Rudolf Klute
Hermann H. Hahn (Eds.)

CHEMICAL WATER
AND
WASTEWATER TREATMENT II

Proceedings of the
5th Gothenburg Symposium 1992
September 28 - 30, 1992
Nice, France

Springer-Verlag
Berlin Heidelberg NewYork London Paris
Tokyo HongKong Barcelona Budapest

Dr. Rudolf Klute
Prof. Dr. Hermann Hahn

Institut für Siedlungswasserwirtschaft
der Universität Karlsruhe
Postfach 6980
D - 7500 Karlsruhe 1

ISBN-13: 978-3-642-77829-2   e-ISBN-13: 978-3-642-77827-8
DOI: 10.1007/978-3-642-77827-8

Typesetting: Camera ready by authors

51/3020-5 4 3 2 1 0 - Printed on acid -free paper

Preface

With joy and pride parents observe the coming of age of their children, conference conveners the acceptance of their programmes, and editors the demand
for their volumes. The scientific advisory board of the Gothenburg Symposia,
the Springer publishing house of the proceedings and the editors are more than
pleased with the fact that the demand for these books far exceeds the supply.

The themes vocalized by the Gothenburg Symposia reflect research and
development needs for the environment more than envisioned at the concep-
tion of the conferences. An environment-oriented analysis of the situation, not
confined to the European community, furnishes the following results: (1) Due
to the very high population density in many areas environmental quality is
endangered; this has become apparent at the very moment in particular in
the aqueous habitat and is corroborated by corresponding regulation proposals
from the European Council. (2) Pollution control concepts and measures are
developed to a varying degree in many countries of the world, reflecting in most
instances the need for environmental protection and the closely related development of (judicial and technical) measures. In most instances these controlling
and protective measures need to be intensified. (3) Thus, nearly all countries
face the problem of developing and/or improving pollution control strategies,
i.e. building new treatment plants, upgrading overloaded or outdated installations and designing new operating and controlling strategies for improved
plant performance. In view of limited financial resources and stringent timing
requirements there is a great need for concepts that allow stepwise realization.
– It is this general theme that the 5th Gothenburg Symposia are devoted to.

As always it is the editors' special privilege and duty to acknowledge the
authors' contributions to this volume. Karin Knisely as well as Adam Leinz
took great care again that a readable book emerged. The cooperation between
the Scientific Committee, the Springer Publishing Company, and Kemira Kemi
of Helsingborg, Sweden made it an enjoyable task for the editors. Last but not
least, the University of Karlsruhe is to be thanked for granting the necessary
leave to one of us.

<table>
<tr><td>Karlsruhe, Germany</td><td align="right">R. Klute</td></tr>
<tr><td>July 1992</td><td align="right">H.H. Hahn</td></tr>
</table>

Members of the Scientific Committee

Prof. T. Asano, USA

Dr. R. Averill, Canada

Prof. P. Balmér, Sweden

Dr. M. Boller, Switzerland

Dr. P. Dolejs, Czechoslovakia

Dr. N. Graham, England

Prof. H.H. Hahn, Germany

Prof. P. Harremoës, Denmark

Dr. R. Klute, Germany

Prof. L. Lijklema, The Netherlands

Dr. M.H. Marecos do Monte, Portugal

Prof. R. Mujeriego, Spain

Mr. J.M. Rovel, France

Mr. J. Sibony, France

Mr. K. Stendahl, Sweden

Prof. N. Tambo, Japan

Dr. G. Tiravanti, Italy

Prof. M. Viitasaari, Finland

Prof. H. Ødegaard, Norway

Contents

Floc Formation

Floc Separation

Chemicals – Dosing Control

Drinking Water Treatment

Wastewater Treatment

Recent Developments in Wastewater Treatment

Floc Formation

Particle and Phosphate Removal Mechanisms with Prepolymerized Coagulants

H. Ratnaweera, J. Fettig, and H. Ødegaard

1. Introduction

The efficiency of wastewater treatment by chemical coagulation depends on the efficiency of each stage: (1) coagulation/precipitation, (2) flocculation, and (3) separation. The efficiency of the coagulation/precipitation stage depends on many parameters including the raw water quality, the chemistry of the coagulant, physical parameters of the process, and the expected quality of the treated water according to the needs. An adequate understanding of the mechanisms of the coagulation/precipitation process is, therefore, essential for increasing its efficiency.

It is widely accepted to divide the mechanisms of the particle removal process into four paths: double layer compression, adsorption-charge neutralization, sweep floc formation, and adsorption-interparticle bridging (Amirtharajah and O'Melia, 1990). In the coagulation with metal salts, the first three mechanisms are of primary importance.

Recently, prepolymerized metal salts have been introduced as coagulants for wastewater treatment. In practice, their performance as coagulants has been found to deviate from the performance of traditional coagulants. Although a number of investigations on the properties of the prepolymerized coagulants can be found in the literature (Ødegaard et al., 1990), the studies on the prevailing coagulation mechanisms are scarce.

Due to the complex nature of wastewater, particle removal cannot be isolated in a study of coagulation. Often, constituents in the wastewater, like phosphates and calcium, interfere with coagulants both chemically and physically. The influence of calcium on particle removal mechanisms is hardly studied with the traditional coagulants or prepolymerized coagulants.

A number of studies have been concerned with the mechanisms of phosphate removal in the coagulation process (Lea et al., 1954; Wuhrmann, 1957; Henriksen, 1963; Stumm and coworkers, 1964–80; Hsu, 1968–76; Recht and Ghassemi, 1970; Lijklema, 1980). However, general agreement on the paths of phosphate removal was not reflected in this literature. In particular, the knowledge of the mechanisms related to the prepolymerized coagulants are poorly understood, primarily due to the lack of research in this area.

It is therefore the purpose of this paper to discuss the mechanisms of particle and phosphate removal mechanisms in wastewater. Based on our own experimental results and on the introductory comments, we hope to contribute to the current understanding of how the degree of prepolymerization of aluminium coagulants effects the coagulation/precipitation stage of wastewater treatment. We try to present a qualitative model of wastewater coagulation with coagulants having various OH/Al ratios.

2. Experimental Procedures and Methods

All experiments reported in this paper can be divided into two major groups. The first group consists of coagulation experiments with model wastewater. These experiments were conducted to investigate the particle and phosphate removal efficiencies of various coagulants under various conditions. The second group of experiments consists of electrophoretic mobility (EM) measurements in coagulation experiments with model suspensions. These were conducted in oder to investigate the destabilizing/coagulating properties of the species resulting from the hydrolysis of metal salts under various conditions.

The soft model wastewater contained 60 mg/l $NaHCO_3$, 400 mg/l NaCl, 100 mg/l NH_4Cl, 50 mg/l K_2HPO_4, 5 mg/l humic substances, 300 mg/l dried milk, 60 mg/l potato starch, 10 mg/l latex particles, and 80 mg/l bentonite. In the hard wastewater, the $NaHCO_3$ concentration was increased to 400 mg/l and NaCl was substituted with 255 mg/l of $CaCl_2$. A detailed description of the preparation and analysis is given elsewhere (Ødegaard et al., 1990).

The model suspensions for the second group of experiments were made with 60 mg/l of quartz particles, which had an approximate diameter of 2 μm. 425 mg/l of $NaNO_3$ and 420 mg/l of $NaHCO_3$ in soft model suspensions and 261 mg/l of $NaHCO_3$ and 338 mg/l of $CaCl_2 \cdot 2H_2O$ in hard model suspensions were included to give an approx. ionic strength of 10 mmole/l. The phosphate concentration was varied, and this was achieved by adding 3.88 or 38.8 mg/l of KH_2PO_4.

The coagulants used for the different experiments were selected from a wide range of commercial and experimental products with prepolymerization ratios (OH/Al ratios) varying from 0 to 2.2. Characteristics of coagulants are given in Table 1.

A modified version of the 'Standard Jar Test' apparatus, the Computer Aided Flocculator (CAF) was used for coagulation experiments. The CAF was purchased from Kemira Kemi AB of Sweden, and modified according to the needs of these experiments. The coagulants were added to a model wastewater stream via a double-piston high-precision pump (type PMP 10, Ismatec Co, Switzerland) and intensively mixed using an in-line mixer (Klute, 1988 and 1990). After that, the jars equipped with automatic stirrers were filled with the suspension. After mixing and settling procedures, the turbidity and the pH were measured on line, and samples were taken from the supernatant for ortho-phosphate analysis. For EM analysis, the samples were pipetted into a flat cell, and values were measured using a particle micro-electrophoretic ap-

Tab. 1. Coagulants used in the experiments

| Name (compound) | Form | Density | Consistance | | | Concentration |
| | | | Al^{3+} | SO_4^{2-} | OH/Al | as g Al/l |
		g/ml	mass %	mass %		
ALG (alum)	Granulate	–	9.1	48.5	0	20–50
Al(NO$_3$)$_3$	Granulate	–	12.7	–	0	1–10
PAX 60P	Liquid	1.33	8.1	< 0.2	1.1	107.8
PAX 19	Liquid	1.17	3.7	< 0.2	1.9	43.3
PAX 22	Liquid	1.39	11.4	< 0.2	2.21	158.2

paratus (Mark II, Rank brothers, UK). All analyses were conducted according to Norwegian standard methods. A detailed description of the apparatus, experimental and analytical methods are given elsewhere (Ratnaweera, 1991).

3. Results and Discussion

3.1 Particle Removal Mechanisms

Figure 1 shows the turbidity removal results from coagulation of the model wastewater with coagulants of different basicities. The coagulant dosage necessary for good turbidity removal decreased with increasing coagulant basicity in both soft and hard waters, except when alum was used in the soft model wastewater.

The wastewater represents a suspension with high particle concentration. Coagulation with alum or ferrichloride in such suspensions is usually explained by the 'sweep floc' mechanism due to the high coagulant concentrations required. Hundt and O'Melia (1988) reported that at pH = 6.5, only 20 % of total aluminium was present as Al(OH)$_3$(s) among the hydrolysis species of a prepolymerized aluminium coagulant. Van Benschoten and Edzwald (1990) found only 10 % of total aluminium as Al(OH)$_3$(s) among the hydrolysis species of another prepolymerized coagulant. However, the hydrolysis products of alum consisted nearly 100 % of Al(OH)$_3$(s). This suggests that if the sweep floc mechanism is exclusively or even primarily responsible for particle removal in coagulation with prepolymerized coagulants as well, the necessary dosage for a given turbidity removal should have been higher than with alum. The observations in Fig. 1 do not support such a conclusion.

The hydrolysis products of coagulants and their EM values are expected be independent of the particle concentration. Figure 2 shows some results from the experiments where the EM of quartz suspensions coagulated with aluminium coagulants with OH/Al ratios of 0, 1.1 and 2.2 were measured. The bottom two lines represent the EM values of soft and hard quartz suspensions without coagulants. The quartz suspension was prepared without phosphates present.

A constant coagulant dosage of 1 mg Al/l was used. The EM of the coagulated model suspensions become more positive with increasing OH/Al-ratio of the coagulants at any given pH.

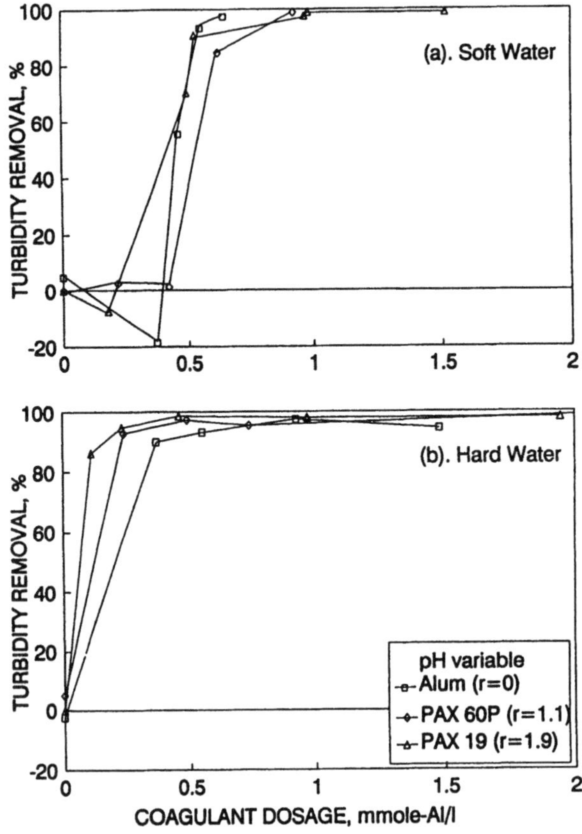

Fig. 1. Turbidity removal of (a) soft and (b) hard model wastewater with $\Delta = F_5^m$ coagulants of different basicities

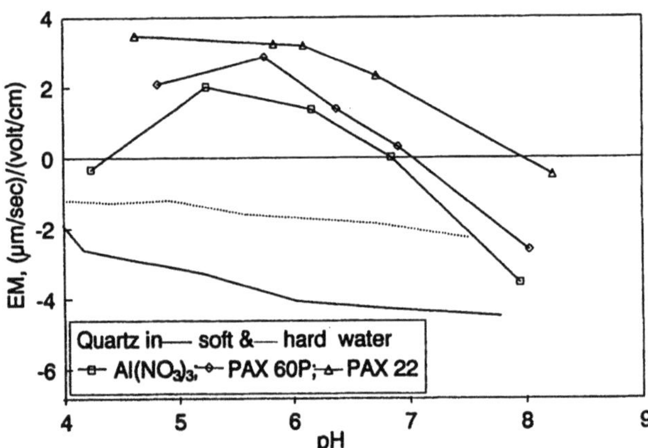

Fig. 2. EM of quartz suspensions coagulated with coagulants of different basicities

Baes and Mesmer (1976), among others, reported on the low stability of polymeric aluminium products, which may convert to $Al(OH)_3(s)$ with aging. However, the hydrolysis products of practical interest to the coagulation process are usually obtained during the first few minutes after hydrolysis. The determination of the accurate concentrations of these species may be complicated. The results reported by Hundt and O'Melia (1988) and van Benschoten and Edzwald (1990) and in this paper should, therefore, represent or relate to the minimum concentrations of polymeric hydrolysis species which may have been present during the first few minutes of coagulation.

Figure 2 suggests that the EM becomes more positive with increasing OH/Al ratio of the coagulant. Thus, if the adsorption-charge neutralization is exclusively or even primarily responsible for particle removal with all coagulants, very high dosages of alum in comparison to prepolymerized coagulants would have been needed in order to achieve an equal turbidity removal efficiency. Although Fig. 1 indicates higher coagulant demands of alum, the differences do not seem to be sufficient to support the above hypothesis.

Obviously, both mechanisms take place simultaneously, and there must also be a shift of the predominant particle removal mechanism with the increasing prepolymerization ratio of coagulants. Alum and other less prepolymerized coagulants remove particles predominantly via the sweep floc mechanism, while with increased degree of prepolymerization the influence of the adsorption-charge neutralization mechanism increases. Particle removal by the adsorption-charge neutralization mechanism is directly proportional to the total positive charge of the hydrolysis species, which increases with the coagulant OH/Al ratio for a given dosage and pH. A slight reduction in the necessary coagulant dosage for a given removal is, therefore, anticipated with the increase of OH/Al ratios of coagulants, as in Fig. 2.

3.2 Influence of Calcium on Particle Destabilization

Figures 1 (a) and 1 (b) illustrate the turbidity removal with different coagulants in soft and hard model wastewater, respectively. The coagulant dosages needed to achieve a certain turbidity removal are very different in the two water types. In order to study the reasons for this difference, the EM was measured in a quartz suspension with 92 mg Ca/l (2.3 mmole/l), and shown as a dotted line in Fig. 2.

When 1 mg Al/l of different coagulants was added to the same suspension of soft model water, the EM (top three solid lines) shifted to the positive side from the negative EM (bottom solid line) before coagulant addition. The potential of aluminium coagulants to raise the EM values was observed throughout the whole working pH range. The effect of the EM shift to the positive side when Ca was added was, however, much less than that when Al was added. Furthermore, the Ca ions did not influence the iep (isoelectric point; the pH where EM = 0) of the quartz suspensions.

The zero point of charge (zpc) corresponds to the pH at which the proton balance at solid surface is achieved (Lyklema, 1975). According to the definition (Hunter, 1987), the amount of specifically adsorbed ions on the particle surface approximates zero when iep = zpc. The iep observed in the model quartz suspensions without Ca and Al ions, therefore, should represent the zpc, since no metal ions were present for specific adsorption then. When the aluminium salts were added, the iep of the quartz suspension was shifted to the pH range between 6.9–8.0 (Fig. 2), indicating a strong specific adsorption of aluminium hydrolysis species, according to the definition of zpc. The same figure also confirms that the Ca ions are not specifically adsorbed onto the particle surface. The Schulze-Hardy rule explains the influence of counter-ions (which are not specifically adsorbed) on the critical coagulation concentration (CCC). Thus, influence of the calcium ion on the EM or zeta potential (both are related to the CCC) can be explained by the Schulze-Hardy rule.

In order to investigate this further, the influence of calcium ions on a system which had both negative and positive EM in a narrow pH range was measured. A quartz suspension which was polluted with a synthetic detergent was found to be suitable for this purpose (Fig. 3). In this suspension, Ca was added to a quartz model water which had an iep of pH = 5.5. Figure 3 shows that as the negative charge of the suspension was reduced (from pH 8 to 5.5), the effect of the calcium ion was also reduced, approaching zero at pH = 5.5. At initial positive EM values of the suspension (at pH < 5.5), the calcium ion did not seem to affect the EM.

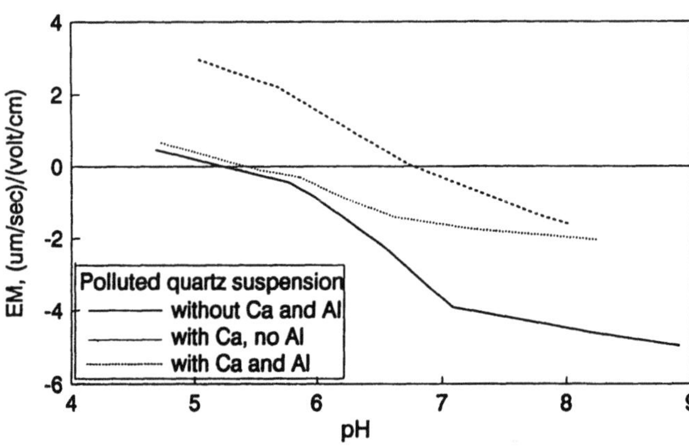

Fig. 3. EM of a polluted quartz suspension and the influence of the calcium ion

The Schulze-Hardy rule states that the CCC (critical coagulation concentration) is inversely proportional to the sixth power of the valence of counter-ions present in the bulk solution. If we assume that the EM is an indirect parameter for determining the CCC, we should see a strong influence of the calcium ion in the range where our model suspension was initially negatively charged, and this agrees with the results shown in Fig. 3.

However, the influence of Ca ions seems to be reduced at lower pH values, starting from pH = 6.6. This phenomenon can be explained using some deviations from the Schulze-Hardy rule given in the literature. They suggest that only at high potentials (highly negative EM values in Fig. 3) is the Schulze-Hardy rule valid with its equation of CCC $\sim 1/z^6$, where z is the valence of counter-ions (Hunter, 1987). At low potentials, CCC $\sim \psi^4/z^2$ (where ψ is the electrical potential at the particle surface). Then the CCC (or EM) can be assumed to be inversely proportional to z^2, but not to z^6. This relationship is therefore valid at situations near coagulation (destabilization) and also at EM near zero in Fig. 3. Thus, a corresponding reduction of the apparent influence of the calcium ions on the EM is probably expected with the decrease of potential (or EM) by a factor of z^6/z^2 (Fig. 3).

In order to investigate the influence of prepolymerized coagulants, further experiments were conducted with different coagulants with and without the presence of calcium ions. The EM measurements are presented in Fig. 4.

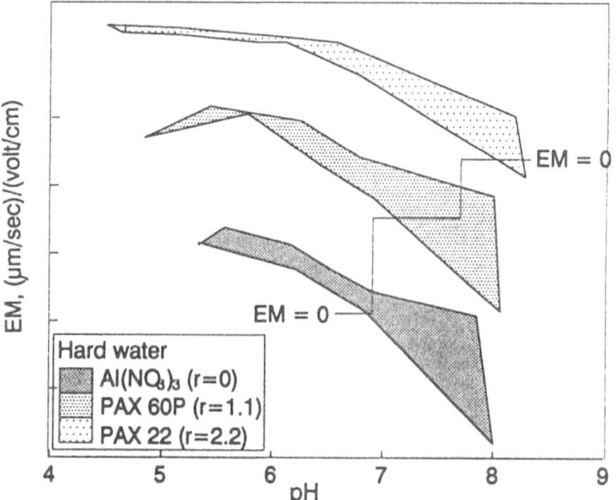

Fig. 4. Effect of calcium on EM. The polygons are constructed with the corresponding EM lines for each coagulant in the presence and absence of calcium and are vertically displaced for easy comparison

The bottom line of each polygon represents EM values for soft quartz suspensions with 1 mg Al/l of the respective coagulant, while the top lines correspond to EM values in hard quartz suspensions with the same amounts of coagulant. The calcium ion seems to influence the EM of coagulated quartz suspensions by extending the EM-shift to the positive side. A significant influence of the calcium ion was observed at high pH values, close to pH = 8. This influence decreased (polygons narrow) with the reduction of pH. With increasing basicity of the coagulants, the influence of calcium on EM at high pH also decreased (the right side of the polygons). This is in accordance with the Schulze-Hardy rule and the discussions in this regard given earlier in this section. When the OH/Al ratio of the coagulants was high, the hydrolysis products tended to reduce the negative surface potential more efficiently than for low OH/Al ratios, especially at high pH values. Then the modified version of the Schulze-Hardy rule for low potentials should be used, and also, depending on the particular case, the results of PAX 22 around pH = 8 (low EM).

When comparing the turbidity removal at pH = 7.2–7.9, a large difference (more than 0.4 mmole/l) in the coagulant dosage needed for a given removal efficiency was observed between soft and hard model wastewater. This difference was insignificant for curves corresponding to pH = 5.0–5.1. It was shown earlier (Fettig et al., 1990) that the particles in model wastewater have a lower negative charge at pH = 5 compared to pH = 7.8. Extrapolating from the aforementioned results with quartz particles, one could assume that the Schulze-Hardy rule for low potentials is valid at low pH. This would explain the smaller or insignificant difference of coagulant dosages between soft and hard wastewater, compared to the dosage of 0.4 mmole/l at pH = 7.2–7.9.

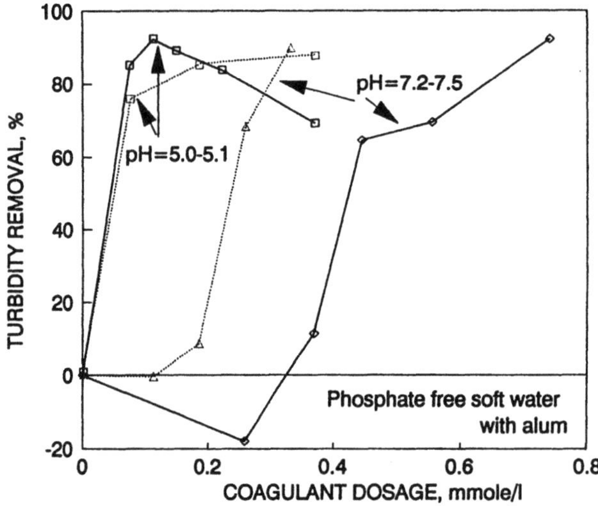

Fig. 5. Turbidity removal by alum coagulation of phosphate free soft (solid lines) and hard (dotted lines) model wastewater at different pH values

3.3 Phosphate Removal Mechanisms

There is no general agreement with respect to which mechanisms are involved in phosphate removal during wastewater coagulation. In this study phosphate removal mechanisms have been divided into three major groups:

(1) Formation of a precipitate, $Me(PO_4)_x(OH)_{3-x}$, as a result of a chemical reaction between Al^{3+} and competitive ions such as PO_4^{3-} and OH^- (chemical precipitate formation);

(2) Adsorption of phosphate ions onto metal hydroxide precipitates. In coagulation with aluminium salts, this mechanism occurs when PO_4^{3-} adsorbs onto amorphous $Al(OH)_3$ precipitates;

(3) Chemical complexation – formation of (initially soluble) complexes by reactions between positive sites on polymeric hydrolysis species and phosphate ions (chemical complex formation).

Some researchers have included the third mechanism as part of either the first or the second. However, it was found necessary here to treat it as an individual mechanism, especially when discussing phosphate removal with prepolymerized coagulants. According to the above definitions of phosphate removal mechanisms, more than one mechanism might occur simultaneously during coagulation.

Lijklema (1980) investigated phosphate removal by alum in model suspensions. Phosphate removal was measured (1) after addition of a concentrated alum solution to a suspension containing phosphates, and (2) after addition of a concentrated phosphate solution to a suspension containing hydrolyzed alum. In the first case, the phosphates were removed according to all three mechanisms defined in this paper. In the second case, however, phosphates must have been predominantly removed by the adsorption or/and chemical complex formation mechanism, since aluminium was already present as a hydrolyzed product at the time of phosphate addition, which would have retarded the chemical precipitate formation mechanism. The portion of phosphates removed by adsorption and chemical complex formation decreased with reduction in pH. At pH = 7 this was only 30 % of the total phosphates removed. It can be expected to be even lower at the low pH values where coagulation normally occurs in practice. Thus, the major phosphate removal mechanism was probably chemical precipitate formation, while the other mechanisms were involved to a lesser extent.

Furthermore, the adsorption mechanism must have been predominant in the second case, since with alum only a very small amount of polymeric hydrolysis species was available for chemical complex formation.

The results of EM experiments with 10 mg Al/l and 8.2 mg P/l are given in Fig. 6. The phosphate removal efficiencies (dotted lines, Fig. 6) seemed to be significantly different for coagulants with different OH/Al ratios, while corresponding EM values were very similar. Since the EM values do not support that phosphate removal took place according to the complex formation mechanism, one can conclude that the complex formation mechanism is not a dominant mechanism in phosphate removal. Using Lijklema's results (1980) it was previously shown that adsorption could not be responsible for a significant portion of the phosphate removal. Thus, the chemical precipitate formation mechanism should be the major phosphate removal mechanism for all coagulants. The predominance among the mechanisms, however, shifts from the adsorption-charge neutralization to the chemical complex formation mechanism with increasing OH/Al ratio of the coagulant. This is simply due to the ability to produce species enhancing each mechanism: $Al(OH)_3$ precipitate or polymeric hydrolysis species.

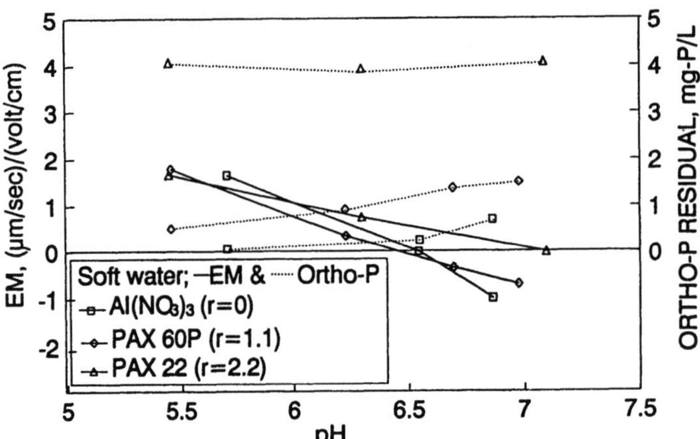

Fig. 6. EM (solid lines) of quartz suspensions with 8.2 mg P/l and coagulated with 10 mg Al/l of coagulants with different basicities. Residual phosphate concentrations are shown with dotted lines

Based on these observations and discussions, the following mechanisms and paths for phosphate removal during metal salt coagulation are suggested:

(1) The predominant phosphate removal mechanism is via formation of a metal hydroxy phosphate precipitate. The adsorption mechanism also contributes, but the proportion of phosphates removed because of this mechanism will always be smaller than that removed by the precipitate formation mechanism. The significance of the phosphate complexation mechanism with polymeric hydrolysis species is also small and comparable to the significance of the adsorption mechanism;

(2) With increasing OH/Al ratio of the coagulant, phosphate removal decreases;

(3) The percentage of phosphate removal by chemical precipitate formation decreases with increasing OH/Al ratio of coagulant;

(4) The percentage of phosphate removed by the adsorption mechanism decreases with increasing OH/Al ratio of coagulant, since this mechanism depends on the production of aluminium hydroxide;

(5) The percentage of phosphate removal by formation of initially soluble complexes increases with increasing OH/Al ratio of coagulant.

4. Presentation of a Qualitative Model of Coagulation

Figure 7 summarizes some of the conclusions of this paper in a qualitative model. In Fig. 7 (a), the variation in the nature of hydrolysis products with the degree of prepolymerization is presented schematically. The coagulants with

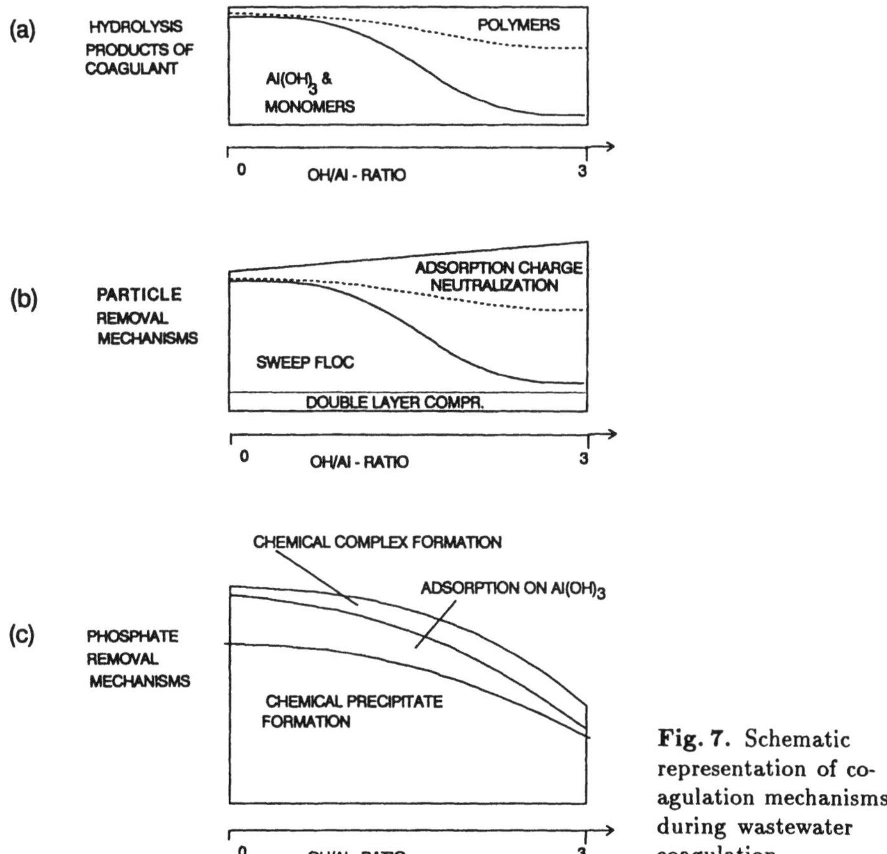

Fig. 7. Schematic representation of coagulation mechanisms during wastewater coagulation

OH/Al ratio = 0 (e.g. alum) have Al(OH)$_3$ precipitate as the major hydrolysis product, together with some monomeric species. With increasing OH/Al ratio, the relative concentration of these species decreases, while the concentration of polymeric species increases. Figure 7 (b) shows the relationship between particle removal mechanisms and the OH/Al ratio of the coagulants. The particle removal efficiency increases slightly with coagulant basicity due to the high destabilizing capacity of the hydrolysis species. The role of the sweep floc mechanism decreases with increasing coagulant basicity, reflecting the decreasing relative amount of Al(OH)3 precipitate among the hydrolysis products (Fig. 7 (a)). Parallel to this, the adsorption charge neutralization mechanism becomes more significant with the increasing appearance of positively charged polymeric species. The difference in coagulant demand between hard and soft water may be attributed to the double layer compression mechanism, which was explained earlier with the Schulze-Hardy rule and its deviations.

The concentration of polymeric species among the hydrolysis products decreases with aging (Baes and Mesmer, 1976). The dashed line in Fig. 7 (a)

illustrates the situation a few minutes after coagulant addition. The involvement of the removal mechanisms relative to each other may also be affected correspondingly, and the dashed line in Fig. 7 (b) reflects this.

The phosphate removal efficiencies and their mechanisms are depicted in Fig. 7 (c). Overall phosphate removal is significantly reduced with increasing coagulant prepolymerization. This effect may be explained with the help of phosphate removal mechanisms. The chemical precipitate formation mechanism is illustrated as the most important phosphate removal mechanism, while the reduction of its efficiency is evident in the reduction of the overall phosphate removal efficiency at high OH/Al ratios. Phosphate removal by adsorption of phosphate ions onto $Al(OH)_3$ precipitates and by complex formation with positively charged hydrolysis species are less significant than the chemical precipitate formation mechanism, although the importance of these two mechanisms is slightly increased at high OH/Al ratios. For the same reasons discussed in the previous paragraph, one may now conclude that most of the phosphates are removed during the first few seconds of coagulation.

The role of aluminium salts and their various paths of consumption during coagulation is shown schematically in Fig. 8. A coagulant is consumed by two major paths during the coagulation of wastewater: (1) a metal ion may be directly involved in chemical reactions to form precipitates or complexes with impurities in the wastewater (for example, formation of aluminium hydroxy phosphate precipitates); (2) a metal ion can undergo rapid hydrolysis, resulting in various polymeric ionic species and $Al(OH)_3$ precipitation. Positively charged hydrolysis species can adsorb onto negatively charged particles and colloids and destabilize them. Given suitable conditions, $Al(OH)_3$ precipitates can remove particles by the sweep floc mechanism. Aluminium hydrolysis species can form complexes with phosphate ions and other anions found in wastewater, resulting in negatively charged ionic complexes which may consequently be removed by other mechanisms such as "sweep floc".

5. Conclusions

- Prepolymerized aluminium coagulants with high OH/Al ratios produce polymeric species rather than $Al(OH)_3(s)$ during hydrolysis. The EM of suspensions coagulated with these coagulants is increased more than that of suspensions coagulated with coagulants of low basicity. A high efficiency is therefore expected in particle destabilization with prepolymerized coagulants.
- In wastewater coagulation, both the mechanisms of adsorption-charge neutralization and sweep floc occur simultaneously. Due to the different hydrolysis products, which include a larger amount of polymeric species, the highly prepolymerized coagulants can be expected to remove particles predominantly by the adsorption-charge neutralization mechanism. Alum produces $Al(OH)_3(s)$ faster and in greater quantity than prepolymerized coagulants,

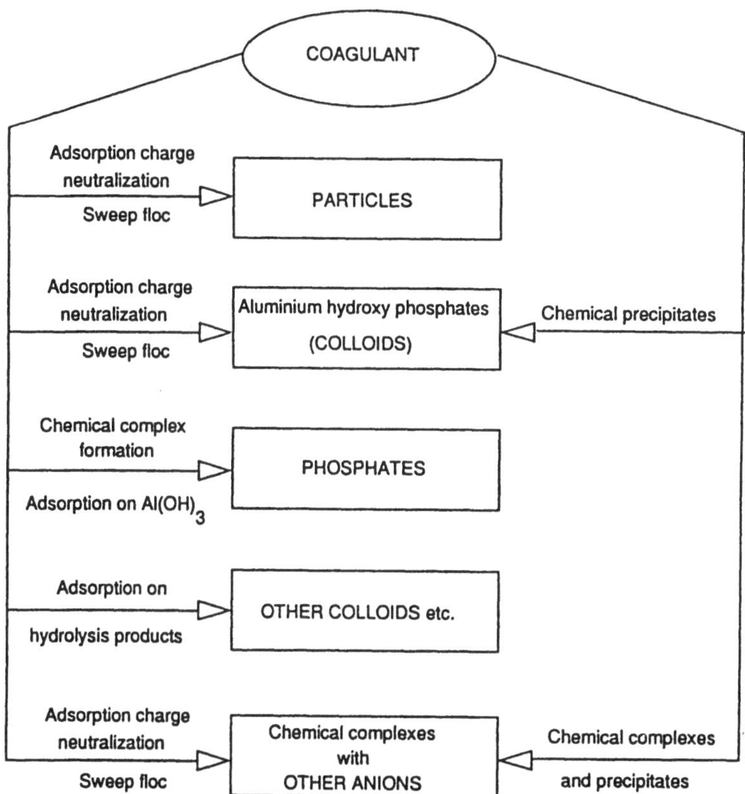

Fig. 8. Schematic representation of the consumption paths and mechanisms of aluminium salts during coagulation of wastewater

and the sweep floc mechanism plays a more dominant role in particle removal when alum is used as coagulant.

– Calcium ions present in the wastewater contribute to particle destabilization by compressing the electrical double layer of colloids as a high valence counter-ion at pH values near and above 7. Unlike aluminium ions, calcium ions are not specifically adsorbed onto particle surfaces, and therefore this phenomenon can be described using the Schulze-Hardy rule (CCC $\sim z^6$, CCC-critical coagulation concentration; z-valence of counter-ions). The positive influence on coagulation is reduced with the reduction of pH if the colloidal charge of the particles becomes less negative, and this can be explained using the modified version of the Schulze-Hardy rule for low potentials (CCC $\sim z^2$).

– Phosphate is primarily removed by chemical precipitate formation during coagulation. Coagulants with low OH/Al ratios efficiently form these precipitate-complexes and consequently they remove phosphate more efficiently than prepolymerized coagulants with high OH/Al ratios.

- Phosphates may also be removed by adsorption of phosphate ions onto metal hydroxide precipitates and by (initially soluble) chemical complex formation with polymeric hydrolysis species, although these mechanisms are of lesser importance. The prepolymerized coagulants remove phosphates via the adsorption mechanism slightly less efficiently than the low basicity coagulants. The complex formation mechanism is, however, favoured by these coagulants.
- A qualitative model of coagulation of wastewater is presented. The role of aluminium salts and their various paths of consumption during coagulation are shown schematically. Despite the large amount of data obtained during this research, it was not found possible to construct a reliable and simple quantitative model for wastewater coagulated with aluminium salts. All of the parameters which are variables in wastewater could not be investigated to a degree that they could be used in a numeric model. The lack of even a qualitative understanding of some phenomena in the coagulation process means that at best only an empirical model could be constructed, which would be of limited interest in practice. However, it is believed that adequate explanations for many phenomena observed in the chemical coagulation of wastewater with metal salts are given and summarized here, and will be fundamental to future modeling approaches.

Literature

Amirtharajah, A., O'Melia, C.R.: Coagulation Processes: Destabilization, Mixing, and Flocculation. In: Water Quality and Treatment. Am. Wat. Works Ass., Pontius, F.W. (ed.), 4th edition. McGraw-Hill, New York 1990, pp. 269–365

Baes, C.F., Mesmer, R.E.: The Hydrolysis of Cations. Wiley and Sons, New York 1976

Fettig, J., Ratnaweera, H., Ødegaard, H.: Status Report for HYPRO-Project. Trondheim 1990

Henriksen, A.: Laboratory Studies on the Removal of Phosphates from Sewage by the Coagulation Process – Part II. Schweiz. Z. Hydrol. 2 (1963) 380–396

Hsu, P.H.: Interactions Between Aluminium and Phosphate in Aqueous Solutions. Adv. Chem. Ser. 73 (1968) 115–127

Hsu, P.H.: Precipitation of Phosphate from Solution Using Aluminium Salt. Wat. Res. 9 (1975) 1155–1161

Hsu, P.H.: Comparison of Iron and Aluminium in Precipitation of Phosphate from Solution. Wat. Res. 10 (1976) 903–907

Hundt, T.R., O'Melia, C.R.: Aluminium – Fulvic Acid Interactions: Mechanisms and Applications. J. AWWA. 80, 4 (1988) 176–186

Hunter, R.J.: Foundations of Colloid Science, Part 1. Clarendon Press, Oxford 1987

Klute, R.: Personal communication (1988)

Klute, R.: Destabilization and Aggregation in Turbulent Pipe Flow. Kinetics of Destabilization with Al(III) in Turbulent Pipe Flow. In: Chemical Water and Wastewater Treatment, H.H. Hahn and R. Klute (eds). Springer, Berlin Heidelberg New York 1990, pp. 33–54

Lea,, W.L., Rohlich, G.A., Katz, W.L.: Removal of Phosphates from Treated Sewage. Sew. and Ind. Wastes 26 (1954) 261–275

Lijklema, L.: Interaction of Orthophosphate with Iron (III) and Aluminium Hydroxides. Env. Sci. & Techn. *14*, 5 (1980) 537–541

Lyklema, J.: Surface Chemistry of Colloids in Connection with Stability. In: The Scientific Basis of Flocculation, K.J. Ives (ed.). Sijthoff and Noordhoff, the Netherlands, 1978, pp. 13–36

Ødegaard, H., Fettig, J., Ratnaweera, H.: Coagulation with Prepolymerized Metal Salts. In: Chemical Water and Wastewater Treatment, H.H. Hahn and R. Klute (eds). Springer, Berlin Heidelberg New York 1990, pp. 189–220

Ratnaweera, H.: Influence of the Degree of Coagulant Prepolymerization on Wastewater Coagulation Mechanisms. Doctoral dissertation, The University of Trondheim, Norway, 1991

Recht, H.L., Ghassemi, M.: Kinetics and Mechanisms of Precipitation and Nature of the Precipitate Obtained in Phosphate Removal from Wastewater Using Al(III) and Fe(III) Salts. Wat. Poll. Contr. Res. Ser. 17010, EKI US-Dept. of the Interior, Washington, DC, 1970

Stumm, W.: Discussion in: Adv. in Wat. Poll. Contr. Res., Proc. 1st Int. Conf. Water Poll. Res. Pergamon Press, London, 2, 1964, pp. 216–230

Stumm, W., Kummert, R., Sigg, L.: A Ligand Exchange Model for the Adsorption of Inorganic and Organic Ligands at Hydrous Oxide Interfaces. Croatica Chimica Acta *53*, 2 (1980) 291–312

Van Benschoten, J., Edzwald, J.K.: Chemical Aspects of Coagulation Using Aluminium Salts, Part I. Hydrolytic Reactions of Alum and Polyaluminium Chloride. Wat. Res. *24* (12) (1990) 1519–1526

Wuhrmann, K.: Die dritte Reinigungsstufe. Wege und bisherige Erfolge in der Eliminierung eutrophierender Stoffe. Schweiz. Z. Hydrologie *19* (1957) 409–427

Harsha Ratnaweera
Norwegian Institute for Water Research
PO Box 69 Korvoll
N-0808 Oslo
Norway

Joachim Fettig
Div. Höxter
University of Paderborn
D-3470 Höxter
Germany

Hallvard Ødegaard
Div. of Hydraulic and Sanitary Engineering
Norwegian Institute of Technology
N-7034 Trondheim
Norway

Model System Studies of Formation and Properties of Flocs Obtained with Cationic Polyelectrolytes

L. Eriksson and B. Alm

Abstract

Flocculation studies were done in dilute suspensions of monodisperse polystyrene latex using an instrument that made it possible to investigate flocculation kinetics, floc size, floc strength, floc structure and reflocculation in the same experiment. The flocculation was of the equilibrium type and the differences in floc properties could be directly related to different characteristics of the cationic polyelectrolytes.

With low charge density polymers, the flocs were shear resistant, had limited ability to reflocculate and consisted of compact microflocs connected into more open macrostructures. They were formed by a bridging mechanism. With high charge density polymers, the flocs had less shear resistance but better ability to reflocculate. The structure had voids on all length scales because of a flocculation mechanism with strong and close contacts between the primary particles. Medium charge density polymers gave smaller flocs due to limited bridging and electrostatic attractions in the total attractive force.

Introduction

Separation of particulate matter from the water phase and dewatering of the sludge are essential steps in most wastewater treatment processes. These operations are generally strongly enhanced if the particles are aggregated into structures that allow rapid settling and a thorough release of the water. For these purposes, both inorganic coagulants as well as organic flocculants have found extensive use.

In such applications a number of desirable characteristics of the flocs can be identified. They should be large and dense to settle properly. They should also be strong enough to withstand disruptive hydrodynamic forces appearing in different stages of the process. It is, for example, necessary with a certain degree of shear to promote mixing, adsorption and flocculation. High shear levels are usually encountered in pumps, in centrifuges and in cakes subjected to pressure differences. It is therefore highly desirable that the system can reflocculate after disruption of the initial flocs. Finally, the floc structure and permeability should be such that cakes with good dewaterability in filtration and pressing are formed.

In practice it may be difficult to achieve good characteristics with regard to all of the aforementioned requirements with a certain flocculant and flocculation process [1]. It is therefore of great interest to understand the relationships between floc properties and the system as well as process parameters. This matter has been the subject of a number of studies both on inorganic coagulants and organic polymers [2–8].

It is generally found that the size of the primary particles determines what type of forces will have predominating effects on the floc characteristics. For particles over the colloidal size range ($> 1~\mu$m), external factors such as hydrodynamic forces will be the most important, while for smaller particles colloidal forces will dominate [7]. It has been clearly shown that the type and amount of flocculant critically determine the structure of flocs formed by colloidal particles with a size comparable to that of the flocculant molecule [9–13]. For larger particles the amount and type of flocculant does not appear to have significant effects on the floc structure [7].

Particles in the colloidal size range are commonly found in wastewaters and wastewater sludges. We may therefore expect that the choice of flocculant will have significant effects on the performance of the separation technique used. This is often found in practice, for example, in conditioning of biological sludges with cationic polyelectrolytes. This type of polymer may also be used for treatment of turbid or coloured waters and for general purpose in combination with inorganic coagulants.

With this background we have studied formation and properties of flocs using monodisperse polystyrene latex as model particles and cationic polyelectrolytes with different charge densities as flocculants. The particle size was chosen close to the upper limit of the colloidal range (diameter 613 nm) in order to be roughly comparable with sizes of important groups of particles such as bacteria and some clays. The model particles are considerably larger than those used in the studies referred to in [9–13].

Experimental

Materials

Latex. Emulsifier-free polystyrene latex was prepared as described in [14], using potassium persulphate as initiator. The mean particle diameter was 613 ± 6 nm as determined from transmission electron micrographs. The surface charge was determined by conductometric titration of acidified latex. The charge density was $10.4~\mu$C/cm^2, which corresponds to a mean distance between charged groups of 1.33 nm assuming hexagonal close-packing of the sites.

Prior to use the latex was cleaned to a conductivity lower than 11.5 μS by extensive washing with Milli-Q water in a serum replacement cell using a membrane with nominal pore diameter 0.22 μm. Freshly cleaned latex samples were used in all experiments.

Polyelectrolytes. The flocculants were polymerized methacrylamido propane trimethylammonium chloride (MAPTAC) or copolymers of MAPTAC with acrylamide (AM). These products were specially synthesized by Allied Colloids. Two series of polymers with molecular weights (M_w) 10^5 and 10^6, as given by the manufacturer, were used. Each series consisted of samples with charge densities (τ) 4, 18, 24, 33, 43, 51 and 100 mole percent charged monomers as given by the manufacturer. These figures were checked by polyelectrolyte titration with potassium polyvinyl sulphate and the agreement was reasonably good.

The water was purified in a Millipore Milli-Q system. The final conductivity was < 10 μS. The NaCl used for varying the ionic strength was analytical reagent grade.

Methods

All experiments were carried out at pH 5.0–6.5 and room temperature.

Flocculation Experiments. The polymers were prepared as 1 % stock solutions. Prior to use the solutions were diluted to 100 mg/l and stirred for 30 min.

The flocculation was done in a 1 l stirred vessel. The desired amount of polymer further diluted to half the final volume (which was 500 or 600 ml) was added as quickly as possible to the latex suspension while the mixture was stirred at 400 rpm with a propeller (diameter: 68 mm). After 1 min the agitation speed was lowered to 270 rpm and a flow of the suspension ($\approx$ 30 ml/min) was taken out from the bottom of the vessel through a 3 mm inner diameter transparent plastic tube using a peristaltic pump. This gives a mean shear rate in the tube of ≈ 80 s^{-1}.

The tube was led through the measuring cell of a Rank Brothers Photometric Dispersion Analyzer, PDA 2000, with which the flocculation was monitored. The signal from the ratio output of the instrument was used as a relative measurement of the floc size. Details about the technique are given by Gregory and coworkers [15–17]. It has been shown that the signal can be used to calculate the mean floc size [18].

Having passed the PDA, the suspension flowed via a microscope equipped with a video camera into a Couette cell with a narrow slit (0.15 mm). The rotational speed in the cell was adjusted to give a shear rate of 2 700 s^{-1}. From the cell the suspension was led through a second PDA with which the effect of the high shear was monitored. Finally, the suspension flowed through a longer piece of tube (1 m) into a third PDA which monitored the extent of reflocculation.

The particle concentration in the main part of the experiments was 134 mg/l, which corresponds to 10^9 particles/ml.

A few trials were also done at 67 and 268 mg/l using polymers with $\tau = 4$, 24 and 100 % and both molecular weights. The stirring rate was 400 rpm for

30 s and then 140 rpm. The propeller diameter was 49 mm. In this case three alternative ways to add the polymer were also tested. The first was the same as above. In the second the polymer diluted to half the final volume was added continuously in the course of 30 s instead of as quickly as possible. In the third a few milliliters (depending on the final concentration) of more concentrated polymer solutions were added instantaneously.

Microscopic Investigation. The video recordings of the flocs were transferred to binary images and further analyzed using the MicroGOP image analysis system. The boundary fractal dimension was determined with the structured walk technique according to [19]. Small samples of settled flocs were also frozen in liquid nitrogen, freeze dried and investigated in a Philips SEM 515 scanning electron microscope.

Results and Discussion

Flocculation Under Standard Conditions

Figure 1 shows an example of the results from a flocculation experiment when 4.6 mg/g latex of a polymer with $M_w = 10^6$ and $\tau = 24\%$ was added at the NaCl concentration 0.05 M. After a short lag period in the results from the first PDA the flocs start to grow with a rate that is represented by the slope of the rising part of the curve. The existence of a lag period indicates that it takes a certain time for the polymer to adsorb [20]. After a maximal value the ratio decreases somewhat as the stirring continues, which may be due to floc break-up or restructuring.

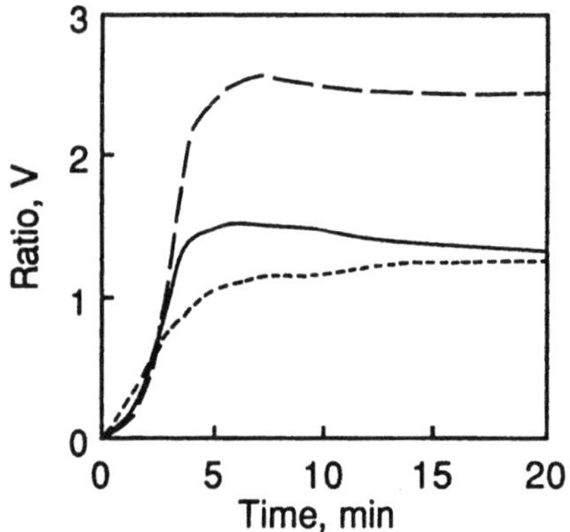

Fig. 1. The PDA-ratio signal as a function of time in flocculation with optimal concentration of a polymer with the characteristics: $M_w = 10^6$, $\tau = 24$ and with addition of 0.05 M NaCl. Results after initial flocculation, high shear and reflocculation

The curve from the second PDA shows that only a limited break-up of the flocs occurred in the shear cell in this case. We also observe that the lag phase is absent. The reason is that the polymer adsorbs faster due to the high shear rate [17]. The flocs can be degraded completely by applying even higher shear rates but so far the comparison between the different polymers has been done systematically only with a constant shear rate.

The significantly higher ratio values from the third PDA show that reflocculation is effective with this polymer. The reason why the reformed flocs are larger than the initial flocs is actually that the conditions in the stirred vessel were not fully optimized. In experiments reported briefly later on in the paper we found that the initial flocs with proper stirring were larger than the reformed flocs. This was also the case when a long piece of tube was inserted between the non-optimized vessel and the first PDA. However, the available results show that the same conclusions can be drawn when comparing the different polymers regardless of the stirring conditions used in the vessel.

Figure 2 shows the maximal ratio values from the time-ratio curves for each PDA obtained at different concentrations of the same polymer as in Fig. 1. The optimal concentration in the initial flocculation was 4.6 mg/g latex. Marginally better optimal values after high shear and reflocculation were found at slightly different polymer concentrations than that for the initial flocculation in this and a few other cases.

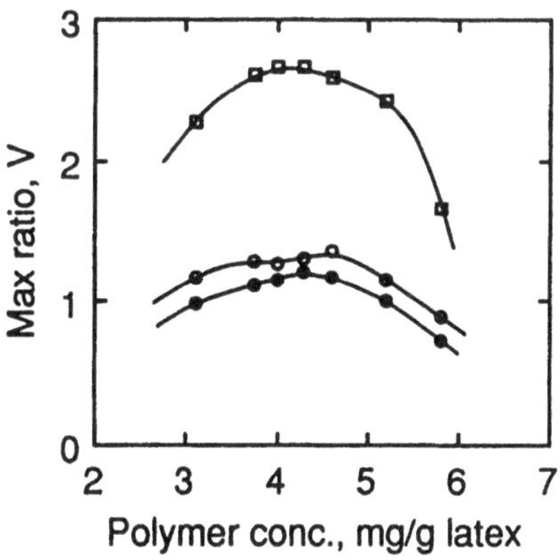

Fig. 2. Maximal ratios as a function of polymer concentration with addition of 0.05 M NaCl. Polymer characteristics: $M_w = 10^6$, $\tau = 24$. Results after initial flocculation, high shear and reflocculation

The ratio values at optimal dosages for the different polymers are shown as a function of polymer charge density (τ) in Figs. 3–4. For simplicity, the same dosage as that in the initial flocculation was used also for the high shear and the reflocculation results.

For both M_w series it is clear that the flocs generally are smaller when 0.05 M NaCl is added. This is most evident at τ around 20 and 100 %. The effect of the charge density also seems to be smaller at high ionic strength. This should be expected since the attraction between polymers and surface are weaker due to shielding of the electrostatic interactions. Furthermore, the polymers will generally be more coiled in solution at high ionic strength, whereas at low ionic strength the higher charged polymers will be strongly elongated due to internal segment-segment repulsion. Both these factors will diminish the effects of different charge densities.

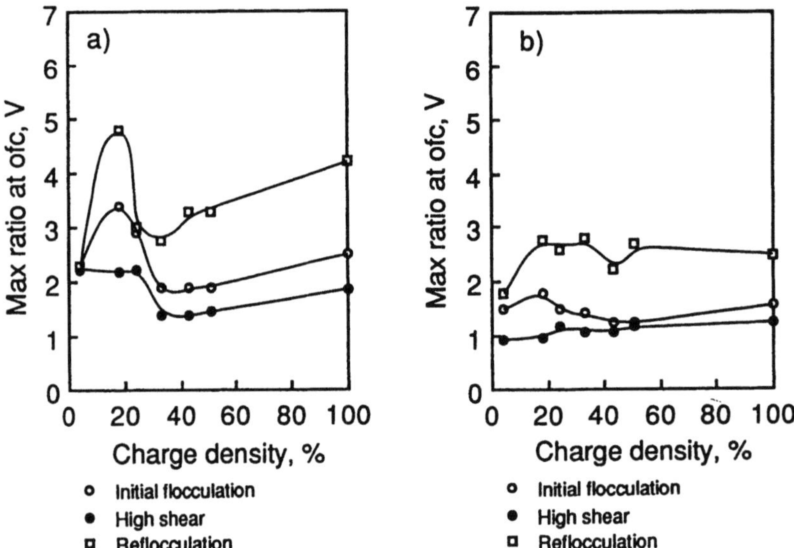

Fig. 3. Maximal ratios at optimal flocculation concentrations as a function of polymer charge density for a series of polymers with $M_w = 10^5$, **a)** with no addition of salt, **b)** with 0.05 M NaCl. Results after initial flocculation, high shear and reflocculation

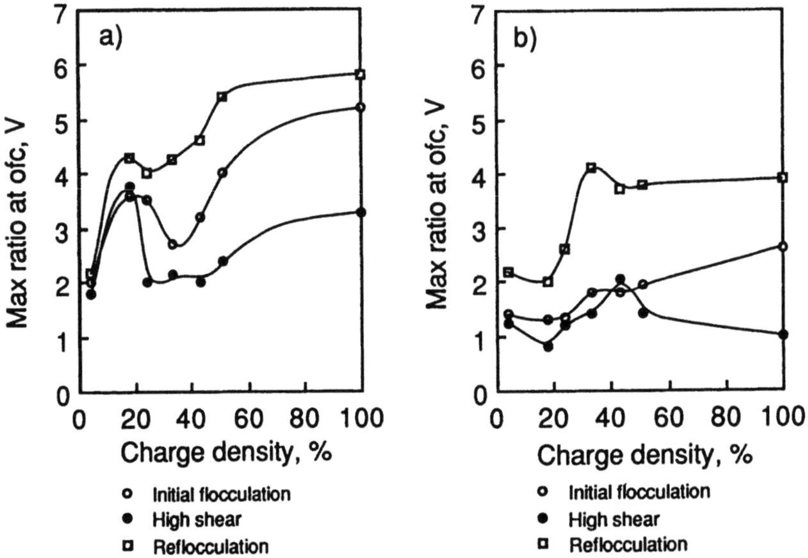

Fig. 4. Maximal ratios at optimal flocculation concentrations as a function of polymer charge density for a series of polymers with $M_w = 10^6$, **a)** with no addition of salt, **b)** with 0.05 M NaCl. Results after initial flocculation, high shear and reflocculation

With no addition of salt (Figs. 3 a and 4 a) there are minima in the ratios at $\tau \approx 35\%$ for both $M_w \cdot s$. In an earlier study we found that the polymers at this charge density at least locally could lay down flat at the surface with a perfect fit of polymer charges to surface charges. The reason is that the mean distance between surface charges matches the mean distance between polymer charges [21]. Ideally this means that at charge neutralization Van der Waals and hydrophobic forces would be the only forces acting between the particles.

At low τ the polymers will adsorb with loops and tails which can form bridges to other particles. The bridging is an added contribution to the attraction together with the forces mentioned above. At high τ there will be a surplus of positive charges at the adsorption site because of the short distance between polymer charges. This will give rise to electrostatic attraction between differently charged mosaic patches on different particles [22] and thus an extra force also in this case. Since other studies have shown that floc size increases with increasing binding strength [23], it is natural to find a minimum in floc size and strength in the medium charge density range.

Of course it is not possible to avoid formation of some loops and tails with high M_w material and consequently we find a higher ratio at $\tau = 35\%$ with the higher M_w. At higher charge densities there is an even larger effect of the M_w which is in accordance with the mosaic patch model [22, 24].

Somewhat surprisingly the ratios go down at $\tau = 4$. Possible reasons may be a decreased adsorption strength due to the few cationic groups, a higher density of the flocs or a different floc structure with a different response of the PDA. Also unexpected is the absence of a M_w effect. This may be due to a higher number of tails with the lower M_w polymer. It was shown in our earlier study [21] that the flocculation with the low particle concentration used here is of the equilibrium type. This means that the polymer approaches equilibrium conformation on the surface before the flocculation occurs. At equilibrium the tails will reach further out from the surface than the loops [25], and thus the higher number of tails may compensate for the larger loops at the higher M_w.

The flocs obtained with the lowest charged polymers are practically unaffected by the high shear and the reflocculation. This shows that the shear resistance is high. With higher τ the size is generally decreased by the shear. This is especially evident with the 100 % charged, higher M_w polymer. However, the reflocculation is also more or less good in all these cases.

The difference in behaviour may be due to a difference in distance between the primary particles in the flocs. With the long bridges at low τ the particles are separated and thus movable relative to each other. This means that the flocs have a certain degree of viscous behaviour and can dissipate energy. Also the flocculation concentrations are relatively high at low τ, which means that many bridges will be formed [21]. At higher τ the particles will stick closer together and have less freedom to move. This results in a more brittle behaviour and thus a higher sensitivity to shear forces when these exceed the binding strength.

It is interesting to note that at high ionic strength the shear resistance is generally better at $\tau > 20\%$ (Figs. 3 b and 4 b). The reason is that the optimal flocculation concentrations are much lower at high ionic strength for the lowest charged polymers. This means fewer bridges which are also weakly bound due to the shielding effect of the salt. At higher τ the optimal dosages are less dependent on the ionic strength [21]. A final remark is that the medium charged polymers perform better relative to the others at high ionic strength than at low. This is probably due to an adsorption conformation with more loops and tails than at low ionic strength, which is the result of the more coiled solution conformation at high ionic strength.

Flocculation Under Various Conditions

It has been shown that the polymer addition procedure is very important for the floc properties in systems with relatively high particle concentrations [26–27]. However, with the low particle concentrations used in this study (67–268 mg/l), we could not see any significant effects when the dosage procedure was varied within the relatively short high speed stirring period. This is further evidence that the flocculation is of the equilibrium type and that the effects reported above truly are caused by differences in polymer characteristics.

The mixing procedure, on the other hand, had a large effect on the results, as already mentioned. This is well known and it is reported in [28] that rather long times with high stirring intensities are needed for good floc formation with polyelectrolytes used in dilute systems. In this study we used short times with the higher stirring rate and medium intensities during the floc growth on purpose in order to be able to follow the growth under as constant conditions as possible and because the measurable size range of the PDA has an upper limit of ≈ 200 μm.

Using the optimal concentration (no salt addition) of the polymer with $\tau = 100\%$ and $M_w = 10^6$ we did a preliminary investigation on the effect of stirring times and rates. Three sets of experiments were done with duration of the high speed stirring (400 rpm) period 15, 30 and 60 s respectively. In each set, there were four runs with the stirring rates 150, 210, 270 and 370 rpm during the floc growth period.

The results from the first PDA were inconclusive. However, the reduction in size after the high shear cell decreases as the stirring rate during the floc growth period increases (Fig. 5). The effects of the time at 400 rpm are smaller and less clear. Nevertheless, the results show that an increase in intensity gives stronger flocs. The reason is presumably that the shear forces the flocs to more compact structures with more primary particle contacts.

A high mixing intensity is thus important in flocculation with highly charged polycations. In the tests with standard stirring conditions the 100% charged polymers gave ratios after 22 min stirring that were 12–27% lower than the maximum. The reflocculation was very good. This indicates that the flocs are amenable to compaction, possibly by breakage and reflocculation. With 4% charged polymers (without salt addition) the decrease was only 0–4%, which

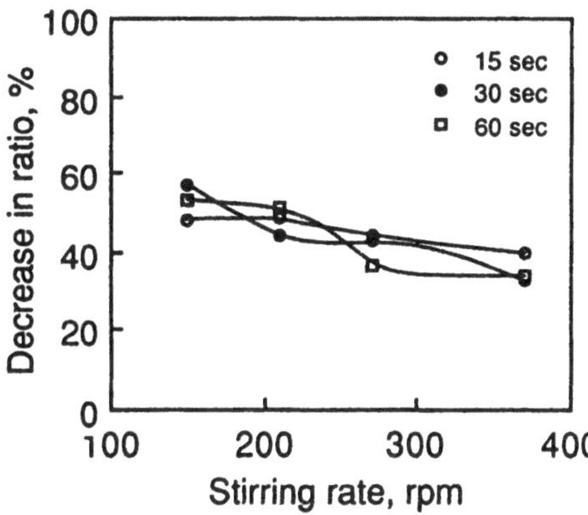

Fig. 5. Decrease in ratio after high shear as a function of stirring rate during the floc growth period and the duration of the high speed stirring period

indicates that the flocs adopt their final structure quickly. As reported above, the shear resistance was very good, but the reflocculation was limited.

Figure 6 shows the increase in ratio from the second to the third PDA. The results after 15 and 60 s stirring at 400 rpm are similar with a strong increase when stirring rates in the floc growth period are higher than 200 rpm. This shows that stronger primary flocs can grow to larger macroflocs. The results after 30 s are much lower except at 270 rpm. A possible interpretation is that the flocs have a multilevel structure and that there is a transition in the structure of the primary flocs that affects their packing properties in the secondary flocs. More direct information on the floc structures are needed to explain adequately these phenomena. However, Clark and Flora also found non-monotonical behaviour in their studies on floc restructuring with varied mixing intensities, which they explained with a multilevel floc model [29].

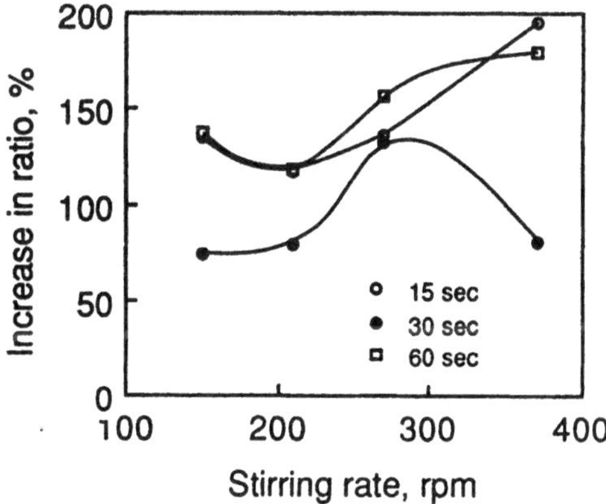

Fig. 6. Increase in ratio in reflocculation after high shear as a function of stirring rate during the floc growth period and the duration of the high speed stirring period

Microscopic Investigation

The standard conditions and polymers with charge densities 4, 18, 33 and 100 % and $M_w = 10^6$ were used for these experiments. The 4 % polymer gave flocs with the appearance of elongated flakes. They had many fissures and nibs and seemed to be shaped without much restructuring in the flow gradients. The flocs with the 33 and 100 % polymers were more regular probably due to breakage and reflocculation. The flocs with the 18 % polymer were something in between.

In Fig. 7 the texture boundary (small length scale) and the structure boundary (larger length scale) fractal dimensions as well as the nr of fissure lines are given. These results show that the small level structure which represents the properties of the primary flocs differs. The low charged polymers give lower fractal dimensions which, in the case of the boundary, may be interpreted as a more smooth line. This may be due to a close packing of the primary particles.

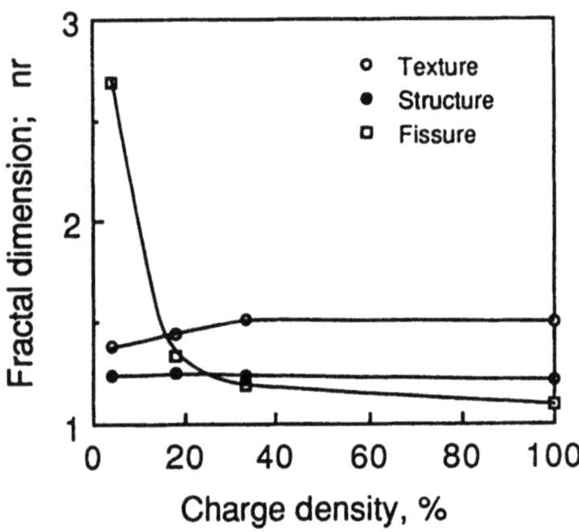

Fig. 7. Texture boundary fractal dimension, structure boundary fractal dimension and nr of fissure lines as a function of polymer charge density

On a longer length scale all polymers give the same results. This further supports the view of multilevel structures. The aggregation of the primary flocs into secondary flocs does not depend on the colloidal forces as far as medium level structure is concerned, but will probably be strongly affected by the stirring analogous to flocculation of larger particles [7]. It is very interesting, however, that the structure on an even larger scale strongly depends on the polymer characteristics as shown by the amount of fissures. This probably has to do with the strength and flexibility of the primary flocs. Thus the colloidal forces will have strong effects on the macroscopic properties in the flocculation of fine particulate (colloidal) systems.

SEM micrographs of flocs obtained with the 4 and 100 % charged polymers are shown in Fig. 8. It appears that the former has small regions with relatively close-packed primary particles, as postulated above, connected into a more open macrostructure. This resembles the results from small angle neutron scattering studies on much smaller silica particles flocculated with low charge cationic polymers where a liquid structure was found on the small length scale and an open fractal structure on larger scales [11]. The formation of small and dense clusters is a consequence of the bridging mechanism allowing the primary particles to move relative to each other to a certain degree, something which also partly explains the good shear resistance as discussed above.

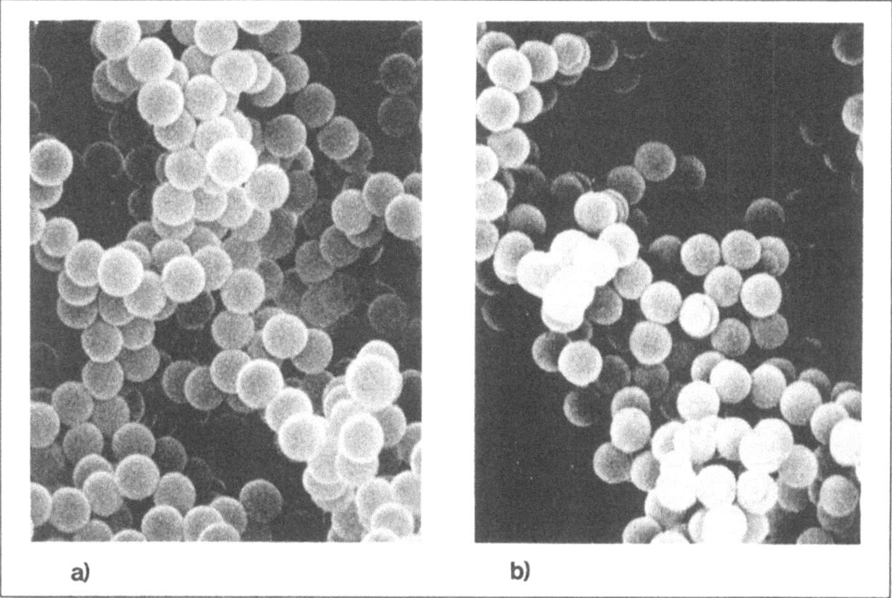

Fig. 8. SEM micrographs of freeze dried flocs obtained with polymers having the characteristics **a)** $M_w = 10^6$, $\tau = 4$, **b)** $M_w = 10^6$, $\tau = 100$

The floc with the 100 % polymer seems to be more open on the smallest length scale which is due to the strong and close contacts between the primary particles which restrict their movement relative to each other. It seems likely that this structure can be further compacted. This is in good agreement with the interpretation of the flocculation results. It also goes along with the texture boundary fractal dimension, since the open structure would give a more irregular boundary and thus a higher dimension.

Conclusions

The results obtained in this study lead to the following conclusions:

1. Flocculation of colloidal particles, even close to the upper size limit, is a multilevel process where the primary and coarse structures clearly depend on the flocculant characteristics. In this study, it is proved only for dilute systems when the flocculation is of the equilibrium type.

2. Low charge density polycations form flocs by a bridging mechanism. The flocs consist of small compact primary flocs with more or less close-packed primary particles that are connected into more open macrostructures. They have good shear resistance due to the strong bridges and because the primary particles can move relative to each other, which allows energy to be dissipated.

3. High charge density polymers give strong and close contacts between the primary particles, which results in an open structure also on small length scales. The flocs are shear sensitive but have good ability to reflocculate. They can be compacted by shear, and high initial mixing intensity is important for the formation of strong flocs.

4. At medium charge density, where the mean distance between charged groups on the polymer is the same as that between charged groups at the particle surface, a minimum in floc size is obtained. This is due to limited bridging and electrostatic attractions in the total attractive force.

5. At high ionic strengths there are fewer differences between flocs obtained with different polymers due to weaker electrostatic interactions and generally more coiled polymer conformations in solution. However, low charge density polymers may give small and weak flocs.

Acknowledgements

This study was financed by the Swedish Board for Technical Development, STU (presently the Swedish National Board for Technical and Industrial Development, NUTEK), and the Association for Surface Chemistry Research. Allied Colloids are thanked for supplying us with the AM-MAPTAC polymers. Supervision of the image analysis work by Sven Forsberg and experimental assistance by Susanna Abrahamsén and Pascal Coenen is gratefully acknowledged.

References

[1] Glasgow, L.A.: Effects of the Physicochemical Environment on Floc Properties. Chem. Eng. Prog. *85* (1989) 51

[2] Tambo, N., Watanabe, Y.: Physical Characteristics of Flocs – I. The Floc Density Function and Aluminium Floc. Wat. Res. *13* (1979) 409

[3] Tambo, N., Hozumi, H.: Physical Characteristics of Flocs – II. Strength of Floc. Wat. Res. *13* (1979) 421

[4] Bache, D.H., Al-Ani, S.H.: Development of a System for Evaluating Floc Strength. Wat. Sci. Tech. *21* (1989) 529

[5] Bache, D.H., Hossain, M.D., Al-Ani, S.H., Jackson, P.J.: Optimum Coagulation Conditions for a Coloured Water in Terms of Floc Size, Density and Strength. Wat. Supply *8* (1990) 93

[6] Glasgow, L.A., Kim, Y.H.: A Review of the Role of the Physicochemical Environment in the Production of Certain Floc Properties. Water, Air, Soil Pollut. *47* (1989) 153

[7] Klimpel, R.C., Hogg, R.: Evaluation of Floc Structures. Colloids Surfaces *55* (1991) 279

[8] Dickinson, E., Eriksson, L.: Particle Flocculation by Adsorbing Poymers. Adv. Colloid Interface Sci. *34* (1991) 1

[9] Wong, K., Cabane, B., Somasundaran, P.: Highly Ordered Microstructure of Flocculated Aggregates. Colloids Surfaces *30* (1988) 355

[10] Cabane, B., Wong, K., Wang, T.K., Lafuma, F., Duplessix, R.: Short Range Order of Silica Particles Bound through Adsorbed Polymer Layers. Colloid Polymer Sci. *266* (1988) 101

[11] Wong, K., Cabane, B., Duplessix, R.: Interparticle Distances in Flocs. J. Colloid Interface Sci. *123* (1988) 466

[12] Cabane, B., Wong, K., Duplessix, R.: Macromolecules for Control of Distances in Colloidal Dispersions. In: Polymer Association Structures, Microemulsions and Liquid Chrystals, M.A. El-Nokaly (ed.). ACS Symposium Series 384, 11989, pp. 312–327

[13] Wong, K., Cabane, B., Duplessix, R., Somasundaran, P.: Aggregation of Silica Using Cationic Surfactant: A Neutron-Scattering Study. Langmuir *5* (1989) 1346

[14] Goodwin, J.W., Hearn, J., Ho, C.C., Ottewill, R.H.: Studies on the Preparation and Characterisation of Monodisperse Polystyrene Latices III. Preparation without Added Surface Active Agents. Colloid Polymer Sci. *252* (1974) 464

[15] Gregory, J.: Turbidity Fluctuations in Flowing Suspensions. J. Colloid Interface Sci. *105* (1985) 357

[16] Gregory, J., Nelson, D.W.: Monitoring of Aggregates in Flowing Suspensions. Colloids Surfaces *18* (1986) 175

[17] Gregory, J.: Polymer Adsorption and Flocculation in Sheared Suspensions. Colloids Surfaces *31* (1988) 231

[18] .Matsui, Y., Tambo, N.: Online Floc Size Evaluation by Photometric Dispersion Analyzer. Wat. Supply *8* (1990) 71

[19] Kaye, B.K.: Image Analysis Techniques for Characterizing Fractal Structures. In: The Fractal Approach to Heterogenous Chemistry, D. Avnir (ed.). Wiley & Sons, Great Britain, 1989, pp. 55–66

[20] Lee, S.Y., Gregory, J.: The Effect of Charge Density and Molecular Mass of Cationic Polymers on Flocculation Kinetics in Aqueous Solution. Wat. Supply *8* (1990) 11

[21] Eriksson, L., Alm, B., Stenius, P.: Formation and Structure of Polystyrene Latex Aggregates Obtained by Flocculation with Cationic Polyelectrolytes I. Adsorption and Optimum Flocculation Concentrations. Colloids Surfaces, Submitted for publication

[22] Gregory, J.J.: Rates of Flocculation of Latex Particles by Cationic Polymers. Colloid Interface Sci. *42* (1973) 448

[23] Smith, D.K.W., Kitchener, J.A.: The Strength of Aggregates Formed in Flocculation. Chem. Eng. Sci. *33* (1978) 1631

[24] Mabire, F., Audebert, R, Quivron, C.: Flocculation Properties of Some Water-Soluble Cationic Copolymers toward Silica Suspensions: A Semiquantitative Interpretation of the Role of Molecular Weight and Cationicity through a "Patchwork" model. J. Colloid Interface Sci. *97* (1984) 120

[25] Fleer, G.J., Scheutjens, J.M.H.M., Cohen Stuart, M.A.: Theoretical Progress in Polymer Adsorption, Steric Stabilization and Flocculation. Colloids Surfaces *31* (1988) 1

[26] Hogg, R.: The Dynamics of Polymer-Induced Flocculation of Fine-Particle Suspensions. In: Flocculation and Dewatering, B.M. Moudgil and B.J. Scheiner (eds.). Engineering Foundation, New York, 1989, pp. 143–151
[27] Gregory, J., Guibai, L.: Effects of Dosing and Mixing Conditions on Polymer Flocculation of Concentrated Suspensions. Chem. Eng. Comm. *108* (1991) 3
[28] Fettig, J., Ratnaweera, H., Ødegaard, H.: Synthetic Organic Polymers as Primary Coagulants in Wastewater Treatment. Wat. Supply *8* (1990) 19
[29] Clark, M.M., Flora, J.R.V.: Floc Restructuring in Varied Turbulent Mixing. J. Colloid Interface Sci. *147* (1991) 407

Leif Eriksson and Barbro Alm
Institute for Surface Chemistry
Box 5607
S-114 86 Stockholm
Sweden

Amphophilic Cationic Polymers as Both Flocculants and Depollutants

E. Ferrer and R. Audebert

Abstract

Linear and cross-linked cationic polystyrene derivatives were prepared and tested as flocculating agents towards some silica suspensions used as model of raw water. For a ratio of substitution of the aromatic ring by an alkyl ammonium group higher than 35 %, the linear polymers are water soluble and flocculate the silica suspensions according to a classical charge neutralization mechanism. The corresponding cross-linked species are very swellable gels which are also flocculants for silica but according to a different mechanism, related to heteroflocculation.

The two types of polymers can also absorb hydrophobic solutes such as polyaromatic hydrocarbons. This phenomenon is independent of the flocculation, and adsorption and flocculation may occur simultaneously. Fluorescence studies showed that this depollution effect involves the hydrophobic part of the polymers. The solute affinity for the polymer behaves like in a liquid/liquid partition and increases with the hydrophobic nature of both macromolecules and solutes.

Introduction

The production of drinking water from ground water generally involves a procedure with two separate steps: i) elimination of the particles in suspension (clay, silica, non-soluble organic matter, etc.) ii) elimination of soluble organic species which are mainly vegetal residues but may also be undesirable chemicals (micropollutants).

The first operation is mainly performed thanks to the addition of aluminium (sometimes iron) salts to the raw water. An increasing interest appears, however, for macromolecular flocculants which are very efficient and do not release aluminium ions. The second step, when necessary, is performed by adding granular or powder carbon black. Practically speaking, the two steps are not completely disconnected since aluminium derivatives lead to the formation of high specific surface mineral precipitates, so a portion of soluble organic species are removed at this stage by adsorption. In fact, this adsorption is much more efficient for humic matter than for undesirable small organic substances [1].

On the other hand, in spite of the attempts to control rejections in agriculture or industry, the number and the level of undesirable substances in raw water is increasing [2]. Powder or granular carbon black are rather expensive and must be used with care because of possible interferences with oxidizing agents or flocculants [3–5] and competitive adsorption between micropollutants and humic substances [6–9]. As for macromolecular flocculants, the compounds presently commercially available are basically very hydrophilic polymers: they are able to remove 60 to 80 % of humic matter, but no interaction was found between them and small hydrophobic solutes [10–12]. Thus a more powerful class of macromolecular flocculants can be expected by selecting new water soluble polymers, preferably cationic, to ensure an easy flocculation of the mineral particles, generally negatively charged in natural water, but with a chemical structure involving hydrophobic parts which might induce the retention of micropollutants via hydrophobic interactions.

We describe here some results obtained with a series of cationic polystyrenic derivatives, some of them linear, other cross-linked, with the above mentioned structure. They were tested as flocculant of colloidal silica suspensions, considered as a very simple simulation of raw water. This flocculation was also done in the presence of polyaromatic derivatives to test the possibility of simultaneous elimination of these model pollutants. Results are analyzed in terms of hydrophobicity of the macromolecular flocculants estimated from fluorescence measurements.

Material and Methods

Choice and Synthesis of the Polymer

We showed previously that cationic polymers derived from polyacrylamide are excellent flocculants of colloidal silica [13–14]. They induce flocculation according to an electrostatic neutralization mechanism [13-15] and they can be used successfully in water treatment [16]. However we verified that they are inefficient for the retention of aromatic hydrocarbons in water. On the other hand, ion exchange resins are depicted as effective for the retention of polychlorobiphenyls and organochlorinated pesticides [17]. Between these two limit structures we expect that water soluble polymers or very swellable gels with both cationic units and hydrophobic styrenic sequences could combine the two advantageous properties of each structure.

The chemical structure of the samples prepared is schematized below where R is a hydrogen atom or an alkyl group.

Gels were synthesized in two main steps [18]. First polystyrene was cross-linked with dichloromethylbenzene according to a double aromatic electrophilic substitution [19, 20]. Quaternization was based upon a modified Tscherniac-Einhorm reaction [21–23] adapted to our polystyrenic substrate [18]. For some samples anionic gels were prepared by substituting the quaternization step by a sulfonation with the complex $SO_3/PO_4(Et)_3$ [24, 25]. Linear cationic polystyrene was obtained by copolymerization of chloromethylstyrene and styrene followed by quaternization of the copolymer [26].

Characterization Techniques

The prepared polymers were characterized by potentiometric titration, IR spectroscopy, NMR, elementary analysis and, also for gels, by the swelling ratio, Q, (weight of swelled gel per unit weight of dry gel) and size exclusion chromatography for linear polymers. Zeta potential measurements were done with a Penkem apparatus and fluorescence studies with a AMINCO SPF 500 fluorimeter.

Nomenclature

Gels are noted PSS-RR-AR or PMA-RR-CR, where

PSS is the sulfonated polystyrene,
PMA the aminomethyl polystyrene,
RR the reticulation ratio (number of crosslinks/monomer, %),
AR the anionic ratio (mean number of anionic site/monomer, %),
CR the cationic ratio (mean cationic site number/monomer, %).

Linear polymers are PMT-CR, PET-CR, PBZ-CR and PBU-CR, where

PMT means trimethylaminated methylpolystyrene,
PET triethylaminated methylpolystyrene,
PBZ dimethylbenzylaminated methylpolystyrene,
PBU tributylaminated methylpolystyrene.

Silica and Micropollutants

The silica used for flocculation studies is a polydisperse precipitated silica (Syton W30) supplied by Monsanto. In electron micrographs, particles appear nearly spherical with a radius ranging from 10 to 100 nm and corresponding to a specific surface area of 35 m^2/g. Among all the different chemical classes considered as micropollutants, we simply chose polyaromatic hydrocarbons which can easily be detected by UV spectrometry.

Experimental Procedure

Flocculation and depollution experiments were always performed by mixing carefully dissolved polymers, or swelled gels, with "synthetic" water, based upon deionized water but containing colloidal silica (5 g of SiO_2 per litre) or the polyaromatic derivative or both components. The pollutants were generally used at a concentration close of fifty per cent of their saturation. In the peculiar case of pyrene, which exhibits a very low solubility in water, it was introduced in a methanol solution. We verified that the presence of this alcohol does not alter the results. Similarly we checked that the pollutants are not adsorbed on silica under our experimental conditions.

The mixtures were simply put in 25 ml top screwed sampling tubes and stirred by slow rotation (60 rpm) for 30 to 60 min. We previously verified that for flocculation of this silica by cationic polymers this procedure leads to the same results as the classical "jar test".

In flocculation experiments a spontaneous phase separation occurs in the vicinity of the optimum flocculation concentration (o.f.c.) but a centrifugation is always needed for a convenient analysis of the pollutant in the supernatant. This determination is much more difficult for absorption experiments (gel plus pollutant solution): we used a gravimetric filtration on a metallic grid with a 63 μm mesh size which prevented the driving of the small pieces of swelled gel. When the treated pollutant solutions were finally recovered, the solute concentration, C_s, was measured by UV spectrometry at 275 nm for naphthalene, 334 nm for pyrene and 400 nm for 1-hydroxyanthraquinone. A comparison with the original concentration, C_o, allows the pollutant absorption, A, to be derived:

$$A\,[\%] = 100 \cdot (C_0 - C_s)/C_0 .$$

Results and Discussion

Flocculation

For ten linear polyelectrolytes differing by the chemical nature of their aminated group or their cationicity ratio (CR): PMT-100; PET-100, -80, -45, -35; PBZ-100, -45; PBU-100, -45, -35 the o.f.c. expressed in g of flocculant per g of silica largely differs. However, when recalculated in term of mole of cationic unit needed per gram of silica, all the o.f.c. are equal to $2.7 \pm 0.3\,10^{-5}$ mole/g. In other words, the o.f.c. is proportional to the equivalent weight of a cationic unit, W^+, (total weight of a macromolecular chain divided by the number of cationic units in the chain). We also followed the evolution of the zeta potential during the flocculation and noticed that the o.f.c. corresponds, practically speaking, to the annulation of the zeta potential. From these results we can conclude [13–15] that a classical charge neutralization mechanism is involved.

On the contrary, flocculation with gels occurs when zeta potential is still negative and the o.f.c. is no longer just related to the number of cationic units introduced, but also depends on the size of the swollen gel particles. As pointed out recently [27] the simple charge neutralization mechanism does not work as soon as the flocculant size is larger than the Debye length in the system. In our case, the gel particles are of some order of magnitude larger than the Debye length. Thus we can expect that the charge neutralization is effective only in a very thin "skin" of the gel and that the weight of silica needed to balance the superficial charges of a given gel particle will be proportional to the accessible surface, S, of the particle and also to the density of cationic units in the organic gel, i.e., inversely proportional to W^+. For a gel with a swelling ratio Q, the surface S varies as $Q^{(2/3)}$. Finally, the o.f.c. is expected to vary as $W^+/Q^{(2/3)}$. This behaviour, depicting a heterocoagulation situation, was effectively observed by comparing three gels with different swelling (PMA-1.5-95 $Q = 300$ g/g; PMA-3-95 $Q = 70$ g/g; PMA-1.5-45 $Q = 40$ g/g): the observed o.f.c. vary, respectively, as 1, 2.2 and 6, and the calculated values according to the above expression, as 1, 2.6, and 6.1, which is an acceptable agreement taking into account the poor precision of the o.f.c. measurements.

This experiment points out the role of the cationic ratio, CR, and of the reticulation ratio, RR, which govern the swelling, Q, hence the accessible surface, S. But another very simple way to change S is to grind the gel. Effectively for gel particles with sizes in the range of some micrometers, the o.f.c. can be of two orders of magnitude larger than those obtained for homologous linear polymers. A progressive grinding of the gels decreases the polymer dose needed for a good flocculation approaching the level obtained for linear polymers.

In conclusion, these polymers appear to be efficient flocculants but the o.f.c. depends very much on the chemical structure. The flocculation properties of the styrenic series of cationic linear polymers are as efficient as those of the classical cationic acrylic derivatives.

Flocculation and Simultaneous Absorption

Flocculation and simultaneous absorption experiments with gels were done by adding pyrene to the aqueous phase before the flocculation procedure. Up to a pyrene concentration of 0.15 ppm used for these experiments, no change in the o.f.c. can be detected. On the other hand, the level of the pyrene absorption by the gel is the same with or without flocculation (Fig. 1). Thus the sites of adsorption of hydrophobic solutes like pyrene and the sites of adsorption on the silica particles are quite different. A similar conclusion can be drawn from the experiments with linear flocculants: again the o.f.c. does not depend on the presence of pyrene. The only point is that if an excess of polymer is added to the silica suspension the micropollutant is indistinctly split between free polymer chains in solution and chains in the floc. This phenomenon explains why when increasing amounts of polymer are progressively added to the colloidal suspension containing pyrene, the micropollutant removed in the floc first increases, rises to a maximum and finally decreases when the amount of free polymer in solution overcomes the amount involved in the floc (Figs. 2–4).

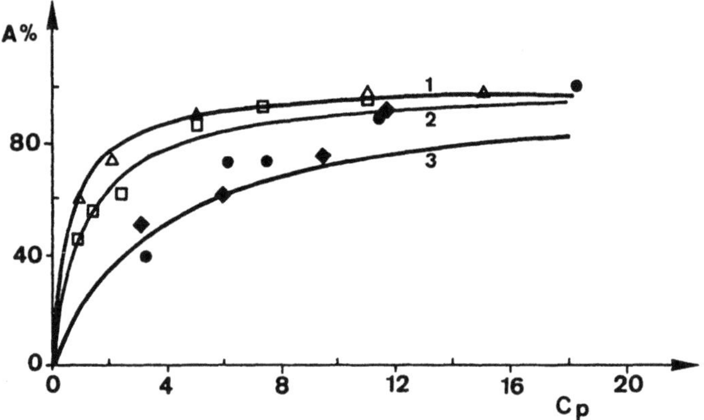

Fig. 1. Pyrene absorption ($A\%$) versus the amount of gel added (expressed in mole of ionic units per liter $\cdot 10^4$).
($\triangle$ – PSS-1-40. Calculated curve N°1; $\square$ – PMA-1.5-45. Calculated curve N°2; ● – PMA-1.5-95. In presence of silica (i.e. with flocculation). Calculated curve N°3; ◆ – PMA-1.5-95. Gel alone. Calculated curve N°3)

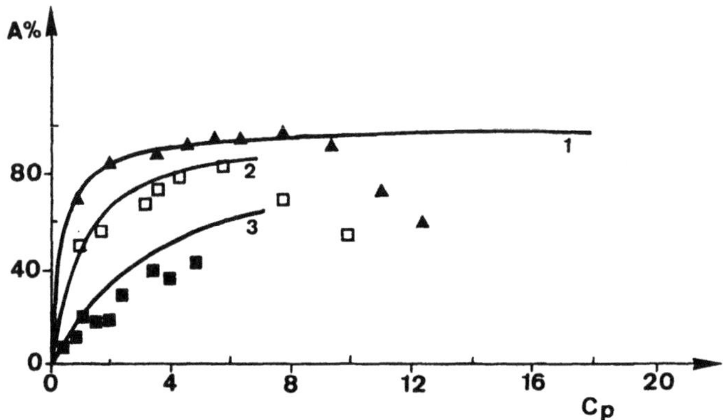

Fig. 2. Pyrene absorption ($A\%$) versus the amount of gel added (expressed in mole of ionic units per liter $\cdot 10^4$) in presence of a "Syton" silica suspension (5 g/l). Three ratios of polymer cationicity are used.
(▲ – Experimental results PET-35. Calculated curve N°1; $\square$ – Experimental results PET-45. Calculated curve N°2; ■ – Experimental results PET-100. Calculated curve N°3)

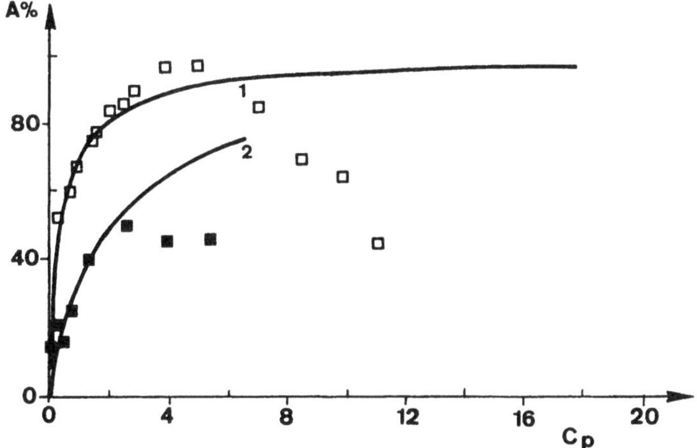

Fig. 3. Pyrene absorption ($A\%$) versus the amount of gel added (expressed in mole of ionic units per liter $\cdot 10^4$) in presence of a "Syton" silica suspension (5 g/l). Two ratios of polymer cationicity are used.
($\square$ – Experimental results PBU-45. Calculated curve N°1; ■ – Experimental results PBU-100. Calculated curve N°2)

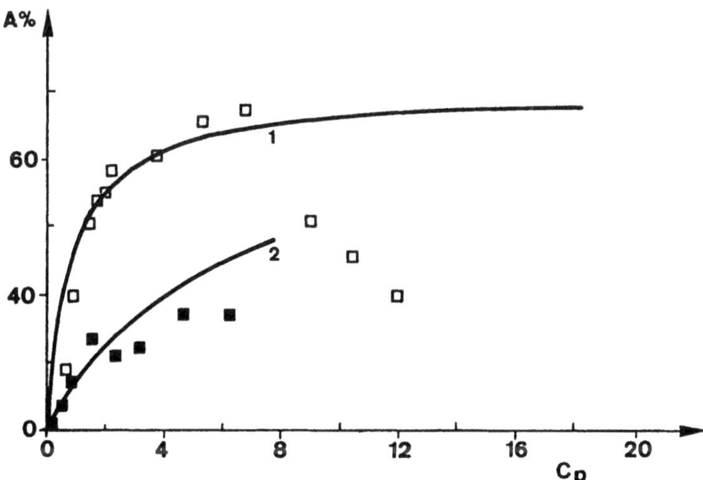

Fig. 4. Pyrene absorption ($A\%$) versus the amount of gel added (expressed in mole of ionic units per liter $\cdot 10^4$) in presence of a "Syton" silica suspension (5 g/l). Two ratios of polymer cationicity are used.
($\square$ – Experimental results PBZ-45. Calculated curve N°1; ■ – Experimental results PBZ-100. Calculated curve N°2)

Absorption

Kinetics of Absorption

The kinetics of absorption are related to the swelling ratio, Q, of the gel in water. For absorption of naphthalene, 10 hours are needed to reach equilibrium with polystyrene ($Q = 0$), but less than 1 hour for a gel with $Q > 40$ g/g. Practically speaking, for all our samples, the absorption equilibrium was reached in one hour, which was the duration selected for our experiments.

Effect of Ionicity and Reticulation Ratio of the Gels

For the limits of the reticulation ratios investigated ($O < RR < 3\%$), the absorption is independent of RR: no difference was found between PMA-1.5-95 and PMA-3-95 towards pyrene. Similarly, the absorption does not depend on the sign of the ionic groups in the grafted gel (at least for our neutral solutes). For instance, the results are quite similar for an anionic gel (PSS-1.5-95) and an analogous cationic one (PMA-1.5-95). This is again evidence that flocculation, which is directly related to the ionicity of the substituent, and the absorption mechanism are completely separate.

Absorption Mechanism

Under our experimental conditions, the amount of flocculant is always in large excess compared to the small amount of pollutant introduced. We observed that the pollutant absorption, A, does not depend on the initial pollutant concentration, C_0. The range of observation is unfortunately narrow for a solute such as pyrene, and difficult to analyse at low concentrations ($0.1 < C_0 < 0.2$ ppm), but was more firmly established for naphthalene ($2 < C_0 < 24$ ppm).

Clearly, A increases when the cationicity ratio (CR) decreases (see Figs. 1 to 4): with the same dose of polymer, 10^{-4} mole/l of PMA, A increases from 20% (CR = 100%) to 50% (CR = 40%). These two figures also show an increase in absorption with the chain length of the ammonium group (1 to 4 carbon atoms). The results for all of our samples confirm a variation in the pyrene absorption in the following order:

$$-N^+(Me)_3 < -N^+(Et)_3 < -N^+(Me)_2Bz < -N^+(Bu)_3\,.$$

For the pollutants, the absorption order is:

$$\text{Naphthalene} < 1\text{-Hydroxyanthraquinone} < \text{Pyrene}\,.$$

For instance, the A values are, respectively, 20, 60 and 95% for a treatment with $3\,10^{-3}$ mole/l of PMA-1.5-45 gel. These results must be correlated with the respective solubilities of these three solutes in water: $2.5\,10^{-4}$ mole/l; $1.6\,10^{-5}$

mole/l; $2.2\,10^{-6}$ mole/l. As for the octanol/water coefficient partition, it increases from $10^{3.38}$ for naphthalene to $10^{5.18}$ for pyrene. In fact, all the phenomena observed can be related to a basic parameter, namely, hydrophobicity.

Pyrene may also be used as a probe for environmental polarity [28] since the shape of its fluorescence spectrum changes according to the microenvironment of the molecule. In particular, the ratio of the first and third fluorescence peaks (I_1/I_3) decreases from a value of about 1.9 in pure water to 0.6 in cyclohexane, and an intermediate value of 1.15 is observed inside a sodium dodecyl sulfate micelle. For the series of linear polymers tested in this investigation, we plotted the evolution of I_1/I_3 versus the absorption ratio (Fig. 5). From the good correlation obtained, we can deduce that the pyrene molecules are mainly located in the vicinity of the hydrophobic groups of the flocculants with an affinity increasing with the hydrophobicity of the flocculant. For the N-substituted derivatives of polystyrene, this hydrophobicity increases in the order:

$$\text{Triethyl} < \text{Dimethylbenzyl} < \text{Tributyl},$$

but is always smaller than the hydrophobicity measured in micelles of classical n-alkyl chain surfactants $(n > 10)$ or in associative polyelectrolytes [29].

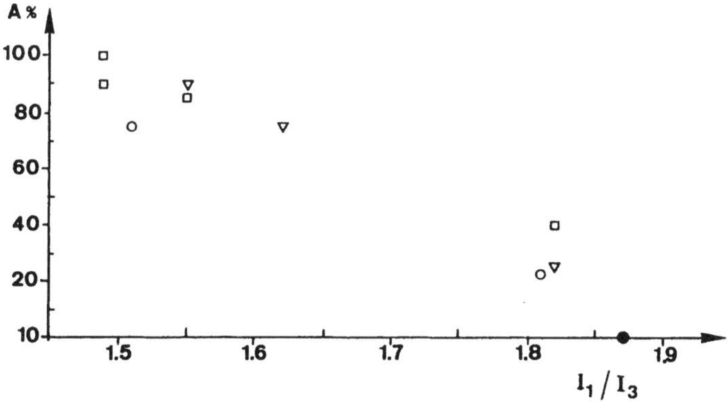

Fig. 5. Pyrene absorption $(A\%)$ versus the hydrophobicity of the polymer expressed by the relative ratio (I_1/I_3) of the first and third fluorescence peak of the pyrene. Pyrene concentration: 0.18 mg/l.
($\bullet$ – Pure water; ∇ – PET type polymers; o – PBZ type polymers; $\square$ – PBU type polymers)

Some desorption experiments were carried out by adding a known amount of pure water to a swelled gel in equilibrium with pollutant solution. Whatever the nature of the pollutant used with the three gels PMA-1.5-95, PMA-1.5-45 and PSS-1.5-40, retention appeared to be quite reversible. On the other hand, as the absorption does not depend on Co since the solution is not saturated, the mechanism of pollutant retention is quite similar to a kind of liquid/liquid partition chromatography, characterized by a partition coefficient, k. $k =$ solute concentration in the polymer phase/solute concentration outside the gel.

Under these conditions, the absorption is related to the polymer concentration C_p (in kg/dm^3) and to the density, d_p, of the polymer ($d_p = 1.05$ kg/dm3):

$$A\,[\%] = 100 \cdot k \cdot C_p/(d_p + (k-1) \cdot C_p)\,.$$

This relation is just obtained by using the k and A definition and expressing the mass conservation of the solute.

This approach where the solute is in equilibrium with the polymer, considered as a liquid matter, and the surrounding medium is consistent with the observed independence of k on the crosslinking ratio, i.e, the swelling of the gel.

This expression of the absorption agrees well with the experimental results (see, for instance, Fig. 1). The corresponding k values are reported in Table 1. It can be seen that k increases with the octanol/water partition coefficient and decreases with the solubility in water. Unfortunately, a large number of experiments, with solutes of various chemical nature, would be needed to establish a correlation.

Tab. 1. The partition coefficient, k, for some polymer/solute systems

Solute polymer	Naphtalene	Hydroxy-1-antraquinone	Pyrene
PMA-1.5-95	1	700	17 000
PMA-1.5-80		1 000	
PMA-1.5-45	800	5 000	70 000
PSS-1.5-95	1		
PSS-1-40	900		130 000
PS-1	2 000		
PET-100			7 000
PET-45			60 000
PET-35			180 000
PBZ-100			6 000
PBZ-45			70 000
PBU-100			15 000
PBU-45			110 000
PBU-35			200 000

For the linear flocculants, the problem is a little bit more complicated: when only a small amount of polymer is added to the suspension it is practically completely adsorbed on the particles; the pollutant is shared between the floc and the supernatant according to the above expression of the absorption and the fit between experimental and calculated values is good (Figs. 1–4). However, for higher particle coverage, as a part of the pollutant is retained by the unadsorbed macromolecules, the absorption apparently decreases and a shift from the theoretical curve appears.

It can be seen from the k values (Table 1), that the unsubstituted aromatic rings of the polymer chain appear to play a major role in the hydrophobic behavior of the polyelectrolytes. k strongly increases when the ratio of ring substitution decreases. As for the alkyl chains substituted on the ammonium group, there must be at least four carbon atoms to enhance noticeably the hydrophobicity (compare, for instance, PBu polymers with others of similar ratio of substitution).

Conclusion

Cationic linear polystyrenes are water soluble for cationicity ratios greater than 35 %. They are good flocculants of silica suspensions and are operative according to a classical charge neutralization mechanism. The homologous cationic gels also flocculate silica, but according to a heteroflocculation mechanism. Their efficiency increases with the cationic site density and greatly decreases with the size of the gel particles.

These organic flocculants can remove small organic molecules and this retention increases with the hydrophobicity of both flocculant and pollutant [30]. This retention can be described according to a liquid-liquid partition mechanism characterized by a partition constant k. According to the k values and fluorescence studies, the hydrophobic behaviour of the polymers is mainly related to unsubstituted aromatic rings and, to some degree, to the n-alkyl chains of the ammonium group substituents.

Under our experimental conditions, flocculation and adsorption are independent. As the level of both pollution and turbidity can drastically changed under practical conditions, the type of polymer to be used must be optimized. A convenient depollution must be obtained at the o.f.c.: if the pollution level is low, the best way is to use a linear polymer with the smallest cationicity ratio compatible with water solubility (CR = 35 % in this study). If the level of organics increases (occasional pollution), a similar increase in polymer is then required, without degradation of the flocculation efficiency. Then a gel can be used and the larger the pollution, the larger the size of the gel particles and their crosslinking.

Acknowledgments

The authors wish to thank MM. POUILLOT and SUTY from ATOCHEM for helpful discussions. The support of this work by ATOCHEM and CNRS (GDR "Traitement chimique des eaux" is greatly appreciated.

References

[1] Rakotonarivo, E., Tondre, C., Bottero, J.Y., Mallevialle, J.: Wat. Res. *23* (1989) 1137

[2] Kraybell, H.F.: J. Am. Water Works Ass. *73* (1981) 370

[3] Gauntlett, R.B., Packham, R.F.: Proc. Conf. on Activated Carbon in Water Treatment, Medmenham, England (April 1973)

[4] Lalezary-Craig, S., Pirbazari, M., Dale, M.S., Tanaka, T.S., Mc Guire, M.J.: J. Am. Water Works Ass. *80* (1988) 3

[5] Mc Guire, M.J., Suffet, I.H.: J. Environ. Eng. Div. ASCE *119* (1984) 629

[6] Robeck, G.G., Dostal, K.A., Cohen, J.M., Kreissl, J.F.: J. Am. Water Works Ass. *57* (1965) 2

[7] Chudyk, W.A., Snoeyink, V.L., Beckmann, D., Temperly, T.J.: J. Am. Water Works Ass. *71* (1979) 529

[8] Frick, B.R., Southeimer, H.: Treatment of Water by Granular Activated Carbon, M.J. Mc Guire and I.H. Suffet (eds.). Adv. in Chemistry Series, 101 ACS, Washington DC 1983

[9] Fettig, J., Southeimer, H.: J. Environ. Eng. *113* (1987) 764

[10] Okubo, T., Ise, N.: J. Phys. Chem. *73* (1969) 1488

[11] Eliassaf, J.: J. Appl. Polymer Sci. *7* (1963) 3

[12] Edzwald, J.K.: In: Organic Carcinogens in Drinking Water, N.M. Ram, E.J. Calabrese and R.F. Christman (eds.). Wiley Interscience Pub., Chap. 8, 1986

[13] Wang, T.K., Audebert, R.: J. Colloid Interface Sci. *119* (1987) 459

[14] Wang, T.K., Piana, G., Lafuma, F., Audebert, R.: Flocculation in Biotechnology and Separation System. Elsevier Pub., 1987, p. 165

[15] Guyot, A., Audebert, R., Botet, R., Cabane, B., Lafuma, F., Jullien, R., Pefferkorn, E., Pichot, C., Revillon, A., Varoqui, R.: J. Chim. Phys. (Paris) *87* (1990) 1859

[16] Wang, T.K., Durand, G., Lafuma, F., Audebert, R.: Influence and Removal of Organics in Drinking Water, M. Suffert (ed.). CRC Press Inc., Boca Raton USA (in press)

[17] Picer, N., Picer, M.: J. Chromatog. *193* (1980) 357

[18] Ferrer, E.: Thèse de doctorat, Univ. P. et M. Curie, Paris 1990

[19] Grassie, N., Gilks, J.: J. Polym. Sci., Polym. Chem. Edn. *11* (1973) 1531

[20] Grassie, N., Meldrum, I.G., Gilks, J.: J. Polym. Sci., B *8* (1970) 247

[21] Tscherniac, J.: Ger. Pat. 134 379, 1902 (Chem. Zentr., II, 1084)

[22] Einhorn, A.: Ger. Pat. 156398, 1905, (Chem. Zentr., I, 55)

[23] Brevet BAYER, Fr 2.190.860, 8 March 1973

[24] Brown, D.W., Lowry, R.E.: J. Polym. Sci. (Chem) *17* (1979) 1039

[25] Turbak, A.T.: I & EC, Product Research and Development *1* (1962) 275

[26] Jones, G.D.: Ind. Eng. Chem. *44* (1952) 686

[27] Nordholm, S., Rasmusson, M., Wall, S.: Simple Analysis of the Electrostatic Screening Mechanism in Flocculation. Paper and Coating Chemistry Symposium, Stockholm, June 1992

[28] Zana, R.: Surfactant Solutions. Surfactant Science Series, Vol. 22, R. Zana (ed.). Dekker Pub., New York 1987

[29] Magny, B., Iliopouls, I., Audebert, R.: Polymer Preprints, ACS Atlanta meeting *32* (1991) 577

[30] Ferrer, E., Audebert, R., Suty, H.: Demande de brevet 89 16118 12/1989 Société ATOCHEM

Eric Ferrer and Roland Audebert
Lab. Physicochimie Macromoléculaire
Univ. Paris et M. Curie ESPCI
10 rue Vauquelin
F-75231 Paris Cedex 05
France

Coagulation and Flocculation in Alkaline Media — The Role of Ca^{2+} and Mg^{2+} Ions

K. Zotter and I. Licskó

Introduction

Background

The physico-chemical, chemical and biological processes involved in water and sewage treatment take place most efficiently in neutral media, i.e. in the pH range around 7.0. In some instances, however, for attaining the desired degree of pollutant removal (e.g. removal of heavy metals, ammonia stripping, lime-soda water softening) it is necessary to change the pH of the water or sewage to be treated. The pH of the effluent discharged is often substantially higher (or lower) than 7.0, so that conventional treatment must be preceded by neutralization, or the possibility and efficiency of removing the particular pollutant(s) in an alkaline or acidic medium by methods other than the conventional ones must be examined. A cost-benefit analysis will indicate whether neutralization-treatment or treatment-neutralization should be the sequence adopted.

Regardless of its numerous drawbacks, the use of lime has found widespread applications in sewage treatment practice. Lime is added primarily for controlling the pH, but it serves also as the precipitating agent for the removal of phosphate ions and dissolved heavy metals. The beneficial effect has long been attributed exclusively to its calcium content. In the combined dosage of iron(III)-chloride and lime, great importance has again been attached to the beneficial effect of the Ca^{2+} ions.

A unit operation of key importance is solid-liquid phase separation, also in the case of sewage treatment in alkaline media. Depending on the quality required for the treated effluent fine phase separation (filtration) may be omitted, but coarse phase separation (settling or flotation) remains essential. The solid part of the pollutants forms in many cases stable colloidal, quasi-colloidal dispersions in the water.

The Al-salts widely used as coagulants in Hungary are unsuitable in alkaline media, owing to the amphoteric properties of the aluminium. The iron(III)-salts are potential coagulants in neutral and alkaline media alike, although the flocculation and settling performance of the iron(III)-hydroxides is less favourable at pH values higher than 10.0 than in the pH range around 7.0.

In water and sewage treatment performed with lime alone, it has been assumed that precipitation and coagulation-flocculation processes occur. The coagulation-flocculation processes have been attributed primarily to the effect of the Ca^{2+} ions.

The groundwaters and treated surface waters used for water supply purposes in Hungary contain – with a few exceptions – Ca^{2+}, Mg^{2+} and HCO_3^- ions in concentrations higher than 1 mmol/l. The addition of lime results not only in a higher pH and a higher Ca^{2+} ion content of the water, but also triggers chemical reactions between the natural components. The higher pH (caused by the addition of lime, but which may be achieved by the addition of, for example, a NaOH solution as well) induces the HCO_3^-, the Mg^{2+}, but even the Ca^{2+} ions to react:

$$HCO_3^- + OH^- \rightleftharpoons CO_3^{2-} + H_2O \tag{1}$$

$$Mg^{2+} + 2OH^- \rightarrow Mg(OH)_2 \tag{2}$$

$$Mg^{2+} + CO_3^{2-} \rightleftharpoons MgCO_3 \tag{3}$$

$$Ca^{2+} + CO_3^{2-} \rightarrow CaCO_3 \tag{4}$$

$$Ca^{2+} + 2OH^- \rightleftharpoons Ca(OH)_2 \tag{5}$$

The equilibrium conditions of the $CO_2 - HCO_3^- - CO_3^{2-}$-system are highly pH dependent ([1, 2]). In the pH range above 8.3, CO_3^{2-} ions appear along with the HCO_3^- ions, their proportion increasing as the pH increases. Thus regardless of whether lime or NaOH is used to attain the desired pH, the reactions (1)–(5) will take place in the overwhelming majority of the surface and groundwaters in Hungary. The efficiency of the water and sewage treatment processes in alkaline media is controlled fundamentally by the properties of the compounds produced by the reactions (1)–(5). The solubility in water of the aforementioned compounds is indicated in Table 1 [3].

Tab. 1. The solubility of alkali-earth metal compounds in water

Compound	Solubility in water g/l
$Ca(OH)_2$	1.85
$CaCO_3$	0.014
$Mg(OH)_2$	0.009
$MgCO_3$	0.106

It can be seen from the tabulated data that calcium hydroxide (which is the most important constituent of lime-hydrate) and magnesium carbonate show the highest, whereas magnesium hydroxide and calcium carbonate the lowest solubility in water. It follows that relatively large amounts of calcium carbonate and magnesium hydroxide must be expected to precipitate if surface or groundwaters commonly encountered in Hungary are involved in water or sewage treatment processes at elevated pH values.

Effects of the Ca and Mg Compounds

In view of the concentration levels of Ca^{2+}, Mg^{2+} and HCO_3^- ions in the water to be treated, considerable amounts of Ca^{2+} ion, precipitated $CaCO_3$ and $Mg(OH)_2$ will be present in the system, regardless of whether lime or NaOH is used to raise the pH. No solid $Ca(OH)_2$ should be anticipated even if the pH is controlled by the addition of lime, since, as revealed by the solubility data in Table 1 and in view of the amounts needed to bring about the desired pH change, all calcium hydroxide added will be dissolved.

In the course of water and sewage treatment in alkaline media – disregarding the pollutants to be removed – the presence and influence of the Ca^{2+} ion and the precipitated $CaCO_3$ and $Mg(OH)_2$ should be taken into account. In order to understand their role in water and sewage treatment, their "performance" in the coagulation-flocculation process is of paramount importance.

Luner and Dence [4], Luner et al. [5], the Interstate Paper Corporation [6], and Oswalt [7] have claimed the addition of lime to be effective in controlling the colour and in treating papermill and cellulose mill effluents. According to Luner and Dence [4] colour control is based on the fact that the functional groups (carboxyl, phenol, enol, etc.) of the lignin substances combine with the calcium ions into poorly water-soluble compounds. The results based on the addition of lime alone and found favourable to minimize colour intensity seem to conflict with those of Black and Christman [8], as well as Christman and Ghassemi [9], who found colour intensity to increase with pH.

Predali and Cases [10] claimed that $(MgOH)^+$ is adsorbed on the $Mg(OH)_2$ particles to lend a positive charge to the magnesium hydroxide, the latter being responsible for the removal of the negatively charged lignin substances and thus for controlling colour intensity. Joffe and Taylor [11] have offered the chemical reactions undergone by the functional groups of the lignin substances and the magnesium ions as an explanation for the development of the poorly water-soluble compounds and for the reduction of colour intensity in the pH range above 10.0.

In alkaline media containing Mg^{2+} ions ($MgCl_2$), the $Mg(OH)_2$ has a structure suited to enhancing considerably the efficiency of clarification [12]. Lecompte [13] added MgO to water containing HCO_3^- ions, resulting in dissolved $MgCO_3$, which proved to be an excellent coagulant in the alkaline medium. Black and Christman [14] and Folkman and Wacks [15] demonstrated the development of magnesium hydroxide in the course of effluent treatment with lime, the magnesium hydroxide being an effective coagulant.

Wentworth et al. [16], Thayer and Sproul [17], and Sproul [18] concluded that in alkaline media viruses cannot be removed at acceptable rates, unless the raw water contains Mg^{2+} ions as well. According to the aforementioned authors, the negatively charged viruses are removed from the water by the positively charged magnesium hydroxides.

Laboratory and pilot-plant experiments by Thomson et al. [19] have lead to the conclusion that in basic media (pH > 10.0) and in the presence of Mg^{2+}

ions (mainly in the forms $MgCO_3$ or $Mg(HCO_3)_2$) flocculation will occur which resembles that triggered by the addition of aluminium and/or iron(III) salts (at pH values close to neutral!). This alkaline coagulation-flocculation treatment is applicable to effluents and surface waters alike.

The foregoing references reflect the conflicting views prevailing on the removal of suspended solids, organic substance and viruses in the course of water treatment in the alkaline range. Some authors regard such treatment as a coagulation-flocculation process, with a readily water-soluble Mg salt as the coagulant, which transforms substantially into a floccular compound (magnesium hydroxide) similar to the aluminium and iron(III) hydroxides. Other authors attribute the removal of pollutants (of the dissolved organic substances in particular) to the chemical reactions between the functional groups of these substances and the alkali-earth metal ions (especially Ca^{2+}) and to the properties of the reaction products.

Removal of Heavy Metals (Zinc)

The main objectives of sewage treatment in alkaline media include the removal of phosphate ions or heavy metals, and, less frequently, the reduction of the levels of organic substances and suspended solids (e.g. in the treatment of bottle washing effluents). An attempt is made here at interpreting the processes involved in the removal of heavy metals, in particular of zinc. This is prompted, on the one hand, by the results of an earlier survey according to which zinc occurs in the highest concentrations in the sludge from communal sewage treatment plants in Hungary [20], and, on the other hand, by the amphoteric properties of zinc which make it readily soluble in acidic and alkaline media alike, so that the development of poorly water-soluble compounds is restricted to a narrow pH range.

The majority of the methods commonly adopted for the removal of heavy metals relies on the precipitation of the metals in an alkaline medium and on separating the resulting solid precipitate containing the heavy metals from the water. Some of the sewage treatment methods referred to as "tertiary treatment" and introduced in the second half of the sixties involved the addition of lime and succeeded in removing heavy metals, too. In their pilot-plant trials Argo and Culp [21] used, besides lime treatment, an activated carbons adsorbent as well. They succeeded in removing cadmium, chromium and zinc at remarkably high efficiencies, but – surprisingly – lead and mercury at 50 % efficiencies only.

In the necessary literature on the removal of heavy metals, there have been similarly conflicting opinions on the optimal pH range of zinc removal. Dean et al. [22] as well as Wing et al. [23] have found zinc removal to be most effective in the 8.5–9.5 pH range. Maruyama et al. [24] have succeeded in demonstrating on pilot scale that zinc removal was excellent even at pH = 11.5, whereas removal efficiency was unsatisfactory in the range below 10.0. The continuous laboratory experiments by Corona et al. [25] have shown zinc removal to be excellent at pH = 12.0. In their laboratory model experiments Patterson et al. [26] have observed in "carbonate" and "hydroxide" systems the removal of four heavy metals, viz. Cd, Ni, Pb and Zn, in alkaline media. At pH values lower than 10.0 and at high CO_3^{2-} concentrations the carbonate system resulted in more effective zinc removal, whereas in the 10.5–11.0 pH range the hydroxide system was more effective. The waters used for industrial purposes in Hungary contain in the vast majority of cases HCO_3^- anions in sufficiently high quantities to ensure the presence of heavy-metal carbonates as the pH is raised. This implies, consequently, that at pH values below and above 10.0 both "carbonate" and "hydroxide" precipitation must be anticipated.

In their pilot-plant trials Idelovitch and Michail [27] added a solution of $MgCl_2$ along with lime to remove trace amounts of heavy metals. At a sewage treatment plant of 4 000 m^3/d capacity Vråle [28] added seawater along with lime because of its high Mg^{2+} content, mainly in order to remove phosphorus. Considerable improvements in treatment efficiency were noted in both cases over the figures obtained prior to the addition of Mg^{2+}.

The laboratory experiments by Joffe and Taylor [11] demonstrated the formation of positively charged cadmium hydroxides but negative zinc and copper hydroxides around pH = 10.5. Consequently, the positively charged magnesium hydroxides proved beneficial to the removal of zinc and copper in this pH range, while having an adverse effect on the efficiency of cadmium removal. In the presence of small amounts of negatively charged lignin substance and magnesium hydroxide the removal efficiency of the three heavy metals was higher.

The data from the literature are seen to be highly conflicting, especially as regards the optimal pH of zinc removal, but also concerning the circumstances under which the experiments were conducted. These vary from the treatment of communal sewage "enriched" with heavy metals pilot plant trials as reported, for example, by Maruyama et al. [24], to distilled water models (e.g. Joffe and Taylor, [11]).

The present study is aimed at obtaining a better understanding of the role in the coagulation-flocculation processes of the Ca and Mg compounds (ions) which are stable in alkaline media, and, furthermore, at determining their contribution to the removal of heavy metal ions, in particular amphoteric zinc. The removal of suspended solids and organic substances by coagulation-flocculation processes in alkaline media was also studied.

Methods and Materials

The impact of Ca^{2+} and Mg^{2+} ions and their simpler compounds on the stability of the colloidal, quasi-colloidal dispersion was studied in laboratory. To a model suspension of 100 mg silicium dioxide in 1 litre of distilled water solutions of $CaCl_2$, $MgCl_2$ and NaCl were added. The influence of the various cations was, of course, studied separately. The chloride Ca^{2+}, Mg^{2+} and Na^+ salts of the cations were used in order to eliminate the influence of the different anions on the Zeta-potential. The concentration of the cations was varied between 1 and 500 mg/l.

The influence of pH on the electric charge of the particles of the colloidal, quasi-colloidal dispersion in the compounds formed from the Ca^{2+} and Mg^{2+} ions was also studied in the model suspension containing silicium dioxide in 100 mg/l concentration. In view of the fact that the surface waters and groundwaters in Hungary contain – with a few exceptions – HCO_3^- ions in concentrations higher than 2 mmol/l, the experiments were carried out in a model system which contained besides the Ca^{2+} and Mg^{2+} ions, HCO_3^- ions as well. The Ca^{2+} and Mg^{2+} ions were added in their chloride salts, whilst the HCO_3^- ions as the Na salt. The pH was increased by adding a NaOH solution.

In the experiments at constant pH, the solutions containing the various cations (Ca^{2+}, Mg^{2+}, Na^+) were introduced under rapid stirring (500–600 rpm) to the silicium dioxide model suspension. After five minutes of rapid stirring a sample was taken for Zeta-potential measurement.

In the experiments at successively higher pH values, the solutions containing the ions Ca^{2+}, Mg^{2+}, and HCO_3^- were added under rapid (500–600 rpm) stirring to the silicium dioxide model suspension. After five minutes of rapid stirring the NaOH solution needed to produce the desired pH change was added and the sample for Zeta-potential measurement was taken after an additional five minutes of stirring.

In the studies related to zinc removal, the model system was composed of $CaCl_2$ solution containing 80 mg Ca^{2+} ion and a $NaHCO_3$ solution containing 220 mg HCO_3^- ion added to one litre of distilled water under rapid (500–600 rpm) stirring. After five minutes a solution was added containing Zn^{2+} ion in quantities from 10 to 50 mg. This was followed after an additional five minutes by the NaOH solution needed for pH control, which contained 5 or 20 % NaOH depending on the change in pH desired.

In several experiments a solution containing Mg^{2+} ion was also added to the system prior to the introduction of the Zn^{2+} ions. The concentration of the Mg^{2+} ions in the experimental system was varied from 10 to 50 mg/l.

In the majority of the experiments liquor from a cellulose mill was also added to the system as a colloid-stabilizing agent, immediately following the addition of the Zn^{2+} ion. The concentrations expressed in terms of TOC of the colloid stabilizing organic substance were 10, 25 and 40 mg/l. In these experiments the NaOH was added (under rapid stirring) five minutes after the colloid stabilizing agent. After an additional 10 minutes of rapid stirring the speed of the stirrer was reduced to 80–100 rpm for 20 minutes, allowing the sample to rest. The sample for pH measurement was taken from the experimental system at the beginning of the slow stirring cycle.

Upon termination of the stirring cycle, 200–300 ml of the experimental mixture were passed through a folded paper filter, 50 ml of the filtrate were conserved with concentrated nitric acid and the concentration of zinc was determined on these samples. The Zeta-potential was measured using a Riddick-type ZETA METER, whilst the concentration of the zinc compounds was determined with PERKIN ELMER 403 and PERKIN ELMER 3030 atom absorption spectrophotometers.

The experiments on the removal of organic substances in alkaline media were made with Danube water to which sulphite waste liquor from a cellulose mill, a solution containing Mg^{2+} ion and a solution of NaOH were added. The original 7 to 10 mg/l TOC of Danube water was increased to 32–35 mg/l by the sulphite waste liquor.

At the time of these experiments the Danube water contained Ca^{2+} and Mg^{2+} ions in concentrations of 75–90 and 15–23 mg/l, respectively. In these experiments the sulphite liquor was added first to the Danube water followed by the solution containing the Mg^{2+}, and, finally by the NaOH solution. The order in which the Mg^{2+} solution and the NaOH solution were added was reversed in another series of experiments. The experiments were performed using the technique described before, the sample for pH measurement having been taken at the beginning of the slow stirring stage. Following slow stirring the experimental system was allowed to settle for 30 minutes, whereafter the concentration of organics remaining in the liquid phase was determined using a Beckmann TOC analyzer.

Results and Discussion

The Zeta-potential of the silicium-dioxide particles shifted from the initial high negative value towards zero as the concentration of the bivalent cations added to the model suspension increased. It can be seen in Fig. 1 that, in accordance with the Schulze-Hardy rule, the high negative charge of the particles in the model suspension was changed by the bivalent cations much more pronouncedly than by the monovalent ones. It can also be seen that the influence of the Ca^{2+} and Mg^{2+} ions on the electric charge was virtually identical. However, in the presence of 500 mg Ca^{2+} or Mg^{2+} ions per litre the Zeta-potential of the particles in the model suspension has not yet reached the zero value which would be most favourable for coagulating the suspension. Increasing the pH in the presence of HCO_3^- ions along with the Ca^{2+} and Mg^{2+} ions in the model system also resulted in significant changes in the electric charge of the particles in the silicium dioxide model suspension. As seen in Fig. 2, in the presence of Ca^{2+} and HCO_3^- ions the negative Zeta-potentials increase with increasing pH. This is due primarily to the fact that as a consequence of increasing pH values, successively larger portions of the HCO_3^- ions were transformed into CO_3^{2-} ions and adsorption of the latter on the silicium dioxide surfaces shifted the Zeta-potential of these particles towards higher negative values. Parallel to this, the CO_3^{2-} ions forming and the Ca^{2+} ions present combine into poorly water soluble $CaCO_3$, the bulk of which consists of particles in the colloidal, quasi-colloidal size range. All these particles carry appreciably higher negative electric charges than the silicium dioxide particles of the original model suspension. The average Zeta-potential of the experimental system was thus shifted vigorously in the negative direction as a consequence of higher pH values.

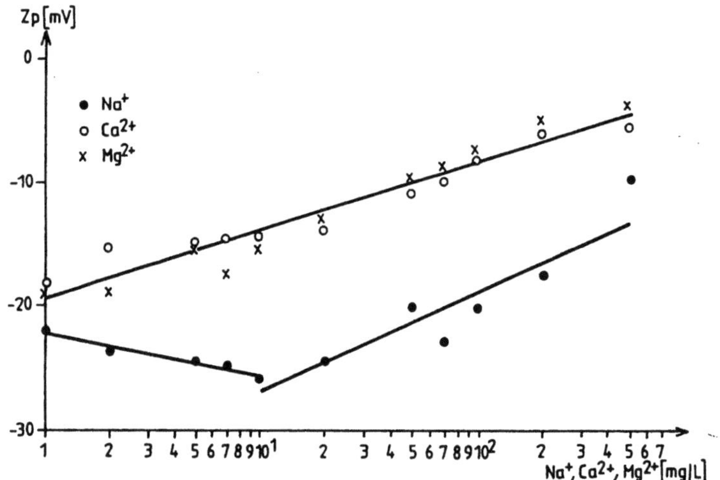

Fig. 1. Change of Zeta-potential of model suspension in the presence of mono- and bivalent cations

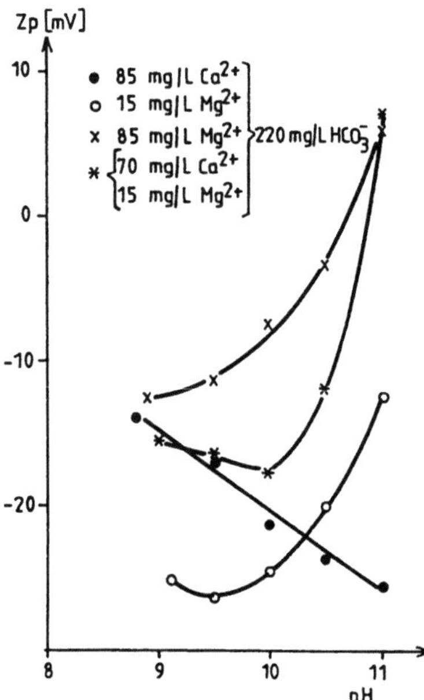

Fig. 2. Zeta-potential of the model suspension in the presence of Ca^{2+}, Mg^{2+} and HCO_3^- ions plotted as a function of pH

The number of solid particles due to the development of $CaCO_3$ was found to increase in the model system as the pH increased. Corresponding to the successively higher negative Zeta-potentials a remarkably stable colloidal, quasi-colloidal dispersion resulted.

In the experimental system containing HCO_3^- ions and little Mg^{2+} (Fig. 2) high Zeta-potentials were measured with the silicium dioxide model suspension in the original condition (without increasing the pH). The average Zeta-potential did not shift towards higher negative values when the pH was increased. In the pH range above 10.0 the electric charge of the particles in the model suspension was shifted towards zero, under the influence of the positively charged $Mg(OH)_2$ formed. In this pH range the development of small flocs was observed and weak coagulation took place in the experimental system. However, the stability of the silicium dioxide model suspension remained substantially unaffected.

Experimental systems with increased Mg^{2+} ion content showed Zeta-potentials in the original condition comparable to those observed in the system containing Ca^{2+} and HCO_3^- ions. Raising the pH shifted the Zeta-potential of the particles in the model suspension towards zero (as experienced earlier) and even into the positive range at pH values close to 11.0 (Fig. 2).

In experimental systems containing Ca^{2+}, Mg^{2+} and HCO_3^- ions in concentrations typical of Danube water, raising the pH resulted first in a mild shift of the Zeta-potential towards zero. However, at pH values higher than 10.0 the Zeta-potential shifted rapidly towards zero and even became positive at values close to 11.0 (Fig. 2). As in the previous experiment, the floccular $Mg(OH)_2$ formed at pH values higher than 10.0 coagulated the hitherto stable silicium dioxide suspension.

As demonstrated by the experimental data shown in Fig. 2, in an alkaline medium the Ca^{2+} ions or their simpler compounds fail to decrease the stability of the silicium dioxide suspension and form even a stable $CaCO_3$ dispersion. In the presence of Mg^{2+} ions, on the other hand, the positively charged magnesium hydroxide which forms induces coagulation processes, as a consequence of which the silicium dioxide suspension is destabilized.

Removal of organic substances during the coagulation-flocculation processes has been assumed to occur in alkaline media, just as in neutral ones. The experimental data are plotted in Fig. 3. Increasing the concentration of the Mg^{2+} ions introduced into the system reduces the concentration of organic substances only up to a particular limit concentration. (In these experiments this was 10 mg/l Mg^{2+}.) It should be noted further that the concentration of organic substances was reduced – at the same pH – within a relatively narrow concentration range of the Mg^{2+} ion added and that a further increase in the Mg^{2+} ion concentration resulted in no further decrease in the concentration of organic substances.

In earlier experiments on the effect of colloid-stabilizing organic substances in coagulation-flocculation processes induced by aluminium and iron(III)-salts as coagulants, the same cellulose-mill sulphite liquor was added as the colloid-stabilizing agent [29, 30]. These experiments also demonstrated that up to a certain limit the amount of coagulant added produced no perceptible reduction in the concentration of the organic substance. Beyond this limit, however, a slight increase in the amount of coagulant caused substantial amounts of or-

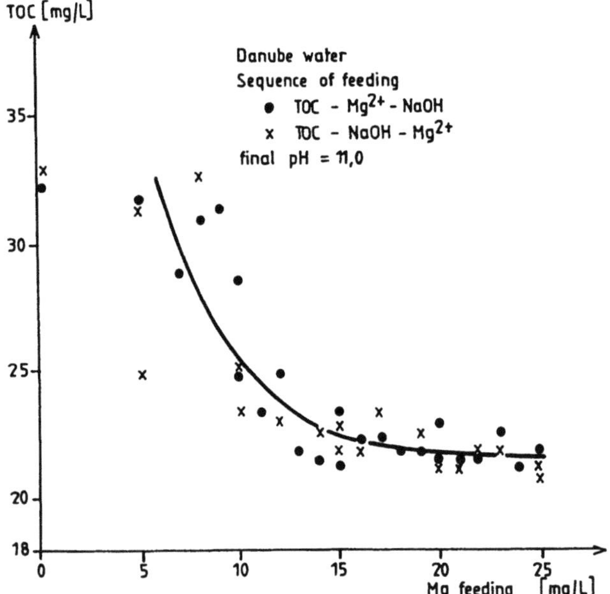

Fig. 3. TOC removal vs. Mg^{2+} feeding at $pH = 11.0$

ganics to precipitate, but the coagulant concentration range within which additional coagulant feed entailed a reduction in the concentration of the organic substance was a relatively narrow one. Beyond a given limit, higher coagulant concentrations failed to decrease the concentration of organic substances any further.

The phenomenon observed in the coagulation-flocculation processes in alkaline media can be interpreted similarly to that observed in neutral media. The negatively charged lignin substances (a component of the sulphite liquor) become adsorbed on the positively charged magnesium hydroxides formed from the Mg^{2+} ions and prevent the magnesium hydroxide sols from aggregating. However, the number of magnesium hydroxide sol particles increased as the concentration of the Mg^{2+} ions increased, and, at a particular Mg^{2+} ion concentration level, all lignin matter available for adsorption was bonded on the magnesium hydroxide surfaces. Still higher Mg^{2+} ion concentrations caused the magnesium hydroxide sols to aggregate and the aggregated magnesium hydroxides – which also contain the lignin substances adsorbed on the surface of the sol particles – can be removed from the water by simple solid-liquid phase separation techniques. As evidenced by the experimental data plotted in Fig. 3, the coagulation-flocculation processes triggered in alkaline media by the magnesium hydroxides decrease the concentration of organic substances in the experimental system by 30–35 %. In neutral media the coagulation-flocculation processes induced by aluminium or iron(III)-salts caused 60–70 % reductions in the concentration of organics in the same experimental system. The difference is due mainly to the difference in pH. In the high pH range the lignin substances undergo structural changes which reduce the amount of organics adsorbing on the surface of the metal hydroxides. Another potential explanation for the difference in organics removal is that the magnesium hydroxides remove other types of organics than the aluminium or iron(III)-hydroxides. However, according to the experimental evidence gathered in neutral and alkaline media alike, the mechanism prevailing up to a certain lignin: metal hydroxide ratio is the impediment to sol aggregation which is removed once this ratio is shifted in favour of the metal hydroxide concentration.

Heavy Metal (Zinc) Removal
in the Presence of Colloid-Stabilizing Substances and Mg^{2+} Ions

The first series of experiments was performed on a model system containing Ca^{2+}, Zn^{2+} and HCO_3^- ions. The pH of the model system was controlled using an NaOH solution. For the subsequent experiments Mg^{2+} ions were also introduced into the model system. The experimental data are shown in Fig. 4.

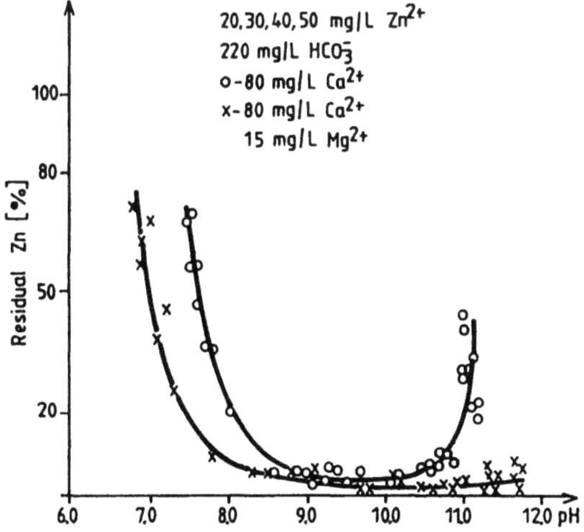

Fig. 4. Removal of zinc vs. pH in model system

The results of the experiments performed in the presence of Ca^{2+}, Zn^{2+} and HCO_3^- ions demonstrate the amphoteric character of zinc, as a consequence of which effective zinc removal is possible within a narrow pH range alone. Regardless of the initial concentration of (dissolved) zinc, the highest zinc removal efficiencies were observed in the 9.0–10.0 pH range. Zinc removal by simple solid-liquid phase separation techniques was found to deteriorate as the pH was raised beyond this limit.

The efficiency of zinc removal will remain satisfactory even in the pH range above 10 if the experimental system contains, besides the aforementioned ions, Mg^{2+} ions as well. The highest zinc removal efficiency was then observed in the 10.5–11.0 pH range. From the experimental data shown in Fig. 4 it is concluded that, in the presence of Mg^{2+} ions (or magnesium hydroxides), the adverse phenomena due to the amphoteric properties of zinc (such as the development of readily soluble zinc compounds at pH values higher than 10.0) tend to decrease. These results imply that in the presence of Mg^{2+} ions co-precipitates containing magnesium and zinc hydroxides develop at pH values higher than 10.0 and these are appreciably less water soluble than the other zinc compounds formed in this pH range. Consequently, the upper limit of the pH range in which poorly water soluble zinc compounds, suited to simple solid liquid phase separation, are formed is modified substantially (extended towards the higher values) in the presence of Mg^{2+} ions. The Ca^{2+} ions had no similar effect.

In the presence of colloid-stabilizing substances (sulphite liquor), the lowest zinc removal efficiencies were observed in the pH range considered optimal formerly (Fig. 4) if the experimental system contained no Mg^{2+} ions (Fig. 5). From the experimental data plotted in Fig. 5 it is concluded that positively charged solid zinc compounds are formed in the 7.5–9.0 pH range, since at constant colloid-stabilizing agent concentrations very satisfactory zinc removal efficiencies were observed at pH values around 8 when the amount of dissolved zinc was increased beyond a certain limit in the experimental system.

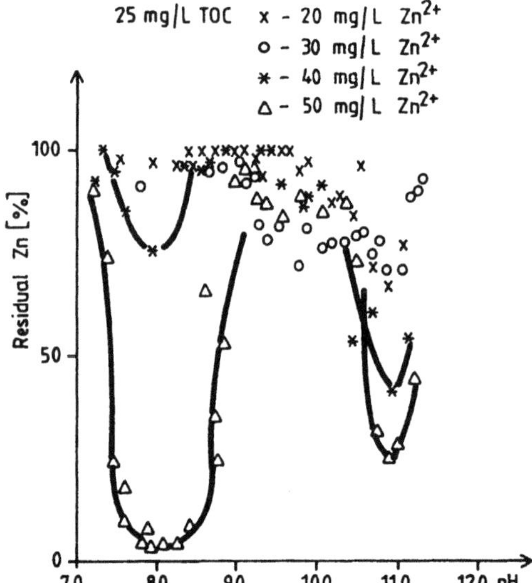

Fig. 5. Removal of zinc vs. pH in the model system containing organic colloid-stabilizing agent

At pH values higher than 9.0, (dissolved and/or solid) zinc compounds are formed which – in the presence of colloid stabilizing substances – are impossible to remove from the water by simple solid-liquid phase separation. The data shown in Fig. 4 imply that the virtually zero zinc removal observed in the 9.0–10.0 pH range (Fig. 5) is due to the fact that the solid zinc compounds formed are stabilized as a colloidal, quasi-colloidal dispersion.

In the next series of experiments an attempt was made to check the validity of this assumption. Relative to the earlier experiments the concentration of the colloid-stabilizing agent was raised from 25 to 40 mg TOC/l and some (10 mg/l) Mg^{2+} ion was also added. Part of the sample taken from the experimental system at the end of the settling period was passed through a folded paper filter, the other part through an 0.45 μm pore size membrane filter. The experimental data are shown in Fig. 6. It can be seen that the membrane filtrate contained less than 10 % of the zinc compounds present in the experimental system over the entire pH range studied (7.5–11.0). More than 90 % of the zinc compounds was retained on the membrane filter, meaning that these are solid forms. On the other hand, over 95 % of the zinc compounds passed through the folded paper filter in the 8.5–10.5 pH range, so that these are present in forms inaccessible to simple solid-liquid phase separation treatment. Precipitation of the zinc compounds is consequently unimpeded, but a stable colloidal, quasi-colloidal dispersion is formed and the precipitated zinc compounds are impossible to separate from the water.

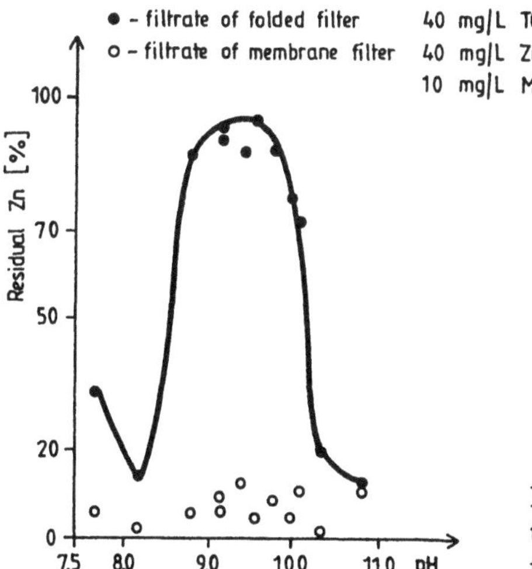

Fig. 6. Zinc concentration measured in the filtrates through folded and membrane filters

It is thus concluded that the zinc compounds formed in the presence of colloid-stabilizing substances at 8.5–10.5 pH values fail to respond to simple solid-liquid phase separation and that the stable, colloidal, quasi-colloidal dispersion must be coagulated to make such treatment possible. The results shown in Figs. 2 and 3 suggest that the addition of Mg^{2+} ions is the simplest approach to induce the stable dispersion to coagulate. The data in Figs. 7 and 8 originate from experiments in which the model system contained Mg^{2+} ions in addition to the Ca^{2+}, Zn^{2+}, HCO_3^- ions and colloid stabilizing agent.

The results shown in Fig. 7 were obtained from experiments in which the colloid-stabilizing agent and zinc were present in lower concentrations (25 mg TOC/l and 30 mg Zn^{2+}/l, respectively). In the experiments shown in Fig. 8, the corresponding concentrations were higher, viz. 40 mg TOC/l and 50 mg Zn^{2+}/l. In both figures the amount of zinc compounds inaccessible to simple solid-liquid phase separation decreases in the 8.5–10.5 pH range as the concentration of the Mg^{2+} ions increases. These measurement data demonstrate further that around pH 11.0 the colloidal, quasi-colloidal dispersion could be precipitated at high efficiencies even at high concentrations of the colloid-stabilizing agent. It should be noted also that around pH = 11.0, even low Mg^{2+} concentrations result in high zinc removal rates.

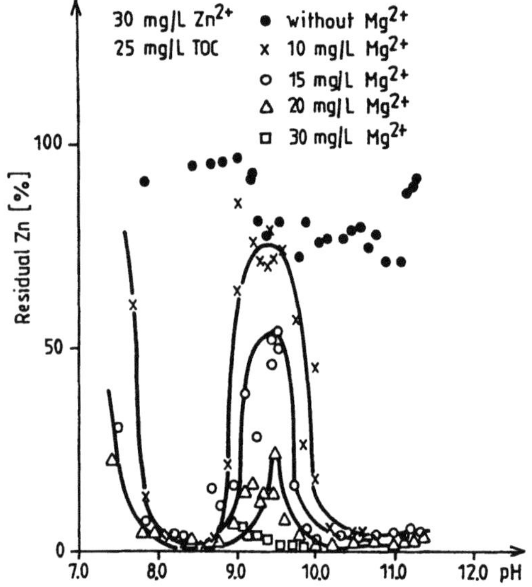

Fig. 7. Zinc removal vs. pH in the model system containing organic colloid-stabilizing agent and Mg^{2+} ions

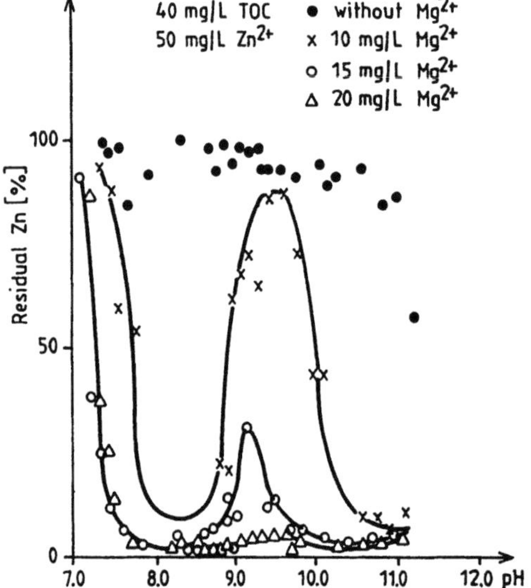

Fig. 8. Zinc removal vs. pH in the model system containing organic colloid-stabilizing agent and Mg^{2+} ions

Both figures imply, moreover, that very high (95–98 %) zinc removal efficiencies can already be attained at relatively low pH values by adding Mg^{2+} ions in sufficiently high quantities. These measurement data suggest the possibility of lowering the pH needed for coagulating the stable colloidal, quasi-colloidal dispersion by raising the concentration level of the Mg^{2+} ion. This assumes special importance in the treatment of wastewaters having high buffering capacities, since important savings in the amount of chemicals needed for pH control can be realized if the optimal pH of heavy-metals (zinc) removal is stabilized by the addition of Mg^{2+} ion at relatively low (8.5–9.0) levels.

Conclusions

In Hungary, two solid compounds, viz. calcium carbonate and magnesium hydroxide, are produced in large quantities during the treatment in an alkaline environment of surface and groundwaters, as well as the wastewaters resulting from them. The calcium carbonate particles carry high negative charges and form a stable colloidal, quasi-colloidal dispersion with water. In contrast, the solid magnesium compounds (magnesium hydroxide) developing in the pH > 10.0 range carry positive charges and their sol particles are susceptible to coagulation. These magnesium hydroxide sol particles can induce the stable dispersion of calcium carbonate particles to coagulate.

Our experimental results have demonstrated that it is not adviseable to add high Ca^{2+} ion concentrations in the alkaline treatment of waters (wastewaters) containing HCO_3^- ions in relatively high amounts. The presence of Mg^{2+} ions is, however, beneficial, since in the pH range above 10.0 positively charged magnesium hydroxide sols are formed which cause the stable colloidal, quasi-colloidal dispersions to coagulate, and which are also capable of removing the colloid-stabilizing organic substances, thus contributing to the coagulation of the colloidal dispersions.

The magnesium hydroxide sols formed at pH values higher than 10 were found to change the Zeta-potential much more radically than the Ca^{2+} and Mg^{2+} ions in neutral media. Around pH = 11.0, the magnesium hydroxide formed with 15 mg Mg^{2+}/l will change the Zeta-potential in a model suspension to the same extent as 500 mg/l of Ca^{2+} or Mg^{2+} ion in a neutral medium.

In the absence of magnesium compounds and in the presence of calcium compounds alone, neither organics nor heavy metals (zinc) are removed in alkaline media containing a colloid stabilizing agent. The conclusion that the optimal pH range of zinc removal can be extended towards higher values by the addition of magnesium compounds is considered to be significant. Comparable importance is attached to the finding that in the presence of colloid stabilizing substances, the lower limit of the pH range of effective zinc removal can be lowered by raising the concentration level of the Mg^{2+} ion. This may be conducive to more economical water and wastewater treatment.

In the 8.5–10.5 pH range and in the presence of colloid stabilizing substances the dissolved zinc compounds were found to transform into solid compounds which form a stable, colloidal, quasi-colloidal dispersion. This dispersion can be induced to coagulate, and thus made accessible to simple solid-liquid phase separation treatment, by the magnesium hydroxide sols present in the water or forming from the Mg^{2+} ions introduced. The dissolved heavy metal (zinc) compounds can be precipitated and the resulting stable dispersion coagulated at the same pH value, provided Mg^{2+} ions are present in sufficiently high quantities. It should be remembered, however, that the transformation of the dissolved heavy metals is a chemical precipitation process, whereas the resulting solid heavy metal (zinc) compounds are changed into forms suited to solid-liquid phase separation by coagulation-flocculation processes. It is important to note further that both are triggered by raising the pH.

Summary

In the course of water or wastewater treatment in alkaline media, the stability of the colloidal, quasi-colloidal dispersion of solid particles is reduced effectively by the positively charged magnesium hydroxide sols formed from the Mg^{2+} ions originally present or introduced artificially into the system. The magnesium hydroxide sols developing at pH values higher than 10 are suited to removing organic substances as well. In the presence of relatively high amounts of HCO_3^- the Ca^{2+} ions are prevented from playing a beneficial role by the stable calcium carbonate dispersion formed at pH values higher than 9.0. In alkaline water and wastewater treatment processes, the beneficial role of lime is attributable to its effect in raising the pH, rather than to the Ca content. The upper limit of the pH range in which the dissolved zinc compounds are precipitated and removed most effectively is shifted towards the higher pH values in the presence of Mg^{2+}, whereas the lower limit of the pH range in which the colloidal, quasi-colloidal dispersion of zinc compounds coagulates most effectively in the presence of colloid stabilizing substances is lowered by raising the concentration level of the Mg^{2+}. In alkaline media the magnesium hydroxide is almost comparable in effectiveness to the aluminium or iron(III) hydroxide coagulants in neutral media.

References

[1] Stumm, W., Morgan, J.J. (eds.): Aquatic Chemistry. Wiley-Interscience, New York 1981
[2] Morel, F.M.: Principles of Aquatic Chemistry. Wiley-Interscience, New York 1983
[3] Perry, J.H.: Chemical Engineers, Handbook. McGraw-Hill, New York London 1950
[4] Luner, P., Dence, C.: Mechanism of Color Removal in the Treatment of Pulping and Bleaching Effluents with Lime – Vol. I. Treatment of Caustic Extraction Stage Bleaching Effluent. Technical Bulletin No. 279 (1970)

[5] Luner, P. Dence, C. Bennett, O. Ota, M.: The Mechanism of Color Removal in the Treatment of Pulp and Bleaching Effluents with Lime. NCASI Technical Bulletin No. 243 (1970)

[6] Interstate Paper Corporation: Color Removal from Kraft Pulping Effluent by Massive Lime Treatment. US EPA-12040-ENC Dec, 1971

[7] Oswalt, J., Land, J.G.: Color Removal from Kraft Pulp Mill Effluent by Massive Lime Treatment. US EPA-R2-73-086 (1973)

[8] Black, A.P., Christman, R.F.: Chemical Characterization of Fulvic Acid. J. AWWA 55 (1963) 897

[9] Christman, R.F., Ghassemi, M.: Chemical Nature of Organic Color in Water. J. AWWA 58 (1966) 723

[10] Predali, J.J., Cases, J.M.: Zeta Potential of Mg Carbonates in Inorganic Electrolytes. J. Coll. Inter. Sci. 45 (1973) 3

[11] Joffee, D.J., Taylor, J.S.: Heavy Metal Removal from Colored and Non-Colored Aqueous Environments. Proc. AWWA Conf. Part 2. 681, San Francisco, California, USA, June 24-29, 1979

[12] Flentje, M.E.: Calcium and Magnesium Hydrates. J. AWWA 17 (1927) 253

[13] Lecompte, A.R.: Water Reclamation by Excess Lime Treatment of Effluent. TAPPI 49 (1966) 121

[14] Black, A.P., Christman, R.F.: Electrophoretic Studies of Sludge Particles Produced in Lime-Soda Softening. J. AWWA 53 (1961) 737

[15] Folkman, Y., Wachs, A.M.: Removal of Algae from Stabilization Pond Effluents by Lime Treatment. Wat. Res. 7 (1973) 419

[16] Wenthworth, D.F., Thorup, R.T., Sproul, O.J.: Polio-Virus Inactivation in Water-Softening Precipitation Processes. J. AWWA 60 (1968) 939

[17] Thayer, S.E., Sproul, O.J.: Virus Inactivation in Water-Softening Precipitation Processes. J. AWWA 58 (1966) 1063

[18] Sproul, O.J.: Virus Inactivation by Water Treatment. J. AWWA 64 (1972) 31

[19] Thompson, C.G., Singley, J.E., Black, A.P.: Magnesium Carbonate. A Recycled Coagulant – Part I.–II. J. AWWA 64/1 (1972) 11, 64/2 (1972) 93

[20] Berta, E., Oláh, J.: Survey of the Heavy Metals in Sludges in Hungary. Hidrológiai Közlöny, 469 (in Hungarian)

[21] Argo, D.G., Culp, G.L.: Heavy Metal Removal in Wastewater Treatment Processes: Part 2. Pilot Plant Operation. Water and Sewage Works (Sept, 1972) 128

[22] Dean, J.G., Bosqui, P.L., Lanouette, K.H.: Removing Heavy Metals from Wastewater. J. Env. Sci. Tech. 6 (1972) 518

[23] Wing, R.E., Swanson, C.L., Doane, W.M., Russel, C.R.: Heavy Metal Removal with Starch Xanthate-Cationic Polymer Complex. J. WPCF 46 (1974) 2043

[24] Maruyama, T., Hannah, S.A., Cohen, J.M.: Metal Removal by Physical and Chemical Treatment Processes. J. WPCF 47 (1975) 962

[25] Corona, J., Llaneza, H.M., Blanco, J.M., Llaneza, R.: Physicochemical and Biological Treatment of Leachates from the Central Urban Solid Waste Landfill in Asturias (Northern Spain). In: Heavy Metals in the Hydrological Cycle, M. Astruc and J.N. Lester (eds.). Selper, London 1988

[26] Patterson, J.W., Allen, H.E., Scala, J.J.: Carbonate Precipitation for Heavy Metal Pollutants. J. WPCF 49 (1977) 2397

[27] Idelovitch, E., Michail, M.: Trace Metal Removal from Secondary Effluent by Two-Stage High Lime. Progress in Wat. Tech. 10, 5/6 (1978) 821

[28] Vråle, L.: Chemical Precipitation of Wastewater with Lime and Seawater. Progress in Wat. Tech. 10, 5/6 (1978) 645

[29] Licskó, I.: Colloid Destabilization in the Treatment of Surface Waters Containing
Colloid-Stabilizing Pollutants. In: Chemical Water and Wastewater Treatment,
H.H. Hahn and R. Klute (eds.). Springer, Berlin Heidelberg New York 1990
[30] Licskó, I.: New Experimental Technique for Measuring the Growth of Aluminium-
and Iron(III) Hydroxide Flocs in the Treatment of Surface Water. Wat. Supply
9, 1 (1991) 45–52

Katalin Zotter and Istvàn Licskó
Water Resources Research Centre, VITUKI,
P.O. Box 27
H-1453 Budapest
Hungary

The Effects of Temperature, pH
and Rapid Mixing Gradient on the
Formation of Particles
in Treatment of Humic Water

P. Dolejs

Abstract

One of the primary objectives of the research described here was to contribute
to the understanding of mutual interactions of three selected factors in the
coagulation of humic waters. The influence of water temperature on coagulation
is still not well understood. Conclusions which are reported in the literature
often appear to be contradictory, because our tendency is to generalize our
results or not to report details (which might be tacit presumptions) about
different conditions in our experiments.

The experiments reported here were carried out at a pilot plant with model
humic water made of peat extract, distilled water and tap water. Particle size
distribution was determined from sedimentation curves as well as after centrifugation of samples. A three-factor central composite design was used to design
and evaluate the experiments. The temperature was varied from 1.7 to 16.3 °C,
the gradient of rapid mixing from 75 to 305 s^{-1} and the pH of coagulation from
5.06 to 6.31.

Results are presented in the form of three-dimensional figures. It was shown
that low temperature suppresses the formation of large flocs and causes higher
residual concentrations of aluminium. To achieve the lowest possible residual
concentrations of aluminium, the pH should be decreased as the temperature
increases. To obtain the largest quantity of large flocs, the tendency is the
reverse, i.e. with increasing temperature the coagulation pH shouldbe slightly
increased. This is also related to the coagulation stoichiometry, which is also
influenced by temperature. More about this was published previously. The alum
dose was kept constant in the experiments described here.

Higher G values during rapid mixing, as in the range used in this study,
increase the percentage of large particles in suspension, but, in practice, do not
influence either the residual concentrations of humic substances or aluminium.
This underlines the importance of a) solution chemistry and b) the perikinetic
phase of coagulation. On the other hand, temperature should not be considered
solely as a factor which can be equated with the perikinetic transport mechanism, but must be evaluated with regard to optimum solution chemistry for
coagulation, which is influenced by several other parameters.

1. Introduction

The effect of temperature in the treatment of drinking water has gained renewed interest in recent years. In the fifties and sixties, Bhoota's master's thesis [1] and the paper written by Camp, Root and Bhoota [2] which was based on it, dominated researchers' thinking. The most often-cited part of Bhoota's conclusions was that "the formation of floc, at all concentrations, is neither retarded nor prevented by the changes in temperature" [1]. This conclusion was based on a comparison of settling curves, produced at different temperatures, which were all adjusted to 10 °C.

This is, of course, absolutely correct from the strictly theoretical point of view, but of very limited use from the point of view of water treatment practice, as it is still not easy for plant operators to adjust the water temperature in their water works. The second point is that for optimal water treatment plant design [3], large flocs (or well-sedimentable flocs) are usually produced with the aim of separating them through settling in sedimentation tanks. In addition, it was shown that low intensity flocculation (optimum for sedimentation) gave faster breakthrough than the situation without preflocculation. The operation of preflocculation in the same manner as before sedimentation in traditional treatment may therefore be disastrous when used aheadof direct filtration [4]. Therefore, the effect of temperature on viscosity can in no way be separated in practice from the real sedimentation velocity in sedimentation tanks, and sedimentation velocity which does not have to be adjusted is the only thing the plant operator is interested in.

This is just one example of a possible source of misunderstanding or misinterpretation of high quality research by professionals in practice and, conversely, a disregard by researchers of what is really needed in practice. Therefore, I consider it quite necessary to express technologically oriented results in terms which reflect the technological similarities between practical processes, installations and laboratory or full-scale experiments. A basic rule of technological similarity in water treatment might be: treatment processes and installations are technologically similar if the suspensions have the same properties from the point of view of separation.

Probably the most comprehensive overview of the effects of temperature on water treatment was published by Heinänen [5]. The whole spectrum of possible ways temperature can affect coagulation is nicely presented, starting with the physical properties of water.

A great deal is known about water's magical properties and how they change with temperature [6, 7]. For this reason, an experimental approach was adopted to track down the interactions of three variables: temperature, rapid mixing and pH. As already mentioned by Hundt and O'Melia [8], different chemicals and different treatment conditions using the same chemical can bring about destabilization of fulvic acid by means of different mechanisms. Therefore, it is important to understand how the parameters function in relation to one another.

The effect of various parameters on coagulation/flocculation and their many mutual interactions was described in review papers by Gregory, Dempsey and Dentel [9–11]. In contrast to most papers which deal with the influence of water temperature on coagulation, including those published in the last few years [12–14], this paper presents experiments which were carried out in a pilot plant.

2. Experimental Methods and Materials

2.1 Pilot Plant

The experiments were performed in a pilot plant shown schematically in Fig. 1. The flow rate in all experiments was 1 l/min.

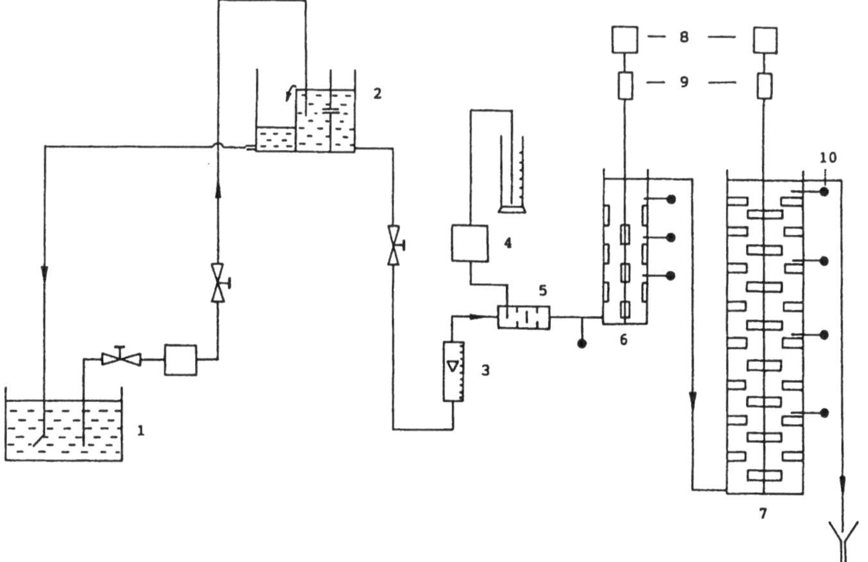

Fig. 1. Schematic diagram of the pilot plant. 1 – Storage tank for the model humic water; 2 – Constant head tank; 3 – Flow meter; 4 – Coagulant dosing pump; 5 – Homogenizer for mixing coagulant with raw water; 6 – Rapid mixing tank; 7 – Slow mixing tank; 8 – Electric motors; 9 – Strain gauge torquemeters; 10 – Sampling point for particle-size distribution determination (4 cm below the surface)

The cylindrical rapid mixing tank had a 9 cm diameter and a theoretical residence time of 168 seconds. The diameter of the slow mixing tank was 14 cm and the theoretical residence time was 1 220 sec.

The root mean square velocity gradient (G) in the slow mixing tank might be in the range of 2–100 s^{-1} and it was set to 7.3 s^{-1} and kept constant during the experiments. The G value for rapid mixing might be in the range of 60–1 200 s^{-1} and was varied in these experiments from 75 to 305 s^{-1}. The power applied to each mixing tank was set and measured with a revolution-meter and continuously monitored using a strain gauge torquemeter, used for the same purpose previously reported by Ødegaard [15].

A homogenizer was placed in the tube just in front of the inflow into the rapid mixing tank. This homogenizer unit consists of three plates which have four, one and four holes, respectively; these plates are placed perpendicular to the flow direction. The coagulant was introduced into the center of the tube, in front of the first plate. The theoretical residence time between the coagulant introduction into the water stream and the outflow from the third perforated plate was 0.65 sec.

The construction of this homogenizer should have reflected the time scale of the reactions which precede particle aggregation, as reported by Hahn and Stumm [16], and in light of the recent work of Klute [17], the homogenizer construction should be regarded as one more variable which deserves considerable attention.

2.2 Model Humic Water

A stock of humic substances concentrate was prepared by extracting peat with demineralized water; this stock was used in all of the experiments. Alkaline extraction was not used in order to avoid changes in humic substances molecules and to simulate natural conditions.

Tap water (having constant quality during the experiments) was diluted about four times with demineralized water for a concentration of $Ca + Mg$ ions $= 0.5$ mmol/l, and the humic substances stock solution was added to get a true colour equivalent of 50 Pt colour units ($COD_{Mn} = 8.1$ mg/lO_2). The initial alkalinity of the model humic water was adjusted to the desired value with $NaHCO_3$ or HCl. After adding all the components to the model humic water, it was gently mixed and cooled in a 120 liter raw water storage tank overnight to reach equilibrium.

The changes in the coagulation pH result from differences in initial alkalinity and the constant addition of coagulant. A 1.5 % solution of $Al_2(SO_4)_3 \cdot 18H_2O$ was used as coagulant. Coagulant was added at a constant rate of 40 mg/l as aluminium sulphate ($Al = 120 \ \mu$mol/l).

2.3 Three-Factor Central Composite Design

It would take one researcher using the classical experimental approach (using only one independent variable) decades to investigate experimentally the effect of several major variables (G and residence time in both rapid and slow mixing, type of homogenization, pH, dose of coagulant, type of coagulant, temperature, type of pollutant to be removed, basic chemical composition of the raw water, etc.) on coagulation/flocculation, as well as their interactions on a pilot-plant scale, and this would only be for five levels of each parameter. Various methods of experimental composite design may substantially reduce the number of experiments needed. A three-factor central composite design was used in this study. A few preliminary tests were used to determine the ranges of the three variables for the series of 15 experiments shown in Table 1.

Tab. 1. Three-factor central composite design for coagulation experiments with model humic water

Nr.	Design levels			Factor levels		
	X_1	X_2	X_3	X_1	X_2	X_3
1.	−1	−1	−1	3	95	0.3
2.	1	−1	−1	15	95	0.3
3.	−1	1	−1	3	285	0.3
4.	1	1	−1	15	285	0.3
5.	−1	−1	1	3	95	0.9
6.	1	−1	1	15	95	0.9
7.	−1	1	1	3	285	0.9
8.	1	1	1	15	285	0.9
9.	−1.215	0	0	1.7	190	0.6
10.	1.215	0	0	16.3	190	0.6
11.	0	−1.215	0	9	75	0.6
12.	0	1.215	0	9	305	0.6
13.	0	0	−1.215	9	190	0.235
14.	0	0	1.215	9	190	0.965
15.	0	0	0	9	190	0.6

Factors: X_1 = Temperature [°C]; X_2 = G (rapid mixing) [s^{-1}];
X_3 = Alkalinity [meq/l]

Using the experimental results, coefficients of the second order polynomial equation (Eq. 1) were determined and the results were drawn by the computer mostly in the form of two-dimensional figures, which were sections on the axis of one X variable illustrating the contour lines of different Y responses between the two remaining X variables. As for depicting the interactions between more than two independent variables, it is sad that our world is only three-dimensional.

$$y = a_0 + a_1 x_1 + a_2 x_2 + a_3 x_3 + a_{12} x_1 x_2 + a_{13} x_1 x_3 + a_{23} x_2 x_3$$
$$+ a_{11} x_1^2 + a_{22} x_2^2 + a_{33} x_3^2 \tag{1}$$

The following responses (Y) were determined:

Y_1 = Macroparticles fraction Y_4 = Non-aggregated aluminium fraction
Y_2 = Microparticles fraction Y_5 = Residual colour
Y_3 = Primary particles fraction Y_6 = pH.

Determination of Y_1–Y_4 is based on the real sedimentation velocity of the particles. The values of Y_1–Y_4 represent the fraction of total aluminium dosed in categories of particles which actually exist, i.e. particles with a sedimentation velocity within certain limits.

The sampling point for determining the particle-size distribution is located 4 cm below the water surface in the slow mixing tank. Particles which sedimented away from the 4 cm depth in 5 minutes sedimentation time are called "macroparticles" (Y_1) in this paper. Their sedimentation velocity is higher than 0.13 mm/s. Such particles can be separated efficiently in sedimentation tanks.

Smaller particles with a lower sedimentation velocity, sedimenting away from the 4 cm depth in 5–60 minutes, are called "microparticles" and their sedimentation velocity is higher than 0.011 mm/s. Such particles can still be separated in clarifiers.

The smallest particles in the experimental system are called "primary particles" in this paper. Their sedimentation velocity is of the order of 0.0001 mm/s and they are separated from the sample, which is taken after 60 minutes of sedimentation time, by centrifugation for 5 minutes at 2 000 g.

The fraction of aluminium and colour in the supernatant after centrifugation is considered to be non-aggregated (and therefore also non-removable by sand filtration). The close correlation between these residual concentrations after centrifugation and real plant data was mentioned in a previous paper [18].

The concentration of aluminium was determined using "aluminon" at pH 3.1. Alkalinity is expressed as concentration of HCO_3^- in meq/l. It was determined by titration with 0.1 mmol/l HCl to pH 4.3. Colour was determined spectrophotometrically at 387 nm [19].

3. Results and Discussion

The experimental results are shown in Tables 2 and 3. For purposes of illustration, Table 3 also gives the values of residual colour calculated back from the regression equation (Eq. 1).

The final results are presented together in three-dimensional figures. Figure 2 shows the relationship between the G of rapid mixing and temperature, with constant initial alkalinity set to 0.6 meq/l, which gives a coagulation pH = 5.85.

At this initial alkalinity (and resulting final pH), the concentration of non-aggregated fraction at a low temperature (1.7 °C) is about twice that at 16.3 °C. Rapid mixing does not have any significant influence on the non-aggregated fraction at this pH. Experimental results indicate that flocs which are formed at a lower pH (initial alkalinity = 0.3 meq/l) might even be more susceptible to breakup and liberation of non-aggregated fraction at lower pH (5.25) under the influence of higher G, see Table 1.

The effect of rapid mixing intensity on the macroparticles fraction is unambiguously positive at all levels of variables investigated. At the lowest temperature and the lowest G of rapid mixing, only 11 % of the dosed Al is in the form of macroparticles. The increase of G from 60 s^{-1} to 315 s^{-1} increases the fraction of macroparticles to 0.23 (or to 23 % of total aluminium dosed). This result should stress the importance of mixing for conventional treatment plants in winter, but on the other hand, its minor importance for direct filtration plants.

Tab. 2. Three-factor central composite design and product data of particle-size distribution for coagulation experiments with model humic water

Nr.	Design levels			Responses			
	X_1	X_2	X_3	Y_1	Y_2	Y_3	Y_4
1.	-1	-1	-1	0.052	0.108	0.638	0.202
2.	1	-1	-1	0.116	0.147	0.593	0.144
3.	-1	1	-1	0.069	0.155	0.523	0.254
4.	1	1	-1	0.204	0.177	0.448	0.171
5.	-1	-1	1	0.146	0.163	0.624	0.067
6.	1	-1	1	0.430	0.080	0.440	0.050
7.	-1	1	1	0.210	0.222	0.502	0.066
8.	1	1	1	0.450	0.130	0.360	0.060
9.	-1.215	0	0	0.093	0.099	0.715	0.093
10.	1.215	0	0	0.448	0.127	0.363	0.062
11.	0	-1.215	0	0.241	0.116	0.578	0.065
12.	0	1.215	0	0.492	0.102	0.344	0.062
13.	0	0	-1.215	0.050	0.153	0.481	0.316
14.	0	0	1.215	0.309	0.168	0.435	0.087
15.	0	0	0	0.328	0.090	0.497	0.085

Factors: X_1 = Temperature [°C]; X_2 = G (rapid mixing) [s^{-1}]; X_3 = Alkalinity [meq/l]. Responses: Y_1 = Macroparticles fraction; Y_2 = Microparticles fraction; Y_3 = Primary particles fraction; Y_4 = Non-aggregated Al fraction

Tab. 3. Three-factor central composite design and product data of measured and calculated values of residual colour and pH for coagulation experiments with model humic water

Nr.	Design levels			Responses		
	X_1	X_2	X_3	$Y_{5\ meas}$	$Y_{5\ calc}$	Y_6
1.	-1	-1	-1	17.1	17.6	5.25
2.	1	-1	-1	11.4	10.8	5.25
3.	-1	1	-1	17.6	17.5	5.25
4.	1	1	-1	9.5	10.2	5.25
5.	-1	-1	1	9.7	8.5	6.20
6.	1	-1	1	7.7	7.4	6.20
7.	-1	1	1	10.8	11.0	6.20
8.	1	1	1	10.1	9.4	6.20
9.	-1.215	0	0	11.4	11.9	5.85
10.	1.215	0	0	6.2	6.8	5.85
11.	0	-1.215	0	6.4	7.8	5.85
12.	0	1.215	0	9.2	9.0	5.85
13.	0	0	-1.215	16.7	16.4	5.06
14.	0	0	1.215	9.0	10.5	6.31
15.	0	0	0	11.0	9.3	5.85

Factors: X_1 = Temperature [°C]; X_2 = G (rapid mixing) [s^{-1}]; X_3 = Alkalinity [meq/l]. Responses: $Y_{5\ meas}$ = Residual colour-measured; $Y_{5\ calc}$ = Residual colour-calculated; Y_6 = pH at the outflow from the slow mixing tank

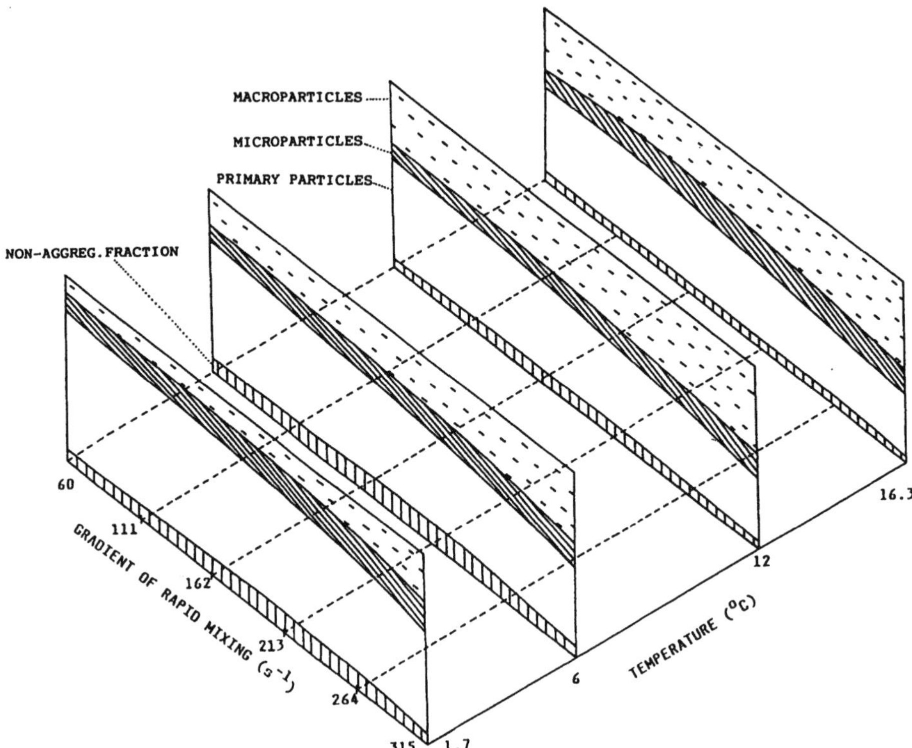

Fig. 2. Influence of temperature and gradient of rapid mixing on particle size distribution (initial alkalinity = 0.6 meq/l, pH = 5.85)

Figure 2 also illustrates the effect of a temperature drop at a conventional treatment plant with insufficient rapid mixing. The percentage of flocs with a high sedimentation velocity (i.e. macroparticles) is as high as 35 % of the total aluminium dosed at 16.3 °C but drops more than three times at 1.7 °C to 11 %.

Figure 3 shows the interaction between the G of rapid mixing and initial alkalinity (which corresponds to the final pH of coagulation), at 3 °C. Here the non-aggregated fraction is also practically independent of the intensity of rapid mixing, but strongly dependent on the chemistry of the system, i.e. pH in this case. A higher rapid mixing intensity might only help to create a larger number of settleable flocs if conditions are nearly favorable.

Figure 4 shows the interaction between temperature and coagulation pH at $G = 111$ s^{-1}. The influence of temperature is in some respects similar to that of rapid mixing.From the lowest to the highest temperature, the non-aggregated fraction decreases by 50 % with increasing temperature, while the fraction of macroparticles quadruples (at "optimum pH").

It is apparent that chemical conditions are of primary importance both for the formation of macroparticles and for residual aluminium. Rapid mixing, on the scale of G values used here, enhances the formation of large sedimentable

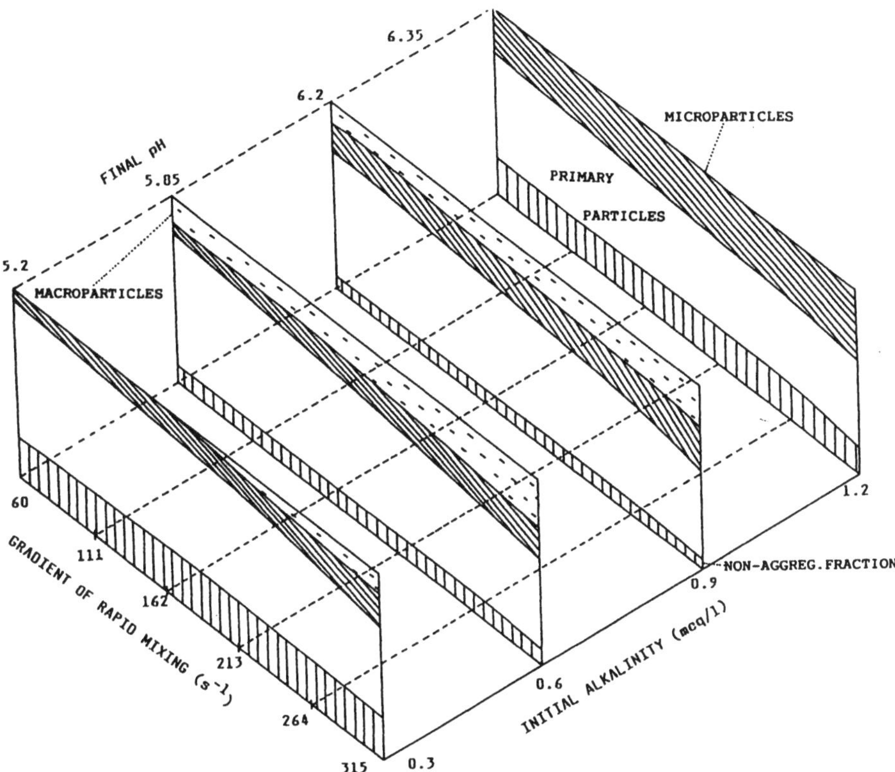

Fig. 3. Influence of initial alkalinity and gradient of rapid mixing on particle size distribution (temperature 3 °C)

flocs, while a temperature increase positively influences both the formation of large flocs and the decrease of the non-aggregated fraction of Al.

A more detailed analysis of the interaction between temperature and pH shows that the term "optimum pH" is relative, see Fig. 5. With respect to the non-aggregated fraction, the optimum pH drops with increasing temperature. The "optimum initial alkalinity" drops about O.055 meq/l per 10 °C, which corresponds to a pH drop of about 0.1 pH units/10 °C. On the other hand, with respect to macroparticles, the tendency is quite the reverse; the "optimum pH" increases with temperature. The "optimum initial alkalinity" increases about 0.105 meq/l for the 10 °C temperature increase, which corresponds to an increase in the "optimum pH" of about 0.12 pH units per 10 °C temperature increase.

This observation is in agreement with previous results [20], which came from a different series of experiments. In addition, it is quite in agreement with the observation that at lower temperatures, the optimum dose (without pH being held constant, which should simulate the real operating conditions at most plants) was lower for both variables (macroparticles and non-aggregated

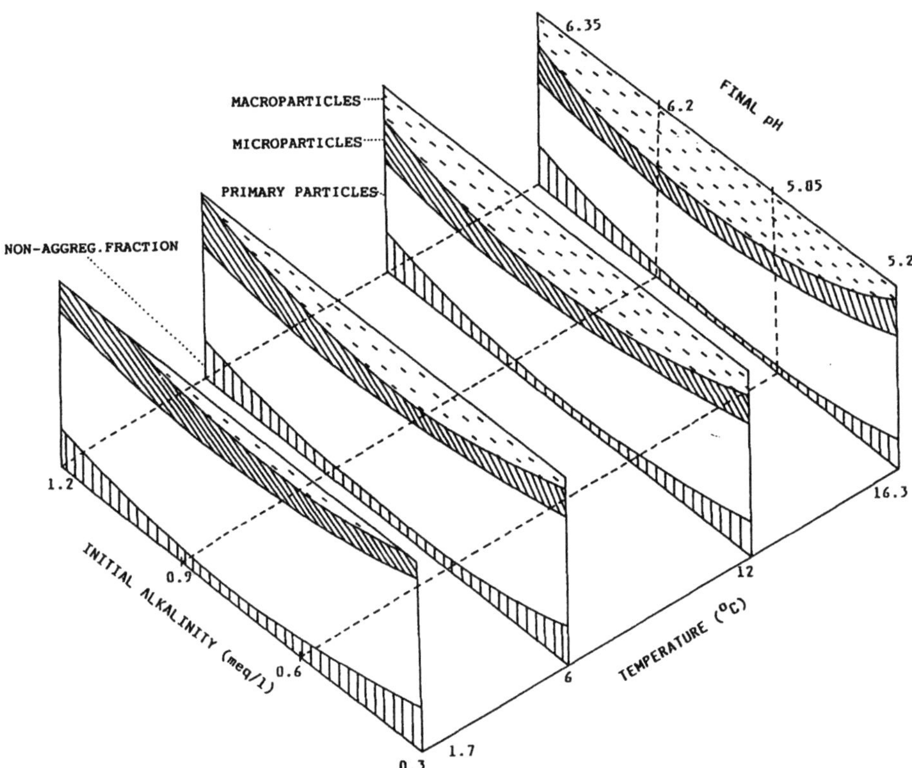

Fig. 4. Influence of temperature and initial alkalinity on particle size distribution (G of rapid mixing $= 111$ s^{-1})

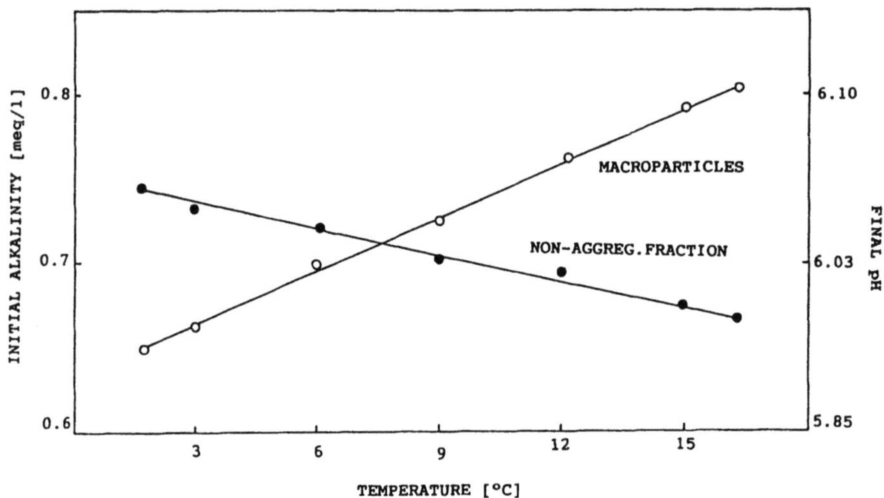

Fig. 5. Influence of temperature on optima of initial alkalinity (i.e. final pH) with respect to macroparticles and non-aggregated fraction, respectively (G of rapid mixing $= 111$ s^{-1})

fraction) [21], because in that paper, that series of experiments was carried out at constant initial alkalinity = 0.6 meq/l, which was below the "optimum initial alkalinity" (i.e. below the "optimum coagulation pH" at constant Al dose) estimated for the same water quality in this paper. It also seems to be in agreement with the results of Fettig, Ødegaard and Eikebrokk [22] who claim that optimum pH for floc separation may be 6 (sedimentation, filtration) or even ≤ 5 (direct filtration).

From the practical point of view, coagulation experiments should examine not only turbidity or colour removal domains or various characteristics of flocs, but quite necessarily also residuals of the coagulant dosed. Previous investigations [4, 22–24] have already addressed the fact that even if optimum destabilization may occur at pH = 4.5, a real plant cannot be operated solely on the basis of coagulation at this pH because of unacceptably high aluminium residual in the treated water. This is also the reason why the build-up of large flocs (macroparticles) is always reported in this paper together with the non-aggregated fraction of aluminium.

4. Conclusions

Chemical conditions are of primary importance for the formation of macroparticles, on the one hand, and residual aluminium, on the other. Rapid mixing, in the range of G values used here, only enhances the formation of large sedimentable flocs, while increasing the temperature positively influences both the formation of large flocs and the decrease in the non-aggregated residual fraction of Al.

The term "optimum pH" is not of universal validity. It changes with temperature and also with respect to the variable under consideration. The effect of temperature on "optimum pH" has opposite tendencies with respect to macroparticles, on the one hand, and the non-aggregated fraction of aluminium, on the other.

5. Literature

[1] Bhoota, B.V.: The Effect of Temperature Upon the Rate of Floc Formation. Master's Thesis, Massachusetts Institute of Technology, 1940, p. 9

[2] Camp, T.R., Root, D.A., Bhoota, B.V.: Effect of Temperature on Rate of Floc Formation. J. AWWA *32* (1940) 1913

[3] Wiesner, M.R., O'Melia, C.R., Cohon, J.L.: Optimal Water Treatment Plant Design. J. Env. Eng. ASCE *113* (1987) 567

[4] Ødegaard, H., Bratteb, H., Eikebrokk, B., Thorsen, T.: Removal of Humic Substances from Water. Report for IHSS 3rd Intl. Meeting, Norwegian Inst. of Technology, Trondheim 1986, pp. 25–26

[5] Heinänen, J.: Effect of Temperature on Water Treatment. Aqua Fennica *17* (1987) 201

[6] Eisenberg, D., Kauzmann, W.: The Structure and Properties of Water. Clarendon Press, Oxford 1968

[7] Letnikov, F.A., et.al.: Aktivirovannaja voda (Activated water). Nauka, Novosibirsk 1975 (in Russian)

[8] Hundt, T.R., O'Melia, C.R.: Aluminium-Fulvic Acid Interactions: Mechanisms and Applications. J. AWWA *80* (1988) 176

[9] Dempsey, B.A.: Removal of Naturally Occurring Compounds by Coagulation and Sedimentation. CRC Critical Reviews in Environmental Control *14*, 4 (1984) 311

[10] Gregory, J.: Fundamentals of Flocculation. CRC Critical Reviews in Environmental Control *19*, 3 (1989) 185

[11] Dentel, S.K.: Coagulant Control in Water Treatment. CRC Critical Reviews in Environmental Control *21*, 1 (1991) 41

[12] Morris, J.K., Knocke, W.R.: Temperature Effects on the Use of Metal-Ion Coagulants for Water Treatment. J. AWWA *76* (1984) 74

[13] Knocke, W.R., West, S., Hoehn, R.C.: Effects of Low Temperature on the Removal of Trihalomethane Precursors by Coagulation. J. AWWA *78* (1986) 189

[14] Hanson, A.T., Cleasby, J.L.: The Effects of Temperature on Turbulent Flocculation: Fluid Dynamics and Chemistry. J. AWWA *82* (1990) 56

[15] Ødegaard, H.: Orthokinetic Flocculation of Phosphate Precipitates in a Multicompartment Reactor with Non-Ideal Flow. Report, The University of Trondheim, NTH, Trondheim 1975

[16] Hahn, H.H., Stumm, W.: Kinetics of Coagulation with Hydrolyzed Al(III): The Rate Determining Step. J. Colloid Interface Sci. *28* (1968) 134

[17] Klute, R.: Destabilization and Aggregation in Turbulent Pipe Flow. In: Chemical Water and Wastewater Treatment, H.H. Hahn and R. Klute (eds.). Springer, Berlin Heidelberg New York 1990, pp. 33–54

[18] Dolejs, P.: A Simple Test for Determination of Optimal Doses in the Treatment of Humic Waters. In: Chemical Water and Wastewater Treatment, H.H. Hahn and R. Klute (eds.). Springer, Berlin Heidelberg New York 1990, pp. 377–390

[19] Dolejs, P.: Spectrophotometric Determination of Colour of Humic Waters. In: Proc. 20th Conf. "Hydrochemia '83", Res. Inst. of Water Management, Bratislava 1983, pp. 361–370

[20] Dolejs, P.: Interaction of Temperature, Alkalinity and Alum Dose in Coagulation of Humic Water. In: Chemistry for Protection of the Environment, L. Pawlowski, A. Verdier and W.J. Lacy (eds.). Elsevier, Amsterdam 1984, pp. 169–178

[21] Dolejs, P.: Effects of Temperature, Coagulant Dosage and Rapid Mixing on Particle-Size Distribution. Env. Protect. Eng. *9* (1983) 55

[22] Fettig, J., Ødegaard, H., Eikebrokk, B.: Humic Substances Removal by Alum Coagulation – Direct Filtration at Low pH. In: Pretreatment in Chemical Water and Wastewater Treatment, H.H. Hahn and R. Klute (eds.). Springer, Berlin New York 1988, pp. 55–66

[23] Vik, E.A., Carlson, D.A., Eikum, A.S., Gjessing, E.T.: Removing Aquatic Humus from Norwegian Lakes. J. AWWA *77* (1985) 58

[24] Vik, E.A., Eikebrokk, B.: Coagulation Process for Removal of Humic Substances from Drinking Water. In: Aquatic Humic Substances: Influence on Fate and Treatment of Pollutants, I.H. Suffet and P. MacCarthy (eds.). Amer. Chem. Soc., Washington DC, 1989, pp. 385–408

Petr Dolejs
Water & Environmental Technology Team
Box 27, Pisecká 2
CS-370 11 Ceské Budejovice
Czechoslovakia

Floc Separation

Effectivity of Liquid-Solid Separation as a Function of Apparatus Characteristics and Wastewater Quality

J. Mihopulos and H.H. Hahn

1. Introduction

Traditionally chemical dosing for the improvement of liquid-solid separation is designed and operated more or less independently of the geometry and hydraulic performance of the actual reactor. The wide-spread use of jar tests in day-to-day operations illustrates this fact [1].

There is not much information available on the relationship between the quality of flocs emanating from coagulation units and the separation process requirements. Exceptions are the recommendations for raw water treatment processes by Klute [2], which are based on general experience and not on the results of specific investigations or analysis.

If quantitative information were available on the influence of the flow structure in any given reactor on flocs, for instance, furthering the aggregation process or destroying already existing flocs, still needs to be determined. A plausible hypothesis is that there are reactor designs that might exert higher stress on aggregates than other designs. Those situations would require differently designed and operated coagulation units in order to produce stress-resistant flocs.

Such possible interactions are described schematically in Fig. 1 [3]. It can be seen that a readily and rapidly aggregating suspension leads to very efficient separation in (nearly) all types of separation units. Contrary to this a less readily aggregating system can only be handled effectively in separators of optimum design.

In order to quantify these relationships, systematic investigations were carried out on the interaction of specifically coagulated suspensions (under defined chemical and physical boundary conditions) with various separation reactors of different but defined geometries. These investigations are described and analysed in the following paragraphs. The study included sedimentation and flotation processes in rectangular and circular basins with moderate to significant depth under systematically varied hydraulic boundary conditions.

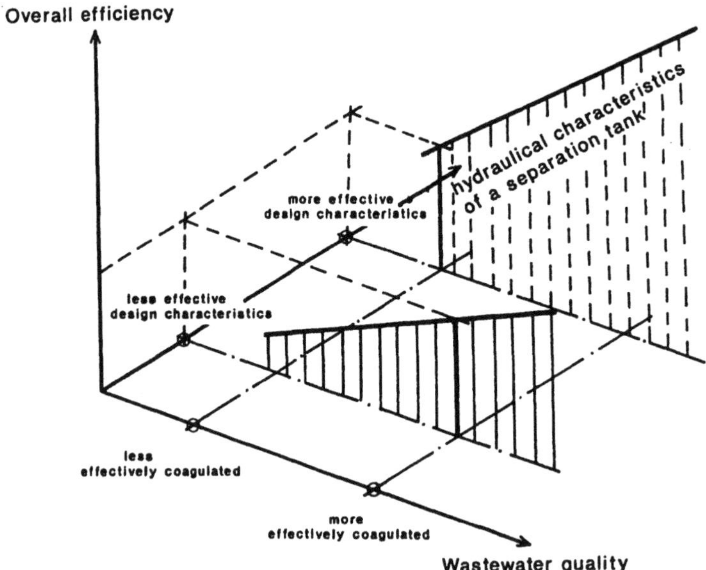

Fig. 1. Schematic representation of postulated relationship between physical and chemical parameters in floc formation and removal

2. Experimental Set-Up

The investigations were performed in scaled-down continuous flow reactors (with a scale of 1 : 40 relative to the size of frequently used real-world reactors – see Fig. 2) for aggregation and separation. Defined suspensions were coagulated under predetermined chemical and physical boundary conditions. The hydraulic regime was closely controlled in both coagulation and separation units through exact temperature control and the maintenance of a constant hydraulic throughput.

Separation was effected by sedimentation or alternatively by flotation. The various shapes of the sedimentation tanks that were employed are shown in Fig. 2. They were designed so that depth to width and length ratios were significantly different, leading to differences in the flow regime for the different reactors even at comparable hydraulic loading. Inflow and outflow structures as well as baffles were also different from design to design.

Flotation cells were designed with similar arguments. The generation of the necessary air bubbles was effected through electroflotation. This allowed easy control of the bubble generation and thus the air-solids parameter which is crucial for flotation processes. Furthermore, electroflotation generates bubbles with a very small diameter; this will then no longer be a rate- or process-controlling parameter. The electroflotation unit used in this study was described earlier (Mayer et al. [4]).

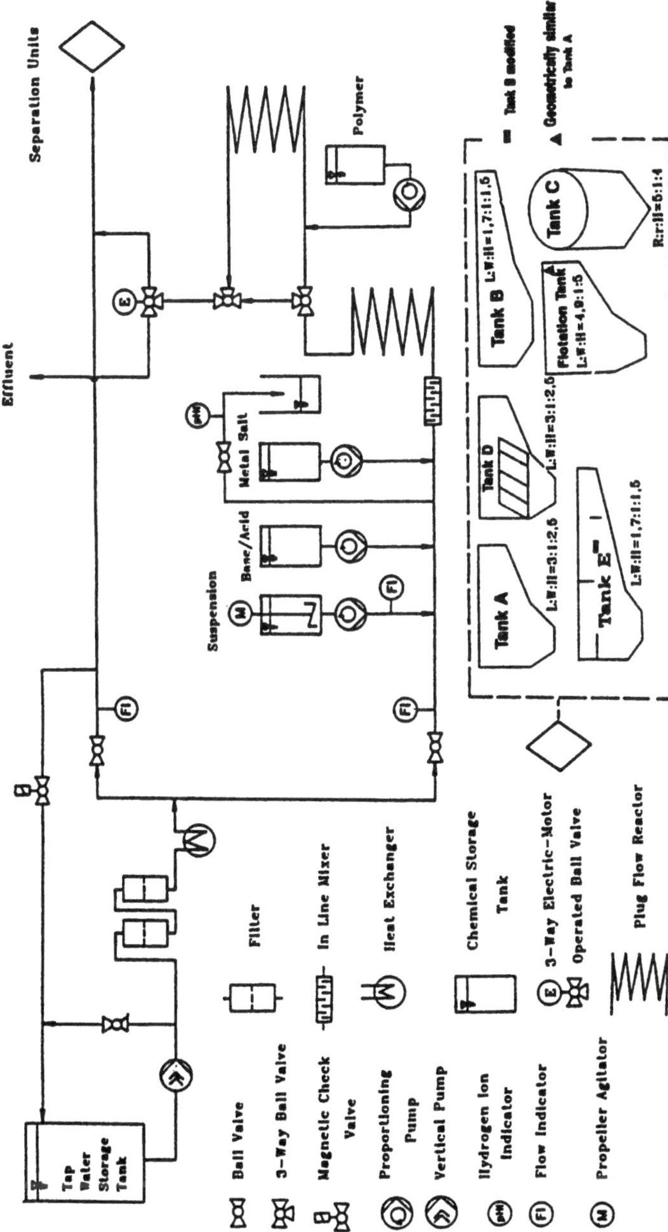

Fig. 2. Experimental set-up

Colloids used for the study had to be as representative as possible for those encountered in water and wastewater treatment. They also should be as defined as possible, both in terms of physical properties (size, density, etc.) and chemical characteristics (for instance, surface charge and its change with variations in the composition of the suspending medium, etc.) Above all, the resulting suspensions were to be as homogeneous as possible. Silicon-based micro-glass spheres suspended in tap water represented the standard suspension used in all investigations (see Table 1).

Tab. 1. Physico-chemical characterization of suspensions used in investigation

Suspension	Suspending medium
Material: Silicon-basis micro-glass spheres (Type 5000) Density: 2.50 g/cm^3 Size: 3–5 μm Concentration: 100 mg/l Chemical Composition: 72.5 % SiO_2 13.8 % CaO 13.7 % Na_2O	pH: from 4.5 to 9.8 depending upon coagulant type Conductivity: 700 $\mu s/cm$ at 12 °C Total Hardness: 21 °dH COD: 7.8 mg/l

With today's insight into agglomeration processes it is possible to 'engineer' flocs having very diverse characteristics, i.e. strong flocs resisting destruction in the process of liquid-solid separation as well as flocs that could predictably to be destroyed under the pre-set operating conditions of the separation units. The chemical boundary conditions used for floc formation are summarized in Table 2). They correspond to floc formation characteristics as they are reported in the literature (cf. Mühle [5]).

Tab. 2. Coagulation-flocculation conditions explored in investigation

Conditions	Material	Concentrations mg/l	pH resulting
Ca^{2+}	$Ca(OH)_2$	28	7.3
Al^{3+}	$Al_2(SO_4)_3$	20	7.3
Polymer	BC11L (kat.)*	5	6.8
Fe^{3+}	$FeCl_3$	20	6.2
Fe + Polymer	$FeCl_3$ + BC11L* (kat.)	12 + 2	6.9

* Fa. Stockhausen GmbH, Germany (Praestol BC11L)

Coagulation and flocculation characteristics were quantified by using the 'collision efficiency factor' as determined with a standard stirrer in a standard aggregation reactor from particle counting analyses using the laser particle analyser CIS Galai II from LOT GmbH [6]. It is understood that this serves only as a relative measure of floc property and floc formation. Other concepts for analysing floc properties, such as floc strength (Kleine [7]) would have required too detailed an investigation and would not have furnished more insight into the coagulating and flocculating properties of the suspensions. This is used as a quantitative measure for the X-axis of Fig. 1.

The quantitative description of the hydraulic performance of the separation units – sedimentation tanks as well as flotation tanks – was one major aim of this investigation. The flow structure, such as conditions of turbulence (microscopic parameters), and the flow regime, such as hydraulic throughput and extent of mixing (macroscopic parameters) were determined by tracer experiments. The effluent concentrations in the investigated reactors under different hydraulic boundary conditions are shown in Fig. 3. Each of these concentration curves can be summarized in terms of a characteristic (dimensionless) number, the 'dispersion number' (Levenspiel [8]).

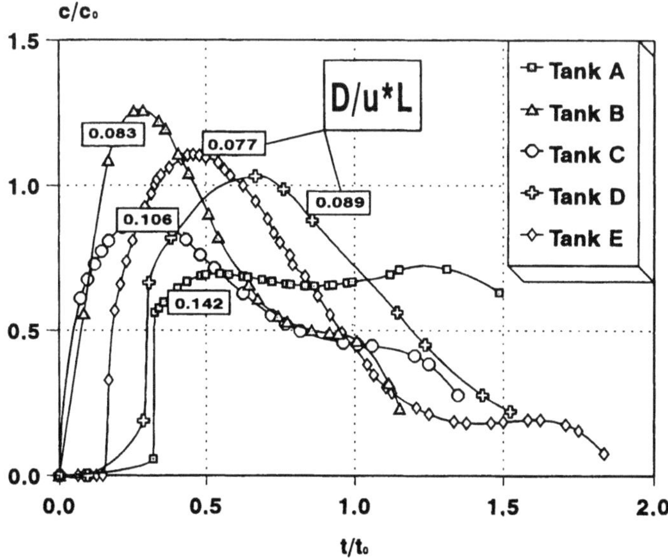

Fig. 3. Flow-through characteristics of separation tanks used in investigation, described as change in tracer concentration in effluents

The various tanks are characterized by different dispersion numbers. If the overflow rate, which was set to 0.58 m/h for the investigations reported in Fig. 3, is changed, then the absolute values of the dispersion number will also change. Some reactors show a more noticeable effect than others (Mihopulos [9]).

Reactor theory yields the basis for an evaluation of the meaning of the dispersion number: very small values indicate that the flow regime ressembles more and more that of a plug flow reactor. The other end of the spectrum of possible flow regimes, the completely mixed flow reactor, is characterized by large disperion numbers, theoretically, infinitely large numbers. It is this dispersion number (D_N) that will be used to quantify the Y-axis of the schematic in Fig. 1.

$$D_N = \frac{D}{u \cdot L}$$

D 'Turbulent' dispersion in the reactor [m^2/s]
u Convectional flow [m/s]
L Characteristic length of reactor [m]

Tanks C and D appear to be less affected by changes in the hydraulic loading than other tank designs. With increasing surface loading, the conditions in tanks A, B and E change from more plug flow to more completely mixed flow.

It is necessary to recall at this point that sedimentation and flotation tank performance are generally formulated on the basis of a 'surface-loading' principle which stipulates plug flow conditions; thus, the closer a reactor approaches plug flow conditions, the more likely a high solids separation performance.

The performance of the separation unit was analysed by introducing a 'delta' impulse, i.e. an infinitesimally small volume of suspension from the aggregation chamber into the sedimentation or flotation unit by means of a three-way switch. The resulting small number of aggregates were then followed on their way through the separation unit by means of counting and image analysing devices. (This type of analysis was also employed to follow the coagulation process.) This provides the basis for the quantification of the Z-axis in Fig. 1.

3. Results

The effect of (waste)water quality and apparatus characteristics on the overall removal efficiency of solids is described by at least a three-dimensional relationship. Thus, the experiments were performed in such a way that one set of variables was varied systematically while the others were kept constant. The parameters of the Y-axis of Fig. 1 were kept constant. In such an experimental series, coagulating/flocculating boundary conditions were then changed systematically, exploring the effect of the independent variable of the X-axis of Fig. 1.

The efficiency of the separation process (as affected by the coagulation process) is described as 'overall solids' retention, derived from particle counting analyses through summation and integration. The efficiency was also evaluated quantitatively by measuring specific (microscopic) suspension parameters such as floc shape, etc. (Mihopulos on image analysis, supported by CIS Galai II Software from LOT GmbH [10]). These latter data were used to explain incongruencies observed in the performance of sedimentation and flotation tanks.

3.1 (Coagulation and) Sedimentation

Figure 4 illustrates the type of results obtained for the series of five sedimentation tanks investigated (cf. Fig. 2). The systematically varied coagulation and flocculation conditions are described as coagulation efficiency factor and also as observed (cumulative) floc size distribution.

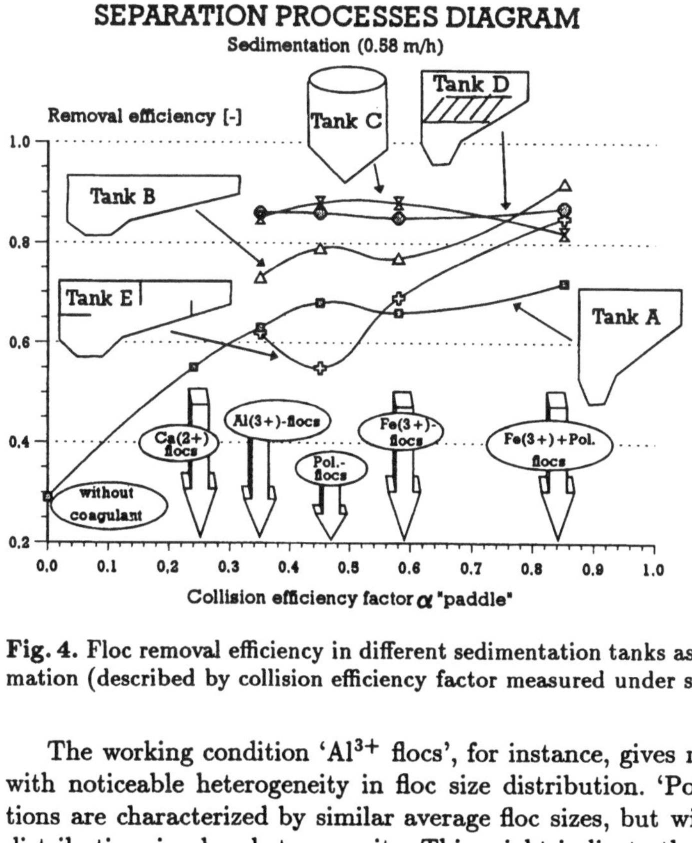

Fig. 4. Floc removal efficiency in different sedimentation tanks as function of floc formation (described by collision efficiency factor measured under standard conditions)

The working condition 'Al^{3+} flocs', for instance, gives medium-sized flocs with noticeable heterogeneity in floc size distribution. 'Polymer floc' conditions are characterized by similar average floc sizes, but with a more narrow distribution, i.e. less heterogeneity. This might indicate that the flocs have a higher stress resistance overall. 'Fe^{3+} flocs' describes a relatively faster or more readily coagulating system with consequently larger average floc sizes and a not so wide distribution. The 'Fe^{3+} flocs plus polymer' is a very quick coagulating/flocculating system with correspondingly large average floc size and moderate heterogeneity (see insert in Fig. 4).

These suspensions are separated with different efficiencies in the tanks investigated:

Tank C, for instance, a so-called 'upflow-clarifier' which has been identified as a reactor with a relatively stable flow structure (a mixture between plug flow and completely mixed flow) even at changing surface loadings, shows very high values of removal efficiency over the whole range of chemical conditions investigated.

In contrast tank A shows a relatively low removal efficiency, even for higher values of collision efficiency, i.e. for better aggregating systems. Tank A is characterized hydraulically by a dispersion number which, for lower surface loadings, is closer to plug flow conditions, while the dispersion number increases noticeably at higher surface loadings.

Tank E shows a significantly different behavior. Flocs formed under less favorable conditions, expressed as lower collision efficiency factor, are removed with quite low efficiency. If coagulation conditions are improved, shown by higher values of the collision efficiency factor, then the separation becomes very effective. This would be an example of improving practical operation: tanks of this type are more dependent in their operation upon an optimized floc formation step than other separation tanks.

In summary then, Fig. 4 shows that: (a) there are separation unit designs that result in good efficiency even when coagulation is less effective; (b) for some types of reactor design the effectiveness of the coagulation process is more significant than for others; and (c) some sedimentation tank designs are more or less independent of the effectiveness of the preceding coagulation stage. It has to be kept in mind that there are only discrete values available for this relationship, the interpolated curves are not te be extrapolated.

3.2 (Coagulation and) Flotation

The investigations on the flotation process used the same colloid-coagulant system as the sedimentation studies (see Table 2 as well as Sect. 3.1). Likewise, the geometry of the reactors employed for (electro-)flotation was kept as similar as possible to the geometry of the sedimentation tanks. The evaluation of separation efficiency was the same as in the case of sedimentation (cf. Sect. 3.1). However, there is one significant difference between sedimentation and flotation: the 'independent' process variable air-solids-ratio. The possibility and necessity of pre-selecting the amount of air provided for the flotation process introduces an additional degree of freedom which has to be kept constant or it has to be controlled in a quantitative way.

Figure 5 shows the effect of the process variable 'air-solids-ratio' (A/S) on the performance of a flotation cell of a geometry similar to the one of sedimentation tank A. It is seen that removal efficiency increases with increasing A/S ratio, as reported by others as well.

There is one exception and that is the polymer-flocculated suspension. This suspension is not removed very well in this flotation unit. There may be several reasons for this:

– the possibly higher density of these flocs, and

Electroflotation / Separation Diagram

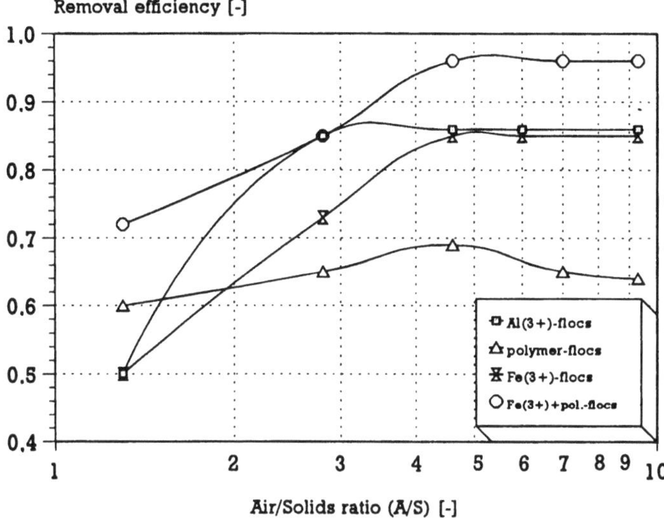

Fig. 5. Floc removal efficiency through flotation as function of A/S ratio for different floc formation conditions

– the relative scarcity of coagulant/flocculant, which probably prevents an effective association between flocs and air bubbles. (This hypothesis should be verified by further investigations.)

The other suspension types are removed with increasing efficiency as the parameter A/S increases. However, there are differences in the observed removal rates: the readily coagulating/flocculating system 'Fe^{3+} plus polymer' is separated best, at lower A/S values as well as at higher ones. The 'least' readily coagulating system 'Al^{3+} flocs' shows removal efficiencies comparable to the better coagulating 'Fe^{3+} floc' system: quite low removal rates at low A/S values and a noticeable increase in removal efficiency with increasing A/S values.

If one keeps the parameter A/S constant, thus eliminating the effect of this parameter upon the removal efficiency of coagulated suspensions, one finds similar relationships between suspension characteristics and removal effect as have been described for sedimentation units (cf. Fig. 4). For purposes of illustration, two series of observations are shows in Fig. 6 for A/S ratios of 2.8 % and 4.6 %, respectively. The abscissa, depicting the independent variable 'aggregation tendency', is quantified again in terms of the observed coagulation efficiency factor. Again, there are only discrete values available for the functional relationship in question. The figure illustrates clearly that:

– The less rapidly coagulating system 'Al^{3+} flocs' is separated in flotation as effectively as the better coagulating system 'Fe^{3+} flocs'. In fact, this system appears more suitable for flotation than the quite effective coagulating/flocculating 'polymer-floc' system (which was removed more satisfactorily by sedimentation).

– The system 'Fe^{3+} plus polymer flocs' shows the best removal efficiency as is also the case in sedimentation units. Higher A/S ratios appear to be more effective at higher overall chemical dosages or at higher coagulation/flocculation tendencies. This suggests that there might be support for the air bubble attachment to flocs by means of chemicals.

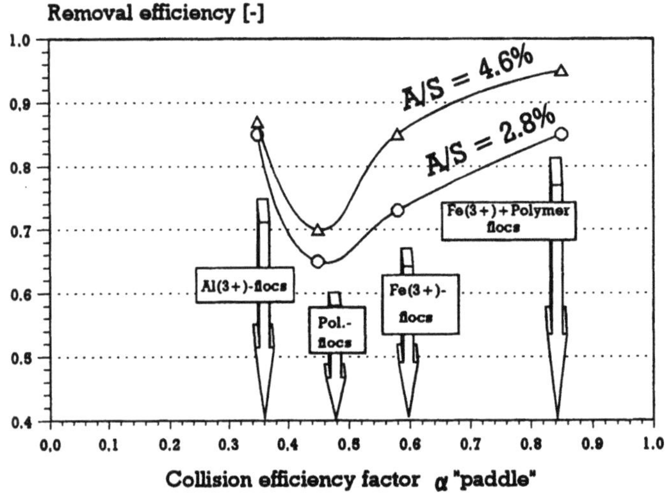

Fig. 6. Floc removal efficiency through floation as funtion of floc formation (described by collision efficiency factor measured under standard conditions)

4. Discussion

The investigations reported here were aimed at clarifying the interaction between floc formation (trend) or floc properties (X-axis), differently designed and operated separation units (Y-axis) and separation efficiency (Z-axis). These interactions can be described in absolute terms. They can also be illustrated by comparing the efficacy of different separation units for specified suspensions.

4.1 Which Type of Floc is Better Removed in Sedimentation, Which in Flotation?

The comparison of sedimentation and flotation efficiency of differently conceived flocs is shown in Fig. 7. The axes of these figures are identical with those of Figs. 4 and 6 and correspond to the X- and Z-axis of the introductory schematic. The data and the corresponding functions show clearly:

- The efficiency of a flotation unit of a geometry similar to that of sedimentation tank A is noticeably higher for the not so readily coagulating suspension (Al^{3+}). There is an indication that the inefficient sedimentation of an Al^{3+}-coagulated system can be upgraded.
- Rapidly coagulating and flocculating suspensions also appear to be removed with higher efficiency in flotation units.
- There is an intermediary situation with respect to the rate of coagulation/flocculation of suspensions; here sedimentation and flotation show similar efficiency in the liquid-solid separation.

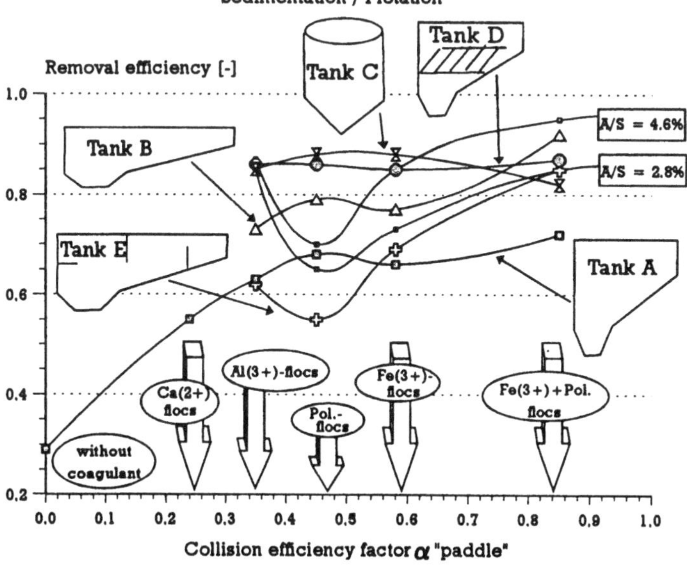

Fig. 7. Floc removal efficiency through sedimentation or flotation as function of floc formation (described by collision efficiency factor measured under standard conditions)

Other authors have also reported that the Al-type floc is removed better in flotation than in sedimentation (Rosén [11]).

The explanation given earlier, that the specific density of aluminium, which is lower than that of iron, is responsible for a better flotation effect is, however, not convincing in light of these data. It appears that floc growth is not rapid enough for the aluminium system to obtain sufficiently high sedimentation rates, yet large enough for good flotation. To what extent the hydraulic regime in tank A is (positively) affected by the bubble movement is still open to speculation; presently available data on particle and floc characteristics suffice for an elucidation of the different removal efficiencies of aluminium flocs in sedimentation and flotation.

If one inspects the insert in Fig. 4, showing the (cumulative) particle and/or floc size distribution function, one finds support for the following statements:

- Both smaller and larger flocs are removed almost equally well by flotation if the bubble attachment is guaranteed.
- Suspensions with heterogeneous size distribution characteristics are removed well in flotation units.
- Systems with sparingly dosed chemicals might not be separated well in flotation if there are not enough chemicals securing bubble attachment in the case of unfavorable surface tension relationships.

4.2 Which Aggregation Process Provides Less Fluctuating Separation Efficiency in Which Separation Reactor?

In Sect. 2 the question of the 'stability' of a flow regime has already been addressed. It has been shown that there are tank designs for which characteristics do not change to any practical extent if the hydraulic loading is increased, while the performance of other tank designs seems to be more subject to such changes under boundary conditions. Similarly, a relative stability or instability in separation efficiency is to be expected, resulting from changing chemical boundary conditions, i.e. changing coagulating/flocculating tendencies.

Therefore, the fluctuation of separation efficiency in different sedimentation tanks has been quantified and described as a function of the aggregating properties of the suspension (see Fig. 8). Such analysis shows clearly:

- There are tanks with lower fluctuation in efficiency and those with a higher fluctuation.
- Tanks C and D are more favorable from this point of view than tanks A, B and E. This is well explained by the hydraulic characteristics described in Sect. 2.
- Aluminum-coagulated suspensions show more sensitivity to different tank designs than other suspensions.
- Better coagulating/flocculating systems lead to more stable removal efficiencies.

If one tries to correlate this last statement again with the (microscopic) information on particle and floc size distribution, one finds confirmation of these conclusions. The higher the rate of coagulation/flocculation, the larger the probability that the minimum required floc size for a given separation process is attained. A comparison of sedimentation and flotation units cannot yet be made due to the lack of data on flotation.

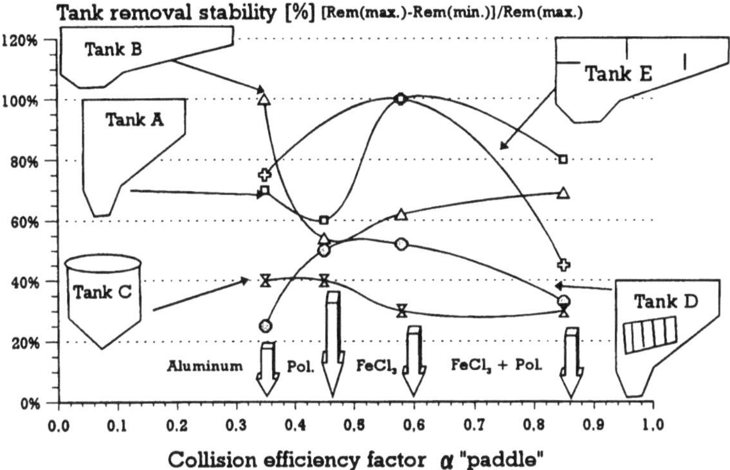

Fig. 8. Stability of floc removal efficiency through sedimentation as function of floc formation (described by collision efficiency factor measured under standard conditions)

4.3 Can One Expect to Find a Quantitative Relationship Between Water Quality Parameters, Reactor Characteristics and Removal Efficiency Values?

'Correlation functions' between suspension characteristics, specific tank designs as expressed in terms of reactor theory and separation unit efficiency are shown in Fig. 9. It is seen that both independent variables, 'alpha' and 'D_N', have noticeable effects on the removal efficiency in sedimentation tanks. It can also be deduced from this graph that an insufficient removal rate resulting from non-optimal hydraulic design of the tank can be compensated by improved operation in terms of improved aggregation characteristics (within certain limits), and vice versa.

One has to keep in mind that these 'correlation functions' do not describe a continuously investigated domain of aggregation and separation. Thus, there is no guarantee that one can 'interpolate' between the correlation functions shown for individual tank designs. For practical purposes, this means that the characterization of a tank by the 'dispersion number' alone, as is done in this study, may not be sufficient.

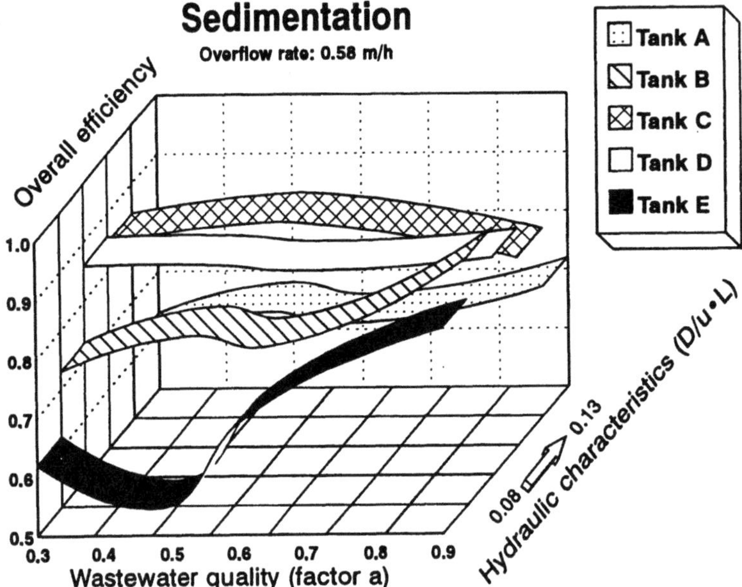

Fig. 9. Experimental results describing the relationship between physical and chemical parameters in floc formation and removal

5. Conclusions

From the aforementioned empirical insights, the following picture emerges for the interaction of coagulation/flocculation with the sedimentation or flotation processes:

1) The geometry of the reactor itself has – even if all other parameters are kept constant – a marked effect upon the removal of aggregated suspensions. This has been shown in detail for sedimentation tanks.

2) There are reactor types (reactor geometries) that show a rather constant flow regime even if the hydraulic loading is increased. Again, this has been investigated and shown in detail for the sedimentation apparatus.

3) Not readily aggregating suspensions are generally removed to a lower degree (in sedimentation units) than rapidly aggregating systems.

4) Removal efficiency becomes more stable, i.e. shows smaller fluctuations, if the aggregating tendency of the system is high(er) and if the hydraulic characteristics of the separation tank remain stable over wider ranges of hydraulic loading. This has been shown in detail for sedimentation units. It may be speculated that the process of flotation compensates for some of these negative effects.

5) The use of flotation for liquid-solid separation may compensate for some of the aforementioned shortcomings, in particular

- with respect to less readily aggregating suspensions if they are coagulated with Al^{3+};
- (probably) with respect to less favorable flow regimes resulting from an increase in hydraulic loading.

6) Systems that sediment readily, possibly due to a large average aggregate size and/or a high aggregate density, also show very good flotation characteristics (i.e. flotation does not require smaller and/or lighter flocs).

7) Not so readily aggregating systems, such as a suspension coagulated with Al^{3+} and showing a lower collision efficiency factor, are separated with very good results in flotation; systems flocculated with polymer and a resulting medium collision efficiency factor are not very suitable for flotation, possibly due to a lack of chemicals to support the effective attachment of air bubbles to solids.

8) A heterogeneous floc size distribution appears to have an effect upon the efficiency of the sedimentation process if the aggregation process has not gone very far, i.e. is characterized by a lower aggregation rate; the influence of this phenomenon is reduced if systems are aggregated to a larger extent. The process of flotation does not appear to be very much affected by this boundary value.

9) The better the hydraulic performance of the tank (expressed, for instance, as dispersion number) and the higher the aggregating tendency of the suspension (expressed as collision efficiency factor), the smaller the differences in efficiency between sedimentation and flotation processes.

We have tried to quantify these relationships (Fig. 9). Yet, the data on which these correlations have been based do not cover the entire range of possible boundary values; therefore, the 'correlation functions' shown in Fig. 9 should rather be taken as illustrations and not necessarily as generally valid functions.

If these interactions are to be included in design and operation of aggregation and separation units, then laboratory investigations with site-representative wastewater in project-specific reactors should be carried out.

6. Literature

[1] Black, A.P., Harris, R.J.: New Dimensions for the Old Jar Test. Wat. and Wastes Eng. (1969) 49

[2] Klute, R.: Abstimmung der Flockeneigenschaften auf den Abtrennverfahren. Proceedings of the "Flocken-Symposium" in Karlsruhe. Wasser (DVGW) *42* (1989)

[3] Hahn, H.H.: Hydraulics of Coagulation Chambers. A proposal for the SFB 210/A/30, Strömungsmechanische Bemessungsgrundlagen für Bauwerke. Karlsruhe 1989

[4] Mayer, J., Zhang, L., Hahn, H.H.: Liquid-Solid Separation by Electroflotation: An Attractive Alternative to Dissolved Air-Flotation. In: Chemical Water and Wastewater Treatment, H.H. Hahn and R. Klute (eds.). Springer, Berlin Heidelberg New York 1990, pp. 151–167

[5] Mühle, K.: Zusammenwirken anorganischer Flockungsmittel und organischer Flockungsmittel. In: Vor-, Simultan- oder Nachfällung? Entscheidungskriterien für Planung, Entwurf und Betrieb, H.H. Hahn und R. Pfeifer (eds). Schriftenreihe des ISWW, Universität Karlsruhe, Bd. 61, 1991 pp. 173–194

[6] CIS Computerunterstützte Partikelanalyse (Manual). LOT GmbH (former ORIEL), Im Tiefen See 58, D-6100, Darmstadt (1980)

[7] Kleine, U.: Der Einfluß der Flockenbildungsbeanspruchung auf die Festigkeit und das Sedimentationsverhalten von Flocken bei der Zentrifugalabscheidung. Dissertation, Department of Chemical Engineering, University of Karlruhe, 1992

[8] Levenspiel, O.: Chemical Reaction Engineering, an Introduction to the Design of Chemical Reactors, Chapt. 9. Wiley and Sons, New York 1962

[9] Mihopulos, J., Hahn, H.H.: Bemessung von Flockungs- und Separationsreaktoren mittels Test-Apparaturen und Nomogrammen. A Report for the Projekt Period 1988–1990, R. Friedrich (ed.). SFB 210/A/75, Karlsruhe

[10] Mihopulos, J.: Digitale Bildverarbeitung zur Analyse von Sedimentationsvorgängen. Proceedings from the 2nd Colloquium in Strömungssichtbarmachung und Digitale Bildauswertung in Karlsruhe 1991, SFB 210 (in print)

[11] Rosén, B.: Dissolved Air Flotation. In: Chemical Water and Wastewater Treatment, A. Grohmann, H.H. Hahn, und R. Klute (eds.). Fischer, Stuttgart/New York 1985

John Mihopulos and Hermann H. Hahn
Institut für Siedlungswasserwirtschaft
Universität Karlsruhe
Am Fasanengarten
D-7500 Karlsruhe 1
Germany

An Analysis
of Floc Separation Characteristics
in Chemical Wastewater Treatment

H. Ødegaard, S. Grutle, and H. Ratnaweera

Introduction

Chemical treatment of wastewater consists of three unit processes: coagulation/precipitation, flocculation and floc separation, the latter normally being carried out by sedimentation. The reaction step (coagulation/precipitation) is obviously of fundamental importance to the total treatment result,but the separation steps (flocculation and floc separation) have far greater importance from an economic point of view, since separation normally takes hours to complete while the reaction is completed within seconds.

Traditionally, salts of aluminium and iron have been used as coagulants in wastewater treatment where the treatment objective primarily has been particle and phosphate removal. In most countries chemical treatment is established either in combination with biological treatment (simultaneous precipitation) or after biological treatment (post-precipitation). In Norway, the experiences with purely chemical treatment plants have been good, as demonstrated by a recent survey carried out by Ødegaard [1], see Table 1.

Tab. 1. Treatment results from 87 Norwegian chemical WWT plants in 1990 [1]

Parameter	N^1	n^2	Influent mg/l	Effluent mg/l	Removal %
BOD_7	23	183	167 ± 95	27.2 ± 12.7	80.9 ± 9.6
COD	87	1193	463 ± 251	104 ± 38	74.9 ± 7.9
P_{tot}	87	1270	5.24 ± 2.53	0.26 ± 0.16	94.0 ± 5.5
SS	78	931	233 ± 186	16.6 ± 9.6	90.6 ± 7.9

[1] Number of plants; [2] Number of monthly averages

Since chemical treatment of wastewater is primarily a separation process, Table 1 is an excellent demonstration of the fact that a very substantial part of the pollutants can be separated directly from the raw wastewater without having to be biologically treated in advance.

Lately the need to remove nitrogen from municipal wastewater has increased as a result of the threat of eutrophication in coastal waters. Biological nitrogen removal requires considerable bioreactor volumes. In many cases in Scandinavia a pre-precipitation, post-denitrification process will be chosen because of its compactness and, consequently, the low investment cost.

Two goals can be set for the chemical pre-precipitation step in such a case:

a. The chemical coagulation/precipitation reaction step has to be carried out in such a way that it enhances the subsequent biological N-removal.

b. The separation of the coagulated/precipitated flocs has to be as efficient and inexpensive as possible.

It has been shown that prepolymerized aluminium coagulants may offer several advantages with respect to the first point when used as coagulants in pre-precipitation plants with N-removal in the bio-step [2].

It has also been shown by several authors [1, 3–6] that the use of organic polymers as flocculation aid may enhance floc separation efficiency considerably. The use of polymers has therefore become more common in recent years. In the survey of Norwegian chemical plants carried out by Ødegaard [2], it was found that more than 10 % of the total number of chemical plants in Norway (and several of the big ones) now use a polymer as flocculation aid at an average dosage of 0.48 ± 0.23 mg/l.

While studies of reaction mechanisms may be done in batch or semi-batch laboratory apparatuses, floc separation characteristics cannot properly be studied in batch reactors. The goal of this study was therefore to develop an apparatus convenient for comparing floc settling and thickening characteristics between various combinations of main coagulants and flocculants and to use this apparatus for such an evaluation.

Experimental Methods and Materials

Experimental Apparatus

The floc separation apparatus that was developed is shown in Fig. 1. It is a small (2 l/min) continously operating chemical treatment plant based on in-line mixing of chemicals, pipe flow flocculation and floc blanket sedimentation. The plant can easily be operated in a laboratory. Each of the treatment steps is designed in such a way that the actual unit process can be controlled.

The water to be treated is pumped from a raw water tank into the plant by a peristaltic pump, whereby the flow (2 l/min) is controlled by a rotameter. In the raw water tank good mixing is provided for by the use of a circulating pump. This arrangement prevents air being sucked into the raw water.

The chemicals (coagulants and flocculants) are added to the water by means of *in-line mixers* having a design proposed by Klute [7] and described in [2]. In this mixer water and chemicals are totally mixed within 0.1–0.3 seconds.

Flocculation is carried out in pipe flow flocculators. Mixing and flocculation proceed in two steps. In the first step (microflocculation), the main coagulant is added through a mixer ahead of the first pipe flow flocculator consisting of a 20 m long, 8 mm inner diameter, soft plastic pipe, giving a detention time in the pipe of 30 seconds and a turbulent velocity gradient, $G = 740\,\mathrm{s}^{-1}$. In this microflocculation step, small, dense flocs are created. The microfloc containing water is then transported through a new in-line mixing unit where the flocculant (if any) is added. When no flocculant is used, this mixer is taken away. When a cationic polymer is used as flocculant, the polymer also acts as coagulant and is therefore added before the microflocculation step.

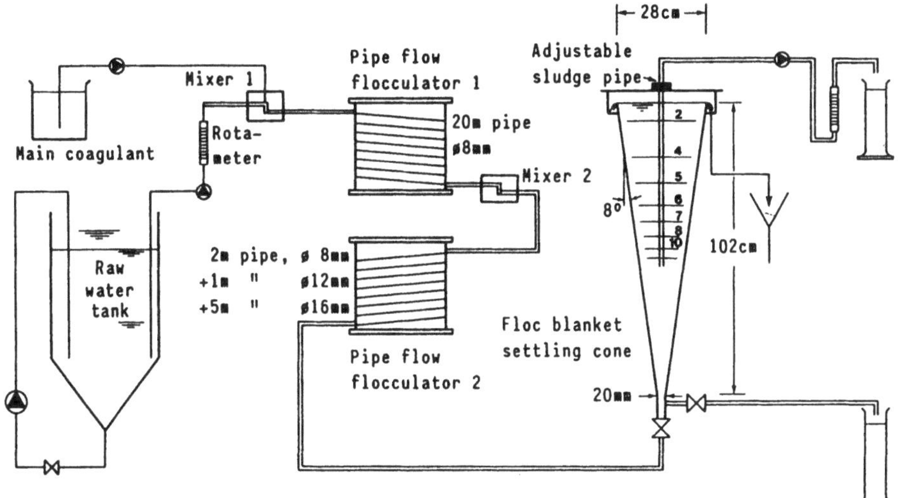

Fig. 1. Chemical floc settling apparatus

After addition of the flocculants, the water flows through the second pipe flow flocculator (macroflocculation step). This pipe flow flocculator consists of three parts in series, the first being a 2 m long pipe with inner diameter 10 mm, followed by a 1 m pipe with 12 mm diameter and finally a 5 m pipe with 16 mm diameter. This gives a total detention time of 40 seconds (at 2 l/min) and a tapered G-value of 350, 190 and $70\,\mathrm{s}^{-1}$ in the three parts of the pipe.

This particular pipe flow flocculator design was chosen after a series of preliminary tests. These tests showed that a detention time of at least 30 seconds was necessary for the microflocs to be created before addition of the flocculation polymer.

The *dosing pump* used for the main chemicals is a very fine peristaltic pump (ALITEA AB-XV) and that for the polymers is another peristaltic pump (12 000 VARIOPERPEX).

The *settling reactor* is designed as a cone of about 1 m height and with a cone angle of 8°. The floc suspension flows into the reactor at the bottom through a pipe with the same inner diameter as the flocculator, thereby minimizing floc break-up at the entrance of the reactor. Because of the balance between the force of upstream flow and the gravitational force on the flocs, a floc blanket will establish itself at a height in the reactor characteristic for the settling velocity of the flocs in the particular suspension. For different overflow rates ($v = Q/A_s$), a mark can be set at various heights for a given flow, thereby making it easy to read the overflow rate for the suspension to be tested at a given time.

The floc blanket settling reactor can be operated in different ways. One way is to let the sludge accumulate in the reactor. This will obviously cause the floc blanket to move slowly upwards after some time as a result of the space needed by the sludge that is produced. If the floc blanket overflow rate plotted as a function of a defined time (for instance 15 min), a floc blanket overflow rate curve characteristic for the suspension in question can be established and compared with other similar curves made under other circumstances. An example of such a comparison is shown in Fig. 2 from some earlier experiments.

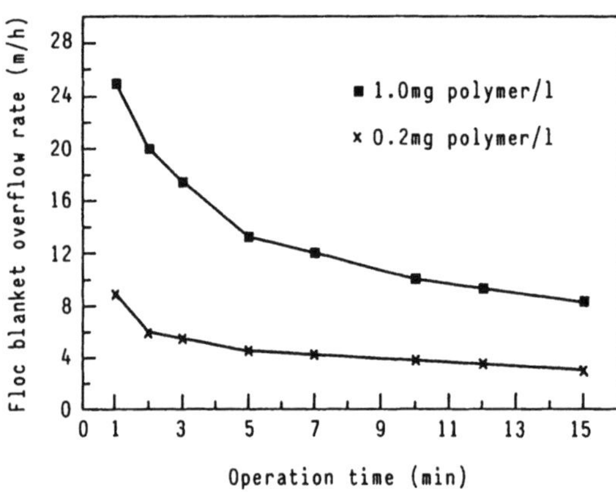

Fig. 2. Floc blanket overflow rate versus time of operation. No sludge removal. Main coagulant: PAX 60. Flocculant: Anionic polymer Praestol 2540

Obviously this method only gives a qualitative picture of the settling rate to be expected in the two situations. The curves in Fig. 2 converge to the x-axis, suggesting that another way to use the floc blanket settling reactor is to establish a steady state situation – after an initial time for the floc blanket to be established – by withdrawing sludge at the same rate as it is being produced. It must be possible to adjust vertically the pipe through which the sludge is to be pumped out, in order to reach the correct position with respect to the floc blanket level.

This method was used in the experiments that will be reported here. The outpumping of sludge by a perstaltic pump was started after 10 min of sludge blanket formation and 5 min was used to find the correct flow which would result in the establishment of a steady state floc blanket. The reactor was operated at this steady state for 5 min. At the end of this period, the height of the sludge blanket was measured for the particular suspension in question. Based on this level and the sludge water flow rate, the operating floc blanket overflow rate could be calculated.

Of course, the floc blanket settling rate determined in this way can only be used for purposes of comparison. One should be very careful in using the absolute values as a measure of what overflow rates can be expected in full scale reactors. It turned out, however, that these relative values are a valuable tool for evaluating the real settling characteristics of different suspensions.

Sludge settling tests can be made directly in the settling cone after the water is turned off or on samples taken out from the settled sludge in the cone. In the experiments to be reported here, the sludge was allowed to settle for 30 min and sludge volume readings were carried out during the settling period. After 30 min settling, a 1 l sludge sample was tapped from the bottom of the settling cone directly to a 1 l settling cylinder where the sludge was left for thickening for 60 min. Sludge volume in the cylinder was read throughout the thickening period. In the thickened sludge, the dry solids concentration and capillary suction time (CST) were measured.

The Testwater

A model wastewater was used in order to be able to control the influent composition and also because the tests were part of a bigger programme (HYPRO) where the coagulation/precipitation mechanism studies were carried out on the same model water [2]. The wastewater was constructed to ressemble a typical soft, Norwegian wastewater. The components of the model wastewater are given in Table 2. This composition results in the values of the group parameters given in Table 3.

Tab. 2. Components of the model wastewater

Component	Concentr.	Component	Concentr.	Component	Concentr.
$NaHCO_3$	200 mg/l	K_2HPO_4	50 mg/l	Potato starch	60 mg/l
NaCl	200 mg/l	Humic subst.[1]	4.0 mg/l	Dried milk	300 mg/l
NH_4Cl	100 mg/l	Latex	2.2 mg/l	Bentonite	80 mg/l

[1] Roth nr. 7824

Tab. 3. Chemical composition of model wastewater

Component	Concentr.	Component	Concentr.	Component	Concentr.
$COD_{unfiltered}$	486 mg/l	pH	7.8	P_{tot}	10.5 mg/l
$COD_{filtered}$	324 mg/l	Alkalinity	3.3 mmol/l	PO_4-P	9.1 mg/l
$TOC_{filtered}$	64 mg/l	Turbidity	150 NTU	NH_4-N	24.8 mg/l

A more detailed description of the model wastewater can be found in [2].

Coagulants and Flocculants Tested

Several combinations of coagulants and flocculants were tested. First we tested the traditional coagulants (aluminium sulfate as the commercial product ALG, and iron chloride as the commercial product JKL) versus the prepolymerized coagulants (products PAX 60 and PAX 21) without the use of polymers as flocculation aid. Secondly we tested the same coagulants with an anionic polymer (Praestol 2540 – shortened P-2540) as flocculation aid. This particular polymer had previously been proven to be suitable.

When using an anionic polymer as flocculation aid after a cationic main coagulant, the system operates best at a slight overdose of the main coagulant, which gives the flocs predominantly positively charged sites to which the anionic polymer can attach. Another idea would be to slightly underdose the main coagulant and to use a cationic flocculant. In this case the cationic polymer acts both as a flocculant and as a coagulant and destabilizes because of adsorption/charge neutralization and to polymer bridging. Different cationic polymers were tested with regard to this concept.

A third possibility would be to use the cationic polymer alone as coagulant and flocculant. Obviously this would only be of practical interest in those cases where only particle removal and minimized sludge production is required. Removal of soluble phosphate is not possible without a phosphate precipitating chemical (as Al- or Fe-salts) present.

Finally, some experimental runs were carried out on a situation where a cationic polymer (Agefloc B50) was mixed with a solution of aluminium coagulant (ALG or PAX 60) and this mixture was then used as the main coagulant. An anionic polymer (P-2540) was additionally used as flocculant in these experiments.

The characteristics of the chemicals used in the different experiments are shown in Tables 4 and 5.

Results and Discussion

When the results are evaluated, one must be aware of the fact that there might be an interaction between the overflow rate and the separation efficiency, and this interaction is difficult to account for. The settling rates presented here are all in situations where the separation efficiency was reasonable, that is, 90 % based on turbidity. Prior to these settling experiments, a row of jar-tests in an apparatus described in [2] were conducted in order to find the most interesting range of dosages to be used in the main experiments.

Tab. 4. Characteristics of main coagulants used

Name	Formula	Form	Density g/cm^3	Al %[1]	Fe %[1]	SO$_4^-$ %[1]	OH/Me
JKL	FeClSO$_4$	Solution	1.50	–	11.9	16.6	–
ALG	Al$_2$(SO$_4$)$_3$	Granulate	–	9.1	–	46.0	–
PAX 60[2]	Poly-AlCl$_3$	Solution	1.30	7.8	–	< 0.5	1.1
PAX 21	Poly-AlCl$_3$ (bauxite based)	Solution	1.32	7.3	0.9	< 0.5	1.0

[1] By weight; [2] Also used PAX 61 with almost same characteristics as PAX 60. Both referred to here as PAX 60

Tab. 5. Characteristics of polymers used

Name	Type	Form	MW g/mol	Charge meq/l
Praestol 2540[1,3]	Anionic	Granul.	13 mill.	−3.5
Praestol 655 BC[1,3]	Cationic	Granul.	7 mill.	+3.7
Agefloc A50 HV[2,4]	Cationic	Solution	300 000	+12.8
Agefloc B50[2,4]	Cationic	Solution	35 000	+10.6
Agefloc WT2206[2,5]	Cationic	Solution	> 1 mill.	+6.4
Agefloc WT40[2,5]	Cationic	Solution	200 000	+10.6

[1] Stockhausen; [2] CPS; [3] Polyacrylamid; [4] Dimethylamin/Epichlorhydrin-copolymer; [5] Polydimethyl-diacryl-ammonium chloride

Floc Blanket Settling Rate (Acceptable Overflow Rate)

Figure 3 shows the registered floc blanket settling rate (equal to the acceptable overflow rate) when using different main coagulants with and without the addition of anionic flocculant.

It is demonstrated that the influence of the main coagulant dosage within the actual range was insignificant, but that the influence of the polymer dosage was considerable. When anionic flocculant was not added, the settling rate in all cases was between 1.5 and 2.5 m/h, which seems reasonable when compared to experiences in practice. At low coagulant dosages, poor coagulation and, consequently, poor floc separation (C_e > 10 NTU) was experienced. In the cases where anionic flocculant was added, the main coagulant dosage ensured good coagulation/precipitation as well as good floc removal (PO$_4$-P in effluent < 0.5 mg/l and turbidity in effluent < 5 NTU).

It is also shown that the addition of anionic polymer improves the settling rate of the flocs and thereby increases the acceptable overflow rate in all cases,

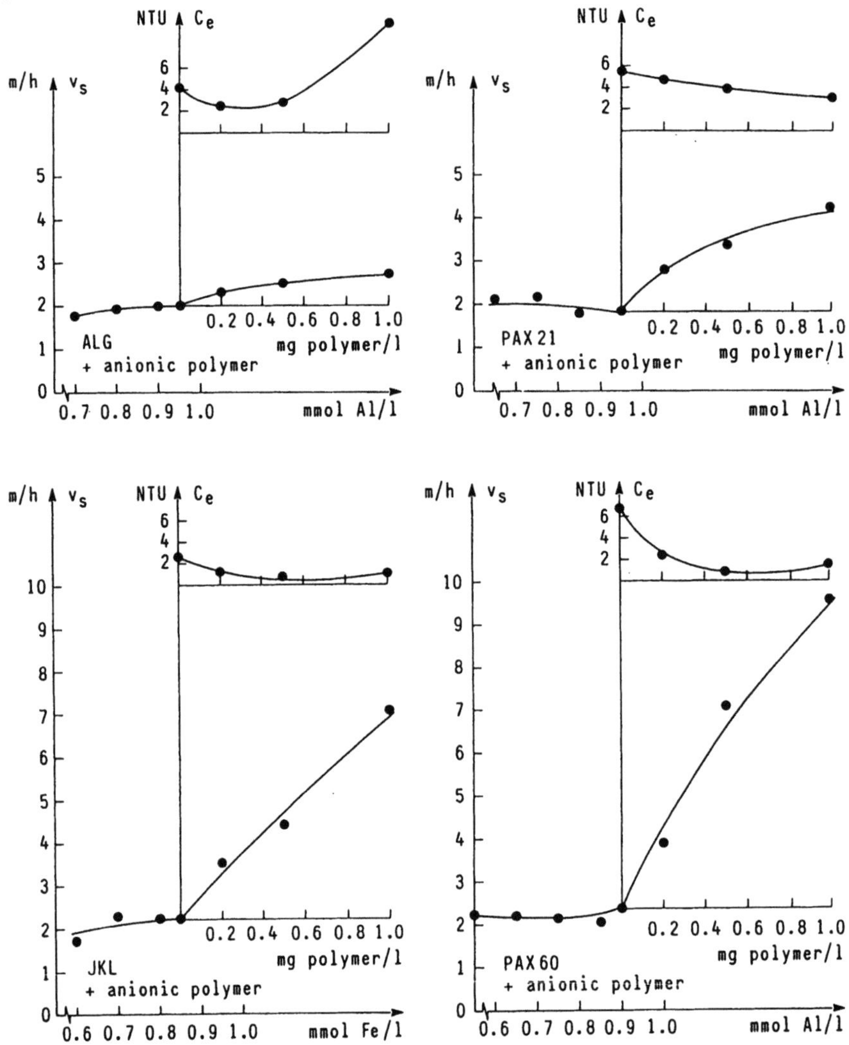

Fig. 3. Floc blanket settling rate, v_s, for different dosages of main coagulant and anionic polymer as flocculant. The left-hand part of each figure shows the influence of the main coagulant dosage (without polymer) while the right-hand part shows the influence of the anionic polymer dosage at a given main coagulant dosage (origin for this curve). The turbidity on the effluent is shown on the top of each figure

but that the increase is relatively small when ALG is used as main coagulant. Especially when PAX 60 was used, the improvement was considerable. It is also interesting to note that in most cases the effluent turbidity also decreased with increasing polymer dosage, resulting in a situation where very low effluent turbidities were obtained at very high overflow rates when an appropriate dosage of anionic polymer was used.

A floc blanket overflow rate of 8 m/h resulted, for instance, with a JKL dosage equivalent to 0.85 mmol Fe/l with the addition of 1 mg/l of the anionic polymer Praestol 2540, and the same floc blanket overflow rate was found at a PAX 60 dosage equivalent to 0.95 mmol Al/l and 0.5 mg/l of the same anionic polymer. Why the results with PAX 21 (based on bauxite), with very similar characteristics to PAX 60, were worse than with PAX 60, is difficult to explain. Not only was the acceptable overflow rate lower, but so was the turbidity removal. Clearly the benefit from polymer addition was lowest when ALG was the main coagulant.

Figure 4 shows the floc blanket overflow rate for experiments where a cationic polymer was added to enhance coagulation and flocculation in a situation where PAX 60 was used as the main coagulant. One wanted to answer two questions: first, can equally good settling characteristics be obtained with a cationic polymer instead of an anionic one and, second, can the dosage of the main coagulant be lowered and compensated for by adding a cationic polymer.

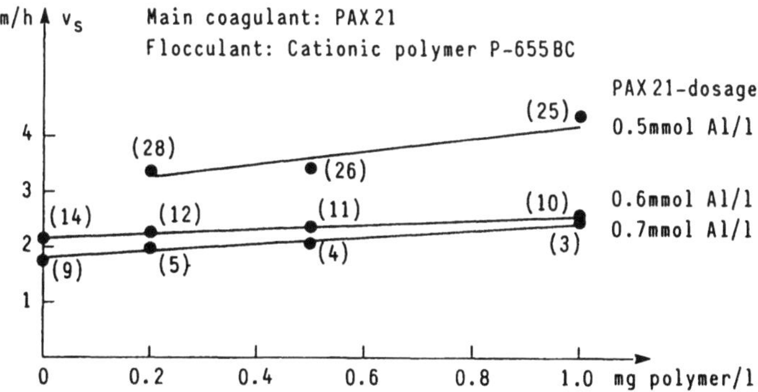

Fig. 4. Floc blanket settling rate versus cationic polymer (P-655 BC) dosage for various dosages of PAX 21 as main coagulant. Effluent turbidities in parentheses

First of all, Fig. 4 shows that the settling rate may also be improved by the addition of a cationic polymer, but that the improvement is far smaller than what was achieved with an equal dosage of anionic polymer. The second question is difficult to answer based on the findings in Fig. 4, since it seems that the lower the PAX dosage, the higher the settling rate. This is, however, caused by the fact that the lower the PAX dosage, the worse the coagulation/precipitation and therefore the floc separation. In Fig. 4 the effluent turbidities at each point is indicated, and it can be seen that it is only at PAX 21 dosages > 0.7 mmol Al/l that coagulation/floc separation was satisfactory.

Figure 5 is intended to clarify the answers to the aforementioned questions. The settling velocities obtained without polymer, with anionic polymer and with cationic polymer are shown – all situations at PAX 21 dosages that gave acceptable coagulation/floc separation (effluent turbidity < 10 NTU). This figure clearly shows that the settling rate was improved with either the addition or anionic or cationic polymer, but that the improvement was far greater using anionic polymer. The figure indicates that a reduction in dosage of the main coagulant (here: PAX 21) of about 0.1–0.2 mmol Al/l could be compensated for by adding a low dosage of cationic coagulant (for instance 0.2 mg/l).

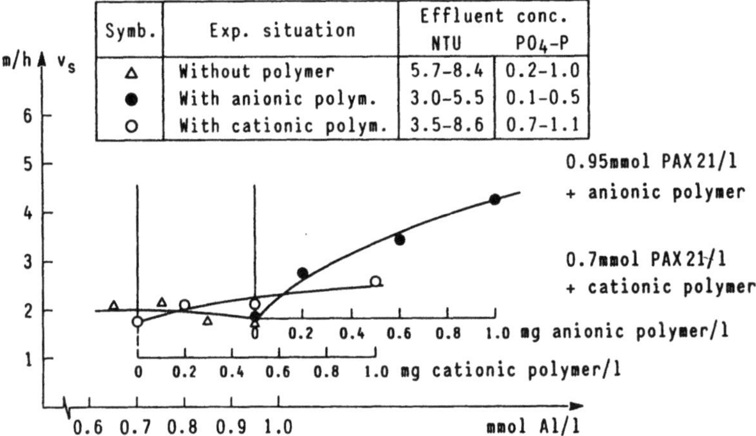

Fig. 5. Comparison of floc settling velocities achieved for different combinations of PAX 21 as main coagulant and anionic or cationic polymer as flocculant for situations giving comparable separation results. For explanation of figure, see Fig. 3

Of course, the reduced main coagulant dosage will also result in reduced phosphate removal and reduced sludge production. This combination of chemicals may therefore be of interest in certain pre-precipitation situations.

If one wants to combine the effect of main coagulant reduction by adding cationic polymer (leading to less sludge production) with the effect of anionic flocculant addition (leading to higher settling velocities), one could add the cationic polymer directly to the main coagulant solution and use anionic polymer in the conventional way as a flocculant. Figure 6 shows the results from such experiments where ALG, or PAX 61, plus a low molecular weight cationic polymer (Agefloc B50) were used as the main coagulants and the anionic polymer (Praestol 2540) was used as flocculant. Two concentrations of cationic polymer were added to the ALG-solution (91 g Al/l), namely, 3 and 6 % based on volume.

It is demonstrated that even if the ALG dosage was reduced from 0.95 mmol Al/l to 0.85 mmolAl/l, considerably higher settling rates could be obtained with a cationic polymer mixed in the ALG solution, especially when anionic polymer dosages > 0.5 mg/l. It is probable that acceptable settling rates could have been achieved at even lower dosages of the main coagulant compensated for by cationic polymer. This was not tested, however.

Without anionic polymer addition, the settling velocity using the cationic polymer-compensated main coagulant was in the same range (actually a little

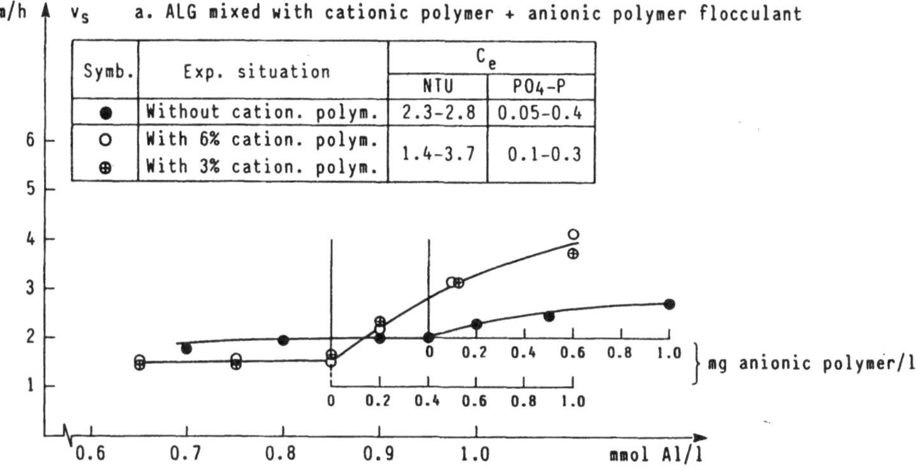

Symb.	Exp. situation	C_e	
		NTU	PO4-P
●	Without cation. polym.	2.3-2.8	0.05-0.4
○	With 6% cation. polym.	1.4-3.7	0.1-0.3
⊕	With 3% cation. polym.		

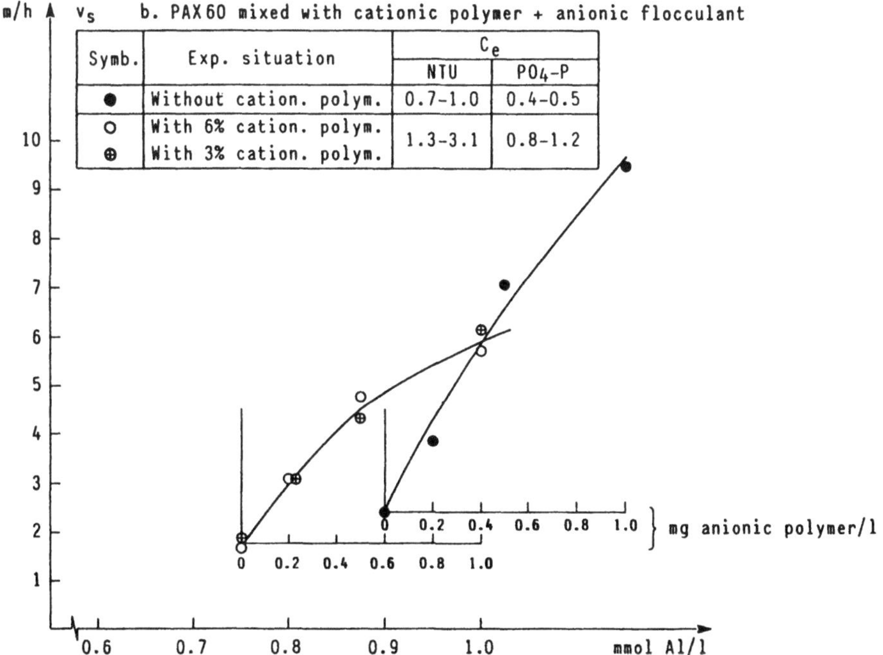

Symb.	Exp. situation	C_e	
		NTU	PO4-P
●	Without cation. polym.	0.7-1.0	0.4-0.5
○	With 6% cation. polym.	1.3-3.1	0.8-1.2
⊕	With 3% cation. polym.		

Fig. 6. Settling velocity when ALG (or PAX 61), plus the cationic polymer Agefloc B50 was used as the main coagulant and Praestol 2540 was used as the anionic polymer flocculant

worse) as the situation without cationic polymer added to the solution. This may indicate that the cost of this relatively complicated combination of chemicals outweighs the benefit. As seen in Fig. 6, no significant difference was found between the 3 and 6 % solutions of the cationic polymer added to the ALG solution. The same situation was also tested with PAX 60 as the main coagulant, see Fig. 6 b. In the cases where the PAX 61 dosage was < 0.8–0.9 mmol Al/l and no anionic polymer was used, the settling velocity was in the 1.5–2.5 m/h range in all cases, but the effluent turbidities were quite poor. These points were not included in the figure. We can see the same trends as with ALG as main coagulant, but the settling rates were considerably higher with PAX 60, as also shown before.

We can see, however, that about the same settling velocity could be obtained with a PAX 60 dosage of 0.75 mmol Al/l + 3 % cationic polymer and with a flocculant dosage of 1 mg anionic polymer, as with a PAX 60 dosage of 0.9 mmol Al/l and a flocculant dosage of 0.4 mg anionic polymer/l. The potential of such an arrangement must be evaluated from an economic standpoint in each case, where proper attention is given to the reduced sludge production that will result from this coagulant arrangement.

Finally the situation where cationic polymer alone was added deserves comment. This concept is only worthwhile when phosphate removal is unimportant and only particle removal and minimization of sludge production is the goal. Figure 7 shows that it was very difficult to achieve low effluent turbidities (< 10 NTU) and at the same time, a good settling rate when cationic polymer alone was used. It must be said that the floc blanket for this situation was not very defined. The higher the molecular weight of the cationic polymer, the higher the floc blanket settling rate. With a dosage of around 10 mg/l of the cationic polymer Praestol 655 BC, a floc blanket settling velocity of 13 m/h could be obtained at an effluent turbidity of about 10 NTU.

Sludge Characteristics

As mentioned before, after performing the continous settling test, the sludge blanket was allowed to settle and after 30 min settling, the sludge was drawn from the bottom of the settling cone into a cylinder, where it was allowed to thicken for 1 hour. Then the dry solids ($X_{\text{thick.}}$) and the capillary suction time (CST) of the sludge phase were determined. This procedure made it possible to calculate the sludge concentration in the floc blanket (X_{fb}), the sludge volume index of the sludge in the floc blanket (SVI$_{\text{fb}}$), and the sludge concentration in the floc blanket sludge after it had been allowed to settle for 30 min ($X_{\text{settl.}}$). The SVI$_{\text{fb}}$ is defined as SV$_{\text{fb}}$/$X_{\text{settl.}}$, and the SV$_{\text{fb}}$ is given as the sludge volume occupied by the settled floc blanket sludge after 30 min batch settling divided by the sludge volume occupied by the floc blanket sludge at steady state (i.e. at 0 min batch settling of the sludge blanket).

In order to present the data in a compact way, Table 6 gives the averages of those sludge characteristics data for the actual coagulant/flocculant combinations that gave acceptable effluent turbidities, that is, C_e < 10 NTU. In most cases, the data given are averages of 3 or 4 experimental runs with different dosages of coagulant/flocculant. Since the differences between each run were

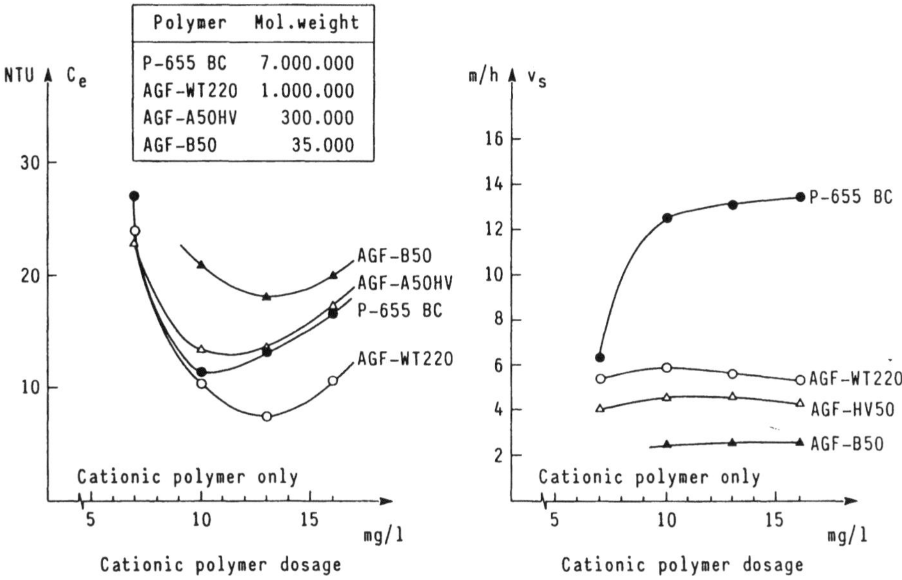

Fig. 7. Effluent turbidity and floc blanket settling rate when different cationic polymers were used as the only coagulant/flocculant

Tab. 6. Sludge characteristics data for different combinations of coagulants/flocculants used

Coagulant/flocculant	v_s m/h	C_{eff} NTU	SVI_{fb} ml/g	X_{fb} g/l	X_{settl} g/l	X_{thick} g/l	CST sec^{-1}	CST/X_{thick} $l \cdot g^{-1} sec^{-1}$
ALG (> 0.7 mmol Al/l)	2.5	6.0	42.2	1.04	5.02	6.08	56.4	9.28
JKL (> 0.8 mmol Fe/l)	2.8	5.1	37.6	1.49	6.36	6.85	50.8	7.42
PAX 60 (> 0.75 mmol Al/l)	2.6	8.1	36.9	1.07	5.31	6.80	58.6	8.61
PAX 21 (> 0.65 mmol Al/l)	2.3	6.5	22.1	0.79	5.87	6.00	59.1	9.85
ALG + anionic polymer[1]	3.4	5.4	32.7	1.27	6.38	8.59	45.3	5.27
JKL + anionic polymer[1]	6.0	1.1	23.9	2.38	9.91	11.73	51.8	4.42
PAX 60 + anionic polymer[1]	8.0	1.4	22.2	2.76	11.59	15.07	51.9	3.44
PAX 21 + anionic polymer[1]	4.3	4.0	20.9	2.39	10.80	11.17	51.9	4.64

[1] Praestol 2540 in dosages varying from 0.2–1.0 mg/l
[2] Praestol 655 BC in dosages varying from 0.2–1.0 mg/l

not dramatic and since the common denominator for all these runs is that the effluent turbidity was low (< 10 NTU), it was found acceptable to present the sludge characteristics data in such a compact way.

In addition to the sludge separation characteristics, the average effluent turbidities and the average floc blanket settling rate are also shown in Table 6. These vary more with the dosage of the coagulant than the sludge separation characteristics, as was shown earlier. The average values are, however, included here for a better overwiev.

Generally it can be seen that the sludge separation characteristics did not vary much for the different coagulants tested when no polymers were added as flocculant aid. JKL seems, however, to give slightly better sludge separation characteristics (lower SVI, higher sludge concentrations and lower CST and CST/DS) than ALG. PAX 60 as main coagulant gave very similar results to those of JKL, while PAX 21 gave better SVI but worse settled sludge concentration than PAX 60.

The separation characteristics changed markedly, however, when an anionic polymer was added. The least improvement upon anionic polymer addition was found in the case of ALG, and the greatest improvement was found when JKL and PAX 60 were used as the main coagulant. It is also interesting to note that this optimal combination of chemicals from a sludge-characteristics point of view also gave the best floc separation (lowest effluent turbidity). Overall, PAX 60 plus anionic polymer gave somewhat better sludge separation characteristics (lower SVI, higher $X_{settl.}$, higher $X_{thick.}$ and lower CST/DS) than JKL with anionic polymer as flocculant. PAX 21 gave about the same results as with JKL, while the sludge separation characteristics when a cationic polymer was used in combination with PAX 21 were significantly worse than those in a situation where anionic polymer was added.

Increased polymer dosage also increased the settled sludge concentration and, to a lesser extent, the floc blanket sludge concentration, as shown for the case of PAX 21 + anionic polymer in Fig. 8.

Real Wastewater Experiments

Some tests were carried out on wastewater from the VEAS treatment plant in Oslo. The raw water was very dilute and it turned out that good coagulation (< 10 NTU and < 0.1 PO_4-P in effluent) could be reached at low dosages, 0.25–0.30 mmol Al/l in the case of PAX 60 and 0.35–0.40 mmol Fe/l in the case of JKL as main coagulant. The same anionic polymer (Praestol 2540) was also used in these experiments, and only the combination of metal-salt (JKL or PAX 60) and anionic polymer was tested. Figure 9 shows the results with respect to floc blanket settling rate. It can be seen that the results are comparable to those obtained in synthetic wastewater (see Fig. 3) and that even higher floc blanket settling rates were achieved. On this dilute wastewater, however, only a low dosage of polymer (< 0.5 mg/l) is needed to improve settleability drastically. In fact, the full-scale plant at VEAS has been operated at a JKL dosage of approx. 0.4 mmol/l with the addition of 0.2 mg/l of an anionic polymer at a maximum overflow rate very close to that determined in this experiment (about 4 m/h).

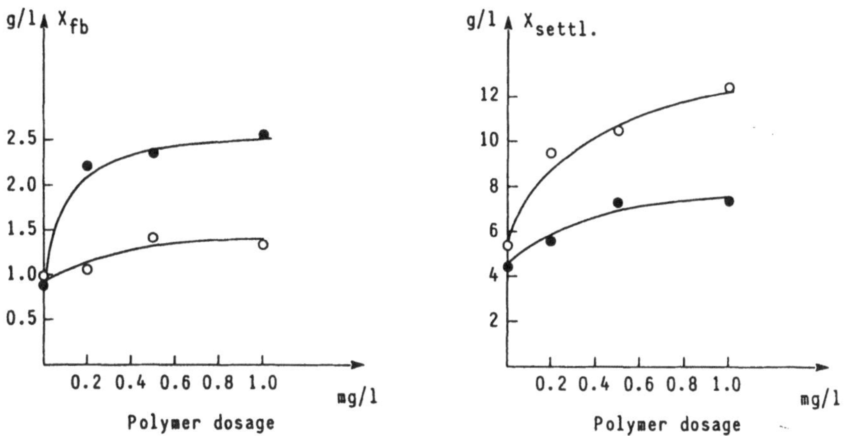

Fig. 8. Influence of anionic polymer (Praestol 2540) dosage on dry solids content in floc blanket (X_{fb}) and settled floc blanket ($X_{settl.}$)

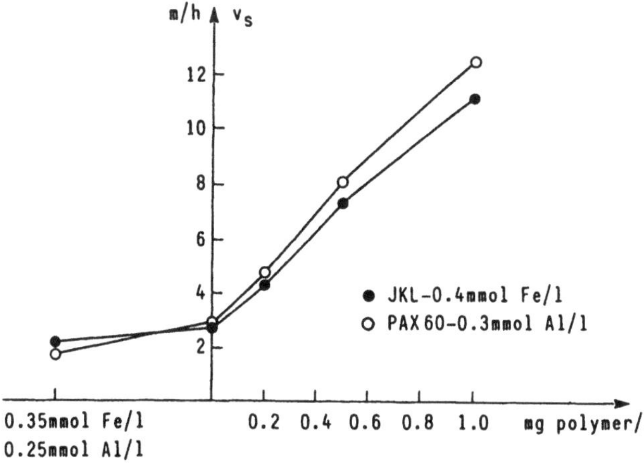

Fig. 9. Floc blanket settling rate, v_s, determined on the wastewater from VEAS treatment plant

Economic Evaluation

Based on the findings here and the experiences with polymer addition for the enhancement of flocculation/floc separation in Norwegian full-scale plants, it is reasonable to expect that the acceptable overflow rate can be doubled when a suitable polymer at correct dosage is used as flocculant compared to a situation where polymer is not used. If one designs for an overflow rate of 3–4 m/h instead of 1.5–2 m/h, there is a considerable savings in investment cost, especially in Norway where the plants normally are covered.

For a typical covered Norwegian chemical plant, the total marginal cost for the settling tanks is about NOK 12 000/m² for a 10 000 person equivalent plant

($Q_{ave.} = 1.5 \cdot 10^6$ m^3/year, $Q_{design} = 200$ m^3/h) and about NOK 9 500/m^2 for a 50 000 person equivalent plant ($Q_{ave.} = 7.3 \cdot 10^6$ m^3/year, $Q_{design} = 900$ m^3/h). Energy cost (primarily for heating and ventilation) is typically 100 NOK/m^2·year. Investment cost for a double line polymer dosing plant with necessary control equipment is about 130 000 NOK per line for a 10 000 pe plant and 180 000 NOK per line for a 50 000 pe plant (Skjelfoss, 1992).

Table 7 shows a calculation example based on the cost of polymer (Praestol 2540) at 35 NOK/kg, a polymer dosage of 0.5 mg/l and an expected improvement of the design overflow rate from 1.5 to 3.0 m/h. When an interest of 12 % and a life expectancy of 20 years is used for capitalization calculations, we can see from Table 7 that the use of polymer can save money.

Tab. 7. Evaluation of yearly cost savings using polymer as flocculant in a Norwegian chemical treatment plant (investment cost capitalized)

Size of plant	Settling tank area saving m^2	Reduced investment NOK/year	Reduced energy NOK/year	Extra dosing equipment NOK/year	Polymer cost NOK/year	Savings NOK/year
10 000 pe	67	107 656	6 700	34 814	37 500	53 292
50 000 pe	300	381 615	30 000	48 204	182 500	235 661

Summary and Conclusions

This study was made to evaluate a laboratory set-up for the analysis of floc and sludge separation characteristics and to compare, with the help of this apparatus, the separation characteristics of different combinations of coagulants and flocculants for chemical wastewater treatment.

The experiences with the apparatus were quite good. Some experimentation on the actual wastewater is necessary to determine the correct time when steady state is reached and the correct withdrawal of sludge from the sludge blanket in order to establish steady state. With our water, the floc blanket was established, at 10 min, and thereafter withdrawal of sludge took 5 min to adjust. Then the apparatus was operated at steady state for 5 min before the floc blanket overflow rate was measured. We would recommend that this steady state period be increased to 10–15 min.

As long as the reaction step was functioning well and a clear water phase was created between the flocs, the settling reactor performed very well and a well defined sludge blanket was formed. When, however, the reaction step did not function optimally, the floc blanket was blurred and readings were difficult. The apparatus should therefore be run only after optimal reaction conditions have been established for the coagulant/flocculant situation to be investigated. This can be done in jar tests or in preliminary experiments in the continous settling apparatus.

It must be emphasized that this equipment is not intended for determination of design overflow rates. It is only suitable to use for the comparison of different coagulant and flocculant combinations on a given wastewater in order to evaluate floc separation characteristics. The choice of coagulants and flocculants in chemical wastewater treatment can have a very significant influence on the floc and sludge separation characteristics and thus also on the cost of chemical treatment. In particular, the correct use of flocculant aid (polymer addition) can make it possible to operate at considerably higher overflow rates in the settling tank and thereby decrease the investment cost more than the resulting increase in operation cost due to higher chemical expenditure.

The following conclusions can be made from this study:

1. The acceptable overflow rate found in the test apparatus used in this study corresponds well with the *maximum acceptable* overflow rates found in full-scale plants in practice.

2. The acceptable overflow rate as well as the sludge separation characteristics do not vary significantly among the commonly used coagulants when these are used without flocculant aid, even though iron-chloride (JKL) seems to give slightly better floc and sludge characteristics than aluminium sulfate (ALG).

3. A significant improvement in both floc amd sludge separation characteristics can be obtained with the addition of 0.2–1.0 mg/l of a suitable anionic polymer. In the case of JKL and PAX 60 as the main coagulants, the average floc blanket overflow rate (overflow rate giving < 10 NTU in effluent) could be increased by 2–3 times compared to a situation without polymer addition.

4. JKL and PAX 60 as main coagulants in combination with anionic polymer seemed to give the best separation results. A floc blanket overflow rate of 7 m/h was, for instance, measured with a JKL dosage of 0.85 mmol Fe/l and the addition of 1 mg/l of the anionic polymer Praestol 2540, and the same overflow rate was measured at a PAX 60 dosage of 0.95 mmol Al/l and an addition of 0.5 mg/l of the same anionic polymer.

5. Addition of cationic polymer could also improve floc and sludge separation characteristics, but not to the same degree as the addition of an anionic polymer.

6. The addition of cationic polymer to the main inorganic coagulant solution in order to reduce main coagulant consumption was also shown to improve floc and sludge characteristics when an anionic polymer was used as flocculant aid. The improvement was not, however, found sufficiently great to justify this rather complicated use of chemicals.

7. From this study it can be concluded that the combination of an optimized dosage (dependent upon the turbidity and PO_4-P removal requirements) of the metal coagulant and a dosage of 0.2–1.0 mg/l of a suitable anionic coagulant should be evaluated with respect to floc and sludge separation characteristics when floc separation in chemical wastewater treatment is to be optimized in a given case. The apparatus developed in this study can be

used for this evaluation, possibly with some modifications.

8. The main coagulant that is to be recommended also depends on the treatment goal. If the chemical treatment is used as the first step in a pre-precipitation plant with subsequent biological N-removal, a prepolymerized coagulant can be recommended because it gives good separation characteristics, but also because the alkalinity consumption is low and because the PO_4-P removal can be controlled in order to prevent phosphate limitation in the subsequent biological nitrification step.

References

[1] Ødegaard, H.: Norwegian Experiences with Chemical Treatment of Raw Wastewater. Proc. Int. Conf. on: Wastewater Management in Coastal Areas, Montpellier, France, April, 1992

[2] Ødegaard, H., Fettig, J., Ratnaweera, H.: Coagulation with Prepolymerized Metal Salts. In: Chemical Water and Wastewater Treatment, H.H. Hahn and R. Klute (eds.). Springer, Berlin Heidelberg New York 1990, pp. 189–220

[3] Ødegaard, H.: Orthokinetic Flocculation of Phosphate Precipitates in a Multi-compartment Reactor with Non-ideal Flow. Prog. Wat. Tech., 1979, Suppl. 1, pp. 61–88. Pergamon Press

[4] Weholt, Ø.: Capacity Increase in Settling Units by the Use of Polymers. Proc. VAR 84-conference, H. Ødegaard (ed.). TAPIR Publ., Trondheim, 1984 (in Norwegian)

[5] Fettig, J., Ratnaweera, H. and Ødegaard, H.: Synthetic Organic Polymers as Primary Coagulants in Wastewater Treatment. Water Supply, Vol. 8, Jønkøping, 1990, pp. 19–26

[6] Sagberg, P., Sæter, R. and Berge, A.B.: Increasing the Surface Load at a Direct-Precipitation Plant, VEAS, Norway. In: Chemical Water and Wastewater Treatment, H.H. Hahn and R. Klute (eds.). Springer, Berlin Heidelberg New York 1990, pp. 271–282

[7] Klute, R.: Destabilization and Aggregation in Turbulent Pipe Flow. In: Chemical Water and Wastewater Treatment, H.H. Hahn and R. Klute (eds.). Springer, Berlin Heidelberg New York 1990, pp. 33–54

[8] Skjelfoss, E.: Personal communication, March 1992

Hallvard Ødegaard and Svein Grutle   Harsha Ratnaweera
Div. of Hydraulic and Sanit. Engineering   Norwegian Institute for Water Research
Norwegian Institute of Technology   P.O. Box 69 Korsvoll
N-7034 Trondheim – NTH   N-0808 Oslo 8
Norway   Norway

Characterization of Sludges
from Wastewater Coagulation
with Polymeric Aluminium Salts

S.J. Langer, F. Seitz, and R. Klute

1. Introduction

The application of hydrolizing aluminium and iron salts in chemical water and wastewater treatment is continuously increasing. Dosage of these chemicals will result in a considerable reduction of phosphorus concentrations in the effluent of phosphorus elimination and sewage treatment plants and hence will satisfy stringent effluent requirements, e.g. in Germany of 1.0 mg P/l for larger sewage works (> 100 000 p.e.) and of 0.3 mg P/l for discharges into protected rivers and lakes. In addition to precipitation reactions with dissolved pollutants the hydrolizing metal salts destabilize suspended and colloidal substances, bringing about their aggregation into larger flocs, which can be separated efficiently. If one takes into account that suspended and colloidal material in raw sewage contributes significantly to the total pollution load, measured for example as BOD or COD (Ødegaard, 1988) the application of these chemicals becomes attractive for mechanical sewage works with increasing effluent demands as well as for temporarily overloaded treatment plants. Furthermore, an extensive and specific elimination of pollutants already in the primary clarifier of a mechanical-biological wastewater treatment plant offers interesting aspects with regard to denitrification without additional tank volume (Karlsson, 1988; Henze and Harremoës, 1990).

After dosing, the aluminium ions are transformed from the trivalent state into a sequence of hydroxocomplex species characterized by positive surface charges, before they finally precipitate as neutral metal-hydroxide. The positively charged hydroxocomplex species destabilize the particulate material by adsorption coagulation and can be considered as highly effective, whereas the metal-hydroxides show a low efficiency because the destabilization is achieved by enmeshment into the precipitate. The destabilization potential is dependent on the valence of the hydroxocomplex compounds and therefore polymeric species carrying higher charge numbers apparently are more effective.

The prehydrolysis of aluminium salts allows to prebuild the active polymeric species before actual use. There are indications that these species are relatively stable, thus explaining the high efficiency of the polymeric aluminium salts when applied for water and wastewater treatment. However, up to now only limited information is available on amount and characteristics of the resulting sludges.

Diamadopoulos and Benedek (1984) investigated aluminium hydrolysis effects on phosphorus removal from wastewaters and found that the degree of hydrolysis affected the supernatant suspended solids concentration, the settling rates, and the dewaterability of the sludge. An increasing OH:Al molar ratio (basicity) resulted in a higher specific resistance and hence a poorer dewaterability. On the other hand, Fiessinger and Bersillon (1977) found that the specific resistance of a sludge from surface water treatment improved with increasing basicity of the coagulant. Dempsey and co-workers (1985) concluded from coagulation experiments of clay-fulvic acid suspensions that alum and polyaluminium chloride were not significantly different in the volume of sludge generated by a given concentration of aluminium. The flocs formed by polyaluminium chloride appeared to be very sensitive to shear. This was apparent in settling tests. In these experiments, polyaluminium chloride produced the fastest-settling floc when rapid mix conditions were correctly controlled.

The goals of this investigation were to compare aluminium chloride, polyaluminium chloride, and a silica-containing polyaluminium chloride as coagulants for a municipal wastewater and to specify the characterics of the resulting sludges with regard to sludge volume, total solids concentration in the sludge, sludge thickening behavior, and sludge dewatering.

2. Materials and Methods

2.1 Wastewater Characterization

Raw wastewater from a municipal sewage treatment plant was used in the coagulation experiments. It is characterized by the parameters listed in Table 1. The phosphorus was analyzed as total phosphorus in an unfiltered sample ($P_{tot.}$) and after membrane filtration (0.45 μm). It is assumed that the filtered phosphorus value is equivalent to the concentration of ortho-P. The charge concentration per unit volume of wastewater was determined by titration of wastewater with a cationic (100%) polyelectrolyte until the isoelectric point was reached. This zero point of charge was detected by an instrument that is based on the streaming current principle (SD 100; Fa. Kolb, Switzerland).

Under average conditions, the wastewater can be regarded as a typical municipal wastewater with medium concentration levels and a high hardness (3.7 mmol/l). Although the use of diluted wastewater after substantial rainfall was avoided, the wastewater conditions changed somewhat, as indicated by the standard deviations in Table 1.

In order to identify the effect of the chemicals added, a sample of the raw wastewater was treated in the same way as in the coagulation experiment. It was stirred and then allowed to settle for 30 minutes. The percent reduction of any wastewater parameter due to coagulation was calculated in relation to the unflocculated, but settled wastewater, not to the raw water data.

Tab. 1. Characterisation of the raw wastewater

Parameter	Unit	Raw wastewater		Settled wastewater	
		$\bar{x}$	σ_n	$\bar{x}$	σ_n
pH	-	7.40	±0.37		
Conductivity	mS/m	138	±23		
COD	mg/l	654	±192	516	±59
TOC	mg/l	139	±60	127	±55
$P_{tot.}$	mg/l	10.7	±2.3	8.7	±2.4
Turbidity	NTU	120	±36	76	±10
Suspended solids	mg/l	239	±29		
Charge concentration	µeq/l	103.5	±21.3	84.3	±12.7
After membrane filtration (0.45 µm)					
DOC	mg/l	79	±31		
ortho-P	mg/l	6.6	±1.3		
Charge concentration	µeq/l	58.1	±10.8		

2.2 Selection of Chemicals

The experiments were performed with three different coagulants based on aluminium salt. Aluminium chloride and PAX-W, both commercial products from Kemira Kemi, were selected to evaluate the effect of prepolymerization both on the treatment efficiency and on the characteristics of the resulting slurry. In addition, a coagulant consisting of polyaluminium chloride with silica compounds (UPAX 6) was selected. UPAX 6 is an innovative product from Kemira Kemi and not yet commercially available. By selecting this coagulant for the experiments, the effect of the silica compounds on coagulation efficiency and sludge properties should be investigated.

Some parameters describing the chemicals are listed in Table 2. As intended, the coagulants differ in basicity. According to the manufacturer, the OH/Al-ratio of PAX-W is 0.75. This ratio cannot be determined for UPAX 6, since the silica present has reacted and formed soluble complexes with an unknown portion of the aluminium.

Since the iron content is negligible in all products, the chemical dose is expressed as mmol/l Al. The experiments were performed at identical Al-doses for all chemicals, since the main objective of this study was to describe the sludge properties under identical chemical conditions.

2.3 Experimental Set-Up

The coagulation of the wastewater was performed in batch-experiments. In order to obtain a sufficient quantity of slurry for the analysis of sludge properties,

Tab. 2. Chemical composition of the coagulants

	Unit	$AlCl_3$	PAX-W	UPAX 6
Al^{3+}	weight %	9.14	5.9	6.70
Al^{3+}	mol/kg	3.39	2.2	2.48
Fe^{3+}	weight %	0	0.6	0.05
Fe^{3+}	mol/kg	0	0.1	0.01
Cl^-	weight %	36.1	18.3	15.2
Basicity	mol OH/mol Al	0	0.75	unknown
Si-content	mol Si/mol Al	0	0	0.17

a batch of 18 l wastewater was flocculated. The rapid mixing of coagulant and
wastewater was carried out during a 30 second period of mixing at 1 700 rpm.
For aggregation and floc formation, the suspension was agitated by a grid stirrer under mild conditions (15 rpm) for 15 minutes. After this period, the flocs
were allowed to settle. After 30 minutes of sedimentation, samples were taken
from the supernatant for analysis of the parameters described above. Then,
16 l of supernatant were removed to obtain 2 l of slurry.

When flocculating in a batch reactor with a relatively large volume compared to jar-tests, it becomes difficult to achieve satisfactory rapid-mix conditions. In the first experiments, a grid stirrer in a circular reactor was set at
1 700 rpm for 30 seconds, resulting in very poor rapid-mixing, as an experiment
with tracer revealed. The tracer stayed in the upper half of the reactor and it
took as long as 20–30 seconds of mixing until a complete homogenization was
achieved. Efforts to optimize rapid-mixing led to the employment of a screw-
type stirrer, which promotes a convection in the reactor in the vertical direction
and distributes the added tracer or chemical solution, respectively, throughout
the entire reactor volume. In addition, baffles were installed on the reactor
wall. In the tracer experiments, it could be observed that within 5 seconds the
chemicals were added and homogenized in the reactor. As will be discussed
later, this improvement in rapid-mixing resulted in higher removal efficiencies.
However, the sludges did not exhibit any significant changes resulting from the
improved rapid-mix conditions.

2.4 Sludge Characterization

2.4.1 Thickening Experiments. After homogenization, the 2 liters of slurry
were filled into a graduated cylinder that served as a settling column. The
height of the interface that developed between the more or less clear upper
region and the hindered-settling region was recorded over time for 180 minutes. The samples were allowed to compress for one day and the thus defined
minimum sludge volume was recorded. All further analysis of sludge properties
was performed with the resulting, thickened sludge.

2.4.2 Dewatering Experiments. The capillary suction time (CST) was determined with standard apparatus (Triton Electronics Ltd., GB) using original CST papers. A better measure of sludge dewaterability is the specific resistance to filtration which is, according to theory, independent of the solids concentration. The specific resistance to filtration was determined by an experimental procedure similar to the one outlined by Christensen and Dick [5]. A 200 ml sample was filtered through filter paper by applying a pressure difference of 1 atm, the filtrate being collected in a receiver on an electronic balance with computer interface. The data recording, transformation and evaluation could be done on a PC.

3. Performance of Coagulants

3.1 Coagulated and Settled Wastewater

The following plots have been set up to emphasize the difference between prepolymerized aluminium species and regular aluminium chloride. The horizontal axis is scaled as percent reduction of the respective parameter after coagulation with aluminium chloride, while the ordinate shows the corresponding performance of PAX-W and UPAX 6 at the same dose (as mmol/l Al^{3+}). In Fig. 1 a, the reduction in COD is plotted from experiments with the three coagulants at different doses. If the performance of the prepolymerized products were identical to that of aluminium chloride, all data points would be on the line bisecting the angle between both axes. As can be seen in Fig. 1, most of the scattering is above this line, indicating a better COD removal when using prepolymerized coagulants. With respect to the reduction in total phosphorus, the performance of the prepolymerized coagulants PAX-W and UPAX 6 is also slightly better (Fig. 1 b). The superiority of the prepolymerized coagulants becomes even more evident when comparing the turbidity of the settled wastewater (see Fig. 1 c). When using UPAX 6 as coagulant, the resulting supernatant turbidity was in all cases superior to that of aluminium chloride, but also to that achieved with the polymeric PAX-W. Coagulation with UPAX 6 yielded also the lowest suspended solids content in the supernatant (Fig. 1 d).

The presentation of the data in Fig. 1 makes it possible to differentiate between different coagulant doses. One can see that the higher the Al dose, the smaller are the differences in performance of the chemicals. At low doses, the prepolymerized aluminium products are more effective in removing turbidity and particulate matter by coagulation, resulting also in lower COD and P_{tot}. values. The use of prepolymerized coagulants seems to be most advantageous at low chemical doses.

The aluminium content in the supernatant was about 20 % lower when using polyaluminium as coagulant compared to aluminium chloride. A comparison of the coagulants PAX-W and UPAX 6 did not reveal a significant difference in residual aluminium. This suggests that the utilization of coagulant is more

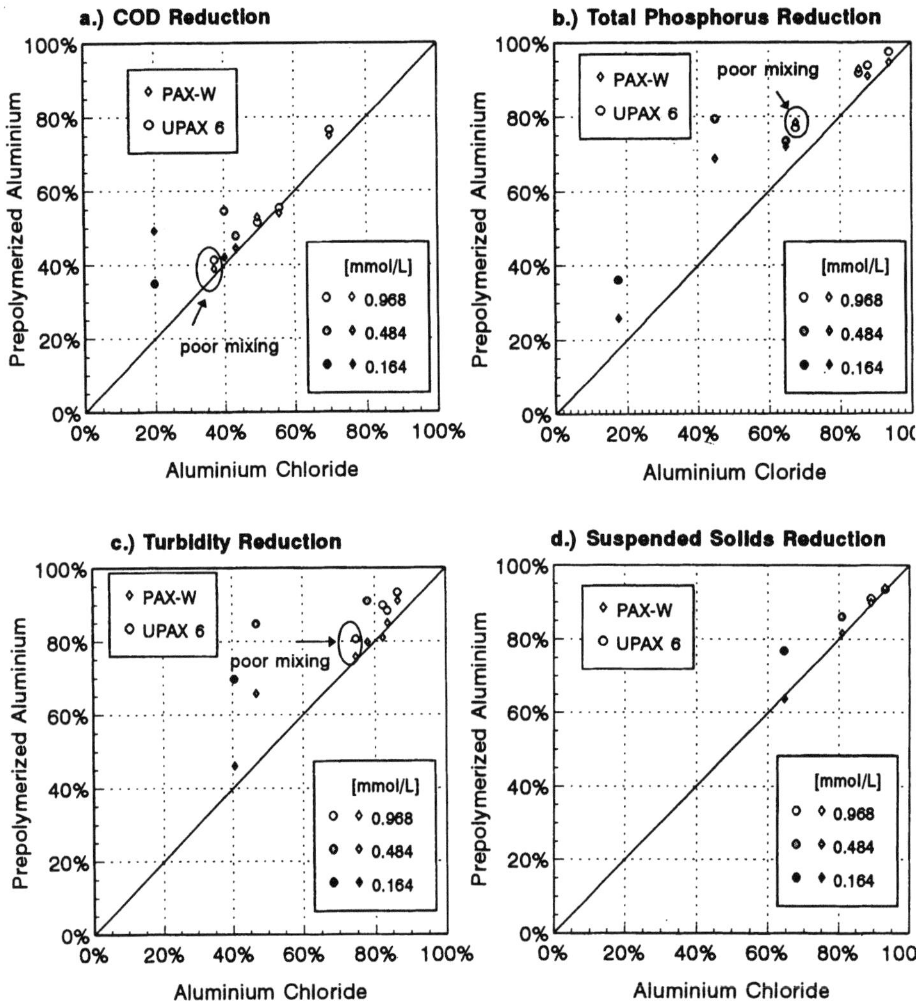

Fig. 1. Comparison of removal efficiencies from coagulation with aluminium chloride and coagulation with the prepolymerized chemicals PAX-W and UPAX6. On the y-axis, the removal rates for PAX-W and UPAX6 are given; for each data point the corresponding removal efficiency for coagulation with aluminium chloride can be obtained on the x-axis

complete when prepolymerized products are used. It is important to note that during all experiments the pH stayed above 7.0, indicating a high buffer capacity of the wastewater. The differences in pH after coagulation with the different chemicals remained negligible.

3.2 Coagulated and Filtered Wastewater

The coagulated and settled wastewater was further membrane filtered (0.45 μm) to characterize the dissolved and submicron matter. In Fig. 2 a, the charge concentration of the filtrate is plotted for the experiments with aluminium chloride versus the results with PAX-W and UPAX 6. The results indicate that dissolved matter and/or submicron colloidal particles in the wastewater are removed slightly better by PAX-W and UPAX 6 than by conventional aluminium salts.

What was not expected is an improved reduction in soluble phosphorus when using PAX-W and UPAX 6 (Fig. 2 b). Especially at low Al doses, the prepolymerized coagulants seem to be more efficient. At the lowest dose, the molar ratio between Al and P (β-value) was about 0.61. The measured reduction for both PAX-W and UPAX 6 was higher, while the reduction due to precipitation with aluminium chloride was only 32 %. No effort to explain this is made since the data basis is weak and experimental error cannot be ruled out completely. However, in all experiments, coagulation with PAX-W yielded the highest reduction in ortho-P.

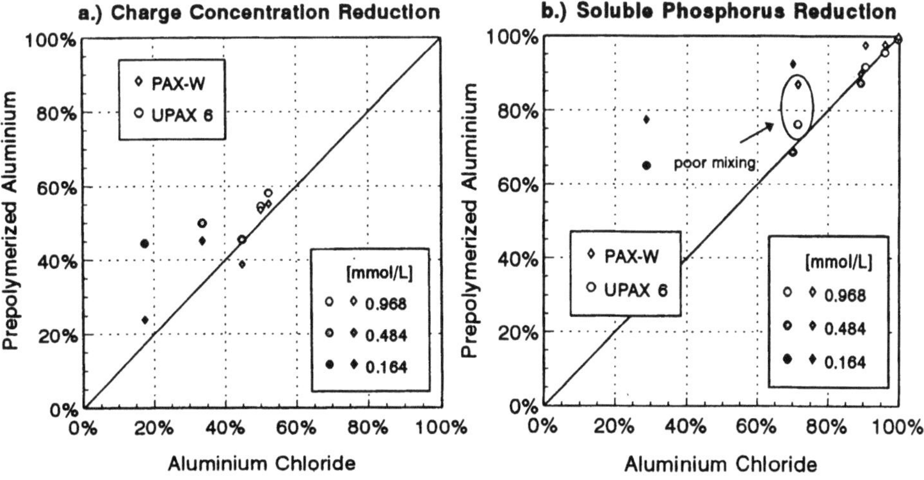

Fig. 2. Comparison of removal efficiencies from coagulation with aluminium chloride and coagulation with the prepolymerized chemicals PAX-W and UPAX 6. The charge concentration (a) and soluble phosphorus (b) were determined after membrane filtration (0.45 μm)

3.3 Effect of Rapid-Mixing

As has been described earlier, the rapid-mix conditions were optimized during the experiments. Thus it is possible to determine how poor rapid-mixing affects the removal rates. As can be seen from Fig. 1, COD, $P_{tot.}$, and turbidity are eliminated to a higher degree when the mixing of coagulant and wastewater is improved. Regarding the removal of phosphorus in the filtered sample, the improvement is even more pronounced. At high Al doses ($\beta > 3$) and proper mixing, the removal rate was at least 90 %, compared to e.g. only 71 % for AlCl$_3$ under poor rapid-mix conditions.

Obviously, poor mixing results in removal efficiencies that can also be achieved at a considerably lower coagulant consumption under optimal mixing conditions (Figs. 1, 2). When comparing the performance of the different chemicals in the experiments with poor mixing, it becomes evident that aluminium chloride yielded the lowest removal rates and had thus the highest sensitivity to poor mixing. In summary, then, when flocculating under unfavourable conditions (poor mixing and/or low coagulant dose) prepolymerized aluminium chloride demonstrates its greatest advantage over regular AlCl$_3$.

3.4 Evaluation of Flocculant Performance

The performance of the different chemicals and the observed superiority of prepolymerized coagulants at low doses for removing turbidity, SS, COD and total-P are well in accordance with the theory of coagulation. Similar results for coagulation with hard wastewater were also presented by Ødegaard et al. (1990), confirming a higher coagulation efficiency with increasing basicity (OH/Al ratio). It is assumed that the highly positively charged aluminium-hydroxo-complex species of the prepolymerized salts are responsible for the observed effect. Furthermore, these species are considered to be more stable after dissolution, as has been pointed out e.g. by Parthasarathy and Buffle (1985).

Recent investigations by Bottero and Bersillon (1989) document a higher specific area of freshly precipitated hydroxide from polyaluminium chloride (OH/Al = 2) compared to aluminium sulfate (OH/Al = 0). Thus, enhanced adsorption or enmeshment in more voluminous flocs might be an explanation for the observed improvement in reduction of particulate and colloidal matter (Fig. 1) and submicron and dissolved species (Fig. 2). The silica content of the UPAX 6 coagulant obviously leads to further improvement. The aluminium silica complexes have a high surface potential and thus tend to adhere strongly to particle surfaces. They might also serve as adsorbent for dissolved macromolecules, as indicated by measurements of the charge concentration after coagulation.

4. Comparison of Sludge Characteristics

4.1 Settling Behavior

The sludges resulting from coagulation with aluminium chloride, PAX-W, and UPAX 6 exhibited significant differences when allowed to settle in a column. The height of the interface that develops between the supernatant and the sludge blanket is plotted versus time in Fig. 3. The results represent typical settling behaviors of the different sludges when operating at high chemical dose (0.97 mmol/l Al). Within three hours, no significant thickening of the UPAX 6 sludge was observed. With the sludges from coagulation with aluminium chloride and PAX-W, a rapid fall of the interface, which goes along with a reduction in sludge volume, was recorded.

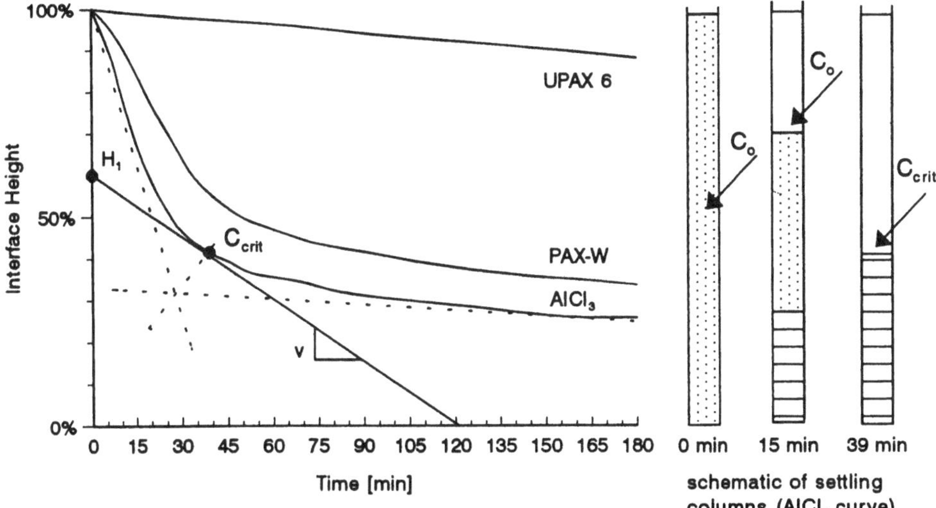

Fig. 3. Typical settling curves of the sludges resulting from coagulation with the different coagulants at high chemical dose (0.97 mmol/l). For the settling curve of the sludge from coagulation with aluminium chloride, the construction of the compression point is demonstrated and the concentrations at the interface at settling times 0, 15 and 39 minutes are illustrated

4.1.1 Theoretical Background. An attempt was made to further evaluate the data from the thickening experiments. Most methods for the interpretation of batch experiments are based on Kynch's theory of sedimentation. A presentation of this theory requires mathematical descriptions, which are not the subject of this paper. In the following, only brief explanations are given to provide the necessary background for the parameters used to distinguish between the sludges from different coagulants. For more detailed information, we refer the reader to the original paper of Kynch (1952) and to a summary of theories and methods for the thickening of sewage sludges presented by Dick (1972) or Braha (1983).

According to the Kynch model, the settling velocity of a particle is a function only of the local solids concentration around the particle. Thus, in a batch test starting at uniform concentration C_0, the solids of a sludge settle at constant velocity as long as the local solids concentration at the interface remains unchanged. At the same time, a layer of higher concentration is propagated up from the bottom of the vessel. At the point where this higher concentration reaches the liquid-"solid" interface, all of the sludge solids are within the compression zone, as illustrated in Fig. 3.

In the plot of interface height versus time, this point can be identified, and is in the following refered to as compression point. Applying the method of Talmadge and Fitch (1955) for the interpretation of data from a single batch test allows this compression point and the corresponding sludge solids concentration at the interface to be determined. Using a simple graphical construction outlined by Metcalf and Eddy (1979), this compression point can be approximated from the settling curve as illustrated in Fig. 3. Tangents are drawn to the settling curves in the hindered settling and the compression regions. The angle between the two lines is bisected. The compression point is then located at the intersect between the settling curve and the bisect. Drawing a tangent to the settling curve through the compression point and determining the intersect H_1 (see Fig. 3) makes it possible to calculate the corresponding concentration $C_{crit.}$ according to Talmadge and Fitch using the equation:

$$C_{crit.} = \frac{C_0 \cdot H_0}{H_1} \tag{1}$$

The tangent can be constructed at any point on the settling curve, and the corresponding concentration C can be calculated by Eq. (1). The slope of the tangent at that point is equivalent to the gravity settling rate v. Multiplication of the concentration and the corresponding settling rate yields the solids flux SF due to gravity:

$$SF = C \cdot v \tag{2}$$

By applying the Talmadge and Fitch method, it was thus possible to plot the solids flux versus the solids concentration without implementing multiple batch tests, as outlined in a method by Dick (1972). The critical concentration at the compression point and the solids flux curves are subsequently used to characterize the difference in thickening behavior of the different sludges from wastewater coagulation.

4.1.2 Critical Concentration. The compression point and the prevailing solids concentration influence the required area of a thickener. In Fig. 4 the critical concentration of the sludges from coagulation with the different chemicals at two coagulant doses are given. For coagulation with UPAX 6 at a dose of 0.97 mmol/l Al, no compression point could be identified. Hindered settling did not take place at the initial solids concentration C_0 and the observed slow fall of the sludge blanket (Fig. 3) is due to compression. The compression point as defined above must be at a solids concentration lower than the uniform concentration at the beginning of the settling. However, when coagulation with UPAX 6 was carried out at low doses, the critical concentration was identified to be about 3 g/l. A comparison of $C_{crit.}$ of the sludges from aluminum chloride with the sludges from prepolymerized aluminium exhibits the lowest concentration for the UPAX slurry, while coagulation with PAX-W and aluminium chloride result in sludges with comparable thickening behavior. As described earlier, $C_{crit.}$ is the solids concentration of the network that is supported from the bottom and propagates to the interface. The creation of such a network requires a significantly higher solids concentration for the sludge from aluminium chloride and PAX-W coagulation. This leads to the conclusion that the flocs of these slurries are denser and form a denser network during the thickening process. Another interesting conclusion that may be drawn from Fig. 4 is that coagulation at lower doses results in sludges that form denser structures. A variation in chemical dose seems to have a strong effect on the properties of the sludge.

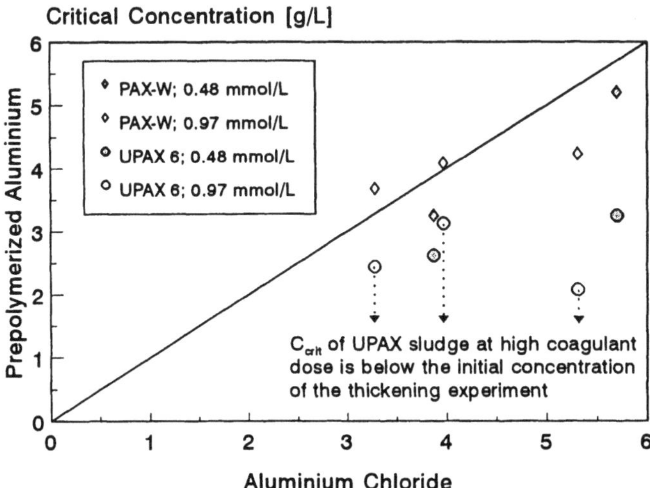

Fig. 4. Comparison of the critical sludge solids concentrations (as defined in Fig. 3) from coagulation with aluminium chloride and coagulation with the prepolymerized chemicals PAX-W and UPAX 6

4.1.3 Solids Flux. This finding can be confirmed by the calculation of the solids flux from the thickening tests. As shown in Fig. 5, a higher solids flux at the same solids concentration is observed at a coagulant dose of 0.48 mmol/l compared to the sludge from coagulation at a higher dose. This was found to be true for all coagulants, although the difference in solids flux was most pronounced when using aluminium chloride (Fig. 5). It appears that an increasing fraction of aluminium hydroxide within the flocs results in a lower solids flux at the same sludge solids concentration. Although at high coagulant dose (0.97 mmol/l) the difference in solids flux between aluminium chloride and PAX-W becomes smaller, the thickening ability of aluminium chloride sludge proved to be superior to the sludge from polymeric aluminium.

4.1.4 Sludge Volume. The sludge volume that could be attained after one day of thickening was recorded and regarded as the final sludge volume. There are significant differences between the sludges that stem from the different coagulants. In Fig. 6, the sludge volumes in liters per m³ of wastewater are compared in the same manner as the removal efficiencies in the previous section.

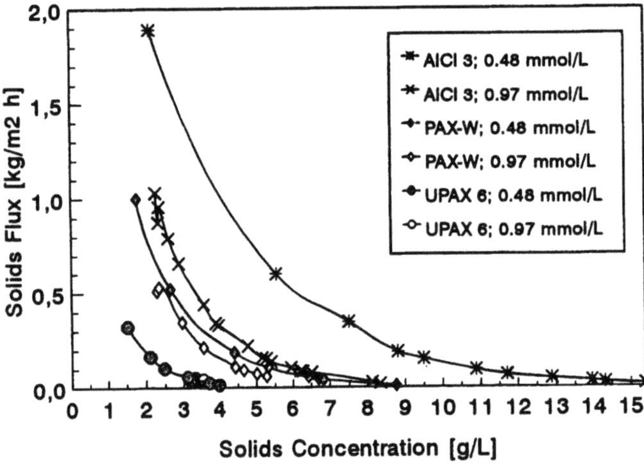

Fig. 5. Solids flux curves as a function of the solids concentration for the sludges resulting from coagulation with different flocculants and chemical doses

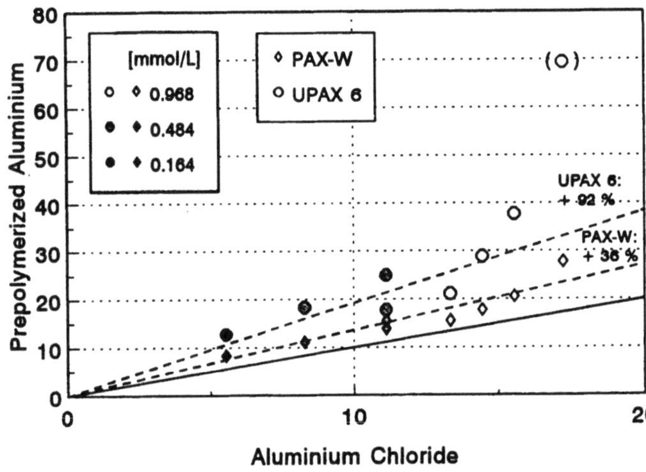

Fig. 6. Comparison of the sludge volume per m^3 of wastewater flocculated resulting from coagulation with aluminium chloride and coagulation with the prepolymerized chemicals PAX-W and UPAX 6

It can be seen that, after coagulation with PAX-W, the resulting sludge volume is 36 % higher than that with aluminium chloride at the same dose; the use of UPAX 6 results in a sludge volume that is almost twice as high.

With regard to practical implications, one must note that the experiments were performed at the same mmol Al/l dose for the different chemicals. However, as indicated in the previous section, a lower chemical dose might be necessary when using prepolymerized aluminium coagulants. In practice, one would have to compare the sludge volumes at the respective proper chemical dose. In this context it is interesting to note that Dempsey et al. (1985) found that substitution of polyaluminium chloride for alum will result in a substantially decreased volume of sludge only when the necessary dose of PACl is lower (in terms of aluminium) than the required dose for alum.

4.1.5 Total Solids Concentration. After recording the final sludge volume, the total solids concentration was determined.

In Fig. 7, the total solids concentrations from coagulation with aluminium chloride are compared to the ones obtained with prepolymerized aluminium. In this plot it is interesting to point out the differences resulting from coagulant dose. A strong relationship between total solids concentration and coagulant dose was found, indicating a change in the composition of the sludge. This is true for all the chemicals under investigation. The observation can be explained by a higher fraction of wastewater suspended solids within the sludge when operating at low coagulant doses. At high doses, the recovery of suspended solids is higher, but the fraction of hydroxides, which are known to be very voluminous, will grow over-proportionally.

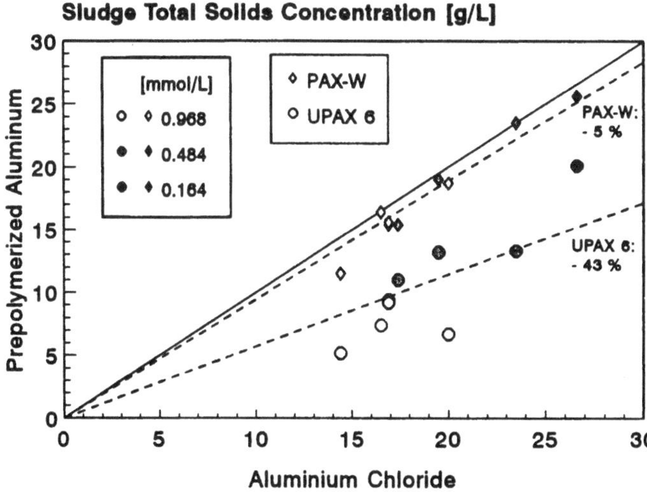

Fig. 7. Comparison of the total solids concentration of the thickened sludges resulting from coagulation with aluminium chloride and coagulation with the prepolymerized chemicals PAX-W and UPAX 6

4.2 Sludge Dewaterability

In order to compare the dewaterability of the different sludges, the capillary suction time (CST) of the thickened sludge was determined. Comparing the CST readings for the sludges from PAX-W and UPAX 6 coagulation with those of the aluminium chloride sludge revealed in all experiments lower CST values for the sludges from coagulation with prepolymerized coagulants. As can be seen in Fig. 8, the CST is a function of the total solids concentration of the sludge. However, at comparable solids concentrations, the CST of the PAX-W sludge is lower than that for the aluminium chloride sludge, indicating a better dewaterability. No conclusion can be drawn for the slurry from UPAX 6 coagulation, since the solids concentrations are not comparable with the others. Therefore, a parameter is needed that describes the dewaterability of the sludge while being independent of its solids concentration.

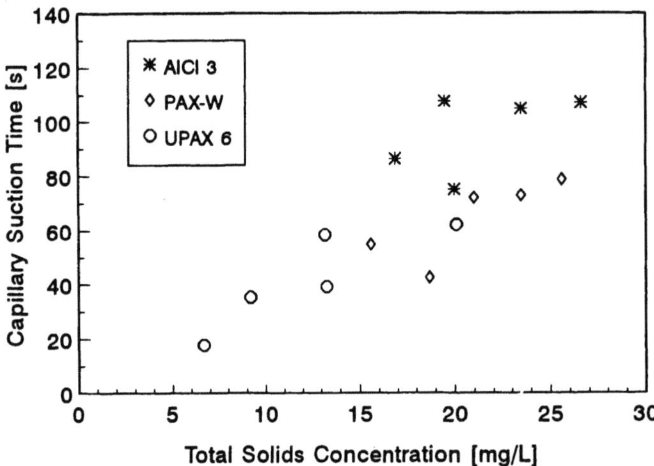

Fig. 8. Capillary suction time (CST) as a function of the sludge solids concentration

In theory, the specific resistance of filtration fulfills this requirement. However, a weak correlation between specific resistance and total solids concentration could be observed. Plotting the specific resistance of the sludges from prepolymerized aluminium versus the corresponding resistance to filtration of the aluminium chloride sludge (Fig. 9) shows that the PAX-W and UPAX 6 sludges have a better dewaterability. Under identical conditions (i.e. Al-dose), coagulation with aluminium chloride results in all cases in a sludge with slightly higher specific resistance to filtration. It is also remarkable that the coagulant dose has a strong influence on sludge dewaterability. Fig. 9 shows that operating at the lowest dose resulted in a sludge with the highest specific resistance.

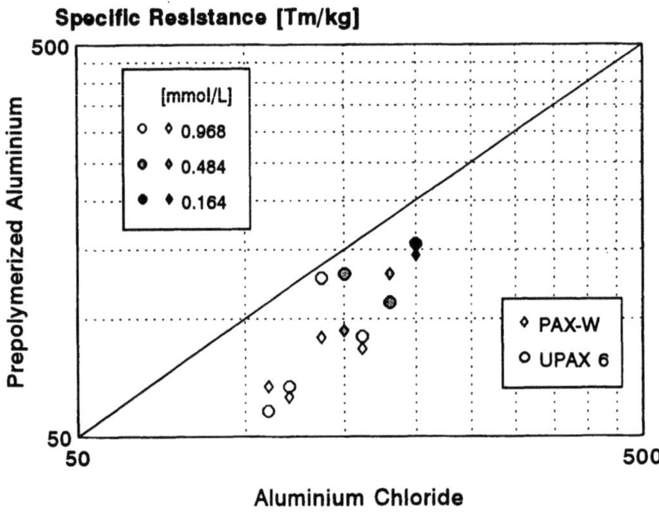

Fig. 9. Comparison of the specific resistance to filtration of the sludges resulting from coagulation with aluminium chloride and coagulation with the prepolymerized chemicals PAX-W and UPAX 6

From the experiments it can be concluded that PAX-W sludge has better dewatering properties than sludge from coagulation with regular aluminium chloride. The same cannot, however, be said for the sludge from coagulation with the silica-containing UPAX coagulant because of its low solids concentration.

Ødegaard et al. (1990) summarized the literature on floc separation, and stated that the effect of using prepolymerized aluminium as coagulant on the sludge dewaterability remains uncertain. Results published by different investigators even seem to be contradictory. Diamadopoulos and Benedek (1984) found that within a pH range of 5–9, the specific resistance increased with increasing basicity of the prepolymerized aluminium chloride. It was also observed, however, that $AlCl_3$ sludge yielded a higher specific resistance than the sludge from polyaluminium chloride with a ratio of $OH/Al = 1.0$. This finding is well in accordance with the results from this study, showing a lower specific resistance for sludge from PAX-W ($OH/Al = 0.75$) coagulation.

4.3 Sludge Production

The sludge volume and solids concentration of the thickened sludge have already been given as a way to distinguish between the different sludges. The product of both parameters yields the mass of sludge solids from the batch coagulation of wastewater. It has been found that the sludge produced from coagulation with the prepolymerized aluminium contained more solids than the sludge from aluminium chloride. This is demonstrated in Fig. 10, whereby all data points are above the line representing identical sludge production. The dotted lines result from the linear regression of the PAX-W and UPAX 6 data. It can therefore be concluded that coagulation with PAX-W and UPAX 6 yields about 30 % more sludge solids.

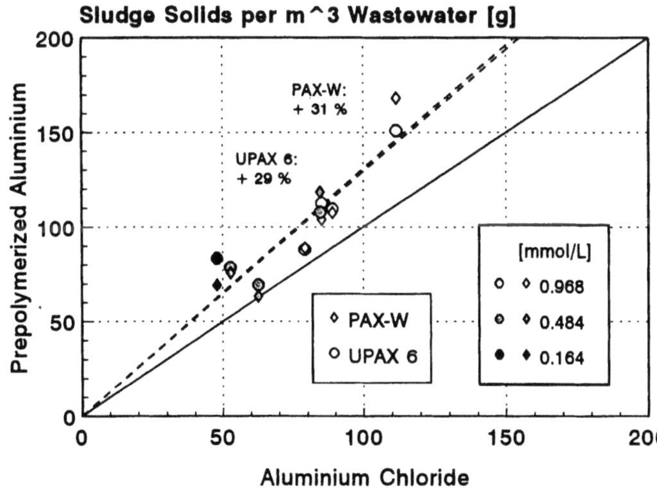

Fig. 10. Comparison of the sludge solids yielded from coagulation with aluminium chloride and coagulation with the prepolymerized chemicals PAX-W and UPAX 6

The question arose, whether the higher sludge production results from a higher removal efficiency of wastewater suspended solids or from an increased recovery of dissolved and colloidal matter by adsorption or enmeshment within the sludge matrix consisting of flocculated and precipitated material. For this reason, in an additional set of experiments, the suspended solids concentration of the raw wastewater and of the supernatant after batch coagulation was measured. Now a simple mass balance could be performed on the solids before and after coagulation. The formulation is as follows:

$$\begin{bmatrix} \text{Wastewater sus-} \\ \text{pended solids} \\ \text{within the batch} \end{bmatrix} + \begin{bmatrix} \text{Precipitates} \\ \text{from coagulant} \\ \text{addition} \end{bmatrix} = \begin{bmatrix} \text{Supernatant} \\ \text{suspended} \\ \text{solids} \end{bmatrix} + \begin{bmatrix} \text{Sludge} \\ \text{solids} \end{bmatrix}$$

In essence, this is a mass balance on suspended solids. However, the sludge solids could only be determined as total solids. Since the sludge volume compared to the wastewater batch volume is small, it is assumed that the concentration of dissolved matter can be neglected. All terms of the balance, except the mass of the precipitates, can be determined directly by measuring solids concentrations and the corresponding volumes. The mass of precipitates is defined as the mass of aluminium dosed into the batch reactor times a factor f. Since the residual aluminium in the supernatant was determined to be below 0.2 mg/l, it was assumed that the coagulant dosed was completely precipitated. Now the mass balance can be solved for f, resulting in Eq. (3):

$$f = \frac{\text{TS} \cdot k + \text{SS}_0(\eta_{\text{SS}} + k + \eta_{\text{SS}} \cdot k)}{C_{\text{Al}}} \tag{3}$$

TS Total solids concentration of the sludge [g/l]
SS_0 Suspended solids concentration of the raw wastewater [g/l]
C_{Al} Coagulant dose as Al [g/l]
k Sludge volume per unit volume coagulated wastewater [-]
η_{SS} Suspended solids removal efficiency [-]

For each batch experiment, the factor $f_{\text{exp.}}$ can now be calculated. The stoichiometry of aluminium hydroxide yields a value of 2.89; if all of the aluminium were precipitated as aluminium phosphate, f would be equal 4.52. For each experiment the theoretical value $f_{\text{theor.}}$ could be calculated and was found to be within the range of 3.2 and 3.7.

Experiences have shown that the coagulation of wastewater with aluminium chloride results in sludge production that is above the mass that can be calculated from stoichiometry. The higher solids production can be explained by a recovery of dissolved anions and organics due to adsorption onto the surfaces of the precipitates and flocs. In addition, enmeshment of colloidal particles within the precipitates might account for the observed increase in sludge solids. Taking into account these uncontrollable processes, the German ATV-standard (1983) quotes a value of 4 gram sludge solids per gram aluminium dosed. The EPA (Barth et al., 1976) suggests that sludge production can be expected to be about 35 % higher than that resulting from stoichiometry.

Figure 11 makes it possible to compare the experimental and theoretical values for f. The sludge solids resulting from aluminium chloride coagulation are well within the expected range. In any case, more sludge was produced by coagulation with polymerized aluminium salts. The mass balance reveals that when prepolymerized aluminium chloride is used, it is not only the higher removal efficiency of suspended solids (as demonstrated in Fig. 1 d) that is responsible for the increase in mass of sludge solids, but also a higher recovery of dissolved matter within the precipitates. Unfortunately, only three experiments are available for the calculation of $f_{exp.}$ for PAX-W and UPAX 6, but considering the higher total sludge solids in Fig. 10 for all experiments, the conclusion seems to be justified that when using prepolymerized coagulants a specific sludge production of more than 4 g/g Al can be expected. According to the experimental results, the sludge production when using PAX-W and UPAX 6 as coagulant will be 5-6 g/g Al, in addition to the suspended solids that are removed from the wastewater.

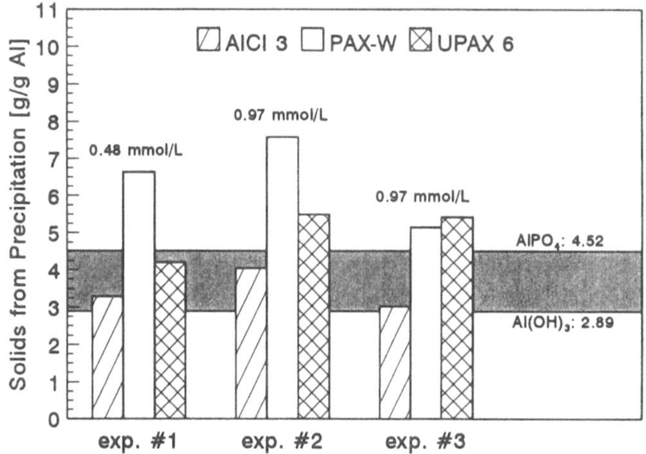

Fig. 11. Specific sludge production (gram per gram aluminium dosed) for the three coagulants

There is another hint that the assumption of increased adsorption of dissolved organics to precipitates from PAX-W and UPAX 6 may be correct. Measurements of the charge concentration within the sludge revealed a higher charge per unit mass of sludge solids for PAX-W and especially for UPAX 6. This corresponds well with lower charge concentrations in the supernatant, as presented in Fig. 2 a.

5. Summary and Conclusions

The coagulation of raw municipal wastewater with prepolymerized and conventional aluminium chloride results in increased removal efficiencies for the organic load (COD), phosphorus and turbidity, when the polymeric products were applied at the same dose. The observed improvement holds true for the entire concentration range examined in this study, but becomes especially evident in the small dose range. A comparison of the results achieved with the two polymeric aluminium chlorides indicates a higher efficiency for the silica-containing product.

The mechanisms involved in the removal of contaminants from wastewater by prepolymerized aluminium chloride are not quite clear, however, measurements of total charge in the supernatant show that the application of these chemicals results in a higher reduction in the negative surface, which can be assumed to lead to improved destabilization and aggregation of the negatively charged colloidal material in the wastewater. The results which were observed make sense when one reflects that the major fraction of the contaminants in municipal wastewater is associated with particles. The superior efficiency of UPAX as coagulant over PAX-W may be based on the silica content of the former, offering additional surfaces for the adsorption of wastewater constituents.

A comparison of the characteristics of sludge resulting from wastewater coagulation with the three different chemicals shows considerable differences with regard to sludge volume, total solids concentration in the sludge, sludge thickening behavior, and sludge dewatering.

The sludge volume that is attained after 24 hours of thickening amounts to an increase of 36 % for PAX-W and 92 % for UPAX 6, in each case compared with the sludge volume for $AlCl_3$ at an identical dose. Because of the higher efficiency of the prepolymerized chemicals, however, smaller doses are required to achieve identical purification results. This will reduce the sludge volume substantially if one takes into account that at high doses the fraction of hydroxides within the sludge will grow over-proportionally. This is supported by the relationship between the total solids concentration and the chemical dose, whereby denser flocs and higher amounts of suspended solids per sludge volume result from small coagulant doses. Based on the data of this experimental investigation, a specific sludge production of 5-6 g TS/g Al due to precipitation (and in addition to the suspended solids removed) can be expected, when prepolymerized aluminium chlorides are added for the coagulation of raw municipal wastewater.

A comparison of the dewaterability of the different sludges shows lower CST values for the PAX-W sludge compared with the $AlCl_3$ sludge at identical suspended solids concentrations, indicating a better dewaterability. This observation was confirmed by specific filter resistance data. The UPAX 6 sludge is characterized by the lowest CST values, which is not surprising, because the low solids concentration of this sludge results in an easier release of water under the influence of gravity compared to the other two sludges.

Acknowledgements

The aluminium salts applied in this investigation have been provided by Kemira Kemi, Helsingborg, Sweden. Helpful discussions with Lars Gillberg, Kemira Water Treatment, regarding properties and application of the prepolymerized chemicals are gratefully acknowledged.

References

ATV-Fachausschuß 2.8. (Dritter Arbeitsbericht): Phosphorelimination. Korrespondenz Abwasser *30* (1983) 191–199

Barth, E., Smith, J., Brunner, C., Farrell, J.: Process Design Manual for Phosphorus Removal. U.S. Environ. Protection Agency Report EPA 625/1-76-001a (1976)

Bottero, J.Y., Bersillon J.L.: Aluminium and Iron (III) Chemistry. Some Implications for Organic Substance Removal. J. Amer. Chem. Soc. *26* (1989) 425–442

Braha, A.: Das Absetzverhalten von Abwässern mit hohem Gehalt an suspendierten Stoffen. Korrespondenz Abwasser *30* (1983) 24–33

Christensen, G.L., Dick, R.I.: Specific Resistance Measurements: Nonparabolic Data. J. Environ. Eng. Div., ASCE *111* (1985) 243–255

Dempsey, B.A., Shen, H., Tanzer Ahmed, T.M., Mentink, J.: Polyaluminum Chloride and Alum Coagulation of Clay-Fulvic Acid Suspensions. J. AWWA *77* (3) (1985) 74–80

Diamadopoulos, E., Benedek, A.: Aluminum Hydrolysis Effects on Phosphorus Removal from Wastewater. J. Water Poll. Control Fed. *56* (1984) 1165–1172

Dick, R.I.: Gravity Thickening of Sewage Sludges. Water Pollut. Control *71* (1972) 368–380

Dick, R.I., Ewing, B.B.: Evaluation of Activated Sludge Thickening Theories. J. Sanit. Eng. Div., ASCE *93* No. SA4 (1967) 9–29

Fiessinger, F., Bersillon, J.L.: Prépolymérisation de l'hydroxyde d'aluminium pour la coagulation des eaux. Tribune du Cebedeau *399* (1977) 52–68

Henze, M., Harremoës, P.: Chemical-Biological Nutrient Removal – The HYPRO Concept. In: Chemical Water and Wastewater Treatment, H.H. Hahn and R. Klute (eds.). Springer, Berlin Heidelberg New York 1990, pp. 499–510

Karlsson, I.: Pre-precipitation for Improvement of Nitrogen Removal in Biological Wastewater Treatment. In: Pretreatment in Chemical Water and Wastewater Treatment, H.H. Hahn and R. Klute (eds.). Springer, Berlin Heidelberg New York 1988, pp. 261–271

Kynch, G.J.: A Theory of Sedimentation. Transactions of the Faraday Society *48* (1952) 166–176

Metcalf & Eddy, Inc.: Wastewater Engineering: Treatment, Disposal, Reuse (2nd Ed.). McGraw-Hill, New York 1979

Ødegaard, H.: Coagulation as the First Step in Wastewater Treatment. In: Pretreatment in Chemical Water and Wastewater Treatment, H.H. Hahn and R. Klute (eds.). Springer, Berlin Heidelberg New York 1988, pp. 249–260

Ødegaard, H., Fettig, J., Ratnaweera, H.C.: Coagulation with Prepolymerized Metal Salts. In: Chemical Water and Wastewater Treatment, H.H. Hahn and R. Klute (eds.). Springer, Berlin Heidelberg New York 1990, pp. 189–220

Parthasarathy, N., Buffle, J.: Study of Polymeric Aluminium (III) Hydroxide Solutions for Application in Waste Water Treatment. Properties of the Polymer and Optimal Conditions of Preparation. Wat. Res. *19* (1985) 25–36

Talmadge W.P., Fitch, E.B.: Determining Thickener Unit Area. Ind. & Eng. Chemistry *47* (1955) 38–41

Stefan J. Langer and Rudolf Klute
Institut für Siedlungswasserwirtschaft
Universität Karlsruhe
Am Fasanengarten
D-7500 Karlsruhe 1
Germany

Frank Seitz
c/o Gauff Ingenieure
Passauer Str. 9
D-8500 Nürnberg
Germany

The DENSADEG – A New High Performance Settling Tank

P. Dauthuille

Abstract

The evolution of settling processes requires increasingly more efficient apparatus. Today the settling tanks must meet the following criteria:
- Great compactness: high settling velocity.
- Excellent treated water quality: use of lamellae.
- Production of concentrated sludge: simultaneous sludge thickening.

The DENSADEG is a new compact clarifier-thickener capable of meeting all these requirements. It fits into the line of sludge recirculation type lamellae settling tanks using chemical reagents. The original design of the reactor induces both a highly dense and homogeneous floc and settling velocities ranging from 20 to 40 m/h. In addition to settling, sludge thickening takes place in the DENSADEG (30 to 500 g/l); the thickened sludge can then be directly handled by a dewatering unit. More than fifty DENSADEG are presently in operation around the world in many areas of applications: drinking water, urban wastewater, sludge treatment, industrial water and storm water treatment.

An example is discussed in drinking water (Morsang-sur-Seine) and in urban wastewater (Gréoux-les-Bains).

1. Accelerated Clarification

Clarification technology has been in a state of constant flux for several decades. The last major innovation dates back to the 1970s with the more or less systematic application of lamellae, tubular or parallel-plate modules to many different types of settling tanks, regardless of their principle of operation.

The lamellae serve two purposes:
- They increase flow rates (or reduce the settling surface area).
- They improve the quality of the clarified water.

The existence of lamellae is generally considered to be the primary criterion for the classification of settling tanks. It would seem better to base classification on the hydraulic design of the system, i.e., whether or not it has an integrated coagulation/flocculation function. Table 1 lists a few examples of settling tanks in the main categories and indicates typical performance levels for clarification of river water.

Tab. 1. Main settling tank categories

Type	Examples	Flocculator	Rising velocity $m \cdot h^{-1}$	Extracted sludge concentration $g \cdot l^{-1}$
Static settling tank	Horizontal flow	In front	0.5–2	1–5
	Vertical flow	In front or integrated	1–4	
Sludge blanket settling tank	Without lamellae (Pulsator)	Integrated	3–5	2–10
	With lamellae (Superpulsator, Lamellar Pulsator)	Integrated	6–10	
Sludge recirculation settling tank	Without lamellae (Turbocirculator)	Integrated	2–3	5–10
	With lamellae (RPS)	In front	10–15	10–20

Today's users require compact systems and accelerated clarification is the key to meeting their needs. Only sludge-recirculation and lamellae modules can offer high settling rates and high sludge concentration levels. To meet all of these requirements, Degremont has developed a new system based on the principle of lamellae settling associated with sludge recirculation and integrated sludge thickening: The DENSADEG.

2. The DENSADEG

The DENSADEG is a new settling tank that associates and optimizes a number of techniques developed by Degremont. It heralds the second generation of sludge-recirculation lamellae settlers.

The main characteristics of the DENSADEG are:

- Innovative design of the reactor, which produces homogeneous and very dense floc.
- Compact design and high settling rate (between 20 and 40 $m \cdot h^{-1}$).
- Integrated thickening of the sludge, which can then be treated directly by a dewatering unit.
- Acceptance of variations in the composition and flow rate of the raw water.

2.1 Principles of Operation

The process is based on three main principles (Fig. 1):

– An integrated coagulation/flocculation reactor of original design.
– Sludge recirculation from the thickening zone to the reactor.
– Lamellae settling.

The DENSADEG consists of three main parts.

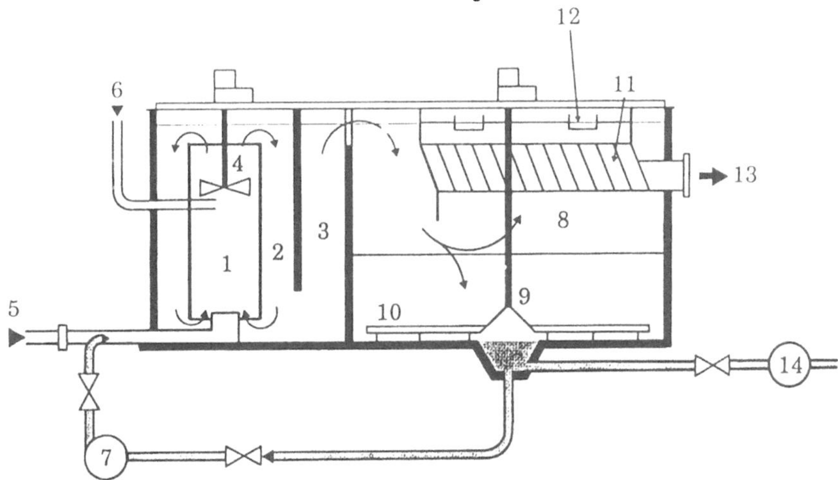

Fig. 1. DENSADEG – Basic diagram (1, 2, 3 – Flocculation chambers; 4 – Propeller; 5 – Raw water inlet; 6 – Reagent feed; 7 – Recirculation pump; 8 – Presettling zone; 9 – Drive shaft; 10 – Scraper; 11 – Modules; 12 – Collection troughs; 13 – Treated water outlet; 14 – Sludge draw-off pump

2.1.1 The Reactor. One of the original features of the process is the reactor, which consists of three successive chambers. Rapid flocculation takes place in the first two chambers and slow flocculation in the third, which operates under plug-flow conditions.

The *agitated reactor* consists of two zones. The cylindrical central zone (1) is equipped with an impeller that recirculates the flow from the second zone (2). Raw coagulated water flows into the base of the agitated reactor (5) and the flocculant (6) is injected at the base of the turbine (4). An external recirculation system (7) adds more sludge to the raw water. Inside the agitated reactor, concentration levels are optimized for the type of treatment (clarification, lime softening, industrial wastewater, municipal wastewater). Upflow conditions prevail in the *plug-flow reactor* (3), which is a zone of *slow flocculation* producing largely homogeneous flocs of considerable size (several millimeters). The innovative design of this highly compact reactor, combined with the high level of sludge recirculation, delivers a particularly dense and homogeneous floc. As a result, the flow enters the settling zone at far greater speeds than in earlier systems.

2.1.2 The Presettler – Thickener – Storage Tank. The incoming floc flows into a presettling and thickening zone (8). Most of the suspended solids in the reactor settle in this zone hence the importance of producing floc of uniform size. The lower cylindro-conical part of this zone is equipped with a picket fence (10) and a bottom scraper (9) for integrated sludge thickening. The thickened sludge is drawn off through a pipe (14). Part of this sludge is recirculated to the reactor's raw water inlet pipe and brought back to the reactor. Recirculation is generally performed by a pump (7). The presettler-thickener produces highly concentrated sludge that can be sent directly to a dewatering unit. The sludge breaks up during recirculation. This considerably increases the density and settlability of the floc. Further, the integrated thickener can also be used as a sludge storage tank between two dewatering operations.

2.1.3 Lamellae Settler. The lamellae settler (11) treats the residual floc leaving the presettler. The chamber is equipped with lamellae or tubular modules for accelerated clarification. The water flows through these modules in the opposite direction to the sludge (12). The low concentration of suspended solids entering the lamellae settler and the excellent floc density promote high settling rates and good water quality (13).

The DENSADEG is characterized by:

- A reactor offering homogeneous optimized flocculation.
- External sludge recirculation.
- An integrated thickener that can also be used for sludge storage.
- A lamellae settler.

The highly compact system offers:

- Very high settling rates (between 20 and 40 $m \cdot h^{-1}$ in the lamellae zone).
- Very high concentrations of extracted sludge.
- Clarified water of high quality.

The DENSADEG is applied in the following areas:

- Clarification of river water.
- Lime softening of river water or drilling water.
- Primary physico-chemical treatment of municipal water.
- Tertiary physico-chemical phosphate removal.
- Settling of storm water.

Some examples are discussed further.

3. Clarification (Potable Water)

The Appraisal of a Two-Year Period of Service at Morsang III

The Morsang-sur-Seine facility is the cornerstone of drinking water production for the Lyonnaise des Eaux network supplying the Southern Paris Area (RPS).

To satisfy the slow but steady increase in consumption arising primarily from the growth of new towns and to promote the security of water resources, it was decided to increase the production capacity of the southern Paris network in the Morsang-sur-Seine facility from 150 000 to 225 000 m^3.

- Annual production   89 000 000 m^3
- Daily production capacity   445 000 m^3
  - Morsang-sur-Seine   225 000 m^3
  - Viry-Châtillon   120 000 m^3
  - Vigneux-sur-Seine   55 000 m^3
- No. of inhabitants supplied   900 000
- No. of communities   130
- Network length   3 500 km
- No. of reservoirs   109
- Storage capacity   160 000 m^3

3.1 The Phase-by-Phase Construction of the Morsang Facility

In 1970, Lyonnaise des Eaux chose Morsang-sur-Seine as the site for its major new drinking water production facility. It was decided to structure the facilities in groups, each in the form of a three-pointed star with a daily production of some 225 000 m^3. Morsang III, a facility consisting of three distinct treatment systems (Fig. 2), completes the first star.

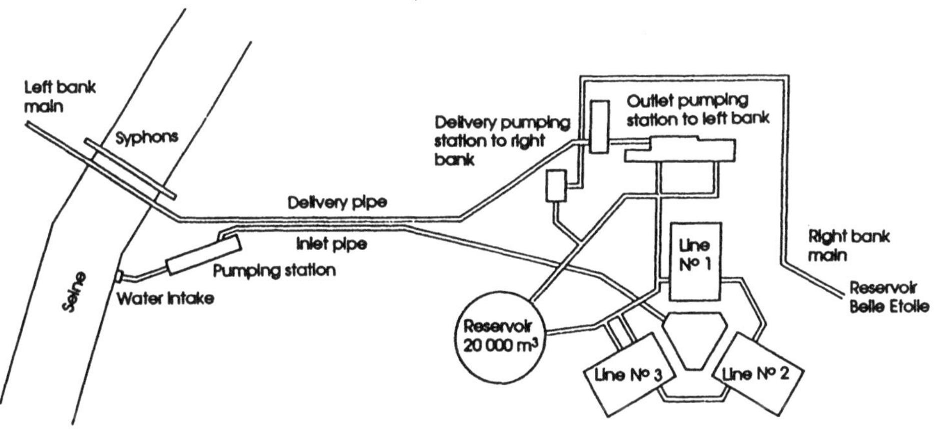

Fig. 2. Morsang-sur-Seine – Layout

The main phases in the construction of Morsang were as follows:

1970: Morsang I. The operations building and the first production unit (75 000 m^3 daily) were built in 1970. The unit was equipped with a Pulsator settling tank (715 m^2), a number of sand filters and a final ozonation process.

1975: Morsang II. The second production unit (75 000 m^3 daily) was built in 1975. It was equipped with a Superpulsator settling tank (416 m^2) and two filter stages (sand followed by activated carbon) with intermediate ozonation.

1988: Morsang III. The third production unit (75 000 m^3) was built in 1988. It was equipped with a DENSADEG settling tank (170 m^2), again with two filter stages (sand and activated carbon) and intermediate ozonation.

In 1970, Lyonnaise des Eaux asked Degremont to manage the design and construction of the different treatment facilities. The environmentally sound design gave the facility a spacious and open layout while the main units, with the exception of the much discussed operations building at the centre of the star (Fig. 3), were placed partially underground.

Fig. 3. Morsang-sur-Seine – The three-pointed star structure

Each phase of the Morsang project marks a step forward in drinking water production technologies. Degremont and Lyonnaise des Eaux have consistently sought to apply their latest technical innovations, e.g., the DENSADEG, at Morsang III.

3.2 Description of Morsang III

3.2.1 Preliminary Treatment. The preliminary treatment process serves the entire Morsang facility. The first stage in the process consists of coarse and fine screenings (15 mm).

The raw water is then treated with chlorine gas for prechlorination in a tank with a minimum contact time of twenty minutes. Injected at doses well below break point, the chlorine helps to reduce oxidizable matter while limiting the development of algae in treatment facilities. The low doses (0.6 g/m^3) also disinfect the water to some extent and restrict the formation of undesirable trihalomethane-type compounds.

The pumping station delivers a maximum throughput of $10\,400$ m^3/hr, which can be modified in steps of approximately $1\,000$ m^3. The raw water is conveyed to a basin at the centre of the star from which it is distributed to the three production systems on a controlled-flow basis.

3.2.2 Reagents. Several different types of reagents are stored in the central operations building.

Coagulants. A variety of liquid coagulants can be injected into any of the three systems. Aluminium sulphate is generally injected into all three settling tanks simultaneously. Technical and/or economic criteria may also make aluminium chloride and aluminium polychlorides an appropriate choice in certain cases.

Flocculation Aids. Two types of reagent have been used at Morsang. The Pulsator and the Superpulsator originally used a solution of activated liquid silica that was prepared on-site. The activated silica was subsequently replaced by aluminium polychlorides, which are used as coagulants.

Since it first opened, Morsang III has used a slightly anionic polyacrylamide-type synthetic polymer (Prosedim AS 24). This product, which was extended to Morsang I and II a few months ago, produces a far more compact floc than activated silica at very low doses (between 0.05 and 0.25 g/m^3). Solid polymer is used to prepare a solution of approximately 2 g/l before it is injected into the settling tank.

Caustic Soda/Sulphuric Acid. On completion of the treatment process, sodium hydroxide (soda lye) or sulphuric acid is added to reestablish the $CaCO_3$ stability of the water and ensure that it is slightly scale-forming when discharged.

3.2.3 DENSADEG

Characteristics

– Nominal flow rate	$3\,540$ m^3/hr
– Maximum flow rate	$4\,000$ m^3/hr
– Settling surface area	170 m^2
– Rising velocity at nominal flow rate	20.8 m/hr
– Rising velocity at maximum flow rate	23.5 m/hr

To optimize the coagulation process, the coagulant is injected into an upstream flash mixer that has a contact time of two minutes. The backwash water from the two filter stages is also recycled in this unit.

A Further Advantage. The DENSADEG does not serve system III alone. The sludge extracted from the Pulsator and the Superpulsator in systems I and II is recirculated in the flash mixer for thickening by the DENSADEG.

3.2.4 Filtration Through Aquazur V. The clarified water is filtered through six sand filters with the following characteristics:

- Unit surface area $49\ \mathrm{m}^2$
- Total surface area $294\ \mathrm{m}^2$
- Filtration rate at nominal flow rate $12\ \mathrm{m/hr}$
- Filtration rate at maximum flow rate $13\ \mathrm{m/hr}$
- Filtering medium   Sand
- Height of medium $1.2\ \mathrm{m}$
- Effective size $1.35\ \mathrm{mm}$

3.2.5 Ozonation. As in the Morsang II, the ozone is injected in two bubble contact towers upstream of the activated carbon filter stage. A number of Degremont ozone generators with a unit capacity of 6.6 kg/hr produce ozone for all three units. Ozone promotes the self-purification action of the carbon by oxidizing the organic matter and oxygenating the water. This synergy significantly increases the amount of organic matter eliminated by the activated carbon, particularly the matter responsible for taste and smell.

3.2.6 Filtration Through Mediazur V. Granular activated carbon filters collect soluble organic matter by adsorption. The biological self-purification action of the carbon increases the effectiveness of the process. Purification takes place on six filters:

- Unit surface area $49\ \mathrm{m}^2$
- Total surface area $294\ \mathrm{m}^2$
- Filtration rate at nominal flow rate $10\ \mathrm{v/v/hr}$
- Filtration rate at maximum flow rate $11.3\ \mathrm{v/v/hr}$
- Filtering medium   Norit Row 0.8
- Height of medium $1.2\ \mathrm{m}$

3.2.7 Storage. When the water has regained its $CaCO_3$ stability, it undergoes final disinfection by chlorine gas prior to storage in a plug-flow reservoir with a capacity of $20\,000\ \mathrm{m}^3$. This reservoir serves two purposes:

- It guarantees a minimum contact time of two hours for post-chlorination.
- It provides a security stock of treated water (sufficient for several hours).

3.2.8 Sludge Treatment. A new sludge treatment unit has been in use since the summer of 1990. Owing to the high concentration of the sludge produced by the DENSADEG, the flow of treated sludge is low and the new unit was therefore designed along particularly compact lines. The unit consists of a sludge storage tank and a filter-press. Lime is used as a reagent. The maximum daily capacity is seven tons of dry matter (30 % dryness).

3.3 Appraisal of the DENSADEG

Morsang III came into operation in July 1988. The operating results from July 1988 to December 1989 are set out below.

3.3.1 Flow Rates Treated. The average flow rate varies between 1.379 m^3/hr (September 1989) and 1 960 m^3/hr (June 1989). The average value for 1989 was 1 623 m^3/hr. The DENSADEG is highly flexible and the operator quickly decided to regulate the flow rate of system III in line with overall demand while retaining a constant flow rate on systems I and II.

Consequently, the flow rate can vary significantly during the course of the day, for example:

– Between 0 and 3 000 m^3/hr in winter.
– Between 1 000 and 3 600 m^3/hr in summer.

The DENSADEG produces on average some 40 % of the total output of the Morsang facility.

3.3.2 Quality of the Clarified Water. The quality of the clarified water is monitored primarily on the basis of its turbidity, which is measured on a continuous basis.

Turbidity at the DENSADEG Outlet. The turbidity of the clarified water depends to a great extent on the turbidity of the raw water (Fig. 4) but the maximum value observed – 1.4 NTU – is within acceptable limits. The average turbidity over the eighteen months concerned was 0.50 NTU (1 mg/l).

Different periods need to be studied to take into account the types of co-agulant injected:

– Aluminium chloride was used between July 1988 and February 1989. The residual turbidity during these eight months was excellent, the average value being 0.32 NTU.
– Results continued to be satisfactory after the change to aluminium sulphate in February 1989, although values were slightly higher than in the preceding period. The average value observed over ten months was 0.60 NTU.

Comparison of Turbidity Levels at DENSADEG and Superpulsator Outlets. The turbidity of the clarified water at the DENSADEG and Superpulsator outlets is compared in Fig. 5.

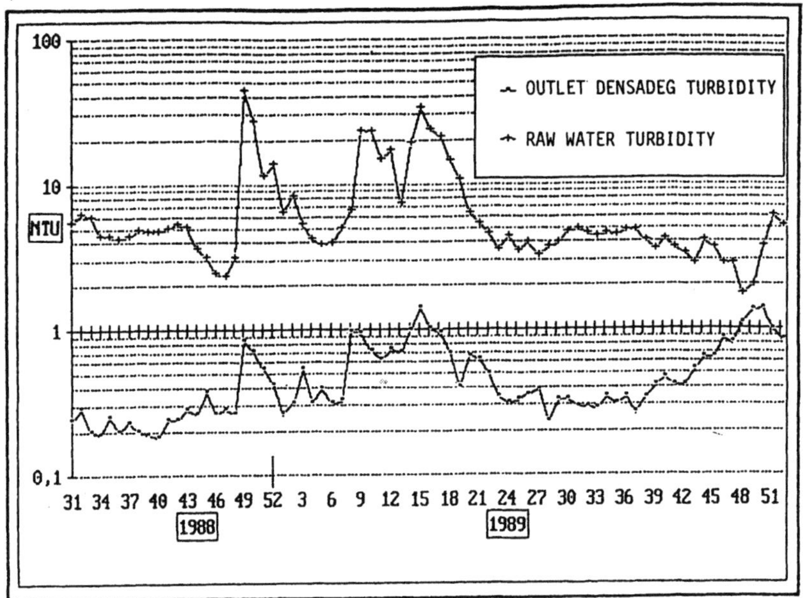

Fig. 4. Comparison of turbidity levels at DENSADEG inlet and outlet

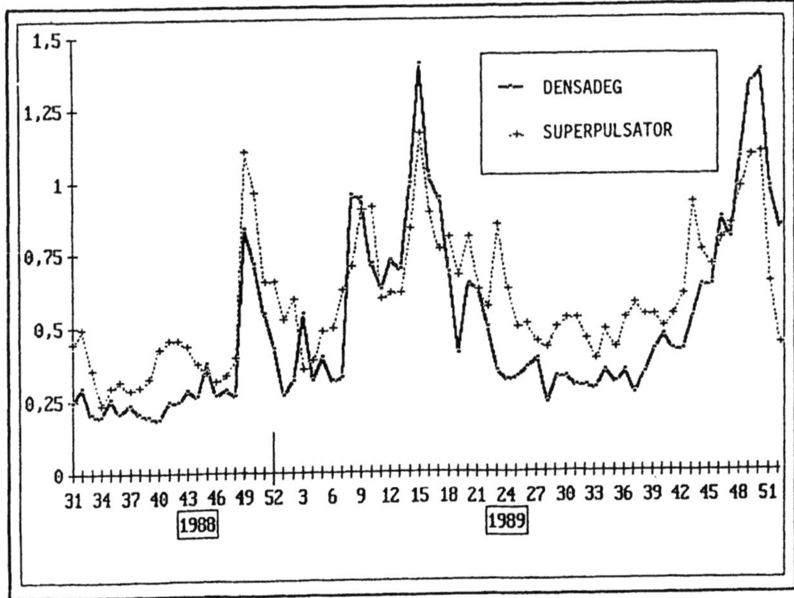

Fig. 5. Comparison of turbidity levels at DENSADEG and SUPERPULSATOR outlets

The flow rate treated on system II amounts, on average, to 75 % of the flow rate treated on system III.

- A direct comparison of results is impossible for the period preceding May 1989 when the two settling tanks were using different reagents (Aqualenc on the Superpulsator and aluminium sulphate or aluminium chloride on the DENSADEG).
- The two systems started to use the same treatment in May 1989 with identical doses of aluminium sulphate and similar doses of anionic polymer. Turbidity was still significantly lower at the DENSADEG outlet (0.40 NTU) than at the Superpulsator outlet (0.62 NTU).

3.3.3 Reagents

Consumption. Coagulants are selected according to effectiveness and cost, hence the use of several reagents during the period in question. The anionic polymer was dosed in relatively low quantities, 0.17 g/m^3 on average, over eighteen months. The conventional dose is between 0.12 and 0.15 g/m^3. A higher dose is applied (up to 0.25 g/m^3) when the water is heavily laden or cold, and settling becomes more difficult.

Dose Control

- The dosage rate of coagulants is determined by the quality of the raw water (twice-daily turbidity measurements, weekly flocculation tests, stochastic model used as a decision-making aid). For the past few months, dose control has relied upon a dedicated sensor called the "Coagulant Control Center" (CCC), which measures the colloidal load continuously. This sensor is used to regulate coagulant injection on the basis of a set-point colloid level.
- The dosage rate of the polymer was originally dependent upon the flow rate of the raw water, but the method was later abandoned. The dose is now proportional to the amount of coagulant injected. The coefficient of proportionality takes into account exceptional situations such as heavily laden or cold water.

It is simple to implement reagent dose control in a settling tank such as the DENSADEG, which thereby becomes totally autonomous.

3.3.4 Variations in the Composition and Flow Rate of Raw Water.
One of the main advantages of the DENSADEG lies in its acceptance of variations in the composition and flow rate of raw water.

Variations in Composition. We saw earlier that the quality of the water leaving the DENSADEG was dependent in part on the quality of the raw water. In this case, however, the difference is solely due to less complete coagulation rather than quantities of flocculated matter. When the Seine is in spate, the operator endeavours to maintain turbidity at around 1 NTU in order to keep doses of coagulant at reasonable levels. Applying higher doses would lower turbidity but the two consecutive filter stages downstream make such an increase unnecessary.

Variations in Flow Rate. The DENSADEG's acceptance of variations in flow rate is illustrated in Figs. 6 and 7.

The first figure corresponds to a typical winter day during which the settling tank is shut down during peak electricity rate periods. On start-up, it immediately regains normal operating speed and the quality of the water produced remains unchanged.

The second figure illustrates the considerable variations in the flow rate applied to the DENSADEG during a typical week of high production levels (June 1989). As before, the impact on the quality of the water is negligible. It was the DENSADEG's total acceptance of changes in flow rate that prompted the operator to select system III to manage variations in the overall output.

3.3.5 Sludge Production. A further advantage of the DENSADEG is that it produces thickened sludge that can be treated directly by a dewatering unit. In view of the low suspended solids content of the water from the Seine several months of the year (e.g., less than 10 mg/l on average from May to December 1989), the sludge extracted should be concentrated to 20 g/l as an annual average. The average value over eighteen months of production is 29.7 g/l. The average concentration is precisely 20 g/l during the many months when the river is not in spate. However, a significant increase in the suspended solids content has an immediate impact on the concentration levels of the sludge extracted. This change is reflected over a period of weeks (52.5 g/l on average between December 1988 and May 1989).

In conclusion, although the water from the Seine contains few suspended solids, the concentration levels of the sludge extracted remain close to 30 g/l. This is the minimum value that is generally obtained on other DENSADEG units where the water has a low suspended solids content. In the case of Morsang, higher values would be obtained by increasing polymer concentration, but this is not necessary.

3.4 Appraisal of Morsang III

3.4.1 Filtration. Conventional post-clarification filtration is performed downstream of the DENSADEG. The quality of the water through the sand varies between 0.06 and 0.15 NTU with an average of between 0.10 and 0.12 NTU. The filters are washed automatically; each has its own programmable logic controller.

3.4.2 Final Quality of Water Supplies. The water from the three production systems at the Morsang facility is mixed prior to quality analysis. A turbidity of 0.06 NTU and an organic matter content of 0.8 mg O_2/l are obtained as yearly average values and reflect the excellent standard of operation of the station as a whole.

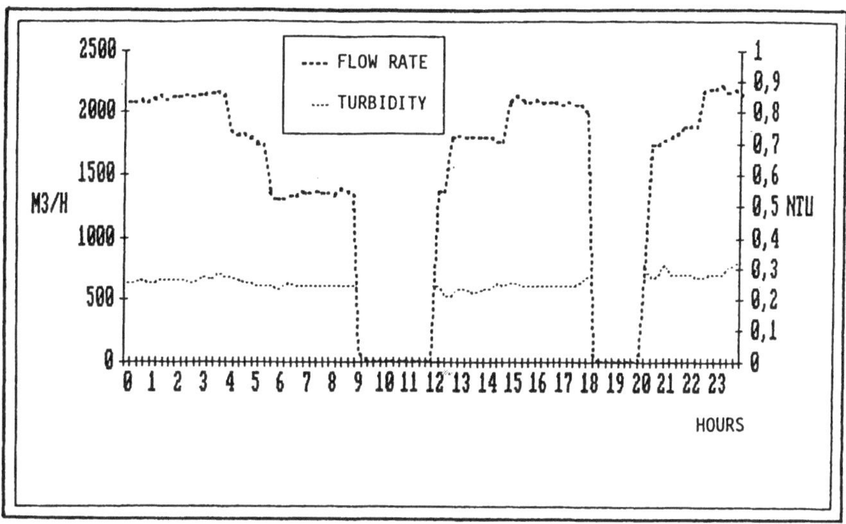

Fig. 6. Variations in flow rate (winter time)

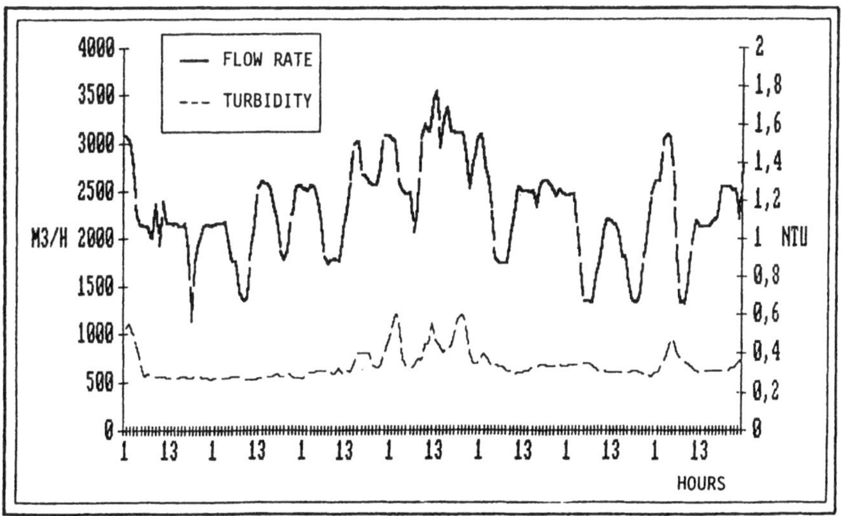

Fig. 7. Variations in flow rate (summer time)

3.5 Conclusion

The Morsang III treatment station represents one major application of a
DENSADEG in the production of drinking water.

Morsang III fully confirms the capacities and high performance of the
DENSADEG:

- Clarified water of excellent quality (0.50 NTU over eighteen months). 0.24 NTU over seven months with aluminium chloride.
- High settling rate (annual average of 9.5 m/hr; peak rate of 23 m/hr.)
- Integrated sludge thickening (30 g/l on average).
- Low reagent consumption (0.17 g/m^3 of polymer on average).
- Acceptance of variations in flow rate.

The highly flexible and fully automated DENSADEG settling tank has all the qualities needed to satisfy the substantial variations in the output of the Morsang facility.

4. Urban Sewage Water

The water treatment of urban sewage water generally employs activated sludge techniques. This proven and reliable process requires large surface areas. In these plants, nuisances (odour, noise) are not easy to remove. In order to obtain compact plants and to ensure a greater treatment reliability, lamellae settlers and fixed film reactors were developed.

The association of a high velocity lamellae settling tank (DENSADEG with reagents) and a biological aerated filter (BIOFOR, upflow process) is a very compact solution for removing suspended solids and carbon. This association is particularly adapted to high load variations, for instance in touristic areas such as seaside or winter sports countries. The example of Grèoux-Les-Bains, a spa in Provence, near the touristic area of the Verdon Cañon illustrates this technology. The treatment plant, implemented by Degremont in 1987, treats pollution corresponding to a variation from 2 000 P.E. to 20 000 P.E.

This plant is composed of:

- Large screening (25 mm)
- Fine screening (10 mm)
- Grit and grease removal
- Flash mixing
- Lamellae settling tank (DENSADEG/21.5 m^2/18 m/h)
- Four biofilters (BIOFOR/4 × 14.1 m^2)

Two periods must be considered:

Slack Season. Daily flowrates are between 900 and 1 000 m^3/d. Different chemical rates are applied (0 to 40 g/m^3 in ferric chloride as commercial product, 0 to 0.5 g/m^3 in polyelectrolyte). The global removals are independent of the chemical rates, the BIOFOR ensure a constant quality of the effluent (Table 2).

Touristic Season. Daily flowrates are between 2 000 and 2 400 m^3/d. Different chemical rates are applied (45 and 120 g/m^3 in ferric chloride, 0.15 and 0.25 g/m^3 in polyelectrolyte). As during the slack season, the global removals are independent of the chemical rates (Table 3).

Tab. 2. Greoux-les-Bains – Main results during slack season

Period	I* g/m^3	II* g/m^3	III* g/m^3	IV* g/m^3	V* g/m^3	VI** g/m^3	VII** g/m^3
Coagulant (FeCl$_3$)	40	15	15	0	0	0	0
Flocculant	0.5	0.5	0.25	0.25	0	0.25	0
Raw water							
SS	102	103	121	102	95	103	133
COD$_{tot}$	183	215	220	231	186	215	250
Outlet DENSADEG							
SS	10	21	29	50	46	47	40
COD$_{tot}$	63	76	102	161	145	151	128
Outlet BIOFOR							
SS	4	4	5	7	8	8	7
COD$_{tot}$	42	36	41	52	36	43	43

* Two BIOFOR into service; ** One BIOFOR into service

Tab. 3. Greoux-les-Bains – Main results during touristic season

Period	I* g/m^3	II* g/m^3	III* g/m^3	IV** g/m^3
Coagulant (FeCl$_3$)	120	45	45	45
Flocculant	0.15	0.15	0.25	0.25
Raw water				
SS	150	151	170	142
COD$_{tot}$	331	365	403	340
Outlet DENSADEG				
SS	15	34	39	45
COD$_{tot}$	130	152	180	214
Outlet BIOFOR				
SS	3	3	5	7
COD$_{tot}$	53	66	51	70

* Three BIOFOR into service; ** Two BIOFOR into service

Different choices are possible in order to guarantee the results:

Slack Season

- DENSADEG alone with reagents (40 g/m^3 as FeCl$_3$ and 0.5 g/m^3 as polyelectrolyte)
- DENSADEG without reagents and two BIOFORS.

Touristic Season

– DENSADEG with reagents with low rates (45 g/m^3 as FeCl$_3$ and 0.25 g/m^3 as polyelectrolyte) and three BIOFORS.

The Grèoux-Les-Bains plant is a perfect example of reliability due to the association of a lamellae settling tank and biofilters. A constant quality of the effluent is obtained and reagents can be saved during the slack season. The average extracted sludge concentration is 45 g/l and the sludge dryness after the belt-filter is at least 25 %. Such a combination can be used to treat the storm water, and a velocity greater than 30 m/h can be applied on a DENSADEG unit.

5. Conclusion

The DENSADEG is part of a new generation of lamellae settlers with sludge recirculation. An innovative reactor combining coagulation and flocculation and an external sludge recirculation system produce floc that is particularly dense and homogeneous. The floc offers high settling rates: between 20 and 40 m $\cdot$ h^{-1}, according to the type of treatment, hence the highly compact design of the DENSADEG. Further, the DENSADEG produces sludge that is sufficiently thick (between 30 and 500 g $\cdot$ l^{-1} depending on the application) to be treated directly by a dewatering unit.

The DENSADEG has many possible fields of application:

– *Drinking water*: Clarification or lime softening (surface water or drilling water); thickening of filter washing sludge.
– *Industrial process water*: As above.
– *Municipal wastewater*: Primary physico-chemical treatment; tertiary physico-chemical phosphate removal; thickening of filter washing sludge; storm overflows.
– *Industrial wastewater*: Clarification prior to discharge; sludge thickening; precipitation reactions. More than fifty DENSADEG are currently in service in almost all the fields mentioned above at flow rates of between 80 and 10 000 m^3 $\cdot$ h^{-1}.

Pascal Dauthuille
Degremont
183 Avenue du 18 Juin 1940
F-92508 Rueil-Malmaison CEDEX
France

Chemicals – Dosing Control

Chemical Dosing Control – Physical and Chemical Boundary Conditions

H.H. Hahn

1. Introductory Remarks – The Problem

Dosing chemicals means starting the intended reaction. If, for instance, phosphate removing processes are aimed for, then the addition of reaction partners, usually in the form of metal salts leads immediately and spontaneously to the formation of (solid) metal phosphate.

If coagulating processes are intended then the addition of chemicals in the form of metal salts initiates a preliminary reaction, that of the coagulant formation. The subsequent step, the attachment of the coagulant onto the colloid, i.e. the destabilization step itself, is by definition distinctly separate but follows immediately and depends in its effectiveness upon the outcome of the previous step. Furthermore, this reaction relies upon the transport of the in-situ formed coagulant to the colloid surface.

Reaction rate and the extent of a reaction depend upon the distribution of the reaction partners in the reactor. Thus, it is essential for the precipitation and/or coagulation process that the addition and resulting concentration pattern of the (needed) chemical is controlled so that a homogeneous distribution in time and space is accomplished.

For an improved understanding and control of chemical dosing one must analyse separately precipitation and coagulation reactions. Yet, for technical applications the first step in successful precipitation operation, i.e. the optimal mixing of chemicals to attain homogeneous distribution, is identical to the first step in coagulation, in particular when metal salts are used that need activation. Thus it is sufficient for this discussion to investigate dosing problems in coagulation. Here, one should differentiate the following situations arising from the interaction of specific boundary conditions with certain process requirements:

- physical boundary conditions that control
  - chemicals distribution (predominantly affecting coagulant formation)
  - chemicals transport (here understood as the preliminary step prior to the attachment of coagulants onto the colloid surface)

and

- chemical parameters characterizing the precipitation or coagulation process in terms of
  - need for coagulant activation

154   H.H. Hahn

- reversibility/irreversibility of destabilization (double layer compaction coagulants are addressed as 'reversibly' destabilizing while high molecular weight organic polymers are addressed as irreversibly attached coagulants).

All process possibilities arising from the coincidence of specific physical process conditions with defined chemical requirements are described in Fig. 1.

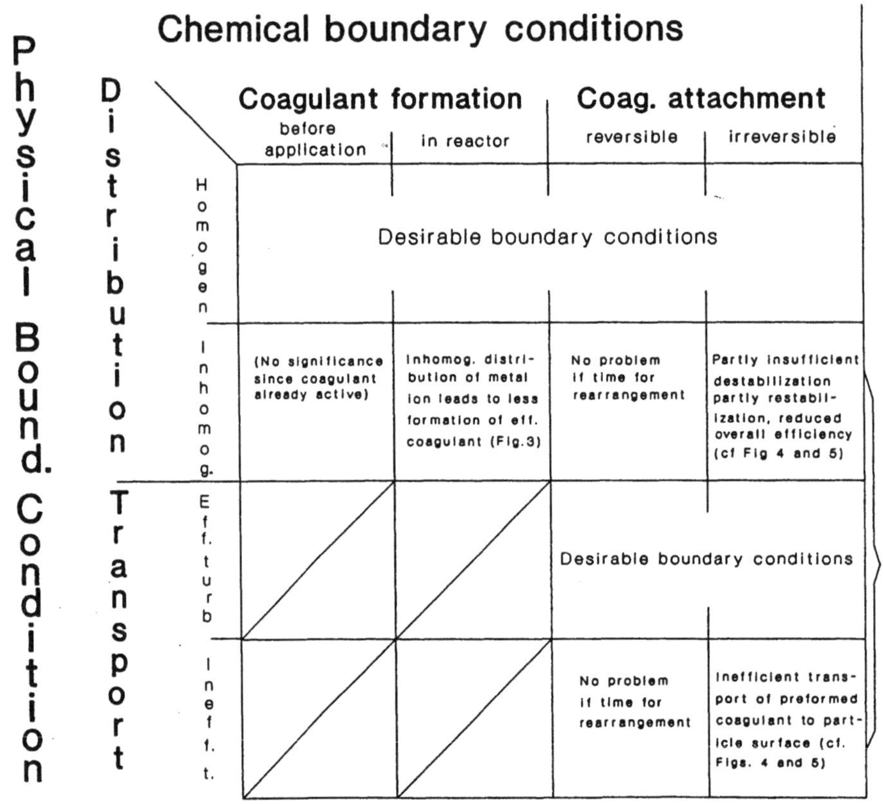

Physical Bound. Condition

Distribution / Transport		Chemical boundary conditions			
		Coagulant formation		Coag. attachment	
		before application	in reactor	reversible	irreversible
Distribution	Homogen	Desirable boundary conditions			
	Inhomog.	(No significance since coagulant already active)	Inhomog. distribution of metal ion leads to less formation of eff. coagulant (Fig.3)	No problem if time for rearrangement	Partly insufficient destabilization partly restabilization, reduced overall efficiency (cf Fig 4 and 5)
Transport	Eff. turb			Desirable boundary conditions	
	Ineff. t.			No problem if time for rearrangement	Inefficient transport of preformed coagulant to particle surface (cf. Figs. 4 and 5)

Fig. 1. Overview of possible effects of physical and chemical parameters in the chemicals dosing phase in the coagulation process

The schematic illustrates clearly that, if the coagulant is active or activated before dosing into the reactor, there is no need for controlling chemicals distribution in terms of coagulant formation. Likewise, if the chemicals distribution is homogeneous then optimal boundary conditions exist for coagulant formation (and attachment).

Problems in chemicals dosing might arise if in-situ formed coagulants, i.e. metal salts, are used in reactor systems with insufficient mixing, causing inhomogeneous concentration patterns (see specifically marked area (1) in Fig. 1).

There might also be a problem in reversible destabilization if the coagulant distribution is not homogeneous and there is not enough reaction time for attaining a new equilibrium by 'desorption' and 're-adsorption'.

Coagulation processes involving mechanisms with irreversible destabilization, i.e. irreversible coagulant attachment, depend to the largest degree possible upon homogeneous chemicals distribution. They also require a most effective transport of these chemicals to the colloid surface. Problems arise where this is not the case, be it in terms of inhomogeneous concentration patterns or low-rate transport of coagulants to the colloids to be destabilized. These phenomena are not easily separated from each other and are therefore analyzed combined (see specifically marked area (2) in Fig. 1).

For the following illustration, case studies are reported that involve only the use of metal ion based coagulants. In one instance these coagulants are in the form of metal salts to be activated in the coagulation process itself and in the other instance they are available as preformed 'polymeric' metal hydroxo complexes. This eliminates additional effects from (surface) chemical reactions resulting from differences in chemical components of the coagulant/flocculant.

2. Physical and Chemical Effects in Coagulation with Aluminium

There are five generally accepted mechanisms of colloid destabilization or aggregation, i.e. double layer compaction through multivalent cations, adsorption coagulation with hydrolized metal ions, patch work destabilization with short-chain polymeric cations, bridge formation with long-chain anionic, cationic or non-ionogenic polymers and, finally, sweep floc aggregation. Aluminium salts may be used to initiate any four of them, except bridge formation. Which of these reaction pathways is followed depends to a large degree upon the chemical boundary conditions, such as concentration gradients and changes in pH regime.

2.1 Chemical Control Parameters

Figure 2 (left part) illustrates what happens if aluminium salts are added as (acid) stock solutions to a raw water or wastewater system, as is usually the case:

- It is plausible that under specific conditions the change in pH and the necessary dilution occur stepwise and in a parallel way, avoiding any extensive hydroxide formation. This means that all reactions occur at the dissolved/non-dissolved boundary, i.e. at the zone of transition from hydrolyzed and hydroxolyzed ions to polynuclear species and finally solid matrices.

- It is conceivable that a change in pH and concentration are engineered so that dilution always occurs prior to the pH increase. This would mean that only in the very last phase is the solubility boundary transgressed towards the formation of hydroxide.
- Finally, it is imaginable that the pH gradient is such that, initially, solid aluminium hydroxide is formed which, in a later phase, upon dilution has to be dissolved as the operational conditions of the system will determine.

In all instances different species are formed with strongly differing coagulation efficiencies.

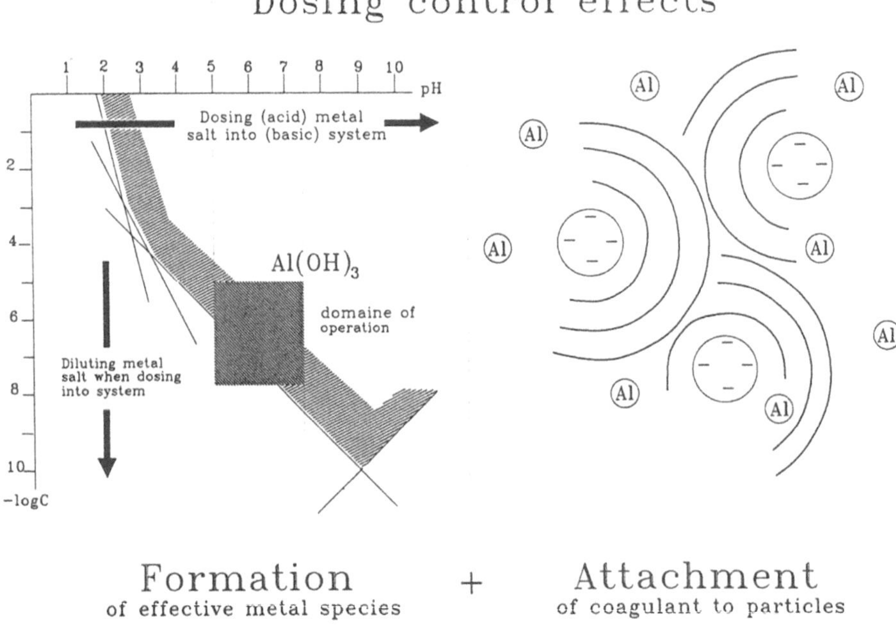

Fig. 2. Schematic illustration of the two reaction steps in the destabilization phase within the coagulation process

In the first case, all reactions that lead to hydroxylation and polymerization are favored. Thus, the (in-situ) production of polymeric aluminium species might reach an optimum under these conditions. The destabilization mechanism most likely to occur is the effective adsorption coagulation or even the more effective patch-work destabilization. (The latter one can be made to dominate if prepolymerized aluminium species are applied.)

In the second case, the solubility boundary of aluminium hydroxide with the surrounding domaine of polynuclear aluminium hydroxide complexes is only reached at the very end, not allowing much time for the formation of the highly efficient hydroxocomplexes, possibly of a polynuclear structure. This pathway of coagulant preparation might furnish effective species, but only to a smaller extent.

In the third instance it is to be expected that a larger amount of the added aluminium ion remains in the form of non-dissolved hydroxide (increasing the tendency of low efficiency sweep-floc coagulation). This will be especially true if the time allowed for coagulant formation and attachment is short and there is no read-justment of the chemical equilibrium to changed boundary conditions.

All arguments presented above contain inherently the assumption that chemical species can be transported without time delay from any dosing point to any location in a reactor. In practice this may be quite different, depending upon the geometry of the entrance point for the chemicals, the geometry of the reactor and above all the (mixing) energy provided for the distribution of the chemicals. It is therefore very instructive to investigate the effect of these physical parameters upon the formation of coagulants and the destabilization of colloids.

2.2 Physical Control Parameters

Dosing of chemicals is intended in most instances to occur under fully developed turbulence, as is indicated in the schematic drawing of Fig. 2 (right part). Yet, this does not guarantee that each increment of chemicals added encounters identical reaction conditions.

If the first reaction pathway for coagulant formation is to be followed in a technical process, then there could still be local differences in aluminium concentration and/or pH value that do not show in an average balance. However, it could lead to locations with high pH milieu with the consequence of local hydroxide formation (reaction pathway 3, as described above), with the logical corollary that there are at the same time other locations where dilution dominates more than intended, resulting in little or no hydroxylation and polymerisation (reaction pathway 2, as described above).

In discussing coagulant formation and its attachment to colloids one has to take the interaction of these two steps into account. This is particularly necessary where reaction times are short and transport distances for chemicals unforeseeably small or large. Again one has to accept that in practice there will be situations of incomplete mixing, i.e. inhomogeneous chemicals distribution. The following situations might arise:

(a) Coagulant formed within close proximity of (stable) colloids will be 'attached' to a larger extent onto those surfaces, leaving other colloids less supplied and less modified in their stability.

(b) If coagulant formation is slow, for whatever reasons, then there might be interactions of the less-active metal ion species with the colloid precluding any further improvement of the coagulant formation reaction and subsequent destabilization.

(c) The time needed for attachment of the coagulant to certain colloids is large compared to the characteristic detention time in the system; the destabilization step will then be incomplete and the provided coagulant not used evenly for all colloids to be destabilized.

High energy mixing and highly turbulent transport conditions in conjunction with optimal geometry of chemicals input and reactor shape provide the technical solution to such problems.

3. Case Studies on Metal Salt Dosing

In the following description of two real-world situations of chemicals dosing, the focus will be on situations where only aluminium species were used as the coagulant. The reason for this selection is that aluminium salts can be used either in the form of in-situ activated coagulants or can also be obtained as prepolymerized polynuclear hydroxospecies. Thus, without introducing any further complicating chemical arguments, both situations outlined in Fig. 1 can be illustrated.

It should be pointed out that equivalent arguments hold for iron salts as coagulants. The sole difference is that prepolymerized iron species are not so readily available for technical processes as are 'aluminium polymers'.

3.1 Problem Area (1): Using a Coagulant that Requires In-situ Activation Under Conditions of Incomplete Mixing

It has been postulated above (Sect. 2.1) that differences in coagulation effectivity might result from different reaction pathways followed in the process of coagulant activation. These arguments have indicated that pathways leading via hydroxide formation to the domain of operation might furnish effective coagulants to a lesser extent.

The data reported by Masides et al. (1986) on the re-use of precipitated aluminium hydroxide may be used to illustrate these arguments. They are shown in Fig. 3 in comparison with data on the coagulation with the usually employed (acid) aluminium stock. The fraction of coagulation-relevant aluminium gained in the process is shown on the ordinate for various operating conditions. These process conditions describe activation or re-activation of the coagulant and the general operation of the coagulation process.

It is recognized from these data that there are significant differences in the amount of coagulation-relevant aluminium species produced under the specified working conditions. An optimum for this system of raw water with its content of coagulation-supporting or coagulation-hindering ions is obtained for a milieu pH of 5.6 with lower output at lower as well as higher pH values. This optimum is similar for both types of chemicals, acid stock and (basic) hydroxide. Furthermore, for activation conditions for the aluminium hydroxide at pH values below 2.25, the output of effective coagulant is comparable.

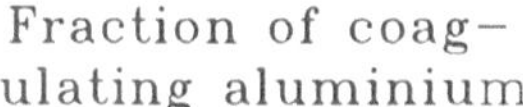

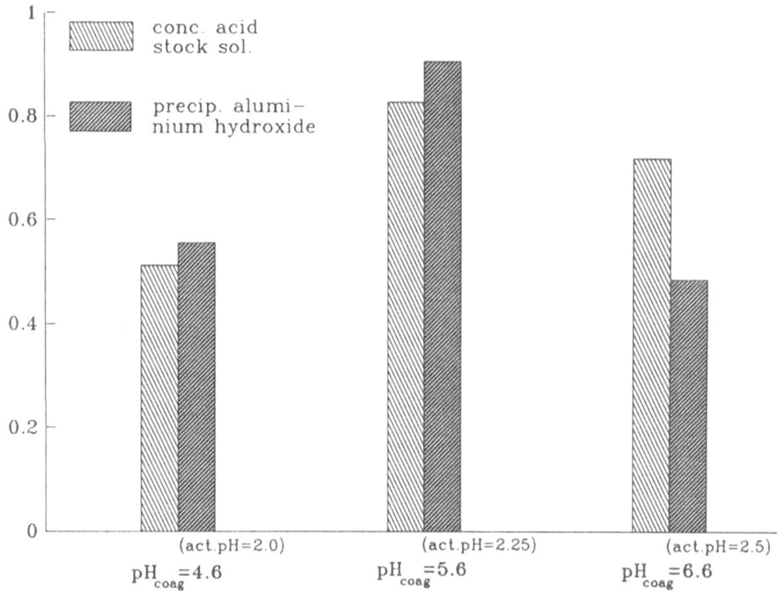

Fig. 3. The effect of pH upon the activation of aluminium species used in coagulation, illustrated by data on the fraction of coagulation-relevant aluminium gained from acidified (recycled) aluminium hydroxide (after Masides et al., 1986)

However, if the basic hydroxide is re-activated at less acidic pH values of 2.5 or more, then the gain in coagulating aluminium species is noticeably lower than that from (acid) aluminium salts. This can only be explained by insufficient dissolution of hydroxide at insufficiently acidic pH values. (It might also be speculated that longer reaction times and/or higher reaction temperatures might in part compensate for less acidic re-activation conditions.)

A different experimental set-up (by Seyfried, 1991) with emphasis on a systematic variation of physical boundary conditions in the activation and destabilization chamber in terms of flow velocity or detention time leads to similar observations. They are shown in Fig. 4, in particular in the reported dosage requirements for non-polymerized aluminium.

It is recognized that there is less aluminium in the form of Al^{3+} needed for a reduction in the (added) turbidity to below 20 % of the initial value if the flow velocity in the mixing (and destabilization) chamber is higher, i.e. the resulting detention time shorter: high mixing intensity secures optimal coagulant formation as the acidic aluminium salt enters a basic milieu of about pH = 7.6 reducing it to a final pH of 6.4. This is a clear example of in-situ coagulant formation along the activation pathway 2 or 3 (as described in Sect. 2.1).

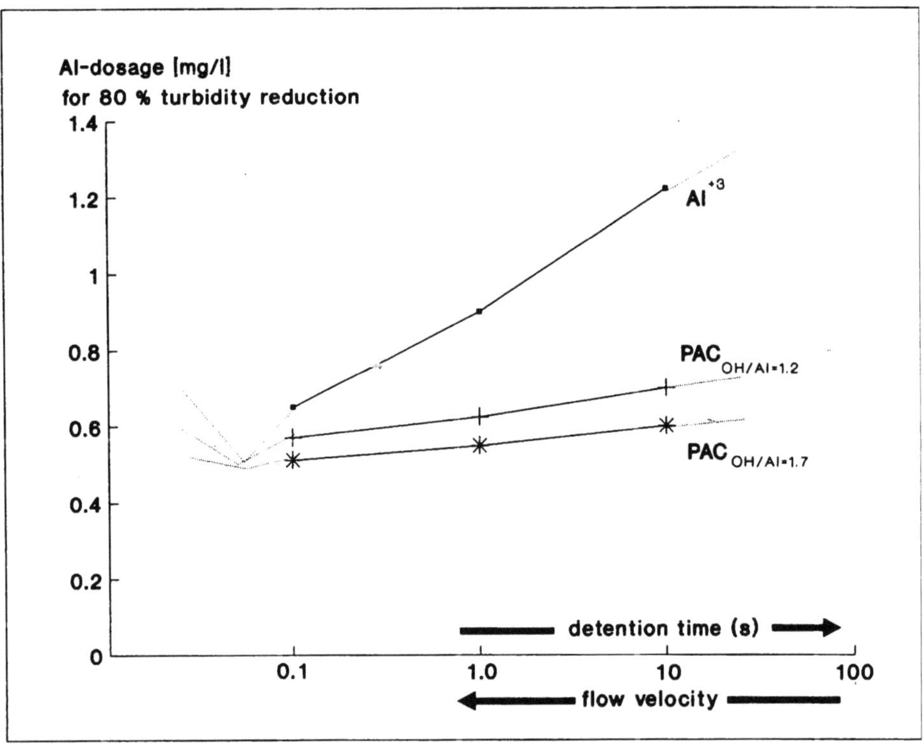

Fig. 4. The role of energy dissipation in destabilization of colloids with non-hydrolyzed aluminium salts and prepolymerized aluminium salts illustrated by experimental data on the required aluminium dosage for turbidity reduction by 80detention time in the destabilization cell (after Seyfried, 1991)

The fact that there is next to no influence of the flow velocity or detention time on the dosage required if prepolymerized aluminium is employed confirms these arguments.

3.2 Problem Area (2): Using Preactivated Coagulants Under Conditions of Insufficient Transport Energy

Let us examine more closely the data of Fig. 4 in terms of the influence of the recorded detention time or the reciprocal of it, the flow velocity (which is taken as a synonym for the mixing intensity), upon the dosage requirements. The data show that there is no visible effect in the case of prepolymerized aluminium. This indicates that the preformed and therefore preactivated chemical is destabilizing the colloids in an optimal way. No increase in energy supplied to the destabilization reactor is needed.

The observations recorded in Fig. 5 (reported by Klute, 1985) describe a different system. The energy provided for destabilization is systematically varied by using different dosing devices. The coagulant used is a preformed and therefore fully activated prepolymerized aluminium cation dosed in the form of a chloride salt. It does not need a chemical activation in the reactor. All changes in destabilization by varying mixing energy must be interpreted as effects upon the transport and attachment step.

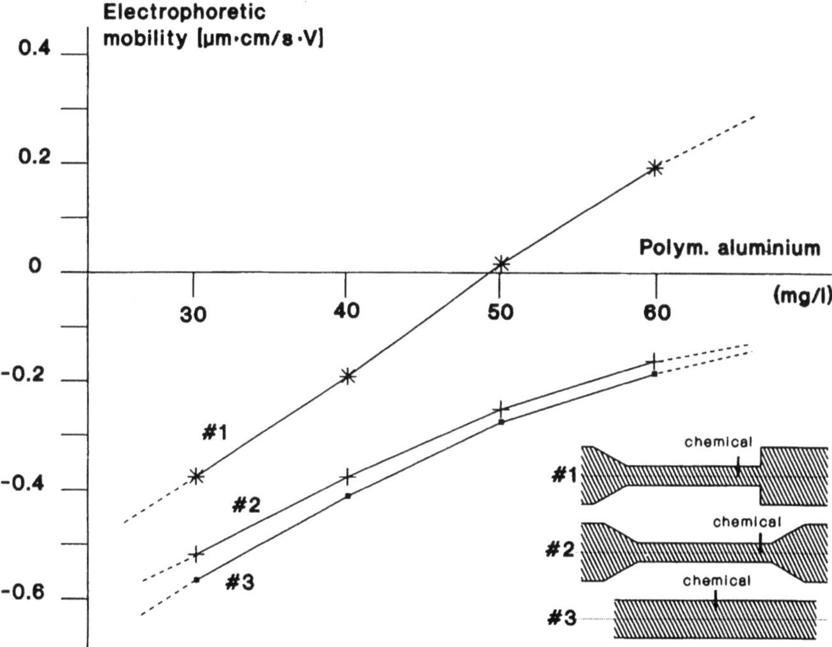

Fig. 5. Destabilization of colloids with prepolymerized aluminium (chloride) using different devices for chemical dosing; the differences observed in resulting electrophoretic mobility, leading to different colloid stabilities, result from differences in transport and attachment of the destabilizing chemicals to the colloid surface (after Klute, 1985)

Device No. 3 would represent the most widely applied dosing into pipe flow. Device No. 2 is thought of as an improvement over device No. 3 in that the flow velocity is increased through a reduction of the pipe diameter. Device No. 1, finally, entails a flow section where a sudden widening of the diameter leads to a type of hydraulic jump with the largest amount of (flow) energy expanded.

It is this latter device with the highest amount of energy dissipation that leads to the best destabilization results: the electrophoretic mobility (ordinate value) assumes the lowest value for any given dosage of polymeric aluminium (PAC) if this device is used. The high effectivity can also be inferred in the following way: higher dosages of PAC are required when using devices No. 2 or 3 and the same electrophoreses values are to be obtained as with device No. 1.

These data are a clear illustration of the fact that 'transport' of preactivated aluminium species to the colloid to be destabilized can also be sub-optimal and therefore a critical technical parameter.

4. Concluding Remarks – What Can be Learned from This?

The discussion confined itself to three very defined situations of using alumini-um-based coagulants with systematically varied chemical and physical bound-ary conditions. It has been shown that there is a significant potential for saving chemicals if both physical and chemical boundary conditions are optimized. While this has been shown or proven only for the case histories described here, the underlying principles are valid for other process configurations as well.

The effects resulting from operating a precipitation or coagulation process under non-optimal boundary conditions are compensated today most frequently by higher dosages of chemicals. This is illustrated by the data in Fig. 5: to attain the desirably low electrophoretic mobility with less efficient dosing and mixing devices one needs to add higher amounts of chemicals. This, together with the fact that no operator of a precipitation or coagulation plant knows the 'true' minimum doses for his process, may explain why such considerations have not received too much attention until now.

The cost of water and wastewater chemicals, even though they appear to rise, may still permit such operation. Yet one has to consider that each incre-ment of chemicals added to the water or waste water stream not only results in costs on the input side but also considerable secondary costs in terms of solids or sludge removed and to be handled. With the present drastic increase in cost for the handling of sludges, reflecting the growing difficulties to handle larger amounts of water and wastewater solids removed, it may become mandatory to use chemicals as sparingly as possible.

The phenomena described above must therefore be translated into recom-mendations for practice in a most global and generalized form. Chemical dosing for precipitation and coagulation processes must take into consideration:

- The actual form in which the chemical is transported, stored and then even-tually needed in the process itself. If activation of the chemical occurs in the course of its actual use, then great care must be exercised in terms of concentration and pH gradients to which the chemicals are subjected.
- For in-situ activation as well as for preactivated chemicals, the transport and attachment to the colloid surface must be optimized in terms of equal opportunity for each colloid to interact with the same amount of destabilizing chemical. Such optimization entails structural provisions such as geometry of the dosing port for chemicals and reactor shape, etc. (cf. Fig. 5) and/or operational measures such as amount of energy dissipated, etc. (cf. Fig. 4). In stating this one has to accept at the same time that these two groups of control variables are to some degree interrelated.

The case studies and the ensuing discussion focussed on coagulation-relevant processes: it was argued for instance, that a certain extent of reaction between metal cation and hydroxide ion was desirable and necessary. It was also shown that the coagulation process was hindered if physical and chemical boundary conditions were such that these reactions did not take place in a predetermined intensity. This statement is valid also for precipitation processes, even though the conclusions are different. Here, for instance, reactions of the metal cation with hydroxide ions compete directly with the desired reaction between the metal cation and the (phosphate) anion. Thus, all previously described arguments have to be applied in such a way that metal ion and hydroxide interaction will not be optimized but minimized. With this difference in mind, all statements can be transferred directly to the improvement of the control of precipitation processes.

5. Literature

Klute, R.: Rapid Mixing in Coagulation. In: Chemical Water and Wastewater Treatment. Fischer, Stuttgart 1985, p. 53
Masides, J., Mata, J., Soley, J.: The Recovery of Aluminum Sulfate Used as Coagulant in Wastewater Treatment. In: Proceedings of the 2nd International Gothenburg Symposium. ISWW Reihe Universität Karlsruhe, Vol. 45, Karlsruhe, 1986, p. 333
Seyfried, A.: Untersuchungen zur Entstabilisierung kolloidaler Partikel mit vorpolymerisierten Aluminiumsalzen. M. Sc. Thesis, University of Karlsruhe, 1991

Hermann H. Hahn
Institut für Siedlungswasserwirtschaft
Universität Karlsruhe
Am Fasanengarten
D-7500 Karlsruhe 1
Germany

Fundamental Considerations in Use of the Streaming Current Detector for Chemical Dose Control

M.L. Elicker, J.J. Resta, J.W. Hunt, and S.K. Dentel

Abstract

The streaming current detector, or SCD, measures the streaming current created in a closed-end cylinder due to particles attached to cylinder and piston walls. The device is used in water and wastewater treatment applications for monitoring or automatic feedback control of coagulant or conditioner dosages. This paper reviews some aspects of proper SCD usage, and then provides an improved theoretical description with experimental validation, and a means for instrument standardization.

The mathematical derivation solves first for flow characteristics in the probe in terms of local fluid velocity. An appropriate version of the Gouy-Chapman equation is used to give the space charge density. The product of these two is integrated over the cross-sectional area between the piston and cylinder to give the streaming current. As with earlier, more approximate equations, the solution predicts the streaming current to be proportional to particle zeta potential and to motor speed. Both proportionalities are experimentally validated although the proportionality with motor speed fails at greater than 360 revolutions per minute. By using different SCD sensor dimensions, the geometric factors in the derived equation were confirmed. Quantitative confirmation of the equation is obtained using surfactant solutions and surfactant-coated particles as a basis.

1. Background

The streaming current detector, or SCD, is an analytic device capable of measuring the streaming current created in a closed-end cylinder due to particles attached to cylinder and piston walls. Particle exchange with fresh sample in the flow-through chamber and electrical isolation of the cell provide the capability for continuous characterization of particle charge in a variety of industrial processes. Hundreds of these units are now in use in water treatment application in the United States for monitoring or automatic feedback control of coagulant dosage.

The device is now being marketed in the EEC and is also being utilized to a greater extent in other applications such as paper manufacture and slurry dewatering.

The instrument is reported to provide a number of benefits in monitoring and controlling coagulant feed rates in water treatment (Dentel and Kingery, 1989), including:

- more immediate control of coagulant dosage, which is advantageous during variable water quality or flow conditions;
- reduced water treatment costs, primarily due to more efficient chemical application;
- more consistent water quality as a consequence of continuous dosage control;
- the possibility for rapid detection of equipment malfunctions related to chemical dosing.

However, one drawback associated with use of the SCD has been a lack of complete understanding of the device's functioning. In this paper we review some aspects of proper SCD usage, and then provide an improved theoretical description with experimental validation, and a means for instrument standardization.

1.1 Description of the Streaming Current Device

The SCD's sensor consists of a reciprocating piston located in a close-ended cylinder (Fig. 1). The piston's up-and-down motion acts to pump fluid sample continuously in and out of the gap area between the piston and cylinder surfaces, with the fluid velocity at any point represented as sinusoidal over time. The sampled water is renewed by continuous flow through entry and exit ports above the piston. Particles and other adsorbable material in the water momentarily attach to piston and cylinder surfaces and, if these attached materials possess some electrical charge on their surfaces, this is imparted to the sensor's inner walls. The charges attached to the piston travel with the piston velocity, and those on the cylinder walls remain stationary. However, the counter-ions beyond the shear plane near these walls assume the velocity of the moving fluid, and this provides a measurable electrical current. Ring-shaped electrodes at the upper and lower ends of the cylinder pick up this signal (which is sinusoidal with time), and phase-sensitive electronic rectification and amplification provide an instrument output which indicates the sign and magnitude of the attached charge. This, in theory, is directly proportional to zeta potential.

Figure 2 depicts SCD placement in a water treatment process sequence for coagulant feedback control. Following flocculation, the flow may proceed to sedimentation, flotation, direct filtration, or a membrane process. The sample flow (approximately 1 L/min) is taken from the rapid mix or immediately thereafter. Sampling further downstream increases the control delay, as does sample flow over an excessive distance. The proper sampling location must be empirically determined.

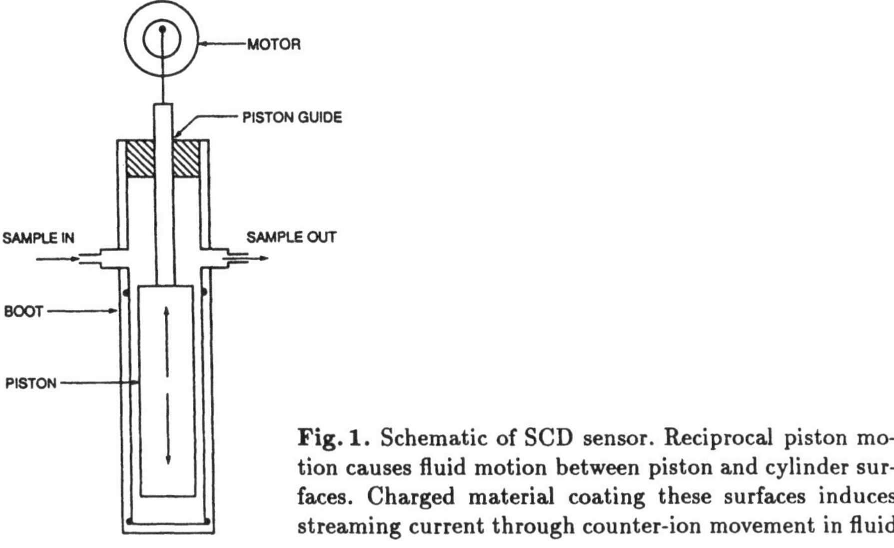

Fig. 1. Schematic of SCD sensor. Reciprocal piston motion causes fluid motion between piston and cylinder surfaces. Charged material coating these surfaces induces streaming current through counter-ion movement in fluid

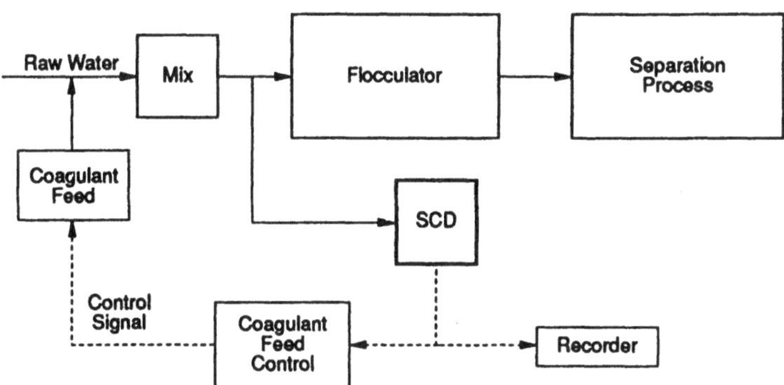

Fig. 2. Typical placement of SCD. Sample flow is approximately 1 L/min in water treatment. SCD output is used by controller connected to metering pump for coagulant feed adjustment. In sludge treatment applications, liquid sidestream is used as sample, and sample flow rate should be higher

1.2 Control Aspects

Figure 3 shows an idealized relationship between particle removal (as turbidity) and such electrokinetic parameters as streaming current or zeta potential. The coagulant causes a particle charge transition from an originally negative value to neutral, and then to positive. The mechanism of charge neutralization theoretically requires a charge near zero to attain optimal removal. In practice, cost considerations may dictate use of a somewhat lower coagulant dose, giving a slightly negative particle charge (point $-A$). If an SCD is employed to determine particle charge, a negative offset of the signal compared to actual particle charge may be noted, also leading to an apparently negative particle charge at the optimal coagulant dose (point $-A'$). In any case, use of the streaming current curve will allow a set value of the signal (such $-A$ or $-A'$) to be targeted as the optimal dosage, either by an operator or by automatic feedback control.

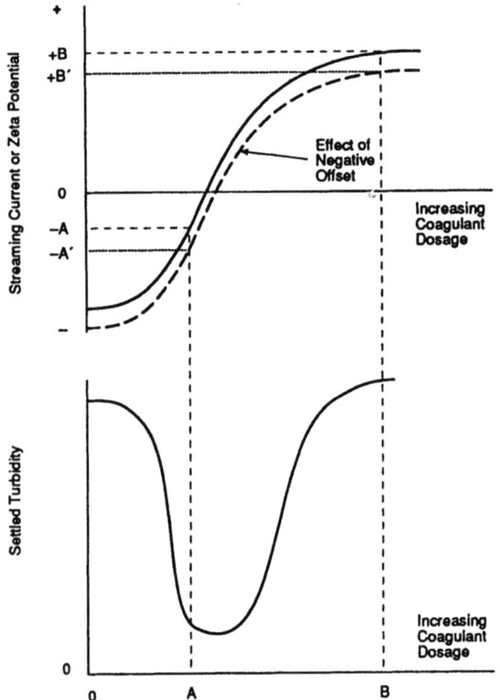

Fig. 3. Possible relationship between particle charge (streaming current or zeta potential) and final turbidity. The darker dashed line shows the streaming current curve if a constant offset from the zeta potential curve exists. Such an offset would dictate use of $-A'$ instead of $-A$ as a set point

Although the residual turbidity vs. dose curve in Fig. 3 might also be used as a basis for control of coagulant dosage, a searching algorithm must be employed which continually imposes dose variations in order to relocate the optimum dose. In fact, a human operator basing coagulant dosage on turbidity essentially utilizes a similar searching procedure, because he or she may not know whether the dosage should be *increased* or *decreased* to decrease the turbidity. For example, if the dose were at point B in Fig. 3, an operator might increase the coagulant dose until observing that the turbidity was increasing rather than decreasing. However, the streaming current shown at +B is more positive than the set point value, immediately indicating that, in this case, the coagulant feed rate should be decreased. It should be noted that the same control difficulties associated with use of residual turbidity will exist for any parameter to be minimized, such as UV absorbance, light blockage, or – in the case of sludge dewatering applications – centrate solids concentration or turbidity. Further aspects of chemical dose control strategies have been discussed in a recent review (Dentel, 1991).

We have evaluated the use of SCDs in sludge dewatering processes for monitoring and control of sludge conditioner dosage. In such applications, the liquid effluent stream (filtrate or centrate) serves as the sample source for the SCD (Fig. 4). Laboratory and plant tests demonstrate a similar relationship to that shown in Fig. 3, i.e. process performance is optimal at doses giving near zero streaming current. However, dewaterability does not deteriorate as severely at higher doses. These results have been observed at both centrifuge and belt filter press installations, and utilizing performance indices such as capillary suction time, specific resistance to filtration, cake solids concentration, and percent solids recovery as exemplified in Fig. 5.

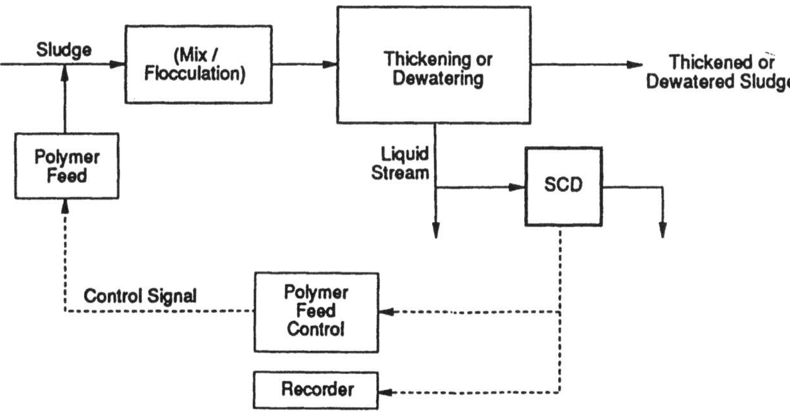

Fig. 4. Typical placement of SCD in sludge treatment applications. Sample flow rate is often higher than in water treatment applications. SCD output is used by controller connected to metering pump for adjustment of chemical conditioner feed rate

1.3 Previous Description of SCD Functioning

Previous work (Gerdes, 1966; Dentel and Kingery, 1988) provided a quantitative model of how the SCD functions, but relied on somewhat simplistic assumptions in development of the descriptive equations. A linear relationship between streaming current and zeta potential was predicted:

$$\bar{i} = \frac{-16\pi w s \varepsilon \zeta}{(1 - \lambda)^2} \tag{1}$$

$\bar{i}$	Time-averaged (rectified) streaming current
w	Motor speed (cycles/sec)
s	Piston stroke
ε	Solution dielectric constant (SI convention)
ζ	Zeta potential
λ	Ratio of piston radius (λR) to inner cylinder radius (R)

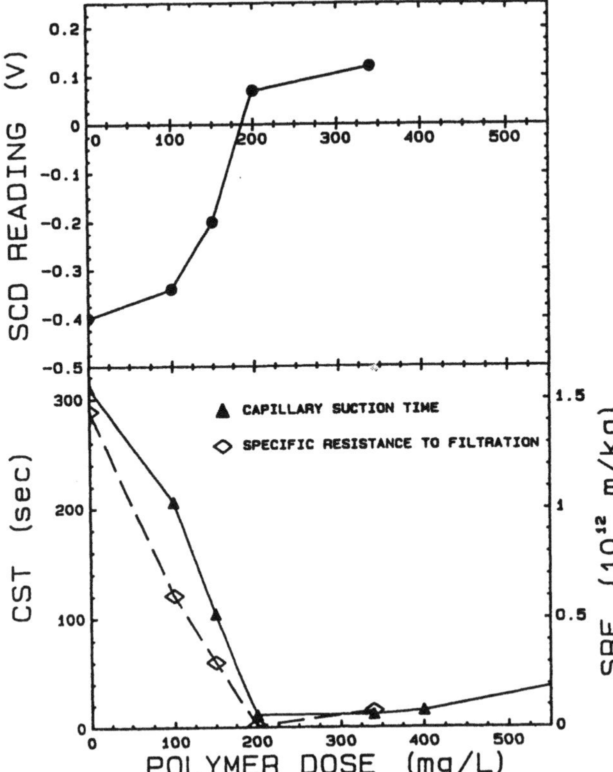

Fig. 5. SCD readings and dewaterability indices as a function of polymer dose to a wastewater sludge. A near-zero SC value typically corresponds to optimal conditioning and dewatering

The linear relationship between measured current and measured zeta potential has been observed. For example, Fig. 6 shows the electrophoretic mobility and streaming current for a Min-U-Sil 5 silica suspension. The particle charge was altered by varying pH and dosage of ferric chloride. The pH values for the two samples used in determining EM and SC for a given data point were approximately, but not exactly matched, leading to some of the apparent variability. Overall, the results illustrate the linear relationship which may be observed between SC and EM as suggested by Eq. (1). However, the above equation has not been quantitatively confirmed to enable a direct conversion from an SCD reading to the corresponding zeta potential. Furthermore, independent SCD readings have been difficult to compare due to non-uniformity of the instruments and lack of a calibration method.

In fact, our results suggest the Eq. (1) to be somewhat inaccurate. The equation was derived assuming a linear velocity profile in the sensor annulus and using a capacitor model of the electrical double layer near the piston and cylinder surfaces. Although these assumptions are reasonable for an initial approximation, a more rigorous mathematical description of the SCD probe can be developed.

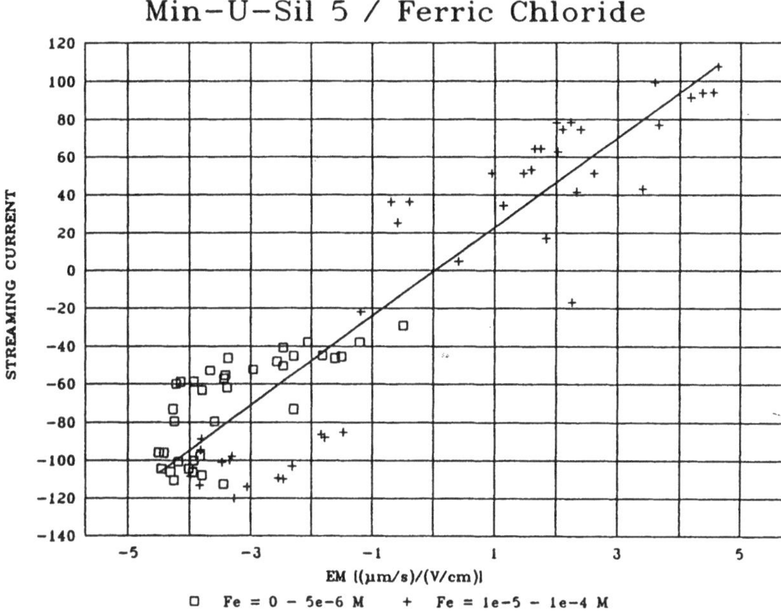

Fig. 6. Proportionality of SCD readings and electrophoretic mobility for 100 g/L silica (Min-U-Sil 5) suspension. SC was determined with a model "Gen I" SCD (Milton Roy Co., Ivyland, PA) for this data. Particle charge was varied by additions of NaOH, HNO_3, and $FeCl_3$

2. Improved Model of the SCD

An analytical mathematical derivation of the flow characteristics in the probe and an appropriate version of the Gouy-Chapman equation are combined in an integration over the cross-sectional area of the annulus; the integrand is the product of velocity (v) and space charge density (ρ):

$$\overline{\imath} = \int_0^{2\pi} \int_{\lambda R}^{R} \rho(r)\overline{v}(r)r\,dr\,d\varphi \qquad (2)$$

In order to obtain $\rho(r)$, the one-dimensional Poisson-Boltzmann equation for symmetrical electrolytes

$$\nabla^2\psi = \frac{d^2\psi}{dx^2} = -\frac{\rho(x)}{\varepsilon} = \frac{2zen_0}{\varepsilon}\sinh\left(\frac{ze\psi}{kT}\right) \qquad (3)$$

is combined with the Gouy-Chapman equation

$$\psi(x) = \frac{2kT}{ze}\ln\frac{\left(\sqrt{u_0}+1\right)+\left(\sqrt{u_0}-1\right)\exp(-\kappa x)}{\left(\sqrt{u_0}+1\right)-\left(\sqrt{u_0}-1\right)\exp(-\kappa x)} \qquad (4)$$

where $u_0 = \exp\frac{ze\psi_0}{kT}$, giving

$$\rho(x) = -2zen_0 \sinh\left[2\ln\frac{(\sqrt{u_0}+1)+(\sqrt{u_0}-1)\exp(-\kappa x)}{(\sqrt{u_0}+1)+(\sqrt{u_0}-1)\exp(-\kappa x)}\right] \tag{5}$$

ψ Electrical potential
x Distance from the surface
z Magnitude of electrolyte charge
e Fundamental electronic charge
n_0 Bulk solution electrolyte concentration
k   Boltzmann constant
T Absolute temperature
κ Inverse double layer dimension

The time-averaged magnitude of the velocity profile $[\bar{v}(r)]$ in the annulus may be related to sensor dimensions and piston velocity as

$$\bar{v}(r) = \left[\frac{1-(r/R)^2+(1+\lambda^2)\ln(r/R)}{1-\lambda^2+(1+\lambda^2)\ln\lambda}\right]2ws \tag{6}$$

where $r =$ distance in the axial direction, and other parameters as defined for Eq. (1). A detailed derivation of this equation is given elsewhere (Elicker et al., 1992).

Numerical integration after combining Eqs. (2), (5), and (6) shows that the current is primarily generated near the inner and outer walls such that approximations for these two regions may be utilized. Thus

$$\rho(x) \approx -\kappa^2\varepsilon\zeta\exp(-\kappa x) \tag{7}$$

and

$$\bar{v}(x) = \left(\frac{d\bar{v}}{dx}\right)_{x=0+} x \tag{8}$$

Near the cylinder wall, $x = R - r$, $dx = -dr$, and with $x \ll R$,

$$\bar{v}(x) = 2ws\left(\frac{(1-\lambda^2)/R}{1-\lambda^2+(1+\lambda^2)\ln\lambda}\right)x \tag{9}$$

$$\begin{aligned}
\bar{i}_{\text{outer}} &= \int_0^\infty \rho(x)\bar{v}(x)2\pi r\,dr \\
&= \frac{2ws(1-\lambda^2)/R}{1-\lambda^2+(1+\lambda^2)\ln\lambda}(-\kappa^2\varepsilon\zeta)\cdot 2\pi R\int_0^\infty x\exp(-\kappa x)dx \\
&= -\left(\frac{4\pi ws\varepsilon\zeta}{1+\frac{1+\lambda^2}{1-\lambda^2}\ln\lambda}\right)
\end{aligned} \tag{10}$$

Near the piston wall, $x = r - \lambda R$, $dx = dr$, and $x \ll R$, so

$$\bar{v}_{\text{inner}}(x) = 2ws\left(\frac{(1-\lambda^2)/R}{1-\lambda^2+(1+\lambda^2)\ln\lambda}\right)(1/\lambda)x \tag{11}$$

$$\bar{\imath}_{\text{inner}} = \int_0^\infty \rho(x)\bar{v}(x)2\pi(x + \lambda R)dx$$

$$= \frac{2ws(1 - \lambda^2)/(\lambda R)}{1 - \lambda^2 + (1 + \lambda^2)\ln\lambda}(-\kappa^2\varepsilon\zeta)2\pi\lambda R \int_0^\infty x\exp(-\kappa x)dx \qquad (12)$$

$$= -\left(\frac{4\pi ws\varepsilon\zeta}{1 + \frac{1+\lambda^2}{1-\lambda^2}\ln\lambda}\right)$$

Thus the currents on inner and outer walls are equal, and the total streaming current is

$$\bar{\imath} = \bar{\imath}_{\text{inner}} + \bar{\imath}_{\text{outer}} = -\left(\frac{8\pi ws\varepsilon\zeta}{1 + \frac{1+\lambda^2}{1-\lambda^2}\ln\lambda}\right) \qquad (13)$$

This solution shows the streaming current to be directly proportional to the zeta potential as in Eq. (1), and thus is consistent with results shown in Fig. 6. However, Eq. (13) predicts a higher current than the more approximate Eq. (1). For typical instrument and sample conditions, the former predicts a higher current by 25–35 %. Further development of this equation is provided in Dentel (1992). For experimental conditions described in this paper, electrophoretic mobility (EM) is proportional to zeta potential, and thus EM and SC are also predicted to be linearly related.

3. Experimental

The SCD used in this research was a model Generation II (Milton Roy Co., Ivyland, PA) but modified in three respects. A variable speed motor was installed to allow investigation of piston velocity effects on instrument output (via ws in the previous equations); the lands were eliminated from the piston by providing bearing surfaces above the submerged portion of the piston; and a modified probe was utilized which placed electrodes on the piston circumference rather than being fitted into the cylinder wall as with prior commercial units (Dentel, 1992). We were then able to vary dimensions of the cylinder (and thus annulus) through the use of removable cylindrical inserts to assess the importance of λ and R. Table 1 gives the five values employed for these parameters. An oscilloscope was used with the instrument to enable analysis of electronic circuitry and for determination of moter speed. A conventional Zeta-Meter (Long Island City, NY) was employed for determination of electrophoretic mobility and zeta potential.

Figure 7 shows measured streaming current values using the four probe inserts and at increasing concentrations of NaCl added to distilled, demineralized water. In this case there are no colloidal or adsorbable components in the solution, and the streaming current thus accrues from the electrical charge on the original probe surfaces. With each probe insert, the streaming current becomes less negative with increasing ionic strength. As predicted by Eq. (13), larger gap dimensions result in a lesser current magnitude (in this case, less negative).

Tab. 1. Dimensions for probes used in experiments

Probe	R [cm]	λ
Standard	0.9376	0.9831
Insert G1	0.9817	0.9389
Insert G2	1.015	0.9079
Insert G3	1.047	0.8803
Insert G4	1.173	0.7859

Equation (13) could thus be tested by changing λ or, using the variable speed motor, the piston rotational speed w.

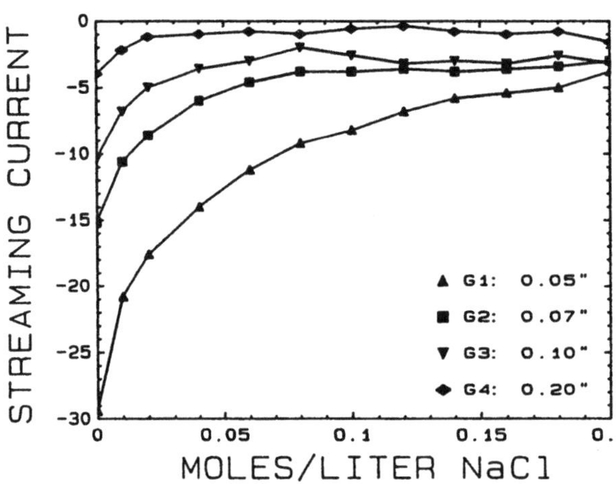

Fig. 7. Streaming current as a function of NaCl concentration in demineralized water, with different probe inserts to provide the annular gap dimensions shown

If Eq. (13) is correct, then effects of probe dimensions should be normalized by multiplying the SC readings with each insert by the appropriate term $[1 + \ln \lambda (1 + \lambda^2)/(1 - \lambda^2)]$. This has been done in Fig. 8. Results are brought into closer, but not exact, correspondence.

The results in Figs. 7 and 8 suggest that the decreased current at increased NaCl additions is due to double layer compression and the resulting decrease in zeta potential in the sensor (Dentel et al., 1989 a,b). A diminished signal could also be caused by increased conductivity at high salt concentrations, and increased back-current through the solution in the annulus rather than through the measuring circuitry. However, the greater gap sizes would accentuate this current loss, which is not consistent with results in Fig. 8. Thus the effect of salt addition on SCD readings is commensurate with a destabilization phenomenon (double layer compression) rather than being an analytical (conductivity) artifact.

Variation of piston velocity via motor speed is predicted by either Eq. (1) or Eq. (13) to give a proportional change in output. However, Fig. 9 shows that this was not the case at higher speeds. The relationship appears to be linear

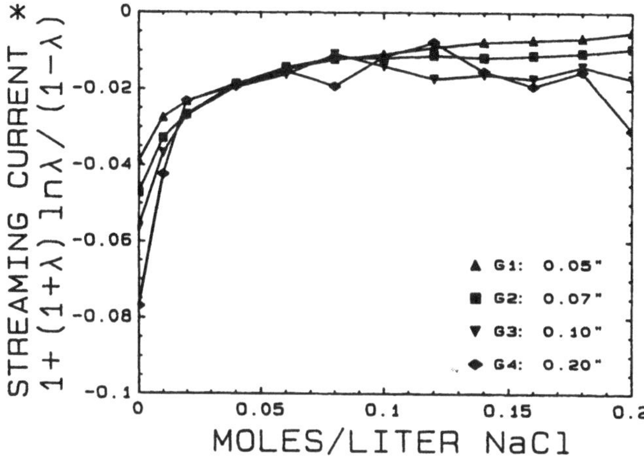

Fig. 8. Streaming current results from Fig. 7, normalized by the geometric factor in the predictive Eq. (13)

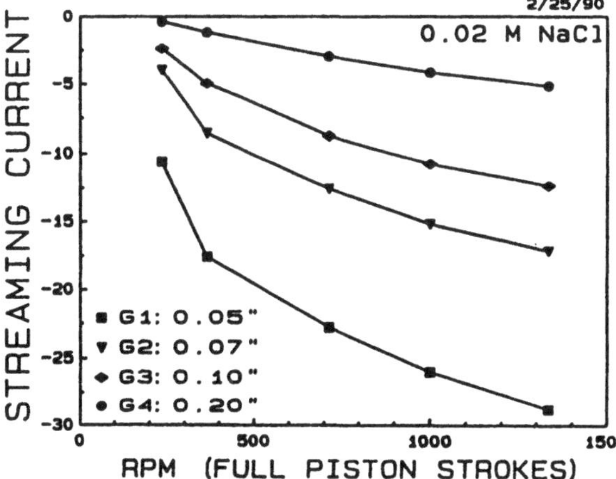

Fig. 9. Effect of increasing revolutions per minute (RPM or ws) on streaming current measurements using four probe inserts and a 0.02 M NaCl solution as sample

(and extrapolates back to the origin) up to about 400 RPM. Coincidentally, conversion of the SC values using the multiplying term used for Fig. 8 collapses the four lines into one, but only in the lower RPM range. Although either turbulence or cavitation could be responsible for the higher-speed non-linearity, calculations of Reynolds numbers or pressure differentials based on our derivation does not identify either of these phenomena as significant. Equation (13) may be considered valid, however, at motor speeds less than 360 RPM at which commercial SCDs operate.

3.1 Standardization Procedure for the SCD

In order to quantitatively assess the validity of Eq. (13), it was necessary to develop a means for standardizing SCD output and associating it with an equivalent zeta potential value. The method for doing this was as follows: the potassium salt of polyvinyl sulfonate (PVSK, Sigma Chemical) was used as a negatively charged surfactant and added to a suspension of TiO_2 particles at pH of 2.5, where the TiO_2 is positively charged. As shown by both SCD readings and electrophoretic mobility or EM (Figs. 10 and 11), the particle charge was changed from positive to negative. At a concentration of 750 mg/L PVSK, the charge was quite insensitive to relative changes in concentration. Moreover, this same addition of PVSK to the SCD *without* TiO_2 showed the same value. This SCD reading can therefore be associated with the corresponding EM (or zeta potential). The ratio of the two numbers provides a conversion factor between SCD reading and zeta potential. We confirmed the same relationship and numerical ratio using a positively charged surfactant (1,5-dimethyl-1,5-diazaundecanemethylenepolymethobromide or DDPM) added to the SCD or adsorbed onto negatively charged silica particles for SC and EM measurements. Either the DDPM or PVSK appear to be suitable as standardization chemicals for the SCD using this method.

The calibration method allows SCDs to be calibrated to a reference sensitivity, using one of the selected chemicals. After reconditioning or replacing an SCD in a treatment facility, it now can be calibrated to agree with previous settings, or, two or more SCDs working in parallel can be calibrated for process uniformity. We have tested this by calibrating two SCDs with substantially different probe designs, then comparing their measurements while varying pH for a TiO_2 sample. The two sets of results were essentially superimposable.

3.2 Quantified Confirmation of the SCD Model Equation

After confirming trends predicted by Eq. (13), and developing a calibration method, Eq. (13) could then be tested in its ability to predict the zeta potential of suspensions based on the SCD reading. This required a detailed analysis of the electronics used to convert the measured current (i) to a phase-sensitive rectified value ($\bar{i}$) and then to an amplified digital or chart-recorded readout (SC). The latter conversion may be written as

$$SC = K_{\mathrm{amp}}\bar{i} \tag{14}$$

with K_{amp} the amplification factor. We determined K_{amp} as the product of gains through several stages of electrical processing, using both a mathematial analysis of the circuit diagram and empirical testing by input of a known alternating current in place of the sensor input. Amplification through the first operational amplifier converts the current to a voltage at an amplification found to be 14.57×10^6 V/A; subsequent stages included adjustable range settings which affect the value of K_{amp}. Thus voltages were determined by oscilloscope after only the first amplification stage and converted to the streaming current.

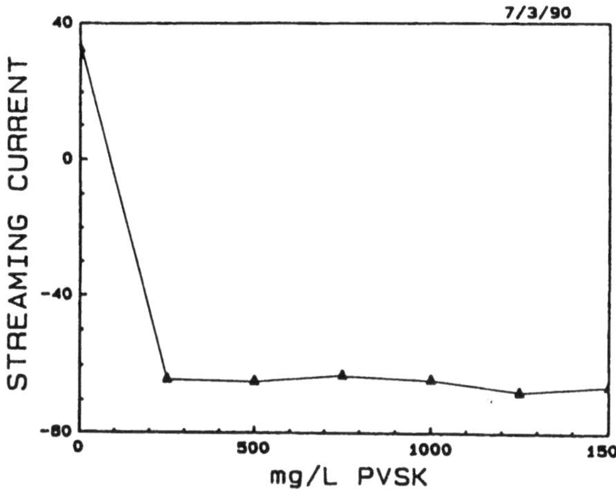

Fig. 10. SCD reading vs. concentration of PVSK added to 300 mg/L DuPont TiO$_2$ (Lot R-107-CD) at pH 2.5

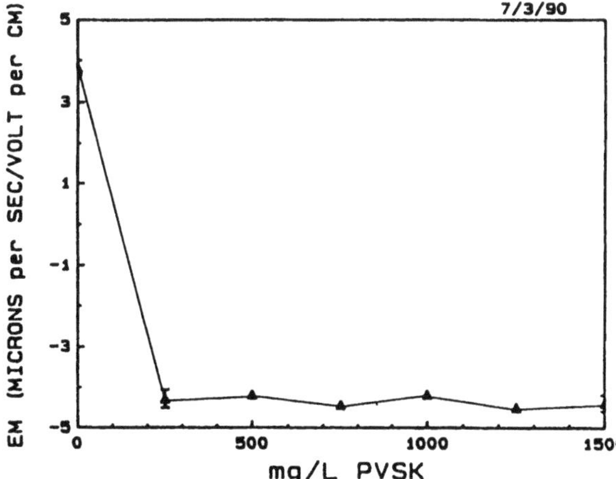

Fig. 11. Electrophoretic mobility (EM) vs. concentration of PVSK added to 300 mg/L DuPont TiO$_2$ at pH 2.5

Tab. 2. Values used for SCD conversions (Insert G1)

Parameter	Value	Units
λ	0.939	–
R	9.82×10^{-3}	m
w	4.83	rev/s
s	6.35×10^{-3}	m
i	17.47×10^{-9}	amp
ζ	-43.02×10^{-3}	volt

Using the PVSK solution as the SCD's sample, the peak-to-peak amplitude of the voltage measured after the first amplification stage was 0.8 V; using the above conversion factor gives a 54.9 nA amplitude at the sensor, corresponding to an average magnitude of 17.47 nA. Using the values in Table 2 resulted in a calculated zeta potential of -43.02 mV. For the same sample, the measured zeta potential was -45.30 mV. (This value does not correspond to EM values shown in Fig. 11 due to different solution ionic strengths).

Similar agreement between zeta potentials determined as electrophoretic mobilities and those found using the SCD has been established using both DDPM (positively charged, adsorbed on Min-U-Sil5 particles for EM measurements) and Min-U-Sil particles alone. The accuracy of the predictive SCD equation appears to be within the inherent uncertainties of EM and SC measurement.

3.3 Effect of Instrument Offset

Figure 6 confirmed the linear relationship predicted by Eqs. (1) or (13), and also that zero streaming current is at zero EM. However, a similar correlation between SCD output and EM may not exist in some samples. Some published data indicate that when a sample possesses a zeta potential of zero, a negative streaming current reading may be observed (Dentel and Kingery, 1987, 1989; Hubele and Bernazeau, 1989; Chowdhury et al., 1990). Since the streaming current reading observed with distilled water is also negative (Fig. 7), this suggests incomplete coverage of these surfaces by particles. The affinity of particles for these negatively charged surfaces is likely to decrease as particle charge decreases, ultimately reaching a minimum, negative streaming current.

Such discrepancies have not been observed with all particle types, and clearly do not arise with charged polymers or surfactants. In any case, the existence of such a zero offset dose not present an operational problem in chemical dose control since the control set point can simply be adjusted downward (dotted line in Fig. 6). Difficulties could arise if this offset is variable in magnitude, however. Our results to date support the generalization that an offset will not exist for sub- colloidal materials or for highly adsorptive or surface-active charged substances.

4. Conclusions

1. Changes in particle charge characteristics, brought about by pH changes or coagulant addition, can be observed with the streaming current detector. The SCD readings are in good qualitative agreement with electrophoretic mobility or zeta potential measurements, and exhibit good quantitative proportionality in many cases. Sensor surfaces themselves display negative charge, and lack of complete particle attachment may lead to an offset from the predicted direct proportionality.
2. The newly developed theoretical description of the SCD enables prediction of sensor output changes with geometrical parameter variations (probe gap width and diameter), and has been experimentally validated. The equation successfully predicts changes in streaming current as a function of piston velocity, but only in low velocity range (up to 360 RPM).

3. A calibration procedure has been developed, using either an anionic or cationic surfactant substance which may be adsorbed onto particles. The procedure may alleviate previous difficulties with non-uniformity of output values. It enables SCD values to be directly tied to equivalent EM or zeta potential values and may be useful for plant use of SCDs in control of chemical dosing.
4. The calibration procedure, combined with analysis of the SCD circuitry, enables validation of the new SCD equation to within about 5 %.

5. References

Chowdhury, Z.K., Wright, J.C., Owen, D.M.: Pilot and Plant Scale Evaluation of On-Line Coagulation Monitoring Devices. Proceedings, AWWA Water Quality Technology Conference, San Diego, 1990

Dentel, S.K., Kingery, K.M.: Theoretical Principles of Streaming Current Detection. Wat. Sci. Tech. *21* (1988) 443

Dentel, S.K., Kingery, K.M.: Using Streaming Current Detectors in Water Treatment. J. Amer. Wat. Works Ass. *81* (1989) 85

Dentel, S.K., Thomas, A.V., Kingery, K.M.: Evaluation of the Streaming Current Detector I. Use in Jar Tests. Wat. Res. *23* (1989 a) 413

Dentel, S.K., Thomas, A.V., Kingery, K.M.: Evaluation of the Streaming Current Detector II. Continuous Flow Tests. Wat. Res. *23* (1989 b) 423

Dentel, S.K.: Coagulant Control in Water Treatment. Critical Reviews in Environmental Control *21* (1991) 41

Dentel, S.K. Wehnes, K.M.: Monitoring Sludge Dewaterability by Streaming Current Detection. Proceedings, Canadian Society of Civil Engineers-American Society of Civil Engineers Joint National Conference on Environmental Engineering, 1988, p. 254

Dentel, S.K. Kingery, K.M.: Final Report: An Evaluation of Streaming Current Detectors. J. Amer. Wat. Works Ass. Research Foundation, Denver, CO, 1987

Elicker, M.L., Hunt, J.M., Resta, J.R., Dentel, S.K.: The Streaming Current Detector: Fundamental Description (submitted). J. Coll. Interface Sci. (1992)

Gerdes, W.F.: A New Instrument – The Streaming Current Detector. Proceeding, 12th Natl. Analysis Instrument Sym., Houston, TX, 1966

Hubele, C., Bernazeau, F.: Automatic Control of the Coagulant Dose in Drinking Water Treatment. Proceeedings, 1st Macau Workshop on Water Treatment (1989)

Mark L. Elicker
Project Engineer
Cascio Bechir Engineers
North Haven, CT 06473
USA

John J. Resta
Program Manager
Hazardous and Medical Waste
U.S. Army Environmental Hygiene Agency
Aberdeen Proving Grounds, MD 21010
USA

James W. Hunt
Department of Civil Engineering
University of Delaware
Newark, DE 19716
USA

Steven K. Dentel
Department of Civil Engineering
University of Delaware
Newark, DE 19716
USA

Particle Separation
in Wastewater Treatment

J. Sörensen and S.-G. Larsson

Introduction

In this paper we wish to show how the F.A.S.T. system works both on a laboratory scale and in full scale plant trials. The system consists of a metal salt dose (M) and a low molecular weight polyelectrolyte (LM), followed by a high molecular weight polyelectrolyte (HM). All conventional metal salt products can be used for the M dose. LM is a cyclic quaternary polyamine. In full scale plant trials the product is dosed as solid grade with only a prewetting, and in laboratory scale trials the product is dosed as a solution. HM is an anionic high molecular weight polyacrylamide dosed as a solution both in full scale and in lab scale trials.

The following metal salts were used in the investigation:

AVR, ALG	Aluminium sulphate
JKL	Ferric chloride
Ekoflock	Prepolymerised aluminium salt

Background

The conclusions drawn by Ødegaard [1] were used as the basis for this study. Since the major portion of the contamination in raw wastewater is associated with particulate matter, direct particle separation is an effective way of lowering wastewater contaminant levels. Particulates have a negative effect on the biochemical oxidation rate both in activated sludge and in biofilm processes, and should therefore be removed prior to biological treatment in order to reduce the organic load and, in addition, to promote rapid biodegradation.

The objective of this investigation was to maximize the removal of particulate matter. A combination of hydrolizing metal salts and of organic polyelectrolytes as mentioned above was chosen to try to accomplish this.

Laboratory Scale Results

The experimental results for the removal of turbidity, phosphorus, and chemical oxygen demand as function of the Ekoflock and the JKL dose, respectively, are presented in Figs. 1 to 4. Obviously, the method gives very good particulate matter removal. Low dosages of metal salts can be used since the metal is only required for precipitation and not for coagulation. The optimum metal salt dose appears to coincide with the turbidity peak on the suspension curve, according to Figs. 1 and 2 (point A – ref.). This occurs with prepolymerised metal salts and ferric chloride, but not with ferrous salts.

As seen in Figs. 1 and 2, the turbidity, after LM and HM polyelectrolyte has been dosed and left for settling, appears to approach the minimum as early as point A (the turbidity maximum with M only). The continuing decrease of turbidity with increasing metal dose, beyond point A, must be due to the reduction of soluble matter.

Figures 1 and 2 show that a very high particle separation is reached with 1–2 gr/m^3 Al or Fe together with 3 ppm LM and 0.2 ppm HM.

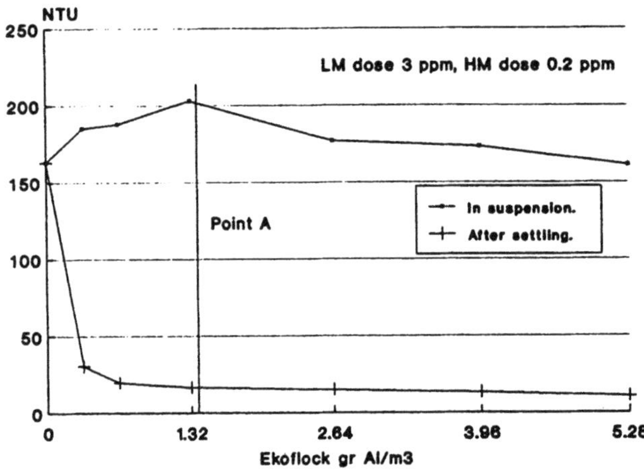

Fig. 1. Turbidity in suspension with M dose only compared with turbidity after LM, HM dose and settling: Ekoflock

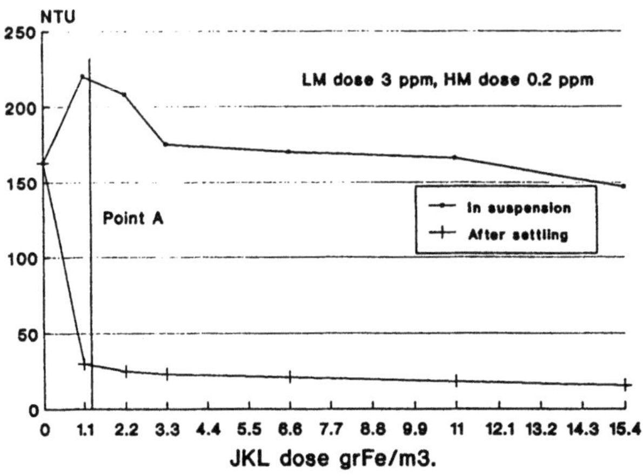

Fig. 2. Turbidity in suspension with M dose only compared with turbidity after LM, HM dose and settling: JKL

AVR seems to be more sensitive in this system, probably due to the pH effect when the metal salt dose level decreases. As seen in Fig. 5, 80 % P_{tot} removal was achieved at a Me/P ratio 0.15.

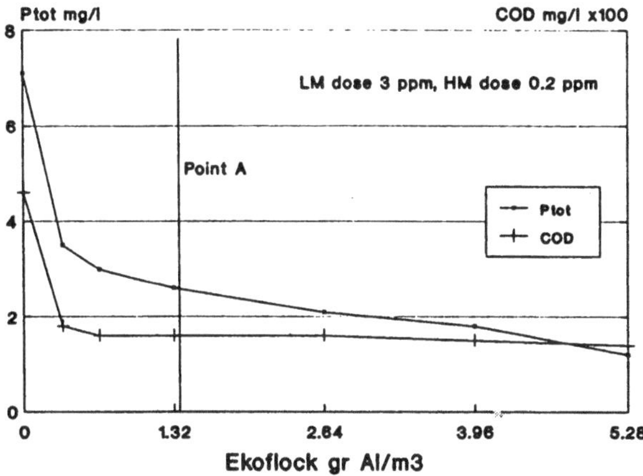

Fig. 3. Removal of phosphorus and COD: Ekoflock

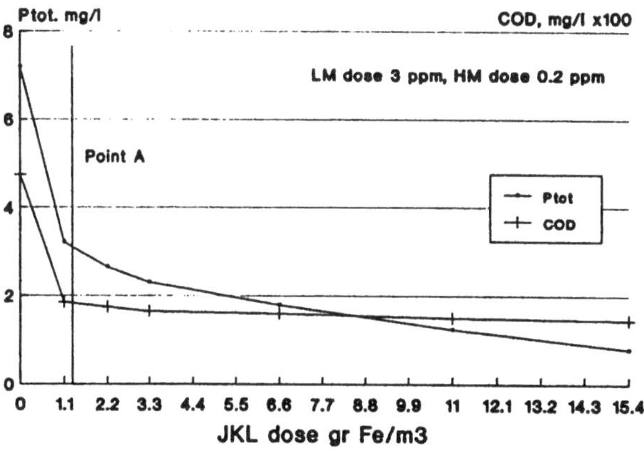

Fig. 4. Removal of phosphorus and COD: JKL

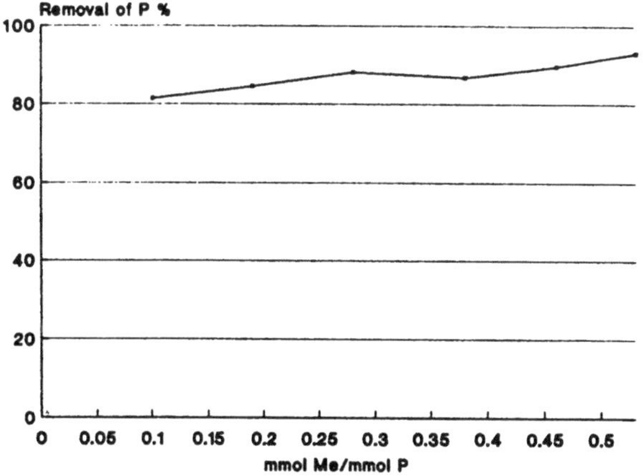

Fig. 5. Removal of phosphorus as function of Me^{3+} dose

As seen in Fig. 6, with a dosing level of 0.875 g Fe/m^3 or 0.015 mmol/l, a turbidity removal of 96 %, COD removal of 57 % and TOC removal of 47 % was achieved.

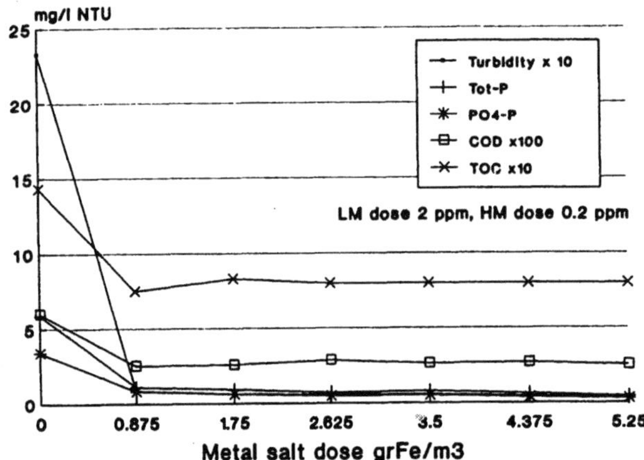

Fig. 6. Variable Fe dose with fixed LM and HM dose

To achieve more than 80 % P removal with the conventional coagulants (AVR, ALG, JKL), Me/P ratios of between 1.7 and 2 mmol/mmol were necessary (Ødegaard [2]).

In this system it seems that the metal salt removal effect increases by approximately a factor of 10, compared with conventional coagulants.

Some results for P_{tot} and PO_4-P reduction with the LM indicate that an organic coagulant contributes to the precipitation of PO_4-P and P_{tot}.

Since the LM contributes, but does not work alone, the only way to prove the above statement is to use the same Me level and vary the LM dose. This is shown in Examples 1 and 2.

Example 1

	SS	COD	P_{tot}	PO_4-P
Untreated water	292	518	9.42	3.16
18 g Fe^{2+}/m^3	116	255	5.06	1.68
18 g Fe^{2+}/m^3, 2 ppm LM, 0.2 ppm HM	72	210	3.47	0.45
18 g Fe^{2+}/m^3, 3 ppm LM, 0.2 ppm HM	64	160	2.28	0.11
18 g Fe^{2+}/m^3, 4 ppm LM, 0.2 ppm HM	64	145	2.17	0.11

Example 2

	Turbidity	COD	PO_4-P
1 g Fe^{3+}/m^3, 2 ppm LM, 0.2 ppm HM	6.2	80	0.32
1 g Fe^{3+}/m^3, 0.75 ppm LM, 0.2 ppm HM	9.8	80	0.65

Full scale trials at several other plants are in progress using pre-polymerised aluminium salts and ferric salts with dosing levels in the range of 2 g Al^{3+}/m^3 and 5 g Fe^{3+}/m^3. These trials have been continued with the realisation of the dose levels and results achieved in the laboratory.

Linköping Wastewater Treatment Plant

The Nykvarn treatment plant in Linköping was commissioned in 1952. Initially composed of only mechanical treatment and sludge digestion, the plant was extended between 1959 and 1961 to include biological treatment. Further additions, in 1972, included chemical post-precipitation, mechanical screening, sludge dewatering facilities and a new laboratory. 1979 to 1980 saw the extension of the biological treatment stage together with the installation of process control and monitoring equipment.

The Nykvarn plant serves a population of approximately 120 000. The total loading on the plant is approx. 167 000 pe, of which 47 000 is industrial discharge. The influent flow is approximately 17 000 000 m^3/y.

Normal Process Operation

Chemical (simultaneous) precipitation with ferrous sulphate is the method employed for phosphorus removal.

The influent water, after passing mechanically cleaned screens, is dosed with ferrous sulphate (approx. 20 g Fe^{2+}/m^3) before passing to an aerated grit chamber. From here the water is pumped via preaeration tanks and two primary settlement lagoons to the biological treatment stage which utilises the contact stabilisation process. Following secondary (intermediate) settlement, the water is dosed with an additional 2–3 g Fe^{2+}/m^3 before passing to the final chemical precipitation stage and from there, as the final effluent, to the receiving waters.

Surplus sludge, together with sludge from the final precipitation is pumped to the influent water for co-settlement in the primary settlement lagoons. All sludge from the primary settlement is pumped to the sludge digestion units. The digested sludge is dewatered with centrifuges and the reject water is recycled to the preaeration tanks.

The F.A.S.T. Process at the Linköping Treatment Plant

The F.A.S.T. process was tested with regard to the following possibilities:
- Improved removal of material in the primary settlement stage.
- Increased biogas production by way of increased organic loading to the digestion units.
- Decreased ferrous sulphate to minimise sludge production without increasing the phosphorus level on the effluent.
- Increased separation of particulate nitrogen in the primary settlement stage*.

---

* Essentially an increase in the nitrogen removal of the plant as a whole since the reject water from sludge dewatering is treated by SBR-technique for nitrogen removal

Tab. 1. Equipment and dimensions of Linköping wastewater treatment plant

Mechanical screens:		2 units	15 mm grating
Aerated grit chamber:	2 u. $\times 200$ m^3	:=	400 m^3
Pumping:	2 u. screw type, cap.		10 000 m^3/h
Preaeration tanks:	4 u. $\times 320$ m^3	:=	1 280 m^3
Primary settlement:	2 u. $\times 3100$ m^3	:=	6 200 m^3
Active sludge aeration lagoons:	$4 \times 500 + 6 \times 515$	:=	5 100 m^3
Secondary settlement:	$6 \times 520 + 4 \times 770$	:=	6 180 m^3
Flocculation unit:	16 u. $\times 75$ m^3	:=	1 200 m^3
Intermediate pumping:	10 000 m^3/h		
Final settlement:	16 u.	total	10 800 m^3
Sludge digestion:	1 u. $\times 3000 + 2$ u. $\times 1700$	:=	6 400 m^3
Sludge dewatering:	centrifuges, 2 u.		
Electrical production from biogas:			460 kW
Discharge limits	BOD$_7$		< 15 mg O$_2$/l
	Total phosphorus		< 0.5 mg P/l
Effluent discharge	BOD$_7$		5–10 mg O$_2$/l
	Phosphorus		approx. 0.35 mg P/l

Methodology

The plant trials were preceded by laboratory bench tests that indicated which polymers, in combination with ferrous sulphate, were suitable and which combinations gave the best results. Repetition of the laboratory tests also gave good results.

Initial full scale trials were performed between 15 May 1991 and 26 June 1991. During this period the dosing of Fe^{2+} and the two polymers was optimized with respect to both quantities and the points of application. The cationic, low molecular weight polymer was dosed as a dry powder and the anionic, high molecular weight polymer as a solution.

Results from this initial trial indicated the following:

– An increase in BOD reduction in the primary settlement from 30 % to 50 %.
– An increase in P removal in the primary settlement from 45 % to 55 %.
– A reduction in Fe^{2+} dose to 12–15 g Fe^{2+}/m^3, with a final effluent quality from the plant of 0.25–0.35 mg P/l.

On the basis of these results, a long-term trial, to last approximately 1 year, was initiated in September 1991.

The following process is presently in operation (February 1992): 10–12 g Fe^{2+}/m^3 is dosed to the influent water after screening. After the grit chamber, 1.5–1.7 g Magnafloc 368/m^3 is dosed as a dry powder (cationic).

The water then passes through the preaeration tanks and flows to the primary settling lagoons. Between the aeration tanks and the primary settlement lagoons, 0.07 g/m^3 of the anionic polymer, Magnafloc 336, is dosed as a solution.

Following the biological active sludge stage, the water is dosed with approx. 2 g Fe^{2+}/m^3 before being pumped to the final chemical precipitation stage. After settlement, the final effluent then passes to the receiving waters.

Results as of February 1992

The results from the application of the F.A.S.T. process at the Linköping sewage treatment plant are presented in Figs. 7–11 and can be summarized as follows:

- Biogas production has increased by 900 m^3/d, see Fig. 7.
- Over the primary settlement stage, the BOD reduction has increased from 30 % to 55–60 %, see Fig. 8, and the total phosphorus reduction has increased from 40 % to 60 %, see Fig. 9.
- Fe^{2+} dose has been reduced from approx. 20 g/m^3 to approx. 12 g/m^3, see Fig. 10.
- The effectiveness of the biological treatment stage has improved as a consequence of the increased proportion of organic material in the active sludge, see Fig. 11.
- Nitrification has been achieved in the activated sludge process.
- Process stability has increased at the plant.

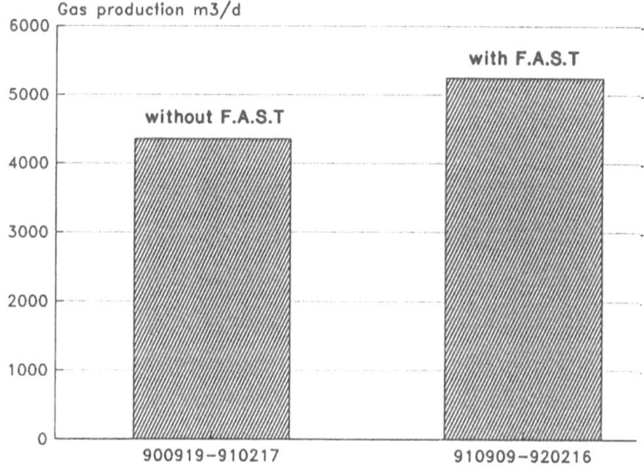

Fig. 7. F.A.S.T. process at the Linköping sewage treatment plant: gas production

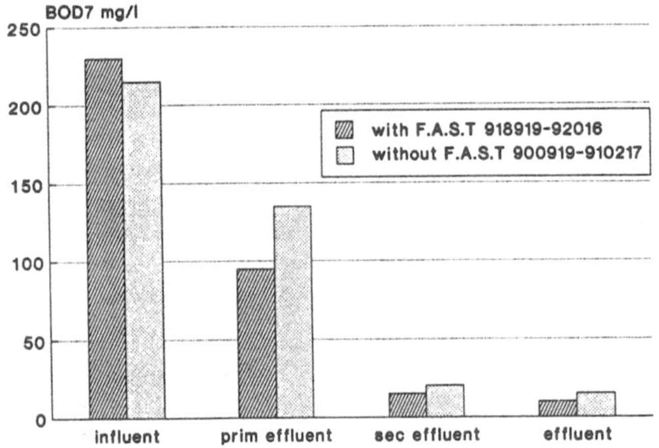

Fig. 8. F.A.S.T. process at the Linköping sewage treatment plant: BOD_7

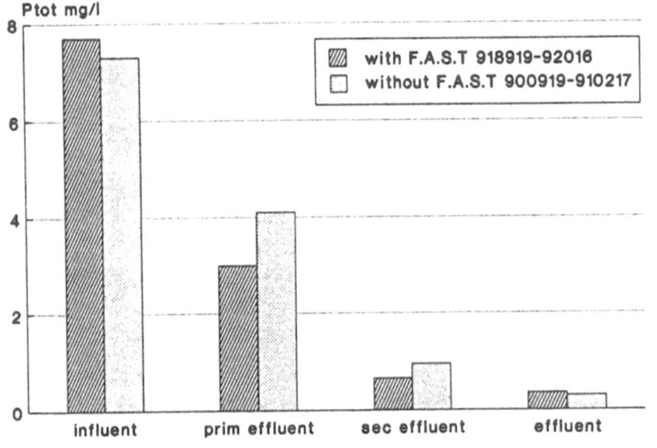

Fig. 9. F.A.S.T. process at the Linköping sewage treatment plant: P_{tot}

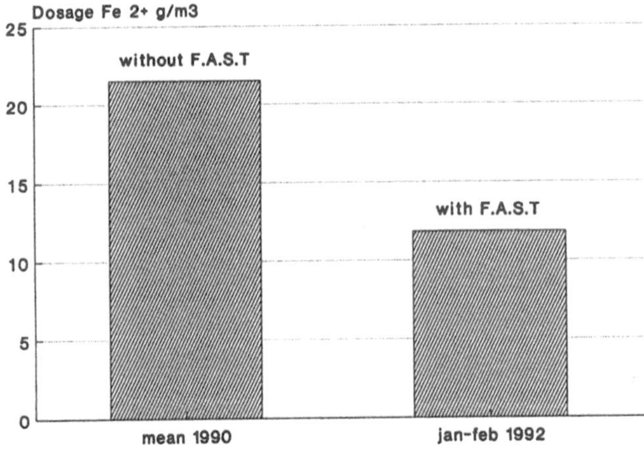

Fig. 10. F.A.S.T. process at the Linköping sewage treatment plant: dosage Fe^{2+}

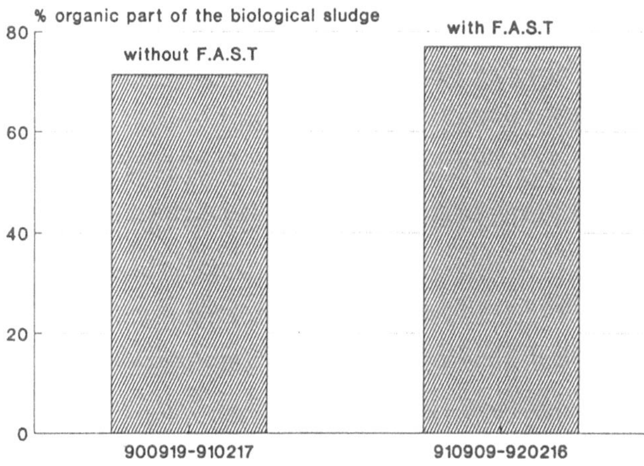

Fig. 11. F.A.S.T. process at the Linköping sewage treatment plant: organic part of the biological sludge

The figures for the period 9/1990–2/1991 result from plant trials carried out when ferric chloride was used in combination with ferrous sulphate. Under normal process conditions, the difference between the effluent values shown in Tables 2 and 3 would be significantly greater.

Tab. 2. Results with normal process operation 9/1990–2/1991

	Influent mg/l	Primary effluent mg/l	Secondary effluent mg/l	Final effluent mg/l
BOD_7	210	141	19	10
COD	535	317	82	54
P_{tot}	7.3	4.1	0.83	0.32
PO_4-P	4.6	2.5	0.49	0.17
Suspended solids	203	137	32	12.4

Tab. 3. Results with the F.A.S.T. process 9/1991–2/1992

	Influent mg/l	Primary effluent mg/l	Secondary effluent mg/l	Final effluent mg/l
BOD_7	239	98	10	5.5
COD	551	260	47	39
$P_{tot.}$	7.7	3.0	0.64	0.36
PO_4-P	5.5	2.1	0.45	0.26
Suspended solids	239	104	13	8.9

Summary of Results from the Plant Trial

The F.A.S.T. process at the Linköping waste water treatment plant.
- Functions well with ferrous sulphate.
- Requires a lower Fe^{2+} dose.
- Increases the separation of organics in the primary settlement stage.
- Reduces the organic loading to the biological treatment stage and increases the active sludge VSS content.
- Results in BOD and total phosphorus levels in the final effluent which meet discharge requirements.
- improves the process stability of the treatment plant.

Conclusions

The results of laboratory tests and plant trials described above indicate that the F.A.S.T. process maximizes precipitation and removal of solids. This in turn has led to significant reductions in other analytical values which would normally be influenced by the presence of solids.

The combination of chemicals used provides for a process which can be controlled in such a way as to allow for the desired amount of PO4 to be passed to the biological stage without loss of impact on solids removal. This has been achieved by judicious variation of the ratio of the chemicals used in the process.

References

[1] Ødegaard, H.: Coagulation as the First Step in Wastewater Treatment. In: Pretreatment in Chemical Water and Wastewater Treatment, H.H. Hahn and R. Klute (eds.). Springer, Berlin Heidelberg New York 1988, pp. 249–260
[2] Ødegaard, H., Fettig, J., Ratnaweera, H.: Coagulation with Prepolymerized Metal Salts. In: Chemical Water and Wastewater Treatment, H.H. Hahn and R. Klute (eds.). Springer, Berlin Heidelberg New York 1990, pp. 189–220

Jens Sörensen   Sven-Gunnar Larsson
AB CDM   Tekniska Verken
P.O. Box 37   S-581 53 Linköping
S-431 21 Västra Frölunda   Sweden
Sweden

Comparison of Various Commercially Available Polyaluminium Chlorides (PAC)

W.C.A. van Dorst and J.W.F.L. Seetz

1. Introduction

Akzo is a large international chemical company producing many chemical products, and thus generating various aqueous waste streams. All of these are treated, some with flocculation agents. Because many different flocculants are commercially available, some of them made by Akzo, we investigated the composition of several polyaluminium chloride (PAC) flocculants and tried to relate this to their effectiveness in water treatment.

2. Production

PAC, chemically represented as $[Al(OH)_x Cl_y(SO_4)_z]_n$, can be produced in several ways. As the production process influences the final composition, the two most important production processes for PAC will be given.

The first process, described in patent literature by Taki [1], is a two-step reaction, as shown below. The key factors are a 2.5 hour reaction time at 112 °C and atmospheric pressure with an excess input of H_2SO_4. Subsequently calcium carbonate or calcium hydroxide is added at 40 °C to increase the amount of OH^-, and to remove the excess of SO_4^{2-} by precipitation of $CaSO_4$.

$$HCl + H_2SO_4 + Al(OH)_3 \longrightarrow \textit{interim gel}$$

$$\textit{interim gel} + CaCO_3 \longrightarrow Al(OH)_x Cl_y(SO_4)_z + CaSO_4 \downarrow + CO_2 \uparrow$$

The second process is described by, among others, Prodeco [2]. This is a single-step reaction at elevated pressure and temperature. Key factors are a reaction time of 3 hours at 160 °C and 5 bar.

$$Al(OH)_3\ (s) + x\ HCl\ (aq) \longrightarrow Al(OH)_{3-x}Cl_x\ (aq) + x\ H_2O$$

3. Analysis

About forty different PACs are known to be commercially available in the European market for water treatment and paper production. Twenty PAC samples were purchased from at least ten different producers. These samples were all analysed for aluminium, hydroxide, chloride, sulphate, sodium, calcium, iron, and potassium. These determinations were made with the normal titrimetric methods.

PACs are usually typified by three parameters, namely the aluminium content, expressed as wt-% Al_2O_3, the sulphate content in wt-% SO_4 and the hydroxide content, expressed as the basicity. This basicity is calculated as $\frac{[OH^-]}{3\,[Al^{3+}]} * 100\%$ and gives the degree of hydrolysis of the Al^{3+} ion. In Table 1 all PAC test samples are listed with the values for these three parameters, and for several other chemical properties.

Tab. 1. Various PAC types, in order of basicity

Producer	Product	A^1	Al_2O_3 wt-%	Bas. %	SO_4 wt-%	M : O : P Ratio			Colour Apha
						M	O	P	
Hoechst	Locron L	V	22.1	89	0	4	3	93	< 10
Hoechst	Locron L	V	22.5	87	0	4	< 3	95	< 10
Breustedt	PAC		12.1	69	0	26	26	49	10
Breustedt	PAC		12.1	66	0	28	28	44	10
Rhône-Poulenc	Aqualenc	W	8.1	58	2.7	39	5	56	100
Atochem	WAC	W	9.9	55	2.6	18	< 3	81	< 10
Montedison	Alpoclar	W	9.9	54	2.7	42	6	52	10
Sachtleben	Sachtoklar	W	10.0	53	2.7	42	7	51	12
Sachtleben	PaperPacN	P	9.7	53	2.7	46	6	48	10
Caldic	PAC		9.7	52	2.6	45	4	51	40
Atochem	WAC	W	9.8	51	2.6	48	7	45	50
Veitsiluoto	Oulupac	W	17.3	48	0	60	30	10	400
Veitsiluoto	Oulupac	W	17.4	48	0	52	31	18	400
Akzo	Redifloc	P	16.8	47	0	62	29	9	300
Atochem	PAC	P	17.3	45	0	61	24	15	400
Akzo	Redifloc	P	16.9	45	0	52	33	14	200
Montedison	Alpoclar200	P	16.5	40	1.2	58	16	26	400
Kemira	PAX-P	W	13.4	36	0	63	27	10	300
Hoechst	Povimal	W	12.5	34	0	77	13	10	150
Kemira	PAX-A	W	14.7	31	0.2	64	20	17	10
Ekokemi	Ekoflock		10.1	29	0.8	71	13	19	400
Ekokemi	Ekoflock		9.8	28	1.2	74	4	22	700
$AlCl_3$	Laboratory	–	9.3	5	0	96	< 3	< 3	500
$Al_2(SO_4)_3$	Laboratory	–	4.1	10	11.7	97	< 3	< 3	< 10
$FeCl_3$	Laboratory	–	8.6^2	10	0	–	–	–	< 10

^{1} Application: P(aper) W(ater) V(arious); ^{2} $FeCl_3$ in weight-% Fe

The PAC samples can be clearly grouped in order of basicity. The first important group with 50–60% basicity is Aqualenc (Rhône-Poulenc) to WAC (Atochem), which has 8 to 10 weight-% Al_2O_3, no sodium but 0.2 weight-% Ca, 2.7 weight-% SO_4. This PAC type is assumed to be produced by the Taki process. There is also a similarity in the degree of association of the molecules, as the measured ratio of monomeric+dimeric : oligomeric : polymeric species does not vary much around 45 : 6 : 50.

The second group with 40–50 % basicity is Oulupac (Veitsiluoto Oy) to Alpoclar 200 (Montedison) with 16.5 to 17.5 weight-% Al_2O_3, no sulphate, sodium or calcium. This type of PAC is assumed to be produced by the Prodeco process. Typically the mono+dimer : oligomer : polymer ratio is 57 : 29 : 13.

The close similarities in concentrations and ratios within these two groups is probably the result of optimization of production process parameters, or of product performance optimization.

The other PACs can nominally be grouped into a High basicity and a Low basicity group. Some of these PACs show a relatively high content of sodium, e.g., PAC (Breustedt), or a relatively high content of Fe, e.g., PAX-A (Kemira) and Ekoflock (Ekokemi). These PAC samples were produced by various processes from various reactants, such as bauxite, e.g. PAX-A, or pickling baths, e.g. Ekoflock.

In order to verify the analytical method of typification, a charge balance was made for every test sample.

$$\sum(\text{positive ions}) \overset{?}{=} \sum(\text{negative ions})$$

These balances showed deviations of only 1–3 % for all PAC samples. This indicates that all important ions have been accounted for with the analytical methods described.

It can therefore be concluded from this analysis that there are clear distinctions between the various available PAC qualities, mainly in the concentrations of Al^{3+} and OH^-, the various co-ions as Ca^{2+}, Na^+, and Fe^{3+}, and the ratios of monomer+dimer : oligomer : polymer.

4. Ratio Monomer+Dimer : Oligomer : Polymer

Literature on the use of PAC in the paper industry indicates that the oligomeric PAC species is the most active species, whereas the monomeric and the dimeric species are not active [3, 4]. The oligomeric species is mostly considered to consist of $Al_{13}O_4(OH)_{24}(H_2O)_{12}^{7+}$ and other ionic complexes containing 6 to 13 aluminium atoms [5, 6]. Larger complexes are considered to be polymeric, and smaller complexes are mono- and dimeric.

The ratio of the sum of monomeric and dimeric species, which are considered equally active, to oligomeric species and to polymeric species has been measured by means of UV kinetic complexometric titrations of the aluminium complex with ferron iodide, as is widely described in the literature [5]. The numerical results are visualized in Fig. 1. It is to be noted that the groups of PAC samples, grouped for chemical similarity in Table 1, appear again, grouped for physical similarity in Fig. 1, with a single exception.

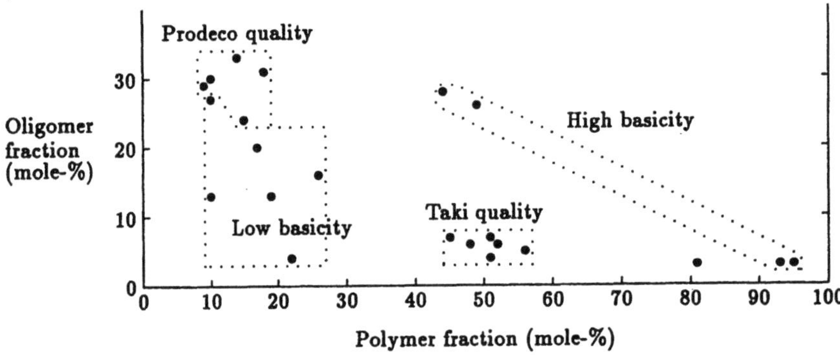

Fig. 1. Oligomer-to-polymer molar ratio [in mole-% Al], measured by means of UV kinetic complexometric titration with ferron iodide

Comparing the high basicity group with the low basicity group it is seen that the basicity correlates with the ratio of the monomeric and polymeric species, so that high basicity relates to a large polymer and a small monomer fraction, and the low basicity relates to a large monomer fraction and a small polymer fraction. The oligomer fraction seems not to be relevant here.

It also follows from the comparison of sulphate-free Prodeco quality with the sulphate-containing Taki quality, that the sulphate content correlates with the ratio of oligomer-to-polymer; low oligomer-to- high polymer when sulphate is present (Taki) and high oligomer-to-low polymer when no sulphate is present (Prodeco). The monomer fraction does not depend on the sulphate content.

4.1 ^{27}Al-NMR Measurements

The results shown above, obtained by the titrimetric method using ferron iodide, appear reproducible, but as it is unknown whether this method introduces systematic errors, such as dilution of the PAC sample to be measured, other methods for the determination of the oligomeric and polymeric species were also studied.

As suggested by the literature and previous research on PAC [7], ^{27}Al-NMR measurements were carried out on non-diluted PAC samples. It was shown that octahedral fragments such as $Al(H_2O)_6^{3+}$, tetrahedral fragments such as AlO_4^{5-} and $Al(OH)_2Al$ molecule fragments were present. On dilution, the tetrahedral fragments, probably the $Al(OH)_4^-$ ions, became more conspicuous, whereas in the non-diluted samples hardly any tetrahedral aluminium fragments were found.

Considering the ratios octahedral : tetrahedral : $Al(OH)_2Al$ fragments, calculated from the structure of the oligomeric Al_{13}-complex, to be representative of the oligomeric molar fraction, these ^{27}Al-NMR measurements on non-diluted PAC samples could have given indications of the ratio mono+dimer : oligomer : polymer. Unfortunately the results were not quantitative enough to calculate this ratio.

4.2 Quasi Elastic Light Scattering Measurements

In a patent by Rhône-Poulenc [8], quasi elastic light scattering (QELS) measurements are used to determine the apparent molecular mass and the hydrodynamic diameter on basis of which claims are made. These same measurements were carried out on several commercial PAC samples. A concentration range of PAC, as is needed for QELS measurements, was made by diluting the PAC samples with their isotonic PAC solutions. Such solutions are made by centrifuging the PAC samples at 55 000 rpm during four days. Then light scattering is measured on the diluted samples.

The PAC samples of PAX-P (Kemira), Redifloc (Akzo) and Oulupac (Veitsiluoto) did not separate on centrifugation so that no measurements could be carried out on these samples. Only the Aqualenc PAC sample made by Rhône-Poulenc itself could be measured completely. The diameter was found to be 22 nm, and the apparent molecular mass is 370 000. For the Sachtoklar (Sachtleben) PAC sample a diameter of 7 nm was estimated, but no apparent molecular mass could be determined.

5. Performance Tests

The flocculating performance of PAC on the laboratory scale is normally tested with the 'jar test'. A detailed description can be found in the Kiwa report SWE–184 [9] and ASTM standard D2035–74 [10].

5.1 Method

The test consists of four jars, each to be filled with 1 litre untreated water. PAC is dosed (25 to 300 μmole or about 1 to 10 mg aluminium per litre) while the water is stirred at 230 rpm for 1 minute to achieve instantaneous mixing. Then slow speed stirring at 43 rpm for 20 minutes allows for floc formation, followed by 30 minutes of no stirring allowing for sedimentation of the flocs which are formed.

At the end of the sedimentation time, a water sample of 200–250 ml is taken from the supernatant water to measure the turbidity, which is an indication of the amount of suspended particles, the UV absorption at 254 nm, which is an indication of the amount of dissolved organic carbon (DOC), and the residual amount of aluminium.

The test sample was taken from the Akzo Hengelo pond, which had an initial pH of 7.1–7.3 and was used at that pH. Usual monitoring analysis showed that the water was normal surface water.

5.2 Comparative Tests of Commercial PAC Samples

Two series of tests were carried out. In the first series, Paper-Pac-N and Sachtoklar (Sachtleben), Redifloc (Akzo), Aqualenc (Rhône-Poulenc), Ekoflock (Ekokemi), Pac (Breustedt) twice, WAC (Atochem) and Pac (Caldic) were tested.

In the second series, performance tests were carried out on Sachtoklar (Sachtleben), Redifloc (Akzo), both for reference with the first test series, Locron L (Hoechst), Oulupac (Veitsiluoto Oy) and Alpoclar (Montedison). For reference, these same tests were carried out on solutions of non-polymerized $AlCl_3$, $FeCl_3$ and $Al_2(SO_4)_3$, produced on a laboratory scale from analytical grade chemicals.

Tab. 2. Products tested in the two series

First series	Second series
Sachtoklar	Sachtoklar
Redifloc	Redifloc
Paper-Pac-N	Locron L
Aqualenc	Oulupac
Ekoflock	Alpoclar
WAC	$AlCl_3$-solution
Breustedt PAC	$FeCl_3$-solution
Breustedt PAC	$Al_2(SO_4)_3$-solution
Caldic PAC	

The same test parameters were used for both series. Only the water temperature was adapted from 15 °C for the first series to 5 °C for the second series in order to establish clearer distinctions between the performance of PAC and other flocculants. The second series was best controlled, and thus gave quantitatively more significant results. Qualitatively, the results of both series were the same, therefore, only the results of the second series are presented.

5.3 Turbidity

The turbidity was measured with a DrLange Trübungsphotometer LTP5, which gives results in Formazine Turbidity Units (FTU). Figure 2 shows a graphical representation of the measurements.

In this figure we clearly see the similarity of the turbidity removal between the various PAC samples. No distinction in effectivity can be found for the presence of SO_4 in the PAC, nor for the other variations in the chemical composition. The performance of the non-PAC flocculants was slightly less than that of the PAC flocculants. For reference, the typical dosage of 100 μmole (3 mg) aluminium per litre is given.

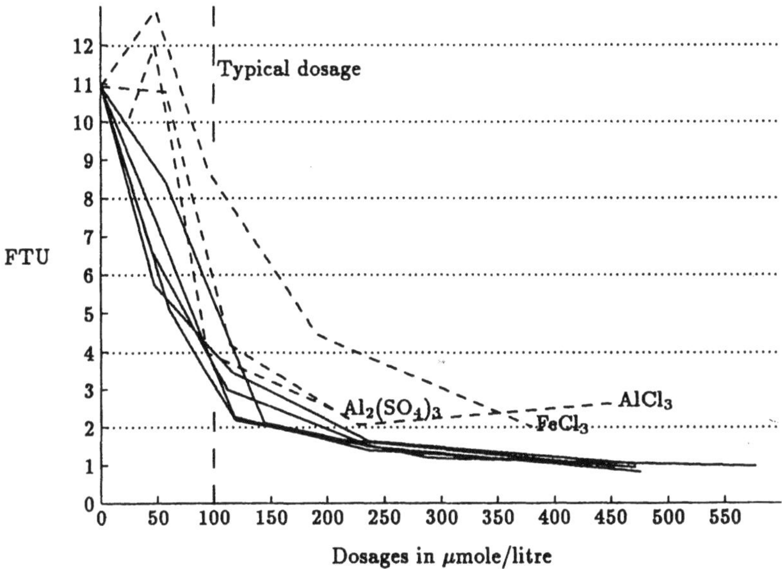

Fig. 2. Residual turbidity [in FTU] in treated ASB pond water for various dosages. Second series flocculation performance tests

5.4 UV Absorption

The UV absorption, which is regarded as an indication of the amount of dissolved organic carbon (DOC), was measured at 254 nm in a 1 cm QS cuvette, using demineralized water as a reference. A graphical representation is given in Fig. 3.

The similarity between the aluminium salts PAC, the $AlCl_3$-solution, and the $Al_2(SO_4)_3$-solution is apparent. The difference in the performance of the $FeCl_3$-solution is also noteworthy.

5.5 Residual Aluminium

In accordance with normal methods [5], the treated water samples are filtered with a Millipore filter (0.45 μm), and then HCl is added until the pH is 1 in order to break up all remaining aluminium flocs. The residual amount of aluminium and iron is then measured by inductively coupled plasma emission spectrometry (ICP-ES). Figure 4 gives the graphical representation of the results.

The similarity among all PACs is again evident in the figure. For dosages up to 200 μmole/litre (4.5 mg Al per litre), the solutions of $AlCl_3$ and $Al_2(SO_4)_3$ show the same performance. For higher dosages, the performance of the $AlCl_3$ solution improves no further, so that the residual amount of aluminium increases again. The performance of the $FeCl_3$ solution is lower than that of the aluminium salts over the whole range.

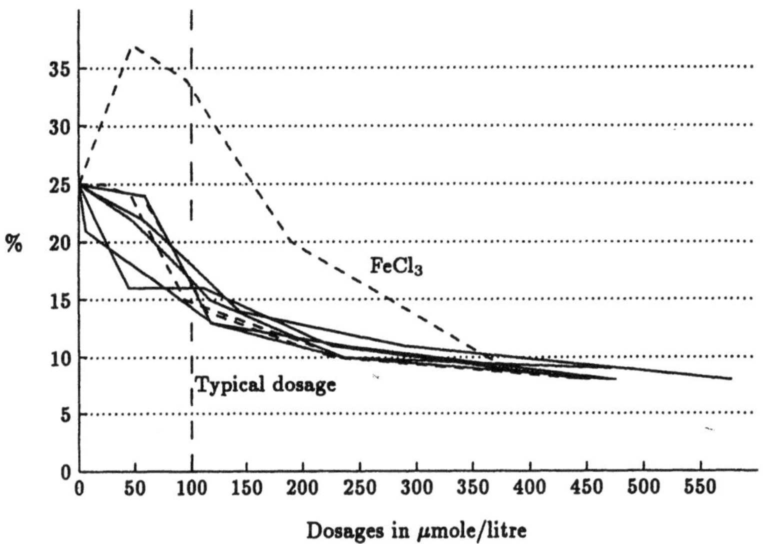

Fig. 3. Residual UV absorption at 254 nm [in % absorption] for various dosages. Second series flocculation performance tests

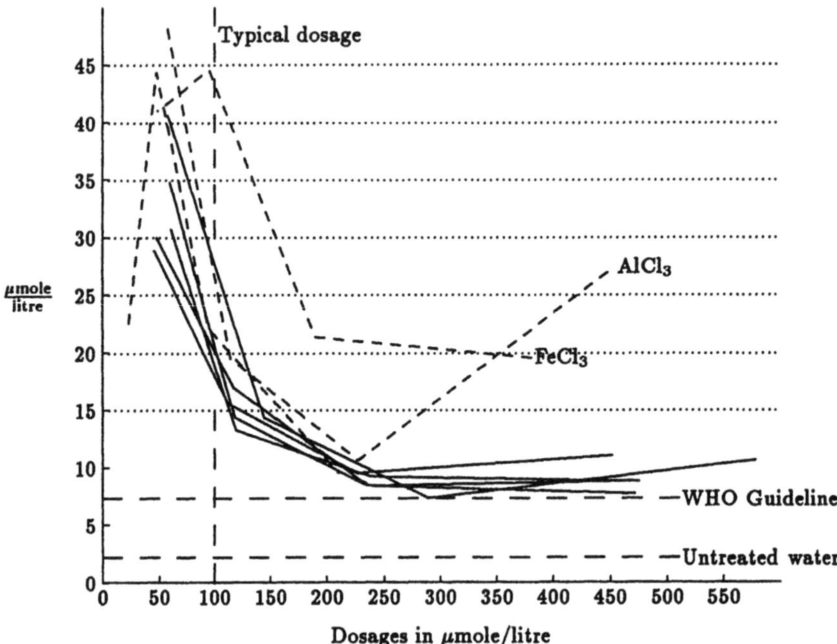

Fig. 4. Residual aluminium and iron concentrations [in μmole/litre] for various dosages. Second series flocculation performance tests

Aluminium dosages in the range of 150–400 μmole/litre (4–10 mg Al per litre) leave a residual amount of aluminium close to the World Health Organization (WHO) guideline of 7.4 μmole/litre (200 μg Al per litre). In real water treatment installations which use PAC, final residual aluminium concentrations as low as 0.4 μmole/litre (10 μg Al per litre) can be obtained, due to longer flocculation and sedimentation times, and to various filtration steps after sedimentation. Normally aluminium-treated water contains 0.5–2 μmole aluminium per litre (10–40 μg Al per litre).

5.6 Results

The results from the removal of the turbidity, the removal of UV absorption, and the residual aluminium concentration in treated water show that no differences in the laboratory scale performance can be found for the various commercially available PAC samples, however different their chemical composition.

This is not in agreement with publications (e.g., see [11]), which claim that sulphate-containing PAC, such as the Taki quality, has a better flocculation performance than sulphate-free PAC. This is also contrary to the claim (e.g., see [3]) that the oligomer fraction causes the main flocculating effectiveness of polyaluminium chlorides.

In the laboratory scale tests, the performance of PAC, compared to other aluminium salts and iron salts, was shown to be better with regard to turbidity removal and residual amounts of aluminium. The performance of PAC and other aluminium salts was better than $FeCl_3$ in the removal of UV absorption.

5.7 Reference Performance Test by Kiwa

To verify the reliability of the tests and results described above, equivalent tests were carried out by the Keuringsinstituut van Waterleidingartikelen (Kiwa), an independent Dutch research institute specializing in water treatment techniques. Kiwa carried out tests on the following unlabelled PAC samples: Sachtoklar (Sachtleben), Redifloc (Akzo), Oulupac (Veitsiluoto), Pac (Breustedt), Pax-P (Kemira) and Ekoflock (Ekokemi).

These tests consisted of jar tests on low and high turbidity water, pH of 7.5–8 and 6.5–7, and a water temperature of 5 °C for each PAC sample. The water types used are representative of the surface water used in The Netherlands for drinking water production. Measurements were done on the turbidity, the UV-absorption and the residual amount of aluminium.

Observation of the test methods carried out by Kiwa showed only small differences, such as the use of pre-sedimentated water, whereas the Akzo test uses direct water sampling and testing. In the Kiwa test equipment, energy is brought in in the course of 10 seconds at high rotational speed and jars fitted with barrages, whereas the Akzo equipment uses slightly lower speeds, no barrages, and a longer mixing time of 60 seconds. In addition, the jars of Kiwa are fitted with sampling nozzles in the lower part of the jars, whereas

Akzo samples are taken from the upper layers. Finally, in the Kiwa method the optimal aluminium dosage is calculated from the measurements and this is compared to the performance found at that dosage, whereas the Akzo method compares the whole range of dosages. These differences make the Kiwa method somewhat more critical.

The Kiwa test shows results similar to those of the Akzo tests presented above. All PAC samples show broadly the same removal of turbidity. Minimal differences are accounted for by the variations in the initial turbidity of the water used. The UV absorption results are even more similar for all PAC samples than the turbidity results. Residual aluminium concentrations are quite comparable. Other variations in the test results can be accounted for by variations in dosages, testing method and analyses. Final conclusions are that no significant difference in performance between the various PAC samples can be found. Therefore, it is concluded that the Kiwa test completely confirms the Akzo tests [12].

6. Conclusions

From the analyses we may conclude there are three groups of PAC with clear distinctions in the chemical composition, namely

1. the Taki quality with 9.8 weight-% Al_2O_3, 2.7 weight-% SO_4 and 53 % basicity,
2. the Prodeco quality with 17.0 weight-% Al_2O_3, and 46 % basicity,
3. other PAC qualities with various aluminium concentrations (8–22 weight-% Al_2O_3), with other components such as Na^+ and Fe^{3+} and with various basicities (28–89 %).

Charge balance calculations show that all relevant ions were found and analysed. The ratio monomeric+dimeric : oligomeric : polymeric species, measured by ferron iodide titrations, shows consistent results. ^{27}Al-NMR and the light scattering method gave extra qualitative information, but they cannot be used for a practical determination of the degree of association of the aluminium species.

From the performance tests it is concluded that the variations in the performances of the available PAC samples are so small that they cannot be measured by laboratory scale tests. The differences in chemical composition of the various commercially available PACs is considered to stem only from the production process, and not from optimization of effectivity.

The differences in performance between PACs in general, and other aluminium salts and iron salts are significant, especially in the removal of turbidity and in the residual amount of aluminium at high dosages.

References

[1] Taki Fertilizer Manufacturing Co., Ltd: Werkwijze voor het bereiden van een stabiele oplossing van een basisch aluminiumzout. Dutch patent no. 164832, April 19, 1973

[2] Prodeco S.p.A.: A Process for the Sizing of Paper. European patent no. 0 063 812, April 27, 1981

[3] Schönherr, S., Frey, H.-P.: Untersuchungen zur Trennung und Charakterisierung kondensierter Aluminiumkationen aus basischen Aluminiumchloridlösungen. Z. anorg. allg. Chem. *452* (1979) 167–175

[4] Schönherr, S., Görz, H., Bertram, R., Müller, D., Gessner, W.: Vergleichende Untersuchungen an unterschiedlich dargestellten basischen Aluminiumchloridlösungen. Z. anorg. allg. Chem. *502* (1983) 113–122

[5] Van Benschoten, J.E.: Speciation and Fate of Aluminum in Water Treatment. Ph.D. dissertation, University of Massachusetts, February 1988

[6] Van Benschoten, J.E., Edzwald, J.K.: Chemical Aspects of Coagulation Using Aluminum Salts: 1. Hydrolytic reactions of Alum and Polyaluminum Chloride. Accepted for Water Research, submitted, June 1989

[7] Angad Gaur, H., Evertse, W.: Onderzoek aan poly(aluminiumchloride)oplossingen m.b.v. ^{27}Al-NMR. Akzo Corporate Research Department Report no. CRL/D 87/17/0516B, Arnhem, March 25, 1987

[8] Rhône-Poulenc Chimie: Chlorosulfate d'aluminium, son procédé de préparation en son application comme agent coagulant. European Patent Office, Numéro de publication EP 0 218 487 B1, October 24, 1990

[9] Kiwa: Bekerglasproef voor coagulatie, Voorschrift voor de uitvoering van de standaard bekerglasproef met sedimentatie. Kiwa nv, Mededeling nr. SWE–184, maart 1978, herziene versie april 1980

[10] ASTM: Standard Recommended Practice for Coagulation-Flocculation Jar Test of Water. American Society for Testing and Materials, ASTM designation D2035–74, 1974

[11] De Hek, H., Stol, R.J., De Bruyn, P.L.: Hydrolysis-Precipitation Studies of Aluminum(III) Solutions 3. The Role of the Sulphate Ion. J. Coll. Interf. Sci. *64*, 1 (1978) 72–89

[12] Heijman, S.G.J., Koppers, H.M.M., Van de Baan, E.: vergelijkend onderzoek van 6 vlokmiddelen op PAC-basis. Kiwa nv, SWO 91.25, februari 1991

W.C.A. van Dorst and J.W.F.L. Seetz
Akzo Salt & Basic Chemicals
Research Department
Boortorenweg 20
NL-7550 GC Hengelo
The Netherlands

Polyaluminium Silicate Sulphate –
A New Coagulant for Potable and Wastewater Treatment

A.K. Arnold-Smith, R.M. Christie, and C. Jolicœur

Summary

Polyaluminium Silicate Sulphate (PASS[(R)]) is a new coagulant for potable water treatment which is increasingly being used in North America, the Far East and Europe.

It is a patented product resulting from a reaction between three chemicals traditionally used in potable water treatment: aluminium sulphate, sodium aluminate and sodium silicate. The empirical formula for PASS[(R)] may be written as

$$Al_a(OH)_b(SO_4)_c(SiO_x)_d \, .$$

Ultrafiltration experiments are described which demonstrate the existence of high molecular weight species (MW > 100 000).

The efficacy of PASS[(R)] as a potable water coagulant is described with reference to laboratory and plant trials involving "problem" waters.

Introduction

Among the processes by which raw water is rendered fit for drinking is that of coagulation. In this process the water is treated with a coagulant to destabilise the colloidally dispersed particles. The destabilised particles then form aggregates and the latter are removed from the water by a selection of flotation, sedimentation and filtration techniques. The coagulation process also assists in the removal of colour, odours and taste from the water.

One of the chemical coagulants most frequently used in the treatment of raw water is aluminium sulphate. The ability of aluminium salts to form polynuclear hydrolysis products is well known. Matijević and coworkers [1–5] and others [6] have shown that the hydrolysis products of aluminium salts are able to coagulate well defined colloidal systems at much lower total electrolyte concentrations than the unhydrolysed aluminium ions. From the results of microelectrophoresis experiments, Matijević suggested that the positively charged aluminium hydrolysis products were capable of being adsorbed by the colloidal particles.

The importance of the role that the hydrolysis products of aluminium salts play in the clarification of raw water has been stressed by Packham [7], Stumm and Morgan [8], and Black [9] in their reviews of the earlier research work in this field. Stumm and O'Melia [10] subsequently proposed that the mechanism of coagulation of a colloidal dispersion by a salt of a hydrolysing metal included several steps:

1. The hydrolysis of multivalent metal ions and subsequent polymerisation to multinuclear hydrolysis species.
2. The adsorption of hydrolysed species at the solid-solution interface to accomplish destabilisation of the colloid.
3. The aggregation of destabilised particles by interparticle bridging involving particle transport and chemical interactions.
4. The aggregation of destabilised particles by particle transport and van der Waals forces.
5. The 'aging' of flocs, which is accompanied by chemical changes in the structure of metal OH metal linkages, and concurrent changes in floc sorbability and extent of floc hydration.
6. The precipitation of the metal hydroxide.

The concept of interparticle bridging was previously introduced by several workers [11–14] to explain the ability of natural or synthetic polyelectrolytes to flocculate colloidal dispersions. The mechanism involves the adsorption of segments of the same polymer chain by two or more particles. Satisfactory flocculation may be achieved when the charges of the polyelectrolytes are of the same sign as those on the particles, or of the opposite sign. Although there is some evidence that polyeletrolytes of higher molecular weight are more effective coagulants than polyelectrolytes of lower molecular weight [15] the reduction of the charge of the colloidal particles may still be an important step in the mechanism of flocculation when negatively charged colloids are coagulated by cationic polyelectrolytes (or vice versa).

Aluminium metal can be dissolved in dilute hydrochloric acid even when the stoichiometric quantity of hydrochloric acid present is insufficient to produce (theoretically) $AlCl_3$ (Eq. 1).

$$2Al + 3HCl + 3H_2O = Al_2(OH)_3Cl_3 + 3H_2 \qquad (1)$$

The solutions are known as basic aluminium chloride solutions. Descriptions of the preparation of basic aluminium chloride solutions from aluminium containing minerals, and of the use of these solutions as coagulants in the treatment of industrial wastewaters and sewage sludges, have appeared in the patent literature [16–19].

More recently, descriptions of the preparations of basic aluminium chloride solutions (eg. polyaluminium chloride or PAC) which contained small proportions of multi valent anions, such as sulphate or phosphate, have appeared in the patent literature [20, 21]. The stoichiometry of the two stage preparation of a typical PAC can be represented as follows:

$$2Al(OH)_3 + 2.5HCl + 1.75H_2SO_4 = Al_2Cl_{2.5}(SO_4)_{1.75} + 6H_2O \qquad (2)$$

$$Al_2Cl_{2.5}(SO_4)_{1.75} + 1.5Ca(OH)_2 = Al_2(OH)_3Cl_{2.5}(SO_4)_{0.25} + 1.5CaSO_4 \qquad (3)$$

In principle, products similar to PAC could also be prepared through the reaction depicted by Eq. (1) using other strong acids, for example H_2SO_4, to yield a basic polyaluminium sulphate (PAS). Several practical routes for the synthesis of basic aluminium sulphates have been proposed involving, either the reaction of alum ($Al_2(SO_4)_3$) with an alkali [23–25], or the reaction of a freshly prepared $Al(OH)_3$ gel with sulphuric acid or alum [26–27]. The stoichiometry of these reactions is illustrated in Eqs. (4) and (5) for products having 50% basicity.

$$2Al_2(SO_4)_3 + 6NaOH = Al_4(OH)_6(SO_4)_3 + 3Na_2SO_4 \qquad (4)$$

$$Al_2(SO_4)_3 + 6NaOH = 2Al(OH)_3 + 3Na_2SO_4 \qquad (5\,a)$$

$$2Al(OH)_3 + Al_2(SO_4)_3 = Al_4(OH)_6(SO_4)_3 \qquad (5\,b)$$

Preparations through the direct alum-alkali reaction (Eq. (4)) are best achieved using weak bases, such as Na_2CO_3 or $Ca(OH)_2$ [28, 29], in order to minimize local alkalinity in the reaction mixture and excessive product hydrolysis (leading to insoluble compounds). In this scheme, substantial levels of inert by-products (eg. Na_2SO_4, $CaSO_4$) are inevitably produced, the solubility of which depends on the alkali used. The alternate reaction scheme depicted by Eqs. (5 a,b) enables the elimination of soluble by-products through washing of the freshly prepared $Al(OH)_3$ gel; the latter operation obviously increases the overall complexity of the process.

Various potential advantages may be anticipated from the use of polyaluminium salts as flocculants due to the intrinsic properties of these compounds. First, because of their low acidity (typically, 50% neutralized), the decrease in pH following a given Al_2O_3 dosage of flocculant salts, and the required pH corrections before and/or after water treatment, will be greatly reduced for polyaluminium salts. Second, although the total number of available positive charges of the flocculant is decreased following partial neutralization, the presence of polymeric units bearing high positive charge leads to stronger electrostatic forces between these polymers and negatively charged species (dissolved or colloidal). Third, the pre-hydrolysed metal species present in basic-aluminium salts may achieve complete hydrolysis faster than the initial Al^{3+} species; this would facilitate low temperature hydrolysis and improve the coagulation/flocculation behaviour in these conditions. Finally, the high molecular weight polymeric species could, temporarily, act as a flocculant aid during the flocculation process, improving floc growth and floc resistance to shear.

Most of these anticipated benefits have been confirmed in laboratory or field trials of polyaluminium salts in water treatment [30–32]. Recent testing and applications of these products have also shown other beneficial features such as low residual Al [33]. On the other hand, shortcomings of polyaluminium products have also become apparent, for example, the difficulty in drying (and reconstituting) these products. Also, polyaluminium salts frequently exhibit a limited shelf life, particularly the sulphate-based products. Moreover, substitution of the classical Fe or Al salt flocculants by the polyaluminium products usually lead to an increase in the cost of the flocculation chemicals (although the actual unit cost may be reduced if handling costs and throughput are taken into account).

To achieve broader acceptance, a polyaluminium product must meet a number of practical requirements. It should exhibit excellent coagulation/flocculation behaviour, to minimize the required dosages. Its preparation should be relatively straightforward, avoiding filtration of a gel or of solid by-products; in addition, the products should be stable for six months or more.

The search for products meeting these criteria has led to the development of a new product: polyaluminium silicate-sulphate (PASS[R]). As the name implies, the product is a basic polyaluminium compound with silicate and sulphate anions [34]. Preparation of this product required new approaches with respect to both the chemicals used and the process.

Synthesis of Poly-Aluminium Silicate Sulphate (Pass[R])

The process used for the synthesis of PASS[R] was developed for the preparation of polyaluminium sulphate (PAS) products [35]. It involves the direct reaction of acid and alkaline solutions of aluminium salts, for example alum and sodium aluminate, under intense mechanical shear (typically 10^5 s^{-1}), yielding the basic aluminium salt and a limited amount of Na_2SO_4. The reaction is illustrated in Eq. (6), neglecting the excess NaOH (approx., 20 mole %) usually present in sodium aluminate.

$$3Na_2Al_2O_4 + 5Al_2(SO_4)_3 + 12H_2O = 4Al_4(OH)_6(SO_4)_3 + 3Na_2SO_4 \qquad (6)$$

In this reaction scheme, the concentrated sodium aluminate solution (typically 10 % Al_2O_3) is introduced slowly into the concentrated alum solution (typically 8 % Al_2O_3). The intense shear prevents the growth of $Al(OH)_3$ gel particles as the aluminate is taken through the neutral pH zone; instead, the gel obtained under these conditions reacts with the alum to form the basic aluminium product. The clarification and equilibration of the resulting milky suspension is obtained by mild post-reaction heating (eg. 50 °C) for one or two hours. The product is a clear solution containing 8–10 % Al_2O_3 (depending on the concentration of the reaction solutions used).

The advantages of this process are readily apparent. First, as can be seen by comparing Eqs. (4) and (6), the use of sodium aluminate as the alkali greatly reduces the level of Na_2SO_4 in the final product. Second, since all reactants and products are liquid solutions, the handling operations are reduced to a miniumum. Third, even for industrial size preparations, the overall reaction scheme is completed in a few hours. With respect to stability towards hydrolysis and precipitation, the PAS products prepared through this route were found comparable to the most stable compounds amongst those prepared via other routes.

In efforts to improve the stability and, as well, the performance of these compounds, a number of PAS-type products were prepared through the process described above, while partially substituting Al, SO_4, or both, for other cationic or anionic species. It was found that incorporating a low level of silicate anions into basic polyaluminium salts was beneficial with respect to their stability and performance as flocculants [35]. The resulting product, a polyaluminium silicate sulphate (PASS$^{(R)}$), is obtained according to the overall reaction depicted in Eq. (7) (50 % basicity, again neglecting the excess alkalinity in the aluminate).

$$5Al_2(SO_4)_3 + 3Na_2Al_2O_4 + 0.80(SiO_2 \cdot 0.3Na_2O) + 12H_2O =$$
$$4Al_4(OH)_6(SO_4)_{2.94}(SiO_{2.31})_{0.2} + 3.06Na_2SO_4 \tag{7}$$

It may be recalled that the combined use of aluminium and silicate salts is not new [36], but the combination of these chemicals to form soluble and stable polymeric species in concentrated solutions had not been reported before. The solubility of silica in acidic solutions (final – pH 3.0–3.5) is very low, and hence, the silicate levels which can be incorporated into polyaluminium salts, as well as the mode of addition, are critical. For performance and long-term stability, a ratio of Si/Al near 0.05 was found adequate, when sodium silicate was added into the alum before the high shear mixing operation.

PASS$^{(R)}$ Physico-Chemical Characterization

The bulk physical properties which characterise the PASS$^{(R)}$ concentrated solutions having a basicity of 50.0 % are summarised below.

Appearance:	clear liquid
Al_2O_3 content:	8.0–10.0 wt %
Sodium sulphate:	5.5–6.6 wt %
pH:	3.0–3.5
Basicity:	50.0 %
Density (25°):	1.25–1.34
Viscosity (25):	11–17 cP

The average chemical composition of the species contained in these solutions can be represented as:

$$Al_a(OH)_b(SO_4)_c(SiO_x)_d$$

For a 50 % basic product with Si/Al = 0.05, the subscripts, when referring to $a = 1$ are $b = 1.5$, $c = 0.735$, $d = 0.05$ and $x = 2.31$. The actual speciation is, however, expected to be quite complex as indicated from earlier studies on various basic polyaluminium compounds [25, 30, 37].

To further characterise the species present in PASS$^{(R)}$ concentrates, various approaches were explored namely ultrafiltration techniques to estimate molecular weight distributions, light scattering to determine apparent average diameters of the polymeric species, and elecrophoresis to monitor the electrostatic charges developed on the floc formed upon hydrolysis. Typical results are reported below, showing the size or molecular weight distributions of the polymeric species present in PASS[R].

Apparent Size/Weight Distribution of Species in PASS[R]

The distribution of apparent polymer size (or molecular weight) in PASS[R] solutions, can be estimated by quantitative ultrafiltration through a series of membranes of decreasing pore sizes. Using a series of membranes with decreasing molecular weight cut-off of 300, 100, 30, 10, and 1 (10^3 Daltons), apparent molecular weight distributions have been obtained as described elsewhere [38] yielding typical results illustrated in Fig. 1. Such distributions can only be taken as indication of the actual species present in the PASS[R] solution. Indeed, size-exclusion filtration of concentrated solutions of highly charged polymers may be affected by membrane polarization and osmatic pressure effects. Also, the PASS[R] polymers being labile (as opposed to co-valent polymers) partial hydrolysis or breakdown may occur during filtration. Hence, the true molecular weight of the species in PASS[R] may be quite different from those specified from the membrane pore size. With proper care, however, the shapes of the distribution, and the aluminium mass balance can be satisfactorily reproduced; as shown by the data reported in Fig. 1 for two different preparations of PASS[R] (8.0 % Al_2O_3) carried out several months apart in different industrial facilities.

As seen from Fig. 1, the apparent size distribution of the aluminium species in PASS[R] shows a substantial level of low molecular weight species; typically, between 30–50 % of the Al present is found within species having MW < 1 000. It is also generally found that another important part of the Al (eg. 15–40 %) is contained within very large species, with MW > 100 000. The Al distribution in the intermediate size species varies significantly (depending on composition, reaction conditions, aging, concentration, etc.), but it is never uniform; there are usually two sharply distinct groups of polymers. A more systematic study of these distributions, and of the influence of various parameters on PASS[R] speciation, will be reported later [38].

Selected Applications of PASS[R]

The polyaluminium silicate sulphate described above has been made commercially available under the trade name PASS[R] and it is being increasingly used in potable water treatment in Europe, North America and Asia. Compared to the conventional flocculants alum or ferric chloride, PASS[R] has shown advantages in, for example, minimum required dosages, pH corrections, low

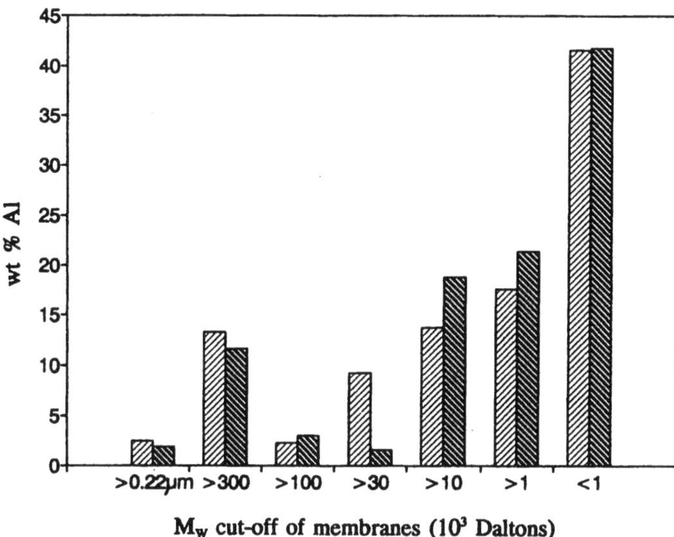

Fig. 1. Distribution of species in PASS[R] (8.0 wt %) according to apparent molecular weight from ultrafiltration experiments. The two data sets characterize preparations several months apart

temperature performance, removal of organic colour, filtration properties and level of coagulant residuals in the treated waters. In most cases, subtle differences were found in the behaviour and mechanistic aspect of PASS[R] and, eg. alum coagulants. Hence, the full benefits achievable from the use of PASS[R] are best evidenced through both laboratory and full scale testing. We report below the results of jar test and field studies aimed at using PASS[R] to reduce the level of residual Fe in treated waters.

Iron Residuals Problem (a Jar Test Exercise)

This works treats a highly coloured, low turbidity soft water which is a blend of a number of reservoir sources. The works has a maximum treatment capacity of 150 Ml/day and this is equally split between sedimentation and dissolved air flotation processes. The two streams of water emerging from these processes are then blended and passed onto rapid gravity filters. Finally, before being pumped into the distribution system, the water is disinfected and pH adjusted to about 9.

The coagulant used at the works at present is ferric sulphate; for the sedimentation half of the works the flocculation process is aided by the use of a polymer extracted from seaweed. Although the removal of colour and turbidity is achieved by these chemicals and the processes mentioned above, the works has difficulty in keeping the iron residuals below the maximum admissable concentration (MAC) of 200 μg/l. With this in mind, and to determine how PASS[R] performs under air flotation conditions, a number of trials employing sedimentation and air flotation jar testers were carried out.

Experimental

The following chemicals were used:

PASS[R] liqiud, 8.0 % Al_2O_3
Na_2CO_3, 5.57×10^{-2} M
Ferric sulphate, 40 % (as supplied by the Water Company)

Jar Test Methods

a) Settlement. 0.5 litre samples of raw water were placed in tall form beakers and mounted on a multiple stirrer unit. The required dosage of PASS[R] and sodium carbonate was then flash mixed with the raw water which was then stirred slowly. The sizes of the resulting flocs were assessed at regular intervals by comparison against a chart. After twenty minutes, stirring was stopped and the flocs were left to settle over a fifteen minute period. The supernatant liquids were then filtered through Whatman GF/C microfibre filters (1.2 μm) and the following measurements were made on the filtrate:

a) Turbidity, using a Hach turbidimeter
b) Colour, using a Lovibond comparator
c) pH, using a Jenway 3020 pH meter
d) Residual aluminium and silica using Palintest chemicals and colourimeter.

b) Air Flotation. 1 litre samples of raw water were placed in tall form beakers and attached to the AZTEC flotation jar test unit. The required dosage of PASS[R], sodium carbonate or freshly prepared ferric sulphate solution was then added to the raw water which was being stirred at high speed.

After half a minute of high speed mixing the water was then stirred slowly and the sizes of the flocs formed assessed at regular intervals by comparison against a chart. Stirring was stopped after twenty minutes and a known volume of air saturated recycled water was injected into the bottom of the beakers. After eight minutes a sample of the treated water was taken from the bottom of the jar and its turbidity measured. The sample was then filtered through a Whatman GF/C microfibre filter. In addition to measuring the pH, turbidity, colour, aluminium and silica residuals the iron residuals were also measured using the spectrophotometer and method employed at the works.

The raw water had the following characteristics at the time of the trial:

pH	6.62	Colour [°Hazen]	175
Hardness [mg/l $CaCO_3$]	< 15	Natural aluminium [μg/l]	30
Alkalinity [mg/l $CaCO_3$]	< 5	Natural iron [μg/l]	185
Turbidity [ntu]	3.4	Natural silica [mg/l]	0.46

The alkalinity and pH of the raw water was unusually low due to maintenance work being carried out on one of the source reservoirs; this requiring total drainage. To overcome this short term problem, alkalinity was being added to the raw water, the amount dependent on the method of treatment – sedimentation or air flotation. The adjusted pH's were:

Settlement   7.4
Air flotation   8.85

For the initial settlement jar test the raw, pH unadjusted water was used, the results obtained being given in Table 1. As can be seen only dosages of PASS[R] above 3 mg Al/l flocced but at these dosages the flocs were fast forming.

Tab. 1. Settlement jar test using raw pH unadjusted water

Raw water: temperature = 9.4°C; pH = 6.62						
PASS[R] [mg Al/l]	1	2	3	4	5	6
Floc size on slow stirring after:						
5 min	–	–	–	C	C/D	D
10 min	–	–	–	D/E	D/E	E
15 min	–	–	–	D/E	D/E	D
20 min	–	–	–	D/E	D/E	D
pH	5.95	5.64	5.46	5.03	4.58	4.35
Filtered turbidity [ntu]	–	–	1.7	0.2	–	–
Colour [°Hazen]	–	–	40/30	< 5	< 5	< 5
Residual Al [μg/l]	–	–	> 500	220	> 500	> 500
Residual silica [mg SiO_2/l]	–	–	0.24	0.26	–	–

Floc size [mm]: A = 0.3–0.5; B = 0.5–0.75; C = 0.75–1.0; D = 1.0–1.5; E = 1.5–2.25

For the 4 mg Al/l dosage level good colour and turbidity removal was effected. However, due to the low resultant pH of 5.03 the aluminium residual was found to be high, being 220 μg Al/l, but still only just above the MAC of 200 μg Al/l. In order to see if this residual level could be lowered and also to more closely examine the dosage levels between 3 and 4 mg Al/l a further settlement jar test was carried out this time using the works settlement pH adjusted water ie. pH 7.4. The results for this jar test are given in Table 2.

As can be seen from Table 2 the 4 mg Al/l dosage level appears to be the lowest at which PASS[R] would floc under the conditions of the trial. At this dosage level although floc sizes after 5 minutes slow stirring were smaller than had been found previously at this stage, comparable sizes were achieved after 15 minutes. As before this yielded a filtered water which had a low associated colour and turbidity. The benefit of having adjusted the pH of the raw water from 6.62 to 7.4 minifested itself in the aluminium residual levels. Thus due to

Tab. 2. Settlement jar test pH adjusted raw water

Raw water: temperature = 9.4°C; pH = 7.4				
PASS$^{(R)}$ [mg Al/l]	3.1	3.4	3.7	4.0
Floc size on slow stirring after:				
5 min	–	–	–	A/B
10 min	–	–	–	D
15 min	–	–	–	D/E
20 min	–	–	–	D/E
pH	6.73	6.67	6.51	6.44
Filtered turbidity [ntu]	3.5	1.7	1.1	0.43
Colour [°Hazen]	> 70	40/30	15/10	< 5
Residual Al [μg/l]	> 500	320	290	60
Residual silica [mg SiO$_2$/l]	0.54	0.08	0.22	0.08

Floc size [mm]: A = 0.3–0.5; B = 0.5–0.75; C = 0.75–1.0; D = 1.0–1.5; E = 1.5–2.25

the post coagulation pH being 1.4 units higher at 6.44 the aluminium residual was found to be 60 rather than 220 μg Al/l as had been the case previously. This new figure is well within the MAC of 200 μg Al/l.

Using the information gained from the settlement jar tests, particularly with respect to the dosing level, a number of air flotation trials were carried out. The first series of these (Table 3) involved variation of the slow stirring speed and the % of air saturated recycle water used to effect flotation. From the results it can be seen that smaller floc sizes are obtained on increasing the slow stirring speed. However, this does not seem to affect the flotation process as demonstrated by the post flotation turbidity for samples using the same recycle volumes.

The results obtained for different recycle volumes seemed to depend on the effect the slow stirring speed had on the size and shear strength of the flocs. For speeds of 30 and 50 rpm, an 8 % recycle, from filtered turbidity and residual aluminium levels seemed to be better than for a 10 % recycle. Comparable results however were found for the slow stirring speed of 40 rpm for both recycle volumes.

The next series of trials (Table 4) involved keeping everything bar the pH of the untreated water constant. This was achieved by varying the quantities of sodium carbonate added to the raw water. From the results obtained previously, a slow mixing speed of 30 rpm was employed whilst the recycle volume of 10 % was chosen as this was equal to that used on the works and would thus enable a direct comparison with ferric sulphate to be made.

As can be seen from Table 4, similar results were obtained for the four pH levels. Thus good colour, post flotation and filtration turbidites were achieved

Tab. 3. Air flotation jar tests involving variation of slow stirring speed and % recycled water used

Raw water: temperature = 9.4°C; pH = 7.4						
PASS$^{(R)}$ dose [mg Al/l]	4	4	4	4	4	4
Fast mix speed [rpm]	400	400	400	400	400	400
Fast mix time [min]	1/2	1/2	1/2	1/2	1/2	1/2
Slow mix speed [rpm]	30	30	40	40	50	50
Floc size on slow mixing after:						
5 min	B/C	B/C	B/C	B/C	B	B
10 min	C/D	C/D	B/C	B/C	B/C	B/C
15 min	C/D	C/D	B/C	B/C	B/C	B/C
20 min	D	D	B/C	B/C	B/C	B/C
Recycle [%]	10	8	10	8	10	8
Turbidity after: flotation [ntu]	2.7	2.4	2.1	3.0	2.9	2.6
filtration [ntu]	0.64	0.5	0.46	0.53	0.59	0.46
pH	6.60	6.62	6.59	6.62	6.59	6.65
Colour [°Hazen]	< 5	< 5	< 5	< 5	< 5	< 5
Residual Al [µg/l]	140	30	100	90	170	60
Residual silica [mg SiO$_2$/l]	0.38	0.12	0.26	0.10	0.14	0.16

Floc size [mm]: A = 0.3–0.5; B = 0.5–0.75; C = 0.75–1.0; D = 1.0–1.5; E = 1.5–2.25

as well as aluminium residual levels well below the MAC. In addition the iron content of the raw water was lowered considerably from 185 to about 30 µg/l which is well below the MAC of 200 µg/l.

The final air flotation jar test carried out was one from which a rough comparison of PASS$^{(R)}$ and ferric sulphate could be made. For PASS$^{(R)}$ the pH of the raw water was adjusted to 7.2 whilst for ferric sulphate the works untreated water, pH adjusted to 8.85 was used. A dosage of 40 ppm of ferric sulphate was used based on a jar test carried out on the day. The results of the comparative jar test are given in Table 5.

In this experiment both chemicals were found to floc quickly but ferric sulphate yielded larger flocs over the period of slow stirring. After flotation the turbidities were found to be comparable. However, after filtration a lower turbidity reading was obtained for ferric sulphate 0.52 compared to 0.68 though this was higher than for the first instance PASS$^{(R)}$ was run under these conditions when a value of 0.46 was recorded. Colour removal for both coagulants was found to be very good.

As expected due to its higher acidity ferric sulphate was found to have lowered the pH by about 4 pH units, from 8.85 to 4.93. This compares with PASS$^{(R)}$ where the reduction was only 1.2 pH units, from 7.2 to 6.0.

Residual aluminium levels were found to be very good for both PASS$^{(R)}$ (30 µg Al/l) and ferric sulphate (20 µg Al/l) and well within the MAC of 200

Tab. 4. Variation of raw water pH

Raw water: temperature = 8.8°C; pH = 6.62			
PASS[R] dose [mg Al/l]	4	4	4	4
Na_2CO_3 dose [mg Na_2CO_3/l]	2.9	3.8	4.4	5.0
pH	7.21	7.73	8.14	8.24
Fast mix speed [rpm]	400	400	400	400
Fast mix time [min]	1/2	1/2	1/2	1/2
Slow mix speed [rpm]	30	30	30	30
Floc size on slow mixing after:				
5 min	C	C	C	C
10 min	C/D	C/D	C/D	C/D
15 min	C/D	C/D	C/D	C/D
20 min	C/D	C/D	C/D	C/D
Recycle [%]	10	10	10	10
Turbidity after: flotation [ntu]	2.5	2.5	2.5	2.5
filtration [ntu]	0.46	0.60	0.45	0.68
pH	5.93	6.20	6.32	6.20
Colour [°Hazen]	< 5	< 5	< 5	< 5
Residual Al [μg/l]	60	110	70	100
Residual Fe [μg/l]	31	34	33	64
Residual silica [mg SiO_2/l]	0.18	0.48	0.32	0.46

Floc size [mm]: A = 0.3–0.5; B = 0.5–0.75; C = 0.75–1.0; D = 1.0–1.5;
E = 1.5–2.25

μg Al/l. The iron residuals for the two coagulants were very different. Whereas PASS[R] had reduced the natural iron content of the raw water from 185 down to 28 μg Fe/l ferric sulphate had actually increased it to 252 μg Fe/l. This later figure being greater than the MAC of 200 μg Fe/l.

A number of conclusions were drawn from this exercise.

PASS[R] produced flocs which responded well to the air flotation technique yielding low pre-filter turbidities. Dependent on the associated slow stirring speed the results indicated that a recycle volume less than 10 % could be employed with PASS[R] usage by virtue of the excellent filtered water obtained. PASS[R] was found to greatly reduce the iron content of the raw water on treatment from 185 down to approximately 30 μg Fe/l. This is obviously well within the MAC for iron of 200 μg Fe/l. Ferric sulphate treatment on the other hand led to iron residuals greater than the MAC. Aluminium residuals for PASS[R] were found to be well below the MAC of 200 μg Al/l and for optimum conditions readings in the range 30–60 μg Al/l were found. PASS[R] was found to require a lower alkalinity adjustment prior to coagulation compared to ferric sulphate and did not lower the pH as much. Thus for both pre and post coagulation relatively large cost savings on pH adjustment could be achieved.

Tab. 5. Comparison of PASS vs ferric sulphate

Raw water: temperature [°C]	8.8	8.8
pH	6.62	8.85
PASS[R] dose [mg Al/l]	4	–
Ferric sulphate dose [ppm]	–	40
Na_2CO_3 dose [mg Na_2CO_3/l]	2.9	–
pH	7.22	8.85
Fast mix speed [rpm]	400	400
Fast mix time [min]	1/2	1/2
Slow mix speed [rpm]	30	30
Floc size on slow mixing after:		
5 min	C	C
10 min	C/D	D
15 min	C/D	D/E
20 min	C/D	D/E
Recycle [%]	10	10
Turbidity after: flotation [ntu]	3.5	3.6
filtration [ntu]	0.68	0.52
pH	5.99	4.93
Colour [°Hazen]	< 5	< 5
Residual Al [μg/l]	30	20
Residual Fe [μg/l]	28	252
Residual silica [mg SiO_2/l]	0.24	0.08

Floc size [mm]: A = 0.3–0.5; B = 0.5–0.75; C = 0.75–1.0;
D = 1.0–1.5; E = 1.5–2.25

Iron Residuals Problem (a Plant Trial)

The raw water, abstracted from two lakes, is quite acidic, the pH varying between 4 and 5.5, and has an associated colour which is normally in the range 20° to 60° Hazen.

The pre-existing treatment process involved initially dosing the raw water in quick succession with the coagulant (ferric sulphate), lime and a chlorine dose. This is followed by a static baffle type in-line mixer after which there was a contact time of 1–1.5 minutes before the pressure filters depending on the flow through the works. Finally after being passed through the pressure filters, the water was dosed with chlorine again and a lime dose sufficient to achieve a final pH of 10.2 and a distribution pH of 8.2–8.5.

The treatment process at the works and the distribution of the treated water is driven by gravity due to the source lakes being at about 1 000 feet above sea level. Between the works and the first service reservoir in the distribution system there are a number of households, thus any problems experienced by the plant unavoidably result in poor quality water being supplied to these households.

The main problem experienced by this works stems from peak demand, mainly during the summer months, exceeding the output from the new works which has a maximum of approximately 6.6 Ml/day or 76 l/s. At such times five filters from the old works have to be used to make up the shortfall. The supply to these filters is taken after the inline mixer. It is estimated that the contact time between dosing and the water reaching the old filters is approximately 4.5 minutes which is obviously longer than that for the new works. During this period of high output (up to 7.4 Ml/day) high iron residuals in the final water, much greater than the maximum admissable concentration (MAC) level of 200 μg Fe/l, are found partly due to employing the old works but also on occasion due to breakthrough on the new works filters. Although this works is to be upgraded in the next few years a solution to the high iron residuals during peak flow is required in the interim and thus the water company, following a successful jar test trial decided to carry out a full scale plant trial using PASS[R] instead of ferric sulphate as the coagulant.

The trial using PASS[R] commenced on the 3rd of September 1991 at which time both the old and the new works were in operation. Over the following few days the filters became conditioned to the new coagulant and its associated operating pH. During, and for a short time after, this period the whole coagulation process was optimised so that the best possible results could be obtained.

On the 9th of September the raw water being treated had the following characteristics:

pH	5.0
Hardness [mg CaCO$_3$/l]	< 5
Alkalinity [mg CaCO$_3$/l]	< 5
Turbidity [ntu]	1.6
Colour [°Hazen]	44
Natural aluminium [μg Al/l]	130
Natural iron [μg Fe/l]	140

As can be seen the geology of the catchment area leads to quite high natural levels of both aluminium and iron in the raw water treated by the works. Upon PASS[R] treatment and just before the old filters received their twice daily wash the final water had the following characteristics:

Turbidity [ntu]	0.1–0.18
Colour [°Hazen]	5
Aluminium [μg Al/l]	180*
Iron [μg Fe/l]	< 10

*Aluminium residual level after the old filters were washed was found to be 100 μg Al/l

Thus the treatment process had effectively removed the natural iron and at potentially the worst time to take an aluminium residual reading a value less

than the MAC level of 200 μg Al/l was obtained. After washing the old filters
the aluminium residual level was found to be 100 μg Al/l which is lower than
the natural level in the raw water of 130 μg Al/l.

By the 19th of September it was no longer found necessary to run the old
works and it was thus switched off. Having allowed the works to adjust to this
change the following readings on the final water were taken on the morning of
the 20th of September.

Turbidity [ntu]	0.08–0.12
Colour [°Hazen]	2.5/5
Aluminium [μg Al/l]	30
Iron [μg Fe/l]	10

As before the natural iron (150 μg Fe/l) had been removed but the improvement in the colour, turbidity and aluminium residuals was very marked
in comparison to the period in which the old filters were being employed. The
aluminium residuals at this point were even below the guidelines of 50 μg Al/l
and much lower than the natural aluminium present in the raw water at the
time of 90 μg Al/l.

The conclusions drawn from the exercise were that during the period when
the old filters were being used PASS$^{(R)}$ was able to yield a final water with low
colour and turbidity as well as an aluminium residual below the MAC level of
200 μg Al/l and only slightly higher on average than the natural level in the
raw water.

Furthermore, when the old filters were not in use the quality of the final
water increased dramatically. This was particularly the case for the aluminium
residuals which dropped down to 30 μg Al/l. At this level they were within the
guideline level of 50 and much less than the natural aluminium level in the raw
water at the time of approximately 90 μg Al/l.

References

[1] Matijević, E., Teǎk, B.: Detection of Polynuclear Complex Aluminium Ions. J.
    Phys. Chem. *57* (1953) 951
[2] Matijević, E., Mathai, K.G., Ottewill, R.H., Kerker, M.: Detection of Metal Ion
    Hydrolysis by Coagulation – III Aluminium. J. Phys. Chem. *65* (1961) 826
[3] Matijević, E., Janauer, G.E., Kerker, M.: Reversal of Charge of Lyophobic Col-
    loids by Hydrolysed Metal Ions – I Aluminium. J. Coll. Interf. Sci. *19* (1964)
    333
[4] Matijević, E., Stryker, L.J.: Coagulation and Reversal of Charge of Lyophobic
    Colloids by Hydrolysed Metal Ions – III Aluminium Sulphate. J. Coll. Interf. Sci.
    22 (1966) 68
[5] Matijević, E., Allen, L.H.: Interactions of Colloidal Dispersions with Electrolytes.
    Env. Sci. Tech. *3* (1969) 264
[6] Teot, A.S., Daniels, S.L.: Flocculation of Negatively Charged Colloids by Inor-
    ganic Cations and Anionic Polyelectrolytes. Env. Sci. Tech. *3* (1969) 825

[7] Packham, R.F.: The Theory of Coagulation Process of Water Treatment - A Review. Wat. Res. Ass. Tech. Publ. No. 12, 1960

[8] Stumm, W., Morgan, J.W.: Chemical Aspects of Coagulation: J. Amer. Waterworks Ass. *54* (1962) 971

[9] Black, A.P.: Elektrokinetic Characteristics of Hydrous Oxides of Iron and Aluminium. In: Principles and Applications of Water Chemistry, S.D. Faust and J.V. Hunter (eds.). Wiley and Sons, New York 1967

[10] Stumm, W., O'Melia, C.R.: Stoichiometry of Coagulation. J. Amer. Waterworks Ass. *60* (1968) 514

[11] La Mer, V.K., Smellie, R.H.: Flocculation, Subsidence and Filtration of Phosphate Slimes. J. Coll. Sci. *11* (1956) 704, 710 and 720; *12* (1957) 230 and 566; *13* (1958) 589

[12] La Mer, V.K., Healy, T.W.: Adsorption – Flocculation Reactions of Macromolecules at the Solid-Liquid Interface. Rev. Pure and App. Chem. *13* (1963) 112

[13] Ruehrwein, R.A., Ward, D.W.: Clay Aggregation by Polyelectrolytes. Soil Sci. *73* (1952) 485

[14] Michaels, A.S.: Aggregation of Suspensions by Polyelectrolytes. Ind. and Eng. Chem. *46* (1954) 1485

[15] Gregory, J., Sheiham, I.: Kinetic Aspects of Flocculation of a Polystyrene Latex by Cationic Polymers. Brit. Polym. J. *6* (1974) 47

[16] Harwood, J.H., Kerr, J., Crundall, S.F. (P. Spence and Sons Ltd): Improvements in the Flocculation of Sewage Suspensions and Sludges. Brit. Patent 630,631 (1949)

[17] Llewellyn, W.B., Odgers, J. (P. Spence and Sons Ltd): Aqueous Solutions of Aluminium Chloride. Brit. Patent 655,617 (1951)

[18] Harwood, J.H., Kerr, J., Crundall, S.F. (P. Spence and Sons Ltd): Improvements in Coagulation and Sludge Conditioning. Brit. Patent 777,041 (1957)

[19] Murray, A., Gleeson, W. (P Spence and Sons Ltd).: Purification and Clarification of Aqueous Liquids. Brit. Patent 823,082 (1959)

[20] Taki Fertilizer Manufacturing Co. Ltd.: Coagulant for Treating Aqueous Media. Brit. Patent 1,222,561 (1971)

[21] Kanamasa, T., Horoshi, S. (Central Glass Co. Ltd.).: Method of Manufacture of Basic Aluminium Chloride Salt. Jap. Patent 73 50,998 (1973)

[22] Smith, F. (The Lord Mayor, Aldermen and Citizens of the City of Norwich): Water Treatment Process and Apparatus. Brit. Patent 1,399,598 (1975)

[23] Hodgson, C.: Chloride and Sulfate Polymer Production. US Patent 4 069 299 (1978)

[24] Nilsson, R.: Aluminium Sulfate Composition Containing Polynuclear Complexes and Uses Thereof. Can. Patent 1 123 306 (1982)

[25] Men, X., Dempsey, B.A., Pile, J.: Production, Characterization and Application of Polyaluminium Sulfate. Proc. Annual. Conf. AWWA, Orlando, Florida, 1988

[26] Gunnarsson, K.L., Nilsson, R.O.: Polynucleate Aluminium Hydroxide Sulfate Complexes Preparation by Treatment of Amorphous Aluminium Hydroxide with Sulfate. Can. Patent 1 203 364 (1986)

[27] Gunnarsson, K.L., Nilsson, R.O.: Basic Aluminium Sulfate Preparation by Addition of Alkaline Earth (hydr)oxide to Aluminium Sulfate. Can. Patent 1 203 655 (1986)

[28] Lindahl, G.M.: Stable Solutions of Basic Aluminium Sulfate Containing Polynucleate Aluminium Hydroxide Sulfate. US Patent 4 536 384 (1985)

[29] Boutin, J., Combet, A.: Basic Aluminium Sulfate Preparation by Reacting Aluminium Sulfate with Alkali Metal or Ammonium (bi)carbonate and Magnesium (hydr)oxide or Carbonate. French Patent 2 563 513 (1985)

[30] Jolicœur, C., Haase, D.: Les aluns basiques dans le traitement physicochimique de l'eau: survol de leurs propriétés et évolution récente. Sci. et Tech. del'eau *22* (1989) 31

[31] Tardat-Henry, M.: Évolution des dérivés de l'aluminium utilisés comme agents coagulants. Sci. et Tech. de l'eau *22* (1989) 297

[32] Mazet, M., Angbo, L., Serpaud, B.: Adsorption des substances humiques sur flocs d'hydroxyde d'aluminium préformés. Wat. Res. *24* (1990) 1509

[33] Simpson, A.M., Hatton, W., Brockbank, M.: Aluminium, its Use and Control in Potable Water. Env. Tech. Letters *9* (1988) 907

[34] Haase, D., Spiratos, N., Jolicœur, C.: Polymeric Basic Aluminium Silicate-Sulphate. US Patent 4 981 675 (1991)

[35] Haase, D., Spiratos, N.: Method for Producing Aqueous Solutions of Basic Aluminium Sulphate. US Patent 4 877 597 (1989)

[36] Smith, F.: Effect of Ageing on Aluminium Hydroxide Complexes in Dilute Solutions. Brit. Patent 1 399 598 (1975)

[37] Van Benschoten, J.E., Edzwald, J.K.: Chemical Aspects of Coagulation Using Aluminium Salts – I. Hydrolytic Reactions of Alum and Polyaluminium Chloride. Wat. Res. *24* (1990) 1519; II. Coagulation of fulvic acid using alum and aluminium chloride. Wat. Res. *24* (1990) 1527

[38] Jolicœur, C. (in preparation)

A.K. Arnold-Smith
R.M. Christie
Alcan Chemicals Limited
Chalfont Park
Gerrards Cross
Bucks, SL9 0QB
United Kingdom

C. Jolicœur
Université de Sherbrooke
Faculté des Sciences
Départment de Chimie
Sherbrooke
Québéc J1K 2R1
Canada

Drinking Water Treatment

Dissolved-Air Flotation:
Theory and Practice

G.J. Schers and J.C. van Dijk

Abstract

This paper* presents a physical-mathematical model for dissolved-air flotation (DAF). The flotation model is an important aid for gaining insight into the effect of design parameters such as bubble diameter, particle size, bubble-volume concentration, contact time, temperature and surface load. The results with the flotation model were compared with the treatment results of existing Dutch flotation plants. The model distinguishes two separate mechanisms: the contact or filtration process and the separation process. The contact process describes the collision between a bubble and a particle on the analogy of the filtration theory. The efficiency of the filtration zone depends mainly on the bubble-volume concentration, contact time and temperature. During the separation process the bubble-particle agglomerates move upward and form a floating layer on the surface of the water. The floating velocity determines mainly the efficiency of the separation zone and appears to be dependent on especially the particle size. Larger particles result in higher floating velocities, through which a high surface load is possible.

1. Introduction

Dissolved-Air Flotation (DAF) is a clarification process, which has been applied successfully in drinking water treatment in the Netherlands for more than 10 years. The process was used prior to that time in Scandinavia and England. As a result of the good results in these countries, the treatment process became interesting for the Dutch drinking water treatment plants. The first flotation plant in the Netherlands was Zevenbergen (1979). In the year 1991 the flotation process is used by five Dutch treatment plants and one Belgian. These are Zevenbergen (NV Waterleidingmaatschappij Noord-West-Brabant), Braakman (Delta Nutsbedrijven NV), Elsbeekweg Enschede (NV Waterleidingbedrijf Oost Twente), Lindenbergh Katwijk (NV Energie- en Watervoorziening Rijnland),

---

* This paper is based on the graduation thesis of G.J. Schers at the Department of Civil Engineering, Technical University Delft, titled "Flotation, Theory and Practice" [1]. The thesis was prepared under the supervision of Prof. ir. J.C. van Dijk

Scheveningen (NV Duinwaterbedrijf Zuid-Holland) and Notmeir (Antwerpse Waterwerken).

The flotation process has been characterized by an empirical approach. In the past, design criteria were determined based on the results of pilot plants or existing full-scale treatment plants. In this paper a flotation model is presented, which is based on the "Conceptual Model for DAF" developed by J.P. Malley and J.K. Edzwald [2–4]. The conceptual model describes the collision between a bubble and a particle in the filtration zone on the analogy of the well-known filtration theory [5–12]. This conceptual model for DAF has been extended by us with a model for the separation of the bubble-particle agglomerates which are formed in the separation zone [1]. The flotation model serves as an important aid for gaining insight into the effect of design and operating parameters. By using this flotation model it is possible to design a more functional flotation treatment plant than by using the historical empirical approach.

2. Principles

Flotation as a clarification process is used at several treatment plants which treat surface water (Fig. 1). The treatment plant prior to the flotation units resembles conventional treatment (i.e. rapid mixing with coagulant addition followed by flocculation).

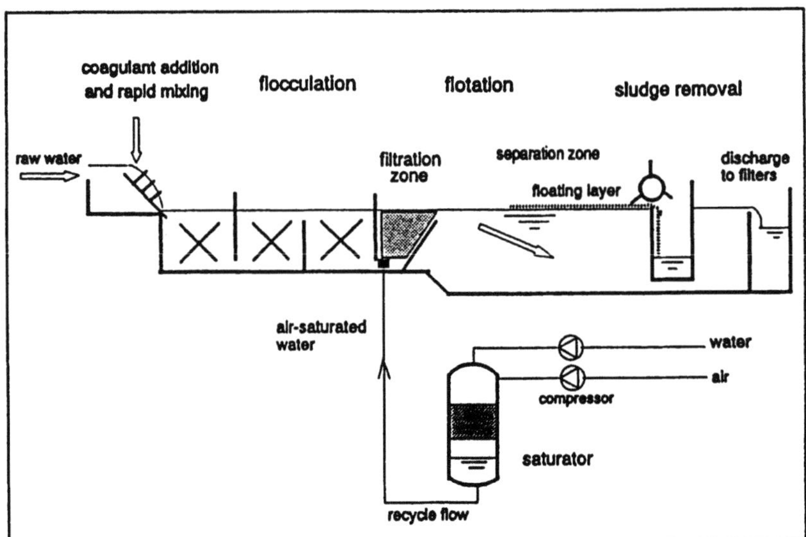

Fig. 1. Schematic diagram of a dissolved-air flotation plant

In flotation practice, the bubbles are introduced exclusively by dissolved-air flotation. The underlying principle is Henry's law: the solubility of air in water is directly proportional to the existing pressure. A process water flow (typically 5–10 % of the raw water flow) is saturated with air under high pressure (typically 500–800 kPa). Either clarified or filtered water can be used for the process water flow. This saturated (recycle) process water flow is introduced into the filtration zone by means of specially designed nozzles and consequently undergoes a pressure drop. This results in the formation of small bubbles with a diameter d_b between 10 and 100 μm. The mean bubble diameter d_{bm} is 30–40 μm for all types of nozzles. The bubble-volume concentration ϕ_b which is used is between 5 and 10 l/m^3. If all bubbles are identical, with a diameter d_b of 30 or 40 μm, the bubble-number concentration N_b (number of bubbles per m^3 total water flow [i.e. the raw water flow plus the recycle flow]) and the distance between the bubbles can be calculated (Table 1).

Tab. 1. Characteristics of the bubbles and the particles ($\rho_d = 1003$ kg/m^3)

Bubble-volume concentration, ϕ_b $\times 10^{-3}$ m^3/m^3	Mean diameter bubbles, d_{bm} μm	Bubble-number concentration, N_b m^{-3}	Distance between the bubbles μm
5	30	$3.5 \cdot 10^{11}$	111
5	40	$1.5 \cdot 10^{11}$	148
8	40	$2.4 \cdot 10^{11}$	121
10	40	$3.0 \cdot 10^{11}$	110
Particle concentration, c_d g/m^3	Mean diameter particles, d_{dm} μm	Particle-number concentration, N_d m^{-3}	Distance between the particles μm
10	100	$1.9 \cdot 10^7$	3 645
10	200	$2.4 \cdot 10^6$	7 290
25	200	$5.9 \cdot 10^6$	5 318

The particles in the surface water are of organic or inorganic origin and vary in size between 0.01 and 100 μm [13]. With proper pre-treatment (rapid mixing and coagulation) these particles grow in size up to 100–1000 μm. The particle density ρ_d is about 1003–1006 kg/m^3. At the existing flotation plants the particle concentration c_d of the surface water usually varies between 10 and 25 g/m^3. If all particles are identical and have the same diameter of 100 or 200 μm, the particle-number concentration N_d (number of particles per m^3 total water flow) and the distance between the particles can be calculated.

Table 1 shows that:

1. The bubble-number concentration N_b is much higher than the particle-number concentration N_d and

2. The distance between the bubbles is smaller than the particle diameter d_d.

This is illustrated in Fig. 2, in which calculations are made with a bubble-volume concentration ϕ_b of 8 l/m^3. This figure shows that a collision between a bubble and a particle is unavoidable. This conclusion will be explained by the following flotation model. The flotation model distinguishes two processes: the contact or filtration process in the filtration zone and the separation process in the separation zone (Fig. 2). The two zones are separated by a partition wall or baffle.

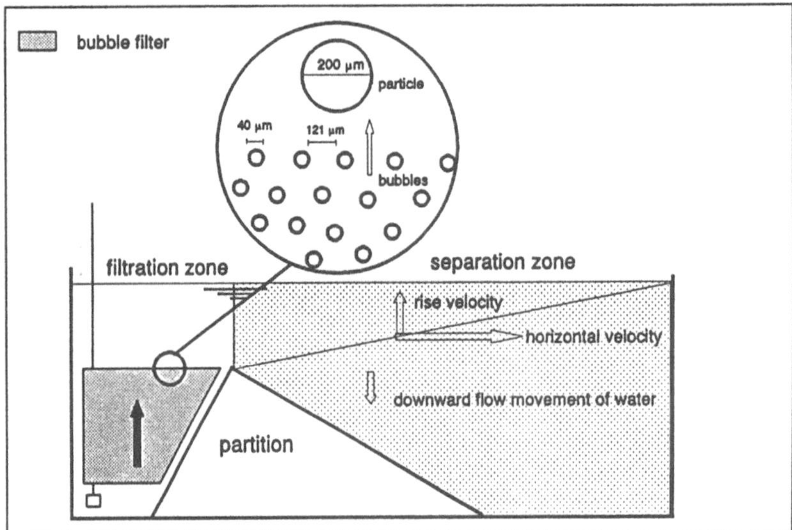

Fig. 2. Illustration of the differences in diameter and in distances of the bubbles and particles

3. Filtration Zone

3.1 Single-Collector Collision Efficiency

In describing the single-collector collision efficiency η between a bubble and a particle, the bubble position is fixed and the fluid flow can be described relative to the bubble in the downward direction (Fig. 3). In this paper the single-collector collision efficiency η is defined as the particle-bubble collision rate divided by the particle-bubble approach rate. Starting point is that the particle is present in the water column above the bubble. The water column has an area equal to the projected area of a bubble perpendicular to the direction of flow. There are 4 particle trajectory mechanisms, which may cause a collision:

1. Diffusion η_D	Particle transport results from the random Brownian motion of (the smaller) particles.
2. Interception η_I	Collision occurs when particle motion along a streamline is sufficiently close to the collector for attachment to occur.
3. Gravity settling η_S	The mechanism of settling is due to the force of gravity and the associated settling velocity of the particles,

which causes it to cross streamlines and reach the collector.

4. Inertia η_{TA} Collision occurs if the (bigger and heavier) particles are not able to follow the streamlines caused by their own inertia weight.

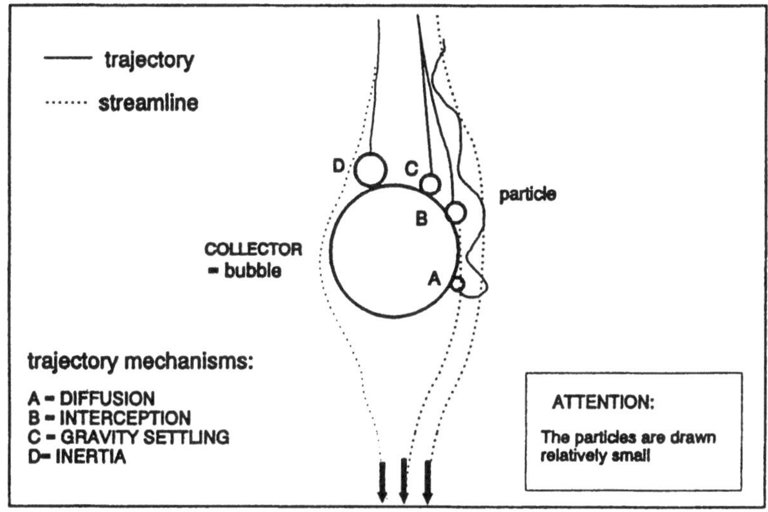

Fig. 3. Particle transport mechanisms for a single (collector) bubble for Stokes flow

The single-collector collision efficiency η can be quantified for these particle trajectory mechanisms. Individual equations for diffusion, interception, gravity settling and inertia follow [2, 3]:

$$\eta_{\text{D}} = 6.18 \left(\frac{k_b T}{g \rho_w}\right)^{2/3} \left(\frac{1}{d_d}\right)^{2/3} \left(\frac{1}{d_b}\right)^{2} \tag{1}$$

$$\eta_{\text{I}} = \frac{3}{2} \left(\frac{d_d}{d_b}\right)^{2} \tag{2}$$

$$\eta_{\text{S}} = \left(\frac{\rho_d - \rho_w}{\rho_w}\right) \left(\frac{d_d}{d_b}\right)^{2} \tag{3}$$

$$\eta_{\text{TA}} = \frac{g \rho_d d_b d_d^2}{324 \, \nu^2 \rho_w} \tag{4}$$

k_b Boltzmann's constant ($1.38 \cdot 10^{-23}$ J/°K)
T Absolute temperature of the water [°K]
g   Gravitational acceleration (9.81 m/s^2)
ρ_w Water density [g/m^3]
d_d Particle Flotameter [m]
d_b Bubble diameter [m]
ρ_d Particle density [g/m^3]
ν Kinematic fluid viscosity [m^2/s]

The total single-collector collision efficiency η_T is equal to the sum of the single-collector collision efficiencies of the different particle trajectory mechanisms:

$$\eta_T = \eta_D + \eta_I + \eta_S + \eta_{TA} \tag{5}$$

Figure 4 shows the effect of the particle size on the single-collector collision efficiency η. The following conclusions may be drawn from this figure:

- the diffusion mechanism controls η_T for particle diameters d_d below 1 μm and
- the interception mechanism controls η_T for particle diameters d_d larger than 5 μm.

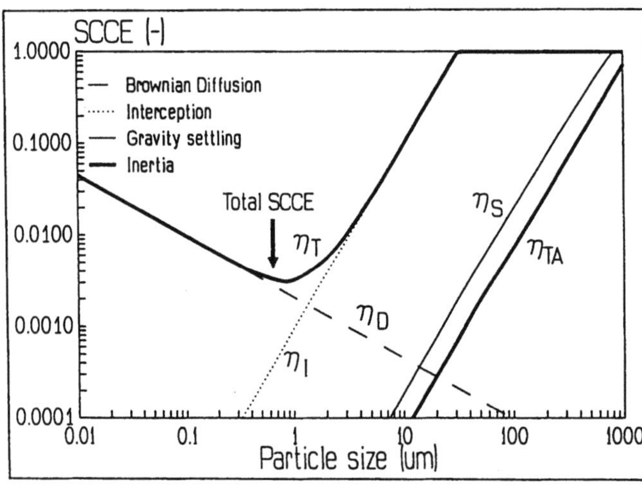

Fig. 4. Effect of the particle size (d_d) on the single-collector collision efficiencies of the transport mechanisms (η_D, η_I, η_S, η_{TA} and η_T)

If the raw water is properly pre-treated, the particle diameter d_d has a value between 100 and 1000 μm. This means that in practice interception is the most important mechanism. Equation 2 shows that, if the bubble diameter d_b is 40 μm and if the particle diameter d_d is larger than 32 μm, the total single-collector collision efficiency η_T is 1. This means that in flotation practice, the total single-collector collision efficiency η_T is always 1, which is also illustrated in Fig. 4.

Besides the particle and bubble diameter (d_d and d_b) the total single-collector collision efficiency η_T depends on the water temperature T, the water density ρ_w and the particle density ρ_d. Figure 5 shows that these last three parameters are subordinate to the influence of the particle and bubble diameter.

The relationship between the particle diameter d_d and the total single-collector collision efficiency η_T, as illustrated in Figs. 4 and 5, is equal to the well-known curves of the filtration theory [5–11]. In the filtration theory, the collectors are the sand grains of the sand filter, while in this flotation model, the collectors are the bubbles. The filtration theory also shows a minimal total single-collector collision efficiency η_T at particle diameters d_d of 1 μm. The

main difference between flotation and filtration is that the numerical values of
the total single-collector collision efficiency η_T for flotation are much higher
than for filtration (Table 2). This results in the following conclusion: "flotation
is an ideal form of filtration".

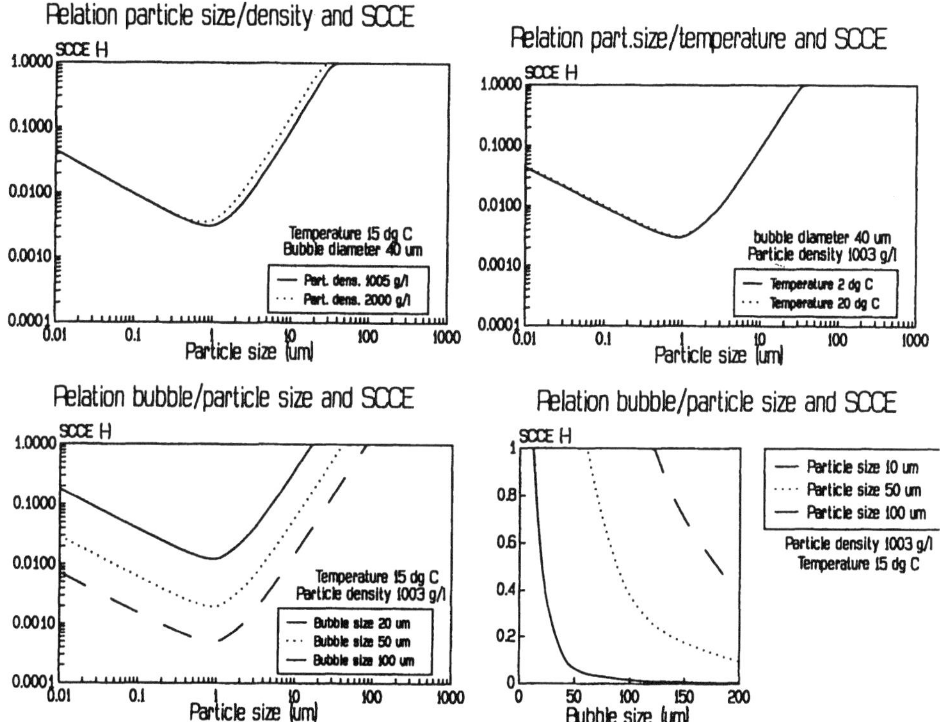

Fig. 5. Effect of the particle size d_d, bubble size d_b, particle density ρ_d, and the
temperature T on the total single-collector collision efficiency η_T

3.2 Bubble-Filter Efficiency

The bubble-filter efficiency X of the filtration zone is described as the fraction
of particles that will be removed by the bubble filter. This means that this
fraction has a permanent attachment to one or more bubbles. To calculate the
bubble-filter efficiency X a mass balance should be made. This results in the
following batch particle removal equation:

$$\frac{dN_d}{dt} = -(\alpha_{db}\eta_T)N_b N_d A_b v_{db} \tag{6}$$

α_{db} Attachment or adhesion efficiency [-]
A_b Projected area of a bubble [m^2]
t Time [s]
v_{db} Bubble particle approach velocity [m/s]

Tab. 2. A comparison between the single collector collision efficiency of the filtration and the flotation models

Flotation model Bubble size, d_b μm	Particle size, d_d μm	SCCE, η_T -
40	10	0.090
40	100	1.000
40	200	1.000
40	500	1.000

Filtration model Specific diameter of filter grains, d_s mm	Particle size, d_d μm	SCCE, η_T -
Slow sand filtration 0.3	10	0.001
($v = 0.4$ m^3/m$^2 \cdot$ h) 0.3	100	0.280
0.3	200	1.000
0.3	500	1.000
Rapid filtration 0.8	10	0.000
($v = 10$ m^3/m$^2 \cdot$ h) 0.8	100	0.028
0.8	200	0.112
0.8	500	0.699

The attachment efficiency α_{db} is the fraction of all collisions that lead to a permanent attachment of a bubble and a particle. The bubble particle approach velocity v_{db} is equal to the bubble rise velocity v_b, which can be described by the Stokes equation [1]:

$$v_b = \frac{1}{18} \frac{g d_b^2}{\nu} \tag{7}$$

The bubble-number concentration N_b is the bubble-volume concentration ϕ_b divided by the volume of a bubble ($\pi d_b^3/6$). Substituting for the projected area of the mean bubble size and the bubble rise velocity (Eq. (7)), and using the above mentioned relationship between ϕ_b and N_b in Eq. (6) yields the following equation:

$$\frac{dN_d}{N_d} = -\frac{1}{12} \frac{\alpha_{db} \eta_T d_b \phi_b g}{\nu} dt \tag{8}$$

In the case of an ideal plug flow, all particles in the filtration zone having the same detention time, Eq. (8) can be transformed in a bubble-filter efficiency equation:

$$X = \frac{N_{d,i} - N_{d,e}}{N_{d,i}} = 1 - \exp\left(-\frac{1}{12} \frac{\alpha_{db} \eta_T d_b \phi_b g}{\nu} \tau\right) \tag{9}$$

τ Net detention time (or contact time) in the filtration zone [s]
$N_{d,i}$ Particle number concentration of the influent [m^{-3}]
$N_{d,e}$ Particle number concentration of the effluent [m^{-3}]

The fluid viscosity ν depends only on the temperature T of the water. The attachment efficiency α_{db} is a parameter which, in the first instance, is determined by the pre-treatment of the raw water. Furthermore, the attachment efficiency α_{db} can be reduced by unfavourable flow conditions in the filtration zone, such as turbulence.

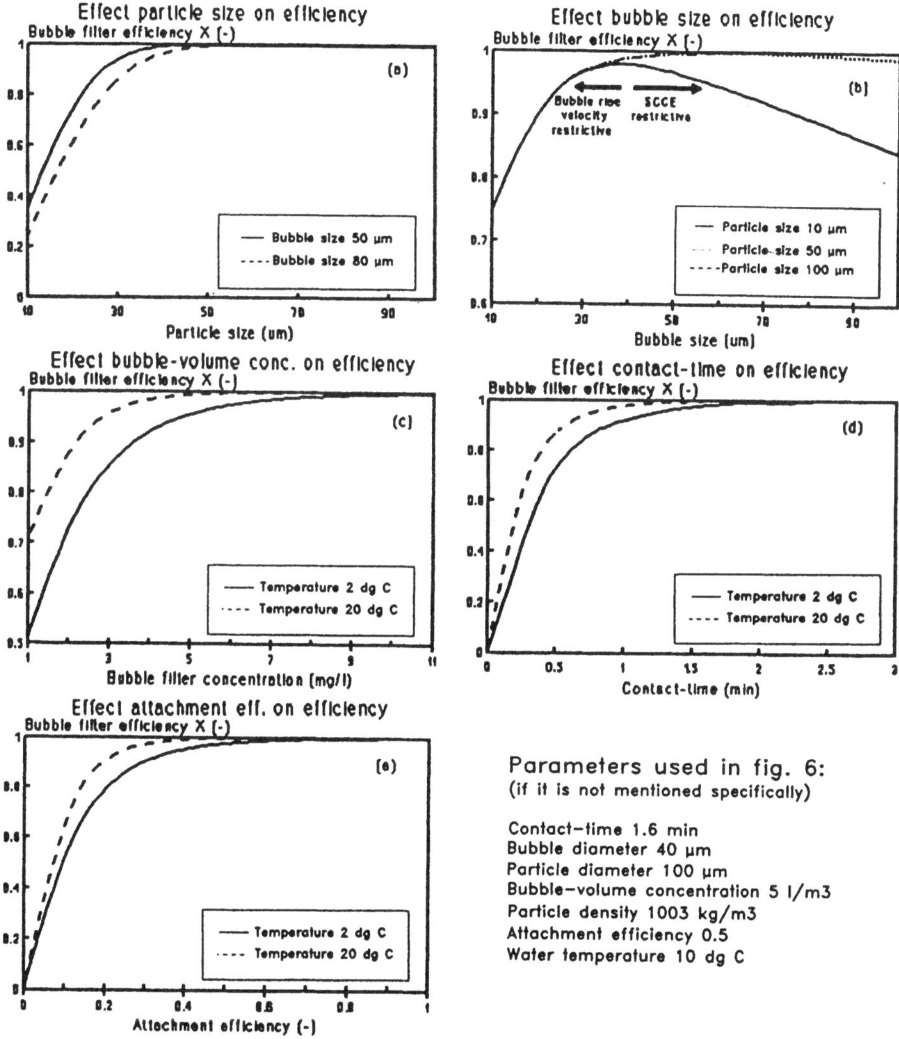

Fig. 6. Effect of the particle size d_d, bubble size d_b, bubble-volume concentration ϕ_b, contact-time τ, and attachment efficiency α_{db} on the bubble filter efficiency X

Theory

Figure 6 shows the influence of several parameters on the bubble-filter efficiency X. If it is not mentioned specifically in Fig. 6, the parameters have the following values:

- contact time $\tau = 1.6$ min
- bubble diameter $d_b = 40$ μm
- particle diameter $d_d = 100$ μm
- bubble-volume concentration $\phi_b = 5 \cdot 10^{-3}$ m^3/m^3
- particle density $\rho_d = 1003$ kg/m^3
- attachment efficiency $\alpha_{db} = 0.5$
- water temperature $T = 10\,^{\circ}$C

Influence of Particle and Bubble Diameter. If the particle diameter d_d is smaller than 0.8 times the bubble diameter d_b. (Eq. (2)), the total single-collector collision efficiency η_T is smaller than 1.0. The lower value of the bubble-filter efficiency X for smaller particles is caused by a reduction in the total single-collector collision efficiency η_T. Figure 7 a shows that the bubble-volume concentration ϕ_b can be limited by using larger particles, if the bubble-filter efficiency X is constant. In practice there is a certain variation in bubble diameters d_b (10–100 μm). Larger bubbles result in a lower bubble-filter efficiency X because the total single-collector collision efficiency η_T is less than 1.0 (Fig. 6 b). Smaller bubbles result in a maximal total single-collector collision efficiency ($\eta_T = 1$), but have a smaller bubble rise velocity v_b, resulting in a reduction in the bubble-filter efficiency X (Eq. (8)).

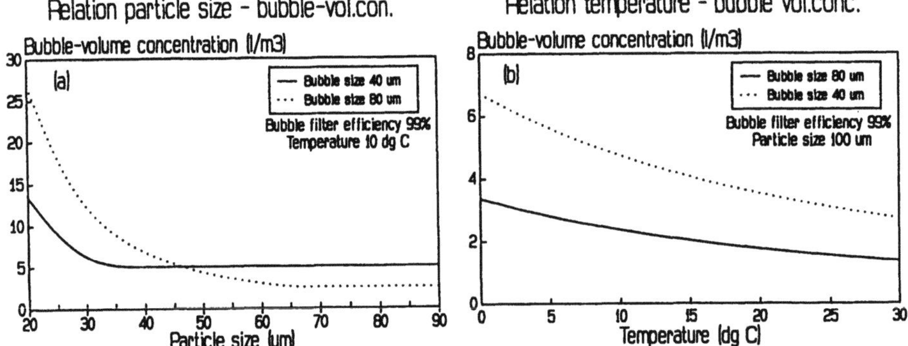

Fig. 7. Effect on the particle size d_d and the temperature T on the bubble-volume concentration ϕ_b by a constant bubble filter efficiency X (99 %)

Influence of Bubble-Volume Concentration. The exponential relationship between the bubble-volume concentration ϕ_b and the bubble-filter efficiency X is illustrated in Fig. 6 c. During operating situations the bubble-volume concentration ϕ_b is very important and can be adjusted to the existing conditions, such as the water temperature T. The bubble-volume concentration ϕ_b depends mainly on the flotation pressure p_f. The flotation pressure p_f is the product of the saturator pressure p_r and the recycle ratio Rec.

Influence of Contact Time. Figure 6 d shows that the selected contact time τ should not be too low and should preferably be higher than 1.5 min. During the design, the choice of the contact time τ or contact volume is of primary importance. Table 3 gives the contact times τ used at the existing flotation plants.

Tab. 3. The contact times at the existing flotation plants

Flotation plant	Design contact time τ min
Zevenbergen	2.1
Braakman	1.8
Enschede	1.4
Katwijk	1.5
Antwerpen	2.1
Scheveningen	0.9

Influence of Temperature. Figures 6 c and 6 d show that low temperatures T (2 °C) decrease the bubble-filter efficiency X up to 20 % compared to extreme summer conditions (20 °C). Figure 7 b shows that the bubble-volume concentration ϕ_b should be adjusted to the water temperature T to fix the bubble-filter efficiency X ($X = 99\%$). The bubble-volume concentrations ϕ_b at a water temperature T of 5 °C must be 1.6 times as high as at a water temperature T of 20 °C.

Influence of Flow Conditions. The efficiency equation (Eq. (9)) can only be applied to an ideal plug flow, where all particles have the same detention time in the filtration zone. In practice it is impossible to have this idealized situation. There are several different theoretical models which describe the flow of water particles through a vessel [1]. In this paper, only the PFDM (Plug Flow with Dispersion Model) will be discussed. In this model the fluid flow can be characterized by the Peclet-number (Pe). Flows with high Pe values (Pe > 20) approximate plug flow (PF). Table 4 shows that low Pe values in the filtration zone result in higher required contact times τ to obtain the same bubble-filter efficiency X. The values of the Peclet-number are mainly determined by the length to width ratio or length to depth ratio of the filtration zone. Higher length to width ratios result in higher Pe values and therefore result in a higher bubble-filter efficiency X. Next to this, the supply and discharge of the feed water and the recycle flow is of importance. These flows should be supplied equally over the cross-section of the filtration zone.

Tab. 4. Effect of the Peclet-number (Pe) on contact time with a constant bubble-filter efficiency X (93 %)

Pe [-]	1	5	10	20	PF
τ [min]	4.5	2.2	1.9	1.3	1.3

Influence of Attachment Efficiency. Figure 6 e shows the exponential relationship between the attachment efficiency α_{db} and the bubble-filter efficiency X. Probably the bubble diameter d_b and the water temperature T also have an effect on the attachment efficiency α_{db} between the bubbles and particles. The magnitude of these effects is unknown, however.

Practice

The results of this model are compared with the treatment results of the existing flotation plants. The numerical values of almost all parameters are known and are listed in Table 5. The only unknown parameter, the attachment efficiency, is determined with Eq. (8) and has a value between 0.2 and 1.0. In most cases it is possible to give narrower limits: 0.3–0.4.

Tab. 5. Values of the parameters at the existing flotation plants

Parameters	Values of the parameters
Single-collector collision efficiency η_T	1
Particle size d_b [μm]	30–40
Bubble-volume concentration ϕ_b [l/m^3]	6–7
Viscosity ν [m^2/s]	1.0–$1.6 \cdot 10^{-6}$
Contact time τ [minutes]	1.0–2.0
Bubble-filter efficiency X [%]	90–100
Attachment efficiency α_{db}	0.3–0.4

Equation 9 shows the influence of the bubble-volume concentration ϕ_b, the contact time τ and the water temperature T on the bubble-filter efficiency X. The theoretical values are compared with the treatment results of the existing flotation plants. The following values are used for the calculations in Tables 6, 7 and 8: $\tau = 1.0$ min, $d_b = 40$ μm, $\eta_T = 1$, $\phi_b = 6 \cdot 10^{-3}$ m^3/m^3, $\rho_d = 1003$ kg/m^3, $\alpha_{db} = 0.3$ and $T = 10\,°C$ ($\nu = 1.3 \cdot 10^{-6}$ m^2/s).

For these values, the bubble-filter efficiency X is 93 %.

Influence of Bubble-Volume Concentration. Table 6 shows that a drop in the bubble-volume concentration ϕ_b causes a reduction in the bubble-filter efficiency X. Experimental results of the DZH [14] with raw water from the Andelse Maas show the same relationship between the bubble-volume concentration ϕ_b and the total removal efficiency R (Fig. 8). The total removal efficiency R is the product of the bubble-filter efficiency X in the filtration zone and the separation efficiency Y in the separation zone. According to Eq. (9), it can be expected that the bubble-filter efficiency X approaches the value 1 for high bubble-volume concentrations ϕ_b. Figure 8 shows that the total removal efficiency R does not approach 1, but is limited to 0.9–0.95 for high bubble-volume concentrations ϕ_b. This can be caused by a reduction in the separation efficiency Y in the separation zone ($Y < 1$).

Tab. 6. Effect of the bubble-volume concentration ϕ_b on the bubble-filter efficiency X

ϕ_b ($\times 10^{-3}$ m^3/m^3)	2	4	6	8	10	15
X [%]	60	84	93	97	99	100

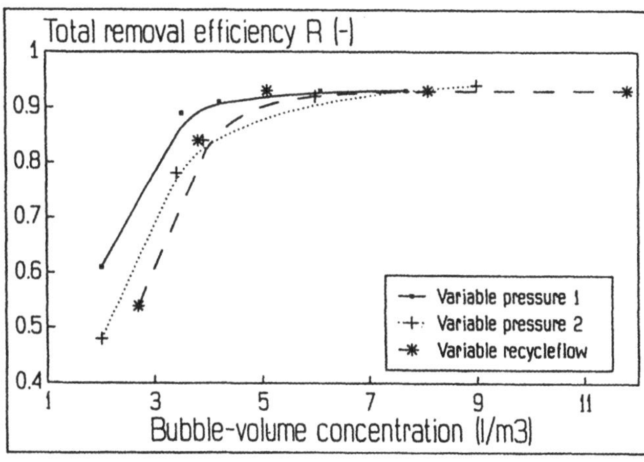

Fig. 8. Effect of the bubble-volume concentration ϕ_b on the total (Fe) removal efficiency R (experimental results of the DZH with Andelse Maaswater [14])

Influence of Contact Time. Table 7 shows that lower contact times τ reduce the bubble-filter efficiency X. Experimental results at the flotation plant Zevenbergen [15] show a comparable relationship, which is illustrated in Fig. 9. This figure shows that for high contact times the total removal efficiency R is lower than 1, which is probably caused by a reduction in the separation efficiency Y. This could be the reason that the total removal efficiency R does not approach the value 1 at high contact times.

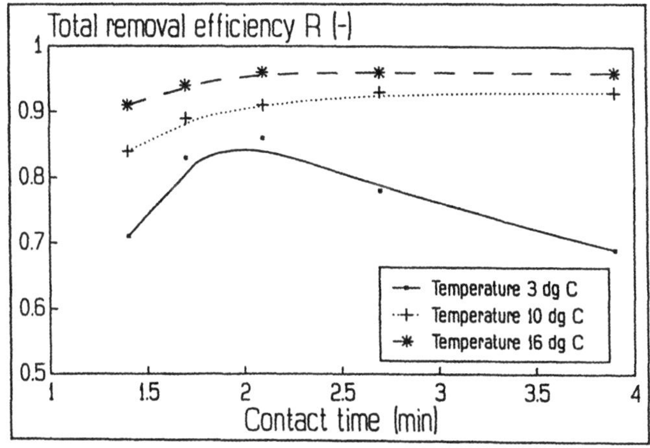

Fig. 9. Effect of the contact-time τ and temperature T on the total (Fe) removal efficiency R (experimental results of the Zevenbergen plan [15])

Tab. 7. Effect of the contact time τ on the bubble-filter efficiency X

τ [min]	0.25	0.50	1.0	1.5	2.0	2.5
X [%]	49	74	93	98	99	100

Influence of Water Temperature. Table 8 shows that the only external condition, the water temperature T, has a significant influence on the flotation process. All flotation plants experience a reduction in the total removal efficiency R in winter. Figure 10 shows the residual Fe concentration ($FeCl_3$ is used as coagulant) in the effluent of the flotation plant Zevenbergen during the year 1990. Next to a reduction in the bubble-filter efficiency X, low water temperatures also lead to lower bubble-particle rise velocities, which reduce the separation efficiency Y, as will be explained in Sect. 4.

Tab. 8. Effect of the water temperature T on the bubble-filter efficiency X

T [°C]	2	10	20
X [%]	88	93	97

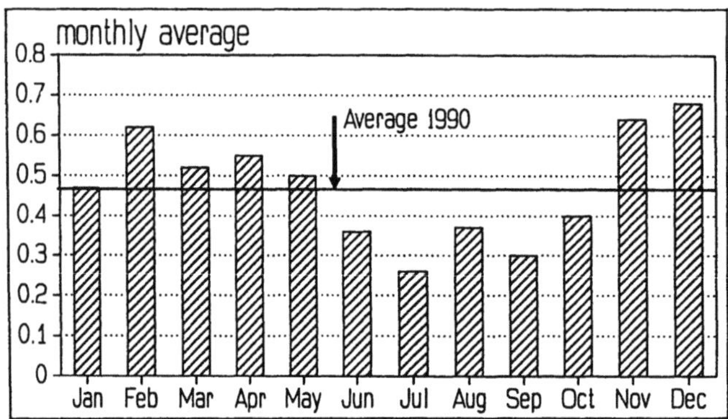

Fig. 10. Residual Fe-concentration [mg/l] of the effluent of the flotation plant Zevenbergen during the year 1990

Table 9 shows the qualitative effect of the design and operating parameters on the contact process in the filtration zone.

3.3 Conclusions About the Filtration Zone

a) Pre-treatment is Important

Particle Diameter d_d. With dissolved-air flotation, the particle diameter d_d should be higher than 30 μm to maximize the total single-collector collision efficiency ($\eta_T = 1$). In practice, with bubble diameters d_b of 10 to 100 μm,

Tab. 9. Summary of model parameters and effects on the bubble-filter efficiency X and on the design and operation of the flotation plant

Parameter	Dependence	Comments
Pre-treatment d_d, particle diameter	1. Nature and origin of the particles 2. Coagulation conditions: – coagulant – flocculation time – energy (G-value)	1. If $d_d > 0.8 \cdot d_b$, SCCE is maximal 2. Proper flocculation leads to larger particles 3. Practice $\eta_T \approx 1$, because $d_d > 1000 - 1000\ \mu$m and $d_b \approx 40\ \mu$m
α_{db}, the bubble-particle attachment efficiency	1. Particle-bubble charge interactions 2. Hydrophilic nature of particles (and bubbles)	1. Increase in α_{db} by a proper coagulant, flocculation and pH adjustment 2. Lower T (hydrophilic), decrease in attachment 3. Larger bubbles, decrease in attachment
T, water temperature	External condition	Higher T, higher v_b, increase in X
Filtration zone d_b, bubble diameter	1. Saturator pressure 2. Nozzles	1. Smaller bubbles, higher particle-bubble interactions 2. Smaller bubbles, increase in η_T (and X) if $d_d < d_b$, but also decrease in v_b results in a drop in X
ϕ_b, bubble-volume concentration	Flotation pressure p_f (Saturation pressure p_r, Recycle ratio Rec)	1. Higher ϕ_b, more collision possibilities 2. High air deficit of the raw water reduces effective ϕ_b 3. Important parameter because it can be changed during operation
τ, contact-time	1. Volume filtration zone 2. Raw water flow 3. Fraction dead space m	Reduce m by paying attention to the supply and discharge of water flow
Flow conditions: Peclet-number	Geometry filtration zone	Higher L/B ratios, higher Pe, increase in X. Prevent short circuiting by a proper supply and discharge of the water flow

this means that particle diameters d_d should preferably be larger than 100 μm. Large particles are not necessary in the filtration zone, but in Sect. 4 it will be shown that large particles are important in the separation zone.

Attachment Efficiency α_{db}. Poor attachment occurs when (stable) particles carry an electrical charge. Next to the formation of a suitable particle size, rapid mixing and coagulation should result in destabilized particles (high α_{db} value).

b) Filtration Zone

Bubble Diameter d_b. In flotation processes, the bubble diameter d_b has an almost constant value (mean bubble size 30–40 μm), so this can hardly be seen as a variable [12]. The designer or the plant operator has little or no control over the bubble size.

Bubble Volume Concentration ϕ_b. The bubble-volume concentration ϕ_b is very important [14] and can be used for optimization during the operation of the flotation plant. In winter higher bubble-volume concentrations ϕ_b are necessary than in the summer. Bubble-volume concentrations ϕ_b of at least 5 l/m^3 are preferable.

Contact Time τ. The contact time τ is determined by the design of the filtration zone and the flow through the flotation plant. A sufficiently high contact time τ (between 1.5 and 2.0 min.) is necessary for a high bubble-filter efficiency X in the filtration zone.

Flow Conditions Pe. Poor flow conditions should be avoided in the filtration zone. An equal supply of the raw water flow and the recycle flow over the cross-section of the filtration zone are necessary. Long and narrow filtration zones or column reactors are preferable.

Water Temperature T. It has been mentioned before that the temperature has an important effect on the bubble-filter efficiency X. Filtration zones should be designed for winter conditions.

4. Separation Zone

4.1 Theoretical Considerations

Removal of the particles takes place in the separation zone. Figure 11 shows an example of a typical separation zone. The condition for the removal of a bubble-particle agglomerate is:

$$\frac{t_b}{t_{st}} > 1 \tag{10}$$

t_b　　　　Detention time of a particle in the separation zone [s]

t_{st}　　　Floating time or time a particle needs to reach the water-surface [s]

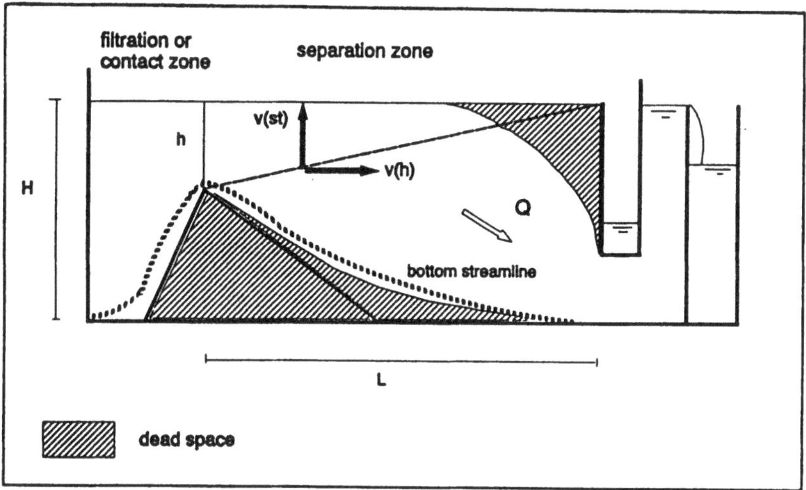

Fig. 11. Cross-section of a typical separation zone of a flotation unit

The floating time t_{st} depends on the depth at which the particle is present in the separation zone. In the critical situation the particle is at the bottom of this zone. The removal condition (Eq. (10)) can be rewritten for ideal plug flow conditions and discrete flotation (constant rise velocity of the bubble-particle agglomerates) into:

$$v_{st} > q_b \frac{1}{(1-m)} \tag{11}$$

v_{st} Bubble-particle rise velocity [m/s]
q_b Surface load of the separation zone [m^3/m$^2 \cdot$ s]
m Fraction dead space in the separation zone [-]

The separation efficiency Y depends on the bubble-particle rise velocity v_{st} and the surface load q_b. The required area of the separation zone is determined by the flow and the bubble-particle rise velocity v_{st}. The bubble-particle rise velocity v_{st} is determined by:

– Stokes equation for laminar flow conditions (Re < 1)

$$v_{st} = \frac{1}{18} \frac{g}{\nu} \left(\frac{\rho_w - \rho_a}{\rho_w} \right) d_a^2 \tag{12}$$

– adjusted Stokes equation for laminar-turbulent flow conditions (1 < Re < 50)

$$v_{st} = \frac{1}{10} \frac{g^{0.8}}{\nu^{0.6}} \left(\frac{\rho_w - \rho_a}{\rho_w} \right)^{0.8} d_a^{1.4} \tag{13}$$

ρ_a Bubble-particle density [g/m^3]
d_a Bubble-particle diameter [m]

In these equations it is assumed that the bubble-particle agglomerates have a spherical shape.

As a result of the higher water viscosity ν, the bubble-particle rise velocity v_{st}, under laminar flow conditions, at $2\,°C$ is 1.6 times lower than at $20\,°C$. The bubble-particle density ρ_a and the bubble-particle diameter d_a can be determined by two better known parameters: the particle diameter d_d and the bubble-particle volume ratio β. In this paper, the bubble-particle volume ratio β is defined as the volume of the bubbles attached to a particle divided by the volume of this particle.

Table 10 shows the effect of the bubble-particle volume ratio β on the flotation driving force ($[\rho_w - \rho_a]/\rho_w$).

Tab. 10. Values for the bubble-particle density ρ_a at different values of the bubble-particle volume ration β (air density $\rho_1 = 1.24$ kg/m^3 and particle density $\rho_d = 1\,003$ kg/m^3)

β	0.003	0.006	0.01	0.05	0.10	1.0	10
ρ_a [kg/m^3]	1 000	997	993	955	912	502	92
$(\rho_w - \rho_a)/\rho_w$	0	0.003	0.007	0.045	0.088	0.498	0.908

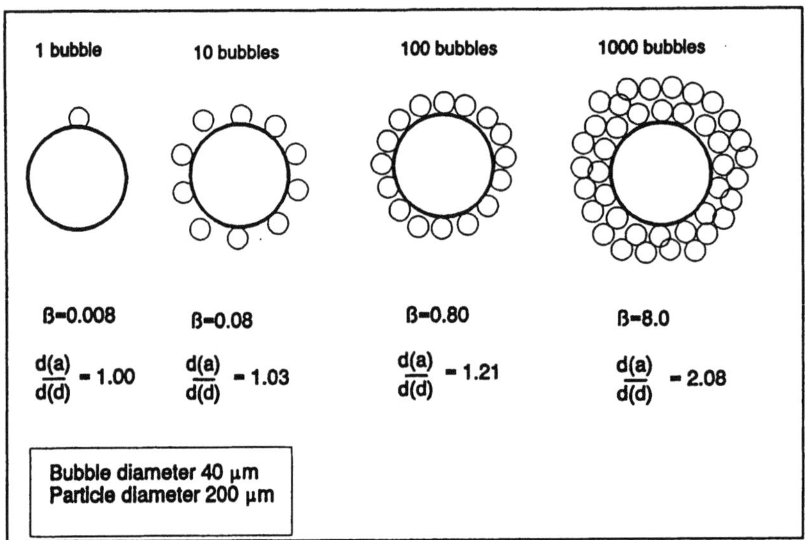

Fig. 12. Illustration of the bubble-particle volume ration β and the diameter ratio between a bubble-particle agglomeration and a particle (d_a/d_d)

Figure 12 shows the effect of bubble-particle volume ratio β on the bubble-particle diameter d_a. In the bubble-particle diameter calculations, the volume of the bubbles has been spread out equally over the bubble surface area. Figure 12 illustrates how a bubble-particle is built up for certain bubble-particle volume ratios β. Physically it is almost impossible for more bubble-layers to be on top of each other, which means that there is a maximal value for the bubble-particle volume ratio β. In the case of particle diameters d_d of 200 μm (Fig. 12), this maximal value is about 1.0.

At bubble-particle volume ratios β of 0.006, the driving force of flotation processes is equal to that of sedimentation processes. If all air ($\phi_b = 5\text{--}10 \, 1/m^3$), which is released at the nozzles, would be used for the separation, the bubble-particle volume ratio β would be 500. For the separation only a small fraction of the bubbles (0.2 %) is used effectively. However, the (high) bubble-volume concentrations ϕ_b that are used are necessary to obtain a sufficiently high bubble-filter efficiency X in the filtration zone. As a result of the excessive amount of air in the separation zone, it is possible to treat water containing high amounts of colloidal impurities and suspended solids, like sewage water flows. Table 11 and Fig. 13 show the effect of the bubble-particle volume ratio β on the bubble-particle rise velocity v_{st} for certain particle diameters d_d. Since there seems to be a physical maximum for the bubble-particle volume ratio β ($\beta < 1$), an increase in the bubble-particle rise velocity v_{st} can be realized mainly by an increase in particle diameter d_d.

Tab. 11. Calculations of the bubble-particle rise velocity v_{st} (10 °C)

Number bubbles ($d_b = 40 \, \mu$m) with attachment		Bubble-particle volume ratio β m^3/m^3	Bubble-particle density, ρ_a kg/m^3	Bubble-particle rise velocity, v_{st} m/h
($d_d = 100 \, \mu$m)	1	0.06	942	0.9
	5	0.32	760	4.3
	10	0.64	612	8.1
	50	3.20	239	29.9
($d_d = 200 \, \mu$m)	1	0.01	995	0.3
	5	0.04	964	2.2
	10	0.16	864	9.0
	50	0.40	716	21.4
	100	0.80	557	34.4
($d_d = 500 \, \mu$m)	5	0.00	1000	0.0
	10	0.01	998	0.8
	50	0.03	976	8.5
	100	0.05	953	15.9
	200	0.10	909	28.0
	400	0.21	850	47.8

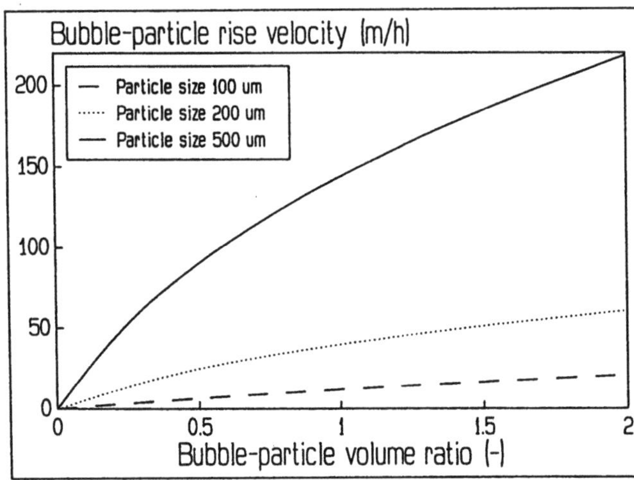

Fig. 13. Effect of the bubble-particle volume ratio β and the particle size d_d on the bubble-particle rise velocity v_{st}

Practice

In practice, surface loads of the separation zone q_b of 10–25 m³/m²·h (Table 12) are used. In the early spring the Zevenbergen plant reaches a maximal surface load of 26 m³/m² · h. Considering a certain dead space fraction (m = 0.15) the bubble-particle rise velocity v_{st} has to be at least 30 m/h for the critical situation to remove this bubble-particle agglomerate (Eq. (11)). The detention times in the separation zone vary between 5 and 10 minutes. Table 13 shows the qualitative effect of the aforementioned parameters on the separation process.

Tab. 12. Surface loads q_b of the separation zone at the existing flotation plants

Floation plant	Maximal surface load q_b m³/m² · h
Zevenbergen	26
Braakman	22
Enschede	20
Katwijk	9
Antwerpen	19
Scheveningen	22

4.2 Conclusions About the Separation Zone

Particle Diameter d_d. As mentioned before the bubble-particle rise velocity v_{st} will be higher if the particle diameter d_d is increased. Pre-treatment should therefore result in large flocs.

Tab. 13. Summary of the model parameters and the effect on the separation efficiency Y

Parameter	Dependence	Comments
$(\rho_w - \rho_a)/\rho_w$, relative difference in densities	1. Bubble-particle volume ratio β 2. Particle and water density	1. Practice: $\beta \leq 1$ 2. Very low bubble efficiency ($\leq 0.2\%$) 3. ρ_w and ρ_d are almost constant and have no effect on v_{st}
d_a, bubble-particle agglomerate diameter	1. d_d, particle diameter 2. Bubble-particle volume ratio β	1. Larger particles result in larger bubble-particle agglomerates and higher bubble-particle rise velocities. Necessity of proper pre-treatment 2. Higher β, higher d_a (physical maximum for $\beta = 1$)
T, water temperature	External condition	Laminar: $v_{st}\,(2\,°\mathrm{C}) = \frac{1}{2} \cdot v_{st}\,(25\,°\mathrm{C})$

Bubble-Particle Volume Ratio β. An increase in the bubble-particle volume ratio β results in higher bubble-particle rise velocities v_{st}. However, this influence is of secondary importance (Fig. 13), because there seems to be a maximal value of the bubble-particle volume ratio β.

Water Temperature T. The water temperature T is an important parameter for the separation zone and determines the water viscosity ν. In the winter the bubble-particle rise velocity v_{st} is lower than in the summer. The flotation plant should be designed for critical winter conditions.

Flow Conditions Pe. The removal condition (Eq. (11)) is only valid for an ideal plug flow. Less ideal flow conditions (low Pe values) decrease the separation efficiency Y [1]. To obtain a flow with high Pe values the separation zone should be long and narrow.

5. Evaluation and Translation of the Model for an Optimal Design

The design of a flotation plant should include a good separation of the filtration and the separation zones.

5.1 Filtration Zone

The starting points of the filtration zone design are:

1. Both a co-current and a counter-current principle are possible. In theory the counter-current option is preferable. However, the surface load q_b of the filtration zone ($50\text{–}100\ \mathrm{m^3/m^2 \cdot h}$) is much higher than the bubble rise velocity v_b (until 2 m/h), so that it does not make much difference whether the design is based on a co-current, counter-current or cross-current principle.

2. The contact time should be longer than 1.5 min.

3. The geometry of the filtration zone should be long and narrow. Good plug flow conditions are obtained by a length to width ratio of 5 or more [1]. This results in the design of a so-called column reactor as a filtration zone.

4. Both flows should be distributed evenly over the area of the cross-section of the filtration zone. At the existing flotation plants, the recycle flow is released in the filtration zone, causing water velocities near the nozzles up to 20–30 m/h. This results in considerable turbulence, as a result of which floc break-up is certainly possible and already existing bubble-particle attachments are broken. This is why, during flocculation, stable (and therefore also small) flocs are usually desired, which can resist the strong recycle flow. To reach the higher bubble-particle rise velocities v_{st} of the larger particles (flocs), the recycle flow should be added in a different way. One possibility is to protect the feed water flow against the recycle flow by placing extra screens around the nozzles. Another possibility is to release the recycle flow in a separate release tank. In the last case, the released recycle flow can be transported to the flotation unit and can be added to the filtration zone with the help of nozzles, which should be divided equally over the area of the filtration zone.

5.2 Separation Zone

The starting points of the separation zone design are:

1. Surface loads q_b of 25–30 $m^3/m^2 \cdot h$ are possible.

2. To reduce high horizontal water velocities, the depth should be approximately 2.0 m, leading to a detention time of at least 5 min.

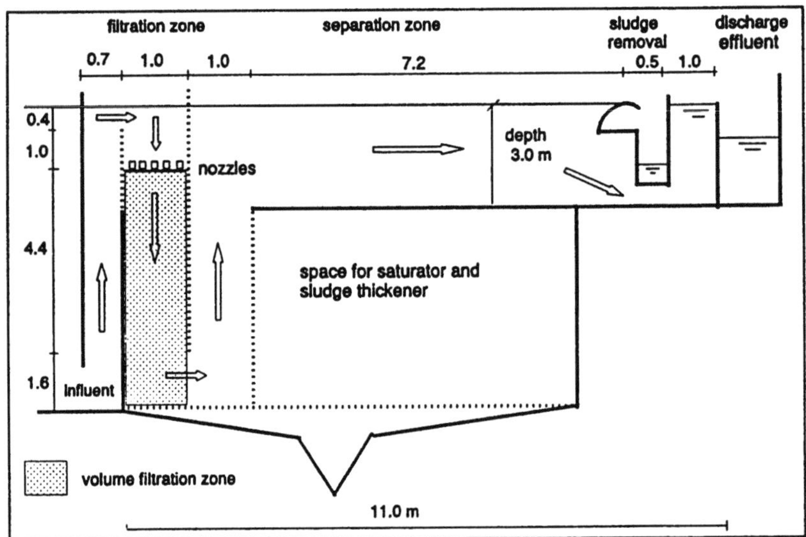

Fig. 14. Cross-section of the concept-design of a hypothetical flotation plant

3. The geometry of the zone should be long and narrow. The use of a partition wall or baffle is not necessary if a contact column is used as filtration zone. The water flow should be supplied and discharged equally over the area of the separation zone to reduce the dead space fraction m.

With these starting points a hypothetical flotation plant was designed conceptually [1], where the existing lamellae separators are to be replaced by flotation units. Figure 14 shows the cross-section of this design.

References

[1] Schers, G.J.: Flotatie, de theorie en de praktijk. Graduation thesis at the Department of Civil Engineering, Division of Sanitary Engineering, Technical University Delft, 1991

[2] Malley, J.P.: A Fundamental Study of Dissolved Air Flotation for Treatment of Low Turbidity Waters Containing Natural Organic Matter. Ph.D. dissertation, University of Massachusetts, Amherst, MA, 1988

[3] Malley, J.P., Edzwald, J.K.: Concepts for Dissolved-Air Flotation Treatment of Drinking Waters. J. Water SRT-Aqua *40* (1) (1991) 7–17

[4] Edzwald, J.K., Wingler, B.J.: Chemical and Physical Aspects of Dissolved-Air Flotation for the Removal of Algae. Aqua *39* (1990) 24–35

[5] Yao, K.-M., Habibian, M.T., O'Melia, C.R.: Water and Waste Water Filtration: Concepts and Applications. Env. Sci. Tech. *5* (11) (1971) 1105–1112

[6] Rajagopolan, R., Tien, C.: Trajectory Analysis of Deep-Bed Filtration with the Sphere-in-Cell Porous Media Model. AIChE J. *22* (3) (1976) 523–533

[7] O'Melia, C.R., Ali, W.: The Role of Retained Particles in Deep Bed Filtration. Progress Wat. Tech. *10* (1978) 167–182

[8] Tien, C., Turian, R.M., Pendse, H.: Simulation of the Dynamic Behavior of Deep Bed Filters. AIChE J. *25* (3) (1979) 385–395

[9] Montgomery, J.M. (Consulting Engineers): Water Treatment – Principles and Design; Chapter 8: Filtration. Wiley, New York 1985

[10] O'Melia, C.R.: Particles, Pretreatment, and Performance in Water Filtration. J. Env. Engin. *111* (6) (1985) 874–891

[11] Payatakes, A.C., Tien, C., Turian, R.M.: Trajectory Calculation of Particle Deposition in Deep Bed Filtration: 1. Model Formulation. AIChE J. *20* (1974) 889

[12] Bennekom, C.A.: De invloed van verschillende nozzletypen op het flotatieproces: 1. Karakterisering van nozzles; 2. Vergelijkende nozzletest in een proefinstallatie. KIWA report SWE-202, 1978

[13] Kitchener, J.A., Gochin, R.J.: The Mechanism of Dissolved Air Flotation for Potable Water: Basic Analysis and a Proposal. Wat. Res. *15* (1981) 585–590

[14] DHV Consultants BV: NV Leidse Duinwater Maatschappij; Flotatie en saturatie. Dossiernumber 1-4378-44-11, October 1987

[15] Groot, C.P.M. de, Breemen, A.N. van: Ontspanningsflotatie en de bereiding van drinkwater. Announcement no. 11 of the Division of Sanitary Engineering, Department of Civil Engineering of the Technical University Delft, June 1987

[16] Huisman, L.: Sedimentation and Flotation. Syllabus college n4, TU Delft, 2nd revised edition, 1982

G.J. Schers
c/o Witteven + Bos
P.O. Box 233
NL-7400 AE Deventer
The Netherlands

J.C. van Dijk
Department of Civil Engineering
Division of Sanitary Engineering
Technical University Delft
P.O. Box 5048
NL-2600 GA Delft
The Netherlands

Experimental Studies on the Kinetics of Flotation

J. Heinänen, P. Jokela, and J. Peltokangas

1. Introduction

Dissolved-air flotation (DAF) is a combination of several physical and chemical unit processes and operations. This study is focused on the phenomena in the flotation tank and especially on the formation and rising of air-bubble floc agglomerates (called agglomerates only). The objective of the study is to investigate the functioning mechanisms of different factors and to establish their interrelationship. The results can be applied in the design of flotation tanks, because, for instance, the rising velocity of agglomerates is related to the overflow rate in the clarification tank.

2. Potable Water Treatment Plants Using DAF

The flotation process has been known for over 100 years, first in the mining industry and then in industrial waste treatment. In potable water treatment, however, DAF is a rather new method. Zabel (1985) reported that in the UK 20 plants were in use or under construction. The first one was started up in 1979 (Anon 1980). Malley (1988) reported that in the USA only 3 plants were working in 1988. We know that in South Africa this process has gained wider acceptance (Botes and van Vuuren 1990) and that there are plants of this type in Sweden. According to Janssens (1992), DAF is now receiving increasing attention also in the Netherlands, Belgium and France.

It is therefore perhaps surprising to know that there are 36 plants which supply water in Finland (serving about one million people), which use DAF as part of their treatment processes. The first plant was started up in 1965 and the latest one in 1990. DAF is very suitable for Finnish raw waters which are usually clear and humic and have low temperatures over a long winter period. Heinänen previously published reports on these water supplies (Heinänen 1988, Heinänen 1990).

The locations and capacities of the plants are listed in Table 1.

Most of the plants have been built within the boundaries of existing water treatment works. In this way a higher capacity and better purification results have been achieved without constructing completely new facilities.

The DAF-plants usually work well, i.e. the purification results are good and mechanical problems rare. But if we look at the dimensioning of the plants, we

Tab. 1. Finnish water supplies using DAF as clarification process

Loacation	Capacity m³/d	Location	Capacity m³/d	Location	Capacity m³/d
Turku	97 000	Seinäjoki	15 000	Uusikaupunki	5 200
Vaasa	59 000	Pietarsaari	12 000	Outokumpu	3 700
Kuopio	41 000	Kajaani	11 000	Toijala	3 500
Kemi	39 000	Jyväskylä, rural	10 000	Kemijärvi	3 300
Oulu, Hintta	26 000	Pieksamäki	9 100	Espoo	2 500
Rauma	24 000	Varkaus	8 600	Vammala	2 100
Oulu, Kurkela	22 000	Kirkkonummi	8 400	Leppävirta	1 800
Savonlinna	21 000	Kuusankoski	6 900	Nokia	1 800
Jyväskylä, urban	20 000	Lohja, urban	6 800	Hamina	1 000
Tampere	20 000	Lohja, rural	5 800	Tornio	700
Pori	20 000	Kaarina	5 700	Kirkkonummi	600
Porvoo	15 000	Iisalmi	5 700	Luoto	200

easily see that nearly all of them are oversized. An example is given in Fig. 1.
The overflow rate in flotation tanks is usually less than 8 m/h, although nowa-
days this figure is more often considered a minimum rather than a maximum.
Overdimensioning does not cause any harm in operation but the costs of investment are unnecessarily high. It also shows a lack of knowledge about the
theory of flotation.

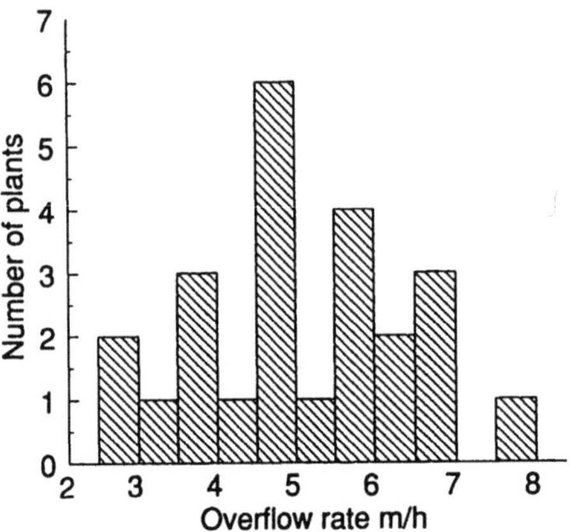

Fig. 1. Distribution of overflow rates in Finnish water supplies using DAF

The coagulants and the pH of coagulation in the Finnish DAF-plants are
summarized in Table 2. The coagulation chemical dosage is related to the
COD_{Mn} of the raw water. In some cases the pH of coagulation is lower in
the summer than in the winter, e.g. 5.8 and 6.2 respectively, when aluminium
sulphate is used.

Tab. 2. Coagulants and the coagulation pH used in the Finnish DAF-plants

Coagulant	Number of plants	pH of coagulation	
		Average	Variation
Aluminium sulphate	19	5.9	5.5–6.3
Polyaluminium chloride	6	5.9	5.4–6.1
Finnferri	4	5.0	4.8–5.2
Ferric sulphate	1	5.0	

3. Kinetics of Flotation

Three Phases of Flotation

The forces controlling *sedimentation* and *flotation* are the same but not similarly directed. *Gravity* affects the particles in both cases. In sedimentation it causes a downward movement of particles. If air bubbles are attached to particles, an opposite force, namely, *buoyancy* overcomes gravity and the agglomerates move upward. In both cases *drag resistance* opposes the movement.

Hahn (1990) suggests that in flotation at least three phases have to be distinguished: [1] bubble formation, [2] bubble attachment and [3] motion of the agglomerates through the flotation tank.

Bubble Formation

In DAF, the recycle water discharges through nozzles to the flotation tank, dissolved air precipitates (or is released) and small bubbles are formed. A certain minimum velocity in the nozzle is required for the air precipitation. Kitchener and Gochin (1981) state that a certain degree of cavitation in the nozzles is required for bubble precipitation.

The bubble size is dependent on various factors. First, the increase in saturator pressure promotes the formation of smaller bubbles up to 500 kPa. Above this pressure the effect is negligible (Kouti, 1983). Many salesmen of nozzles insist that the type of nozzle is the key to flotation. This is not necessarily true, however. Rees et al (1979, cited by Zabel and Melbourne 1980) compared needle valves and the patented nozzles of the Water Research Centre. In both cases the bubbles were within a range of 10–120 μm. The percentage of smaller bubbles was higher when the WRC nozzle was used, but the difference was not significant. The aforementioned range of bubble sizes is also given in many other literature references.

Kitchener and Gochin (1981) photographed bubbles in the vicinity of the nozzles and, in addition to small bubbles, they also found bigger bubbles. For instance, they observed 0.5 mm bubbles at a distance of 7 cm from the nozzle.

Flotation is favoured by small bubbles but apparently bigger ones cannot be avoided.

The constituents of the water also have an effect on bubble size. Cassel et al (1975) found that small amounts of ethanol have a drastic reducing effect. Consequently, one can assume, for instance, that humic substances have a similar effect.

It is surprising to learn how large the concentration of bubbles really is. Edzwald and Wrangles (1990) give an example: if the saturator pressure is 483 kPa and the recycle ratio 5 %, the volume concentration is 2.9 ml/l. If the bubble diameter is 40 μm, the concentration is 10^5 bubbles/ml.

Prerequisites for Agglomerate Formation

It is commonly known that floc formation is possible only if the colloids are destabilized or, in other words, if the electric charge around the particles is compensated. This charge is called the zeta-potential (ZP). Nearly all particles dispersed in natural waters have a negative charge. The compensation is achieved by coagulants, but even after coagulation, the flocs may still have a small charge, usually negative. The air bubbles in DAF have a negative charge, as well. This is from -50 to -300 mV according to Tambo et al (1985). The magnitude of the charge depends on the pH, but there is hardly any practical way to change it.

Before the collision of a floc and an air bubble becomes possible, the repulsive coulombic forces must become as near to zero as possible. Because pH affects both the charge of a floc and that of a bubble, it is obvious that pH plays a key role in floc-bubble attachment. In practice, the charge of the flocs is made as close to zero as possible, even a little to the positive side. Grutsch (1977) suggests the use of titration curves (ZP versus dosage) to find the best coagulant dosage. Because ZP is directly proportional to electrophoretic mobility (EPM), low ZP also means low EPM. Edzwald et al (1990) achieved the best aggregation when EPM was as near zero as possible.

On the other hand, many researchers have found that air bubbles attach to hydrophobic particles better than to hydrophilic ones. This is explained by a layer of water molecules around the particles. Where hydrophobic particles are concerned, the layer is thin and does not cause any hinderance to attachment. In the case of hydrophilic particles, the situation is reversed.

Bubble Attachment

Malley (1988) and Edzwald et al. (1990) developed a mathematical model for DAF performance. This is based on the removal efficiency of particles *by a single bubble*.

Particle deposition on the bubble surface can occur by means of three different mechanisms: Brownian diffusion (the efficiency η_D), interception (η_1) and gravity settling (η_G). These mechanisms are the same as in filtration. In DAF there are bubbles instead of sand corns. η is defined as follows:

$$\eta = \frac{\text{particle} - \text{bubble collision rate}}{\text{particle} - \text{bubble approach rate}} \tag{1}$$

The total single collector collision efficiency (η_T) is considered to be equal to the sum of the individual terms.

The final model is:

$$\frac{dN_p}{dH} = -\frac{3}{2} \cdot \frac{\alpha_{pb} \cdot \eta_T \cdot \phi_b \cdot N_p}{d_b} \tag{2}$$

N_p is the particle (without air bubbles) concentration and H the flotation chamber depth. Instead of H, time can also be used. The left-hand side of the equation represents the decrease in particle concentration when going upwards in the flotation tank or when progressing in time. The greater the term dN_p/dH, the quicker the aggregation occurs.

α_{pb} is the attachment efficiency or the fraction of successful collisions. It depends on particle-bubble charge interactions and the hydrophilic/-phobic nature of the particles as discussed earlier.

In η_T the gravity component is the least important. Larger floc sizes (above 10 μm) contribute to an increase in η_T. However, if the floc mass is too big compared to the number of attached air bubbles, the rising velocity will consequently decrease.

ϕ_b is the bubble volume concentration. High figures favour flotation or, in other words, large amounts of air are favourable.

d_b is the bubble diameter. This term indicates that small bubbles are favourable.

In the aforementioned equation there is no temperature-dependent term. We note, however, that the equation describes the *attachment mechanism*, not the whole DAF process. The next step, the rising of agglomerates, depends on temperature, as will be shown later.

Tambo (1990) has a different approach to the problem. He gives the following equations for a relative concentration of agglomerates:

$$N_{F,i} = \frac{m_F!}{i!(m_F - i)!} \cdot e^{-(1+F)^3 T} \cdot \left[e^{(1+F)^3 T/m_F} - 1 \right]^i \tag{3}$$

The time variation of mean air bubble number on flocs of diameter F is as follows:

$$\bar{i}_F = m_F \left[1 - e^{-(1+F)^3 T/m_F} \right] \tag{4}$$

In those equations:

$F = d_f/d_a$: the nondimensional floc diameter derived from floc and air bubble diameters d_f and d_a. $N_{F,i} = n_{fi}/n_f$: the nondimensional concentration of flocs with i attached air bubbles derived from the concentration of such flocs with and without air bubbles.

m_F is the maximum number of air bubbles that can be attached to a floc having the diameter F. It is defined as $m_F = \alpha_0 F^2$, where α_0 is the initial collision attachment factor.

T is the nondimensional elapsed time.

Hence, Tambo assumes that more than one air bubble can attach to a floc and that there is a maximum value for this number.

An increase in $N_{F,i}$, the concentration of flocs (with air bubbles), means progress with regard to attachment. $N_{F,i}$ is a function of the time, the floc diameter and the number of attached air bubbles. This relationship is rather complicated. Tambo gives a graph (Fig. 2) that shows that with time, the number of flocs having a larger number of attached air bubbles, increases.

The attachment process is relatively rapid compared to the whole DAF process. Tambo found that for coloured water, treated with aluminium, only 15–20 seconds of mixing of flocs and air bubbles at a G-value of about 10 s^{-1} are required to reach the maximum number of attachments.

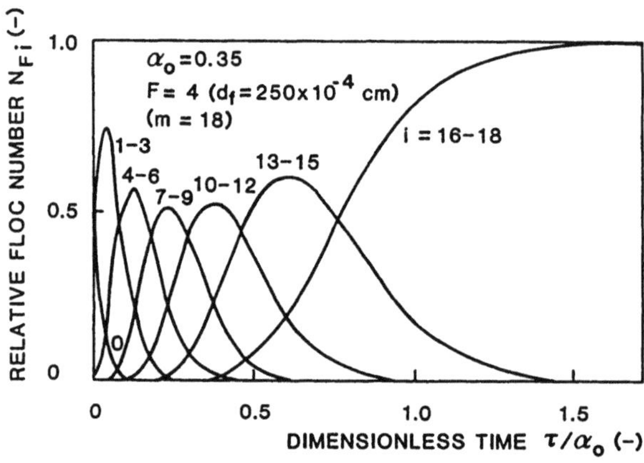

Fig. 2. Progress of air bubble attachment according to Tambo (1990)

Motion of Agglomerates

Bratby and Marais (1975) developed a concept for the motion of agglomerates. Their investigations dealt with waste water but, nevertheless, the same idea can also be applied to lake water. In the upflow section of a flotation tank the hydraulic flow is upward, and in the same direction as the buoyancy. In this stage the flocs have no chance to settle whether they are attached to air bubbles or not. After the threshold (= the back wall of the upflow section) the hydraulic flow turns downward or opposite to the buoyancy. If the velocity of the agglomerates is less than the hydraulic velocity, the agglomerates flow out. If the velocity of the agglomerates is greater, they stay in the sludge layer. Bratby and Marais call the velocity that just keeps the agglomerates in the sludge *a limiting downflow rate*, v_L. It is defined as (total flow into unit)/(plan area of flotation tank). It has the same dimensions as overflow rate, m/h $(m^3/m^2\,h)$.

Bratby and Marais state that v_L depends on the amount of air fixed to the agglomerates. They express it as mg air/mg solid mass of flocs. This is a useful parameter in waste water technology but, when the TS concentration is less than 400 mg/l, the required amount of air is no longer related to the solids concentration (Kiuru 1988). It is the better to express the amount of air as grams per cubic metre inflow.

Tambo et al. (1986) derived an equation for the rising velocity of agglomerates:

$$v = \frac{4g}{3\gamma K} \cdot \frac{i - a(d_f/1)^{-K\rho} \cdot (d_f/d_a)^3}{i + (d_f/d_a)^3} \cdot d_{af}^2 \tag{5}$$

In this equation:

g is the gravitational acceleration.
K, $K\rho$ and a are constants.
i is the number of air bubbles attached to a floc.
d_f and d_a are the diameters of flocs and air bubbles, respectively.
γ is the kinematic viscosity of water.
d_{af} is the diameter of an agglomerate that is approx. equal to d_f.

4. Experimental Materials and Methods

Raw Water Characteristics

The raw water was taken from a small lake, Veittijärvi, close to the research station. From November to February it had the composition given in Table 3.

Tab. 3. Characteristics of the raw water

Parameter	Unit	Number of tests	Average	Variation
Temperature	°C		5.5/2[xx]	
pH		15	6.86	6.68–7.01
Alkalinity	mmol/l	3	0.4	
Colour	mg Pt/l	19	33	25–40
Turbidity	FTU	13	1.6	1.0–2.6
COD_{Mn} (x)	mg O_2/l	27	4.1	3.0–5.7
UV-abs. (250 nm) (x)		27	0.117/0.17[xx]	0.109–0.269[xx]
Aluminium (x)	mg Al/l	7	< 0.05	
Iron (x)	mg Fe/l	11	0.08	0.03–0.16

Analyses marked with (x) were carried out in Kemira's Oulu Research Centre. The other analyses were carried out on site. (xx) the two figures mean before and after ice covering.

As seen from the characteristics, the raw water is clear and slightly humic, much better than is usual for Finnish lake waters. It was used because of its vicinity to the research station. The alkalinity is rather high (compared to usual Finnish lake waters) resulting in a need to dose sulphuric acid in some cases to achieve an optimal pH.

Pilot Plant

The tests were carried out in a pilot plant which is shown schematically in Fig. 3. The following list gives the main features of the plant:

- volume of flocculation tanks 2×1.0 m^3
- volume of flotation tank 1.35 m^3
- volume capacity 5 m^3/h
- retention time in flocculation 28 min
- retention time in flotation 12 min
- overflow rate in flotation tank with 10 % recycle water 10.5 m/h
- volume of the saturator 0.10 m^3.

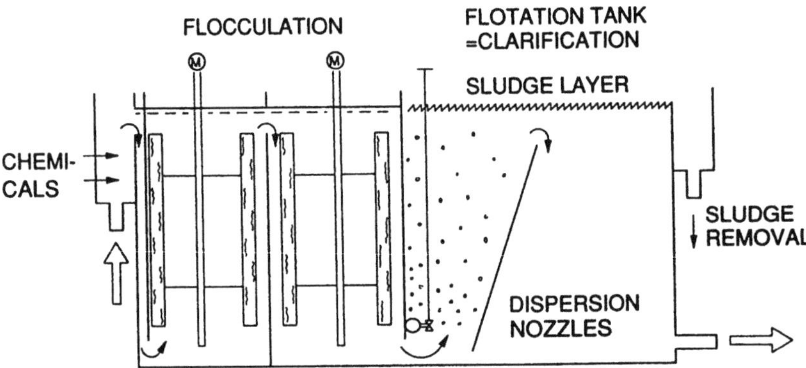

Fig. 3. Schematic diagram of the flotation pilot plant

The flow rate of the influent was measured by an ordinary water meter. The coagulant solutions, as well as sulphuric acid, were dosed by adjustable dosing pumps. The flocculation was carried out by two consecutive paddle mixers.

The inflow to the saturator was measured by a rotameter and adjusted by a segment valve. The pressure in the saturator was 600 kPa. Because the saturator was unpacked, only about 50 % efficiency was achieved in saturation. The amount of air was measured using a method suggested by Vosloo et al. (1986). It was 5–13 g/m^3 influent.

The recycle water was discharged initially through patented nozzles, but air precipitation was not sufficient. Apparently the flow rate was too low for those nozzles. They were replaced by ordinary adjustable needle valves and the precipitation improved.

Sampling and Analysis

Effluent samples were composite samples from a period of one or two hours. The sampling interval was half an hour. Whatman GF/A glass fibre filters were used to get filtered samples. Analyses were the same as for raw water, excluding alkalinity.

Coagulants and Other Chemicals

The following coagulants were applied:

1. Finnferri (abbreviation FF) is a mixture of ferric chloride and sulphate solutions. The bulk solution contained 186 g Fe^{3+}/l or 124 g Fe^{3+}/kg. It was diluted to prepare a proper dosing solution.
2. Kempac 20 is a solution of polymerized aluminium chloride having a common formula $Al_n(OH)_m Cl_{3n-m}$. The bulk solution contained 97 g Al^{3+}/l or 72 g Al^{3+}/kg. It was diluted to prepare a proper dosing solution.

Sulphuric acid was used to adjust the pH.

Rising Velocity and Photography

In order to measure the rising velocity of the agglomerates, a small side flow was taken out of the flotation tank about 75 cm above the dispersion valves. It was siphoned through a thin (Ø 8 mm) copper tube to a lamella made of transparent acrylic plastic. The clearance between the sides of the lamella was 25 mm. The arrangement is shown schematically in Fig. 4.

The agglomerates rising in the lamella were surveyed through a theodolite and the velocities were measured. The measured velocities varied due to the size of the agglomerates and the number of attached air bubbles. An ideal way to present the results would be a velocity distribution curve. However, precision of the observations was not good enough for this system. In the following, *mean* velocities are presented instead (each case includes about 20 observations). This figure cannot be used as such for designing the flotation tank, because the overflow rate should be more related to the *minimum* rising velocity.

The same lamella was used for photography, too. Photographs were taken from a distance of 55 mm (from the lamella side to the lens) using an ordinary camera with a macro lens.

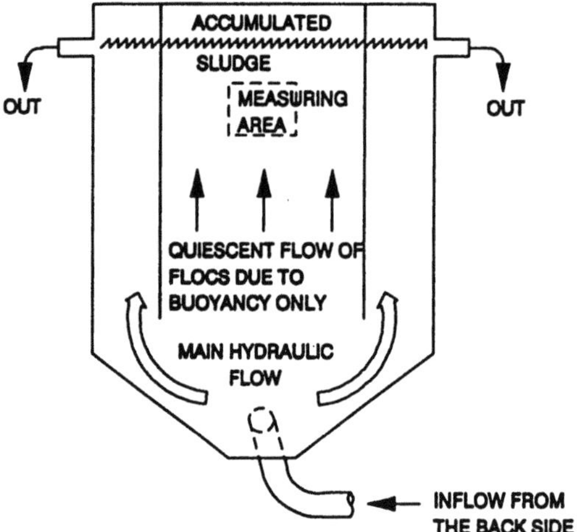

Fig. 4. Schematic diagram of the lamella used in the experiments

5. Results and Discussion

General

At first preliminary tests were carried out. These had two purposes: (1) to gain experience with the pilot plant and (2) to find the right dosages.

Concerning FF, the best results were obtained at pH 5.25 with the dosage 7.3 mg Fe/l. The subsequent jar tests showed that the dosage can essentially be lower if acid is used for pH control. With Kempac, the corresponding figures were 5.9 and 3.7 mg Al/l, respectively. It was not necessary to use acid.

UV-absorption at 250 nm was used for measuring organic matter (in addition to COD). However, residual iron also gives a certain absorbance at the same wavelength. Therefore, if iron is used as a coagulant, it is meaningless to give UV-absorbances unless the sample is filtered.

A water purification process can be considered successful, if

- residual organic matter, measured as COD and UV-absorption, is low,
- residual of the coagulant, iron or aluminium, is low and
- colour and turbidity are low.

In addition to that, flotation is successful if the rising velocity is high, because it corresponds to a high overflow rate. This means a smaller flotation tank and a consequent reduction in costs. These are the criteria used in the following chapters, when the success or failure of the process is estimated.

Tests with Finnferri

Results from tests 1–10 are presented in Table 4 (amount of air, dosage and chemical parameters) and in Fig. 5 (rising velocities and ZP). If the dosage

Tab. 4. Results from the tests with FF as a coagulant and using sulphuric acid for pH control

Parameter	Test numbers									
	1	2	3	4	5	6	7	8	9	10
Air amount $[g/m^3]$	6.5	5.0	5.1	9.9	9.9	11.1	12.3	12.3	12.3	12.3
Dosage [mg Fe/l]	4.8	4.7	4.7	4.8	4.5	4.8	2.1	3.2	3.5	3.3
ZP [mV]	-5	-10	-13			-12	-15	-10	-11	-14
pH	4.15	4.4	4.95	4.05	4.45	5.05	4.55	4.55	4.9	5.5
Colour [mg Pt/l]										
C	70	90	90	80	60	60	60	60	60	60
F	7.5	5	5	5	5	5	5	5	5	10
Turbidity [FTU]										
C		4.4	6.7	7.4	5.0	3.8	5.0	5.0	4.4	5.6
COD_{Mn} $[mg\,O_2/l]$										
C	5.2	2.8	2.6	3.5	2.9	1.8	3.7	3.4	2.8	3.9
F	1.0	1.0	< 1.0	1.6	< 1.0	< 1.0	2.1	1.6	1.2	1.8
UV-abs. ×1000 [250 nm]										
F	85	41	38	120	40	38	76	62	57	108
Iron [mg Fe/l]										
C	4.8	3.9	4.1	4.2	2.4	2.0	1.9	2.3	2.0	3.2
F	1.0	0.27	0.12	1.5	0.35	0.12	0.41	0.54	0.29	0.68

Variables: pH, dosage and air amount. Constants: flow rate 5.0–5.1 m^3/h, overflow rate 10.5–11.4 m/h. C = clarified water or water after flotation; F = filtered water or water after filtration

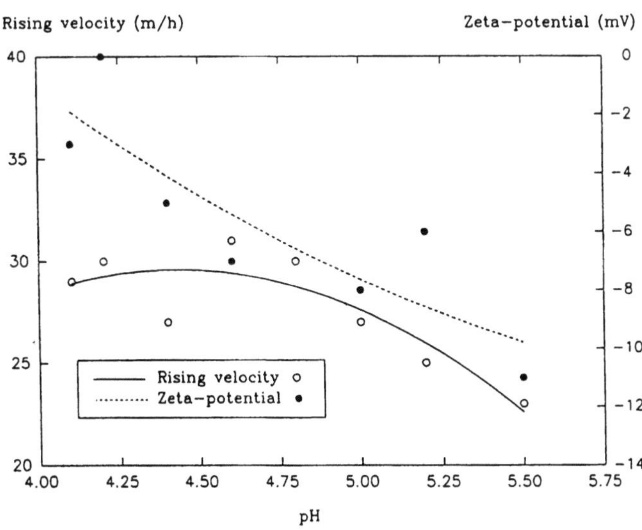

Fig. 5. Rising velocity and ZP as a function of pH. FF as a coagulant. pH controlled by the sulphuric acid

is sufficient, say more than 3.5 mg Fe/l, the results depend on pH more than anything else. The optimum pH is around 5.0 and, in this area, residuals of organic matter and iron have their lowest values. ZP is still negative and the rising velocity, 27 m/h, is not the highest of the observed values. When pH values decrease and, at the same time, ZP approaches zero, the rising velocity increases. However, such circumstances cannot be utilized, because they also mean an increase in the concentration of the residual coagulant.

In tests 1–3 and 4–6, the effect of the amount of air was tested. A higher amount favours better results after flotation but after subsequent filtration, the differences are negligible.

Tests 7–10 were carried out with a relatively high amount of air and a low dosage. Even then, the best results were reached at pH 5.0 but the dosage, 3.5 mg Fe/l, is obviously slightly too low, resulting in a higher residual iron concentration (0.29 mg/l versus 0.12 mg/l when the dosage was 4.7 mg/l). When pH goes up to 5.6, floc formation was no longer observed.

Tests with Kempac

The results from tests 11–17 are presented in Table 5 (amount of air, dosage and chemical parameters) and in Fig. 6 (rising velocities and ZP). The dosage and the corresponding pH are the most important parameters affecting the process, as in the tests with FF. ZP becomes zero at pH 6.15. At this point, the chemical parameters show a successful result. The rising velocity increases up to this point or slightly further but after that, on the acid side, it is nearly constant within the limits of the measuring accuracy.

The dosage in test 16 was obviously too low, resulting in an unsatisfactory purification result.

Test number 17 was carried out with a lower overflow rate (6.7 m/h) than all the other tests (around 11 m/h). pH was at the optimum value. Hardly any difference was found in the results. Consequently, the overflow rate of 11 m/h is not too high. This is a logical conclusion from the rising velocity measurements, too.

In tests 12 and 13 the effect of the amount of air was checked. There was hardly any difference in water quality. The rising velocity is higher when more air is used, but so few tests were carried out that this result cannot be confirmed. Zabel (1985) recommends air in the amount of 8–10 g/m^3, and this is confirmed by these investigations.

Comparison of Iron and Aluminium as Coagulants

Measured in chemical parameters, iron gives the best results when ZP is −17 mV and aluminium when ZP is around 0. The corresponding rising velocity with iron is not the highest, rather this is achieved with aluminium. Aren't these results contradictory?

Tab. 5. Results from the tests with Kempac

Parameter	Test numbers						
	11	12	13	14	15	16	17
Air amount [g/m^3]	13.0	6.9	11.6	12.3	11.6	12.3	10.3
Dosage [mg Al/l]	3.4	3.1	3.1	2.5	1.9	1.3	2.8
pH	6.0	6.05	6.05	6.2	6.35	6.4	6.2
Colour [mg Pt/l]							
C	20	25	30	15	25	60	25
F	5	5	5	5	5	10	5
Turbidity [FTU]							
C	3.6	3.8	4.8	3.6	3.6	6.0	3.3
COD$_{Mn}$ [mg O$_2$/l]							
C	1.5	1.9	2.0	2.5	3.1	5.0	1.5
F	0.4	0.2	0.2	1.3	1.7	2.8	0.2
UV-abs. ×1000[250 nm]							
C	84	99	119	90	115	231	86
F	42	38	45	53	54	92	45
Aluminium [mg Al/l]							
C	2.1	2.3	2.3	1.7	1.4	1.6	1.9
F	0.1	0.1	0.1		0.1	0.2	0.1

Variables: pH, dosage and air amount. Constants: flow rate 5.0–5.1 m^3/h, (tests 11–16) and 3.0 m^3/h (test 17), overflow rate 10.8–11.4 m/h (tests 11–16) and 6.7 m/h (test 17). C = clarified water or water after flotation; F = filtered water or water after filtration

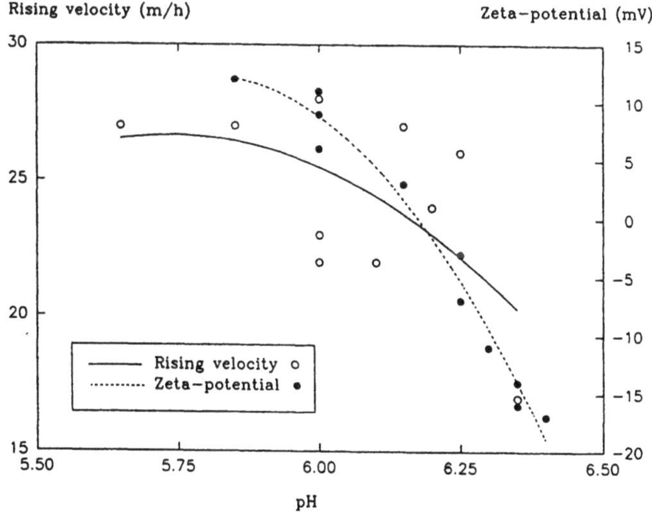

Fig. 6. Rising velocity and ZP as a function of pH. Kempac as a coagulant. pH controlled by the coagulant only

It is known that the isoelectric point of aluminium hydroxide is about 6.3 and, in practice, when humus is involved, the precipitation is carried out at a slightly lower pH. These tests, as well as many others, show that ZP is then 0. Hence the best precipitation point is equal to the isoelectric point.

The isoelectric point of ferric hydroxide is around 8 but, in practice, precipitation of humus with ferric coagulants takes place best at a considerably lower pH, around 5. These tests show that this is not the real isoelectric point (ZP = 0), which is still lower, around 4. But at pH 4 iron does not precipitate completely, resulting in higher residual iron concentrations in the water. For iron, therefore, the isoelectric point is not equal to the best coagulation point.

As stated before, air bubble attachment is best when ZP is zero, and consequently, the rising velocity is supposed to be highest then, too (due to the maximum number of attached air bubbles). Where aluminium is concerned, the observed rising velocity reaches the highest value when ZP is zero. Even where iron is concerned, the rising velocities increase when ZP approaches zero. As stated before, however, such circumstances are not feasible for other reasons. At optimum coagulation pH, the rising velocities are equal, about 27 m/h, for both coagulants.

One might find that chemical parameters do not produce such good results after flotation, and the final polishing is achieved after filtration. It is known that flotation of pure aluminium hydroxide is difficult (Malley 1988) and that humic substances are essential for successful flotation (Kitchener & Gochin 1981). The humic concentration in the raw water used in these tests was very low and, apparently, this is one of the reasons for the unsatisfactory clarification results. Another reason might be the sludge removal. In the pilot plant, it was not possible to remove sludge while the plant was running. It was found visually that some small flocs dropped off from the sludge layer. Similar tendency of floc escape was observed by Rees et al. (1980), when they made experiments using slightly humic raw water.

Behaviour of Agglomerates

Visual observations (in the lamella) showed that the number of attached air bubbles varied from 1 to about 20 per agglomerate and did not increase further in this stage. Apparently the attachment process was completed before water was taken to the lamella for velocity measurements (about 3 min). Furthermore, hardly any separation of air bubbles was observed in this stage.

The photographic methods have not yet been perfected, and, despite hundreds of pictures, only a little information has been obtained. Nevertheless, the pictures in Fig. 7 show distinctly that the flocs have various numbers of attached air bubbles. Some of the agglomerates include tens of bubbles but some have only a few. Of course, the floc size has an influence on that, but if there are some other reasons for the variation, these are not yet clear. In addition, the different amounts of air in the tests can easily be visualized in the pictures.

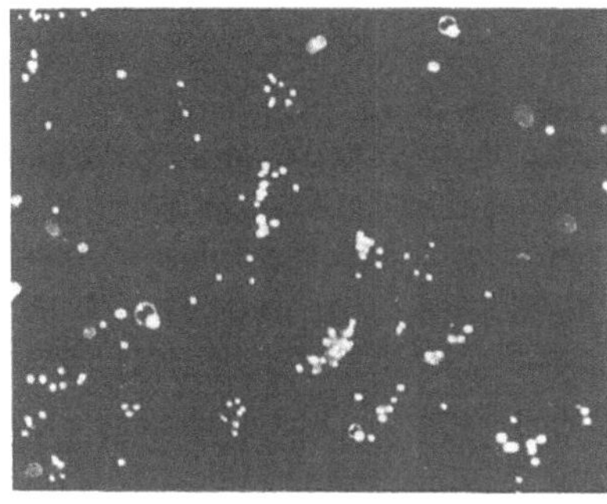 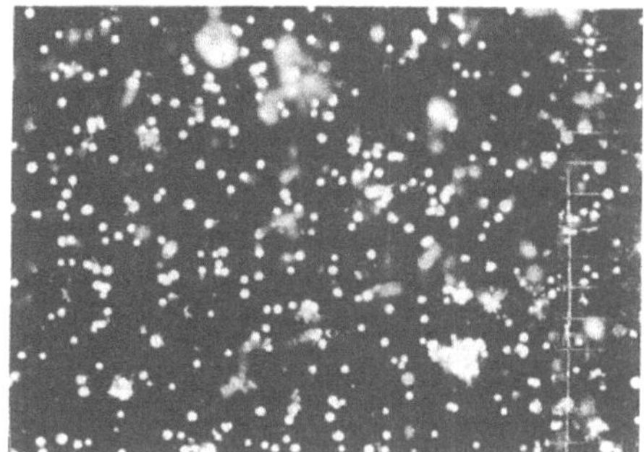

Fig. 7. Agglomerates with different amounts of air   I = 1 mm

6. Conclusions

In the past, when flotation was the object of research, its performance was usually followed using chemical parameters, sometimes also with ZP of EPM measurements. Some other reports on rising velocities also exist. In the investigations reported here, however, all of these elements were incorporated. Only in this way can a comprehensive picture of the whole flotation process be achieved. For the time being, the following conclusions can be presented.

(1) A proper pH value is the most important factor both for coagulation and flotation. It depends on the type of the coagulant and raw water quality.
(2) At the optimal pH, zeta potential is approx. zero when Kempac is used as the coagulant. When Finnferri is used, zeta potential is on the negative side of zero at the optimal coagulation pH.
(3) At the optimal pH, the mean rising velocity is about 27 m/h, regardless of the type of coagulant. When Kempac is used, this is also the highest velocity, but with Finnferri, the rising velocity still increases slightly when the pH decreases.
(4) These investigations encourage the use of higher overflow rates than usual, probably up to 15 m/h.

References

Anon: The First Dissolved-air Flotation Plant for Water Clarification in the UK. Water Services *84* (1007) (1980) 19–23
Botes, V., van Vuuren, L.R.J.: Dissolved-air Flotation for the Removal of Algae and Inorganic Turbidity on a Large Scale. Water Supply *8* (1990) Jönköping, 133–139
Bratby, J., Marais, G.V.R.: Dissolved-air Flotation – an Evaluation of Inter-Relationships between Process Variables and their Optimisation for Design. Water SA *1* (2) (1975) 57–69
Cassel, E.A., Kaufman, K.M., Matipevic, E.: The Effect of Bubble Size on Microflotation. Wat. Res. *9* (1975) 1017–1024
Edzwald, J.K., Wingler, B.J.: Chemical and Physical Aspects of Dissolved-air Flotation for Removal of Algae. Aqua *39* (1) (1990) 24–35

Edzwald, J.K., Malley, J.P., Yu, C.: A Conceptual Model for Dissolved-air Flotation in Water. Water Supply *8* (1990) Jönköping, 141–150

Grutch, J.F.: The Control of Refinery Mechanical Waste Water Treatment Processes by Controlling the Zeta Potential. Symposium on Physical-Mechanical Treatment of Waste Waters, Cincinnati, Ohio, USA, April 5–6, 1977 (EPA) pp. 44–74

Hahn, H.H.: Sedimentation and Flotation: Present Status and Necessary Development: an Introduction. Water Supply *8* (1990) 111–122

Heinänen, J.: Use of Dissolved-air Flotation in Potable Water Treatment in Finland. Aqua Fennica *18* (2) (1988) 113–123

Heinänen, J.: Flotation in Potable Water Treatment. Tampere University of Technology. Institute of Water and Environmental Engineering. Publication A 43. Tampere 1990, 59 pages (in Finnish)

Janssens, J.G.: Developments in Coagulation, Flocculation and Dissolved-air Flotation. Water/Engineering & Management (1) (1992) 26–31

Kitchener, J.A., Gochin, R.J.: The Mechanism of Dissolved-air Flotation for Potable Water: Basic Analysis and a Proposal. Wat. Res. *15* (5) (1981) 585–590

Kiuru, H.: High-grade Wastewater Treatment through Flotation and Filtration. Helsinki University of Technology, Laboratory of Sanitary and Environmental Engineering, Otaniemi, Finland 1988 (in Finnish)

Kouti, P.: Clarification by Flotation. Rakennustekniikka *39* (1983) 559–565 (in Finnish)

Malley, J.P.: Fundamental Study of Dissolved-air Flotation for Treatment of Low Turbidity Waters Containing Natural Organic Matter. Univ. of Massachusetts, Amherst, USA, 1988, 385 pages

Rees, A.J., Rodman, D.J., Zabel, T.F.: Operating Experiences with Dissolved-air Flotation on Various Raw Waters. Aqua (8) (1980) 0170–0177

Tambo, N.: Optimization of Flocculation in Connection with Various Solid-Liquid Separation Processes. In: Chemical Water and Wastewater Treatment, H.H. Hahn and R. Klute (eds.). Springer, Berlin Heidelberg New York 1990, pp. 17–32

Tambo, N., Matsuri, Y., Fukusi, K.: A Kinetic Study of Dissolved-air Flotation. World Congress III of Chemical Engineering, Tokyo, 1986, pp. 200–203

Tambo, N., Igarashi, T., Kiyotsuka, M.: An Electrophoretic Study of Air Bubble Attachment to Aluminium Clay/Color Flocs. J. Japan Wat. Works Ass. *54* (1) (1985) 2–6 (in Japanese)

Tambo, N., Watanabe, Y.: Physical Characteristics of Flocs – I. The Floc Density Function and Aluminium Floc. Wat. Res. *13* (1979) 409–419

Vosloo, P.B.V., Williams, P.G., Rademan, R.G.: Pilot and Full-Scale Investigations on the Use of Combined Dissolved-air Flotation and Filtration (DAFF) for Water Treatment. Wat. Poll. Contr. *85* (1) (1986) 114–121

Zabel, T.: The Advantages of Dissolved-air Flotation for Water Treatment. J. AWWA *77* (5) (1985) 42–46

Zabel, T.F., Melbourne, B.E.: Flotation. In: Developments in Water Treatment – 1, M.W. Lewis (ed.). London 1980, pp. 139–191

J. Heinänen, P. Jokela, and J. Peltokangas
Tampere University of Technology
Institute of Water and Environmental Engineering
P.O. Box 600
SF-33101 Tampere
Finland

Algae, Coagulation, and Ozonation

J.K. Edzwald and A. Paralkar

1. Introduction

In addition to the direct disinfecting and oxidizing benefits of ozone, several secondary benefits have been attributed to its use. These include 1) microflocculation (i.e., the spontanaeous flocculation of particles), 2) a reduction in coagulant dosages, 3) improved water quality (usually measured as turbidity) following filtration through deep bed granular filters, and 4) a reduction in the filter bed head loss gradient. How ozone causes a reduction in the coagulant dose or induces flocculation is not well understood. Several proposed mechanisms are listed and discussed in a recently published book on ozone (Langlais, Reckhow, and Brink, 1991). These include: 1) increased Al complexation with natural organic matter (NOM), 2) increased Ca complexation with NOM, 3) reduction in particle stability due to loss of adsorbed organic matter from natural particles, 4) polymerization of NOM, 5) break-up of metal-organic complexes such as iron yielding *in situ* production of metal coagulant, and 6) reactions with algae.

Much of the previous research on preozonation effects on coagulation and flocculation has been directed at the role of humic substances (Farvardin and Collins, 1989), particularly the effects of ozone on the interaction of humic substances with various mineral particles (Grasso and Weber, 1988; Dowbiggin and Singer, 1989; Jekel, 1990). Some of this work and other research has identified water hardness or calcium as being an important factor. Chang and Singer (1991) investigated several water supplies in the United States and found that the TOC and hardness levels of the raw waters have a great impact in determining whether preozonation affects the flocculation of particles. They concluded that "optimal ozone induced coagulation occurs with hardness to TOC ratios > 25 mg CaCO$_3$/mg C and ozone doses of about 0.4–0.8 mg O$_3$/mg C".

Less research has been done on the subject of this paper, the role of algae in ozone induced coagulation and flocculation. This paper has three goals. First, the effect of ozone on reducing coagulant dosages in treating synthetic and natural waters containing algae is examined. Second, the effect of ozone on the flocculation kinetics of algal suspensions is examined – i.e., microflocculation. Third, ozone effects on algal size distributions, algal particle volumes, particle charge, and release of EOM (extracellular organic matter) are presented.

2. Background

2.1 Ozone Effects on Algae

There have been observations that ozone improves the removal of algae and that the presence of algae in water supplies subject to preozonation improves the coagulation and flocculation of particles (Langlais, Reckhow, and Brink, 1991). How does ozone accomplish these two benefits? One possibility, often cited, is attributed to biopolymers (polysaccharides and anionic polymeric acids such as alginic acid). Alginic acids are used in water treatment as natural polymers so that a model of particle flocculation by interparticle bridging from anionic biopolymers released from algae is plausible. An important related question is whether ozone causes cell lysis releasing intracellular organic matter (IOM) to solution or is it extracellular organic matter (EOM) that is important? In fact, it is frequently stated that ozone causes cell lysis. If the IOM were of sufficient molecular weight and structure, it could flocculate mineral particles but obviously not the destroyed algae. Cell lysis and release of IOM is probably not a likely mechanism since ozone does not destroy (lyse) most algae cells at ozone dosages frequently used to aid coagulation and flocculation such as 3 mg/l or less. This is shown in the results section (Sect. 4.3) of this paper. In fact, high concentrations of released intracellular or extracellular organic mattter is more likely to impair coagulation by increasing the coagulant demand or by adsorbing and stabilizing mineral particles (Akiba et al., (1990), Bernhardt et al. (1986), Hoyer et al. (1987)).

Algae during growth excrete polysaccharides and other polymeric matter. The amount and structure of this EOM attached to the cell surfaces depends on algae type and their physiological growth phase (Lüsse et al., (1985), Bernhardt and Clasen, (1991)). The adsorbed and EOM in solution may consist of polysaccharides, pectins, lipoproteins, and polyamino acids. The EOM can carry a negative charge and if of sufficient molecular weight (MW) may adsorb on adjoining particles producing flocculation via the interparticle flocculation model. Both the molecular weight and concentration of the EOM are important considerations. Bernhardt et al. (1986) report that low MW fractions of EOM do not act as polymeric flocculants. In his experiments low MW fractions were defined as less than 2 KDal (kilodaltons). Bernhardt et al. (1986) report that EOM may aid or inhibit flocculation depending on the concentration. At low concentrations (1 mg/l or less as DOC), the EOM aided flocculation. Higher concentrations inhibited iron coagulation. Hoyer et al. (1987) report that whether ozone aids or inhibits flocculation depends on the algae type, the concentration of EOM, and the absorbed ozone dose. At low ozone doses (1 mg O_3/mg C), ozone may aid flocculation depending on the type of algae present. At higher ozone doses (greater than 1 mg O_3/mg C), flocculation was impaired. This was attributed to a breakdown in the molecular weight of the EOM and an increase of the acidity of this low molecular weight material.

2.3 Flocculation Kinetics

Particle flocculation models are based on the Smoluchowski equations. The appropriate equation describing particle flocculation kinetics for homogeneous suspensions undergoing mixing in a flocculation reactor is

$$\frac{dN}{dt} = \frac{-4\alpha\phi GN}{\pi}$$ (1)

N Particle concentration at any time t
N_0 Initial particle concentration
α Stability factor
ϕ Particle concentration on a volume basis
G Mean velocity gradient
t Time
π Mathematical constant, 3.14

An integrated form of the equation follows

$$\ln(N/N_0) = \frac{-4\alpha\phi Gt}{\pi}$$ (2)

The model directly describes particle flocculation or transport, and also accounts for particle stability through α, the particle stability factor or collision efficiency factor. Low alpha values, 0.01 or less, indicate stable particles that flocculate very slowly whereas higher values approaching 1 indicate destabilized particles that flocculate rapidly. Singer and co-workers (Dowbiggin and Singer, 1989, Chang and Singer, 1991) have used the above model to characterize the effects of preozonation on flocculation kinetics and particle stability. This approach is also used here in examining ozone effects on the flocculation rates of various algae.

3. Experimental Procedures

3.1 General

Four algae were tested: 1) *Chlorella vulgaris* (green alga), 2) *Scenedesmus quadricauda* (green alga), 3) *Cyclotella sp.* (diatom), and 4) *Synura petersenii* (golden brown alga). Experiments with synthetic waters were carried out under controlled conditions of pH (7), inorganic chemistry (NaHCO$_3$ at 10^{-3} M and Ca^{++} added as CaCl$_2$ at 0, 30, or 50 mg/l CaCO$_3$), algae growth phase (log), alga cell concentration, and absorbed ozone dose. Most experiments were done with ozone as an independent varible at 0, 1, and 3 mg/l – some additional tests were done at higher absorbed ozone doses to determine effects on algae cells.

Some of our experiments were done using water collected from the Quabbin and Wachusett Reservoirs. The Massachusetts Water Resources Authority

(MWRA) provides water for Boston (U.S.A.) and surrounding communities serving 2.5 million people. The water supply is taken from the protected upland Quabbin Reservoir and the Wachusett Reservoir without filtration – the Quabbin flows by tunnel into the Wachusett Reservoir. A water plant is being considered on the Wachusett Reservoir site. Due to the Surface Water Treatment Rule (SWTR) dealing with filtration and disinfection and other anticipated regualtory requirements dealing with disinfection by-products, bench-scale and pilot plant-scale studies have been taking place since 1990 to determine effective treatment methods and to determine the role of preozonation in coagulation and filtration.

Typical water quality data are shown in Table 1. These supplies are high quality of low turbidity, low color, and low TOC with low alkalinity and hardness. The Wachusett Reservoir is oligotrophic, but does experience seasonal algae problems with occurrences of certain diatoms (*Asterionella* and *Cyclotella*), blue-greens (*Anabaena*), and golden-browns (*Synura*). Studies of the effects of ozone on coagulant dose and flocculation kinetics were done in our laboratory for water samples collected from both reservoirs and spiked with either *Cyclotella* or *Synura*.

Tab. 1. Typical water quality data for the Quabbin and Wachusett Reservoirs

Parameter	Quabbin	Wachusett
pH	6.1	7.2
Alkalinity, mg/l $CaCO_3$	4.0	7.5
Hardness, mg/l $CaCO_3$	11	14
Turbidity, NTU	0.3	0.6
UV at 254 nm, cm^{-1}	0.032	0.070
DOC, mg/l	2.4	3.0
TOC, mg/l	2.6	3.1

3.2 Jar Tests: Ozone Effects on Coagulant Dose

A cationic polymer was used as the sole destabilizing coagulant. The polymer is a high charge density polymer of the epichlorohydrin dimethylamine type, Clarifloc C 309P (Allied Signal Inc.). Metal salts (alum, iron salts) were not used to avoid enhancement of particle flocculation due to metal hydroxide precipitation. Coagulation following ozone was studied by varying the cationic polymer dose for a control case and two preozonation dosages: 0 mg/l (control), 1 mg/l and 3 mg/l. Ozone dosages are given on an absorbed basis. Particle destabilization and coagulation effectiveness were based on particle charge measuements (electrophoretic mobility (EPM); Rank Brothers, Mark II), residual turbidity, and residual algae (cell counts and optical density (OD) measurements.

3.3 Ozone Effects on Flocculation Kinetics

Flocculation kinetic experiments were used to measure the effect of ozone on alpha (particle stability factor or collision efficiency factor) values for the algae. No coagulant was used. Mixing was controlled (velocity gradients of 10 or 50 s^{-1}) as well as the initial algae cell concentration. Alpha values before and after preozonation were calculated from algae particle number measurements over time (Coulter Counter). In some experiments the effect of calcium was examined by varying Ca^{++} at 0, 30, and 50 mg/l as CaCO$_3$.

3.4 Other

Other measurements were made to provide information on the mechanisms of ozone effects on algae stability. Particle size distributions (PSDs) were measured before and after ozonation using a Coulter Counter. Also, scanning electron micrographs (SEM) were made of the algae before and after ozonation to determine whether ozone destroyed or physically altered the cells.

4. Results and Discussion

4.1 Jar Tests

Results for *Scenedesmus* are presented in Fig. 1. Without preozonation, the polymer dosage required for charge neutralization (EPM data) is 0.3 mg/l. The addition of ozone prevents charge reversal at polymer dosages up to 0.6 mg/l; however, *Scenedesmus* is coagulated, flocculated, and settled with or without ozone at a polymer dosage of 0.1 mg/l. Considering 0.1 mg/l as the effective coagulant polymer dose based on algae removal and residual turbidity, there is no evidence that ozone reduces the coagulant dose. Whether ozone reduces the coagulant dose or not, may be of minor importance here because the dose is low anyway. Similar results were observed for the other three algae tested.

The data of Fig. 1 show that ozone improves performance (slightly lower turbidities, optical densities, and greater removals of algal-size particles), and there is evidence of microflocculation – i.e., at no polymer addition, ozone caused the flocculation of this algae. This is seen in Fig. 1 for the turbidity, optical density, and fraction of algae remaining data at zero polymer dose. It was also visually observed that ozone induced flocculation of *Scenedesmus*. Data for the other three algae produced similar results for *Cyclotella* and *Synura*, but not for *Chlorella*. For *Chlorella*, there was no evidence of microflocculation.

Interestingly, the EPM data show that ozone affected the *Scenedesmus* cells such that additional polymer is needed to reduce the cell surface charge to zero, but does not cause a corresponding increase in polymer dose for effective coagulation. This was also observed for the other algae. For two of the algae (*Synura* and *Cyclotella*), the effect of ozone on surface charge was examined in

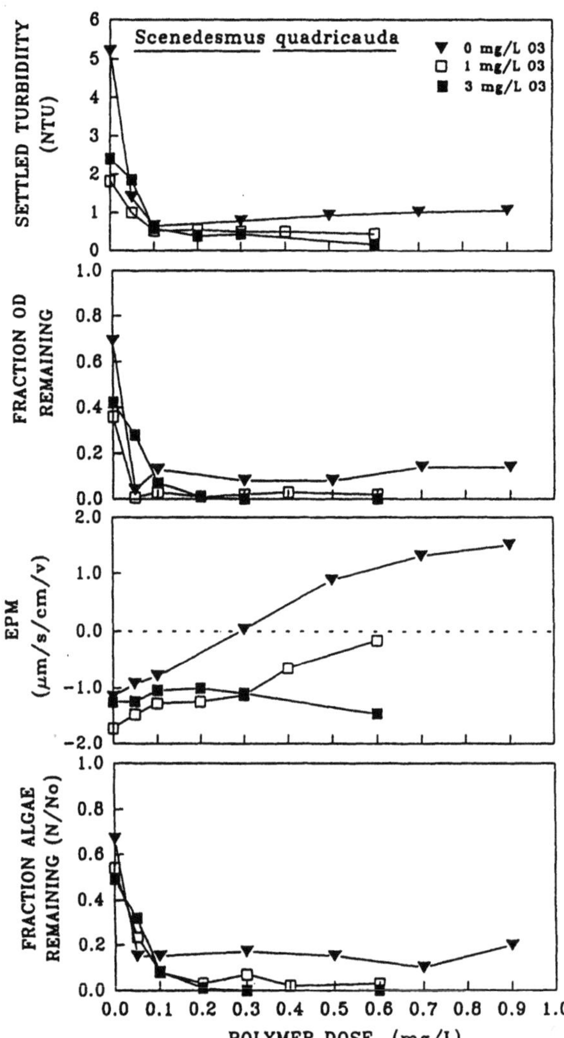

Fig. 1. Effect of polymer dose on the coagulation of *Scenedesmus quadricauda* with and without pre-ozonation

both synthetic and natural waters. The results are presented in Figs. 2 and 3. Two points are discussed. First, the polymer dose required for charge neutralization without preozonation in synthetic waters is much less than for natural waters. For both *Synura* (Fig. 2) and *Cyclotella* (Fig. 3), the polymer dose is about 0.6 mg/l in synthetic waters while a charge of zero is not obtained for the natural waters for polymer dosages up to 3 mg/l. The hypothesis is that the presence of natural organic matter in Wachusett Reservoir water, particularly DOC, exerts a demand for cationic polymer. The second point is that ozone had more of an effect on EPM data for the experiments done in synthetic waters in contrast to the experiments done in natural waters. Ozone causes more polymer to be needed for charge neutralization for synthetic waters – see EPM data in Figs. 1–3. On the other hand, in natural waters (top parts of Figs. 2 and 3) ozone has little effect on the EPM data. It is noted that ozone was applied under batch conditions so the reported ozone dosages are both applied and absorbed dosages.

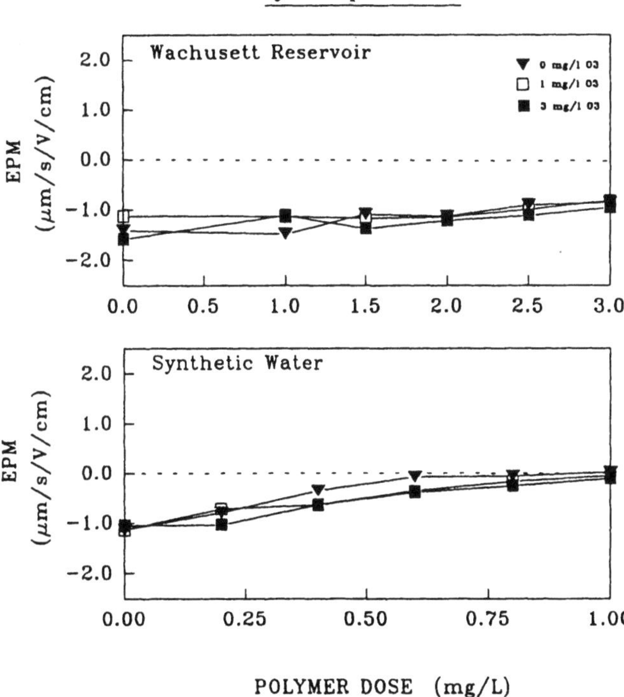

Fig. 2. Effect of pre-ozonation on the surface charge of *Synura petersenii*. Top figure for *Synura* in Wachusett Reservoir water, bottom figure for *Synura* in a synthetic water (Synura at 2×10^4 cells/ml; Synthetic Water: pH 7, 10^{-3} M NaHCO$_3$, 30 mg/l Ca as CaCO$_3$)

4.2 Flocculation Kinetics

Flocculation kinetic experiments were used to measure the effect of ozone on algal particle stability by determining alpha values. Experiments were done in which algae were added at a known (measured) cell number concentration. Over time, the algal suspension was mixed at a known velocity gradient of 10 or 50 s^{-1} and samples were withdrawn for particle size and number measurements. When natural waters are ozonated, ϕ may not remain constant. It may increase following ozonation due to production of particles through precipitation (e.g., oxidation of Fe or Mn producing Fe(OH)$_3$ or MnO$_2$ particles), or it may decrease due to breakup or shrinkage of organic particulates such as algae. In our experiments with synthetic waters, problems from Fe and Mn were avoided. To account for any change in ϕ, Eq. 2 is normalized as follows

$$\frac{\ln(N/N_0)}{\phi} = \frac{-4\alpha Gt}{\pi} \tag{3}$$

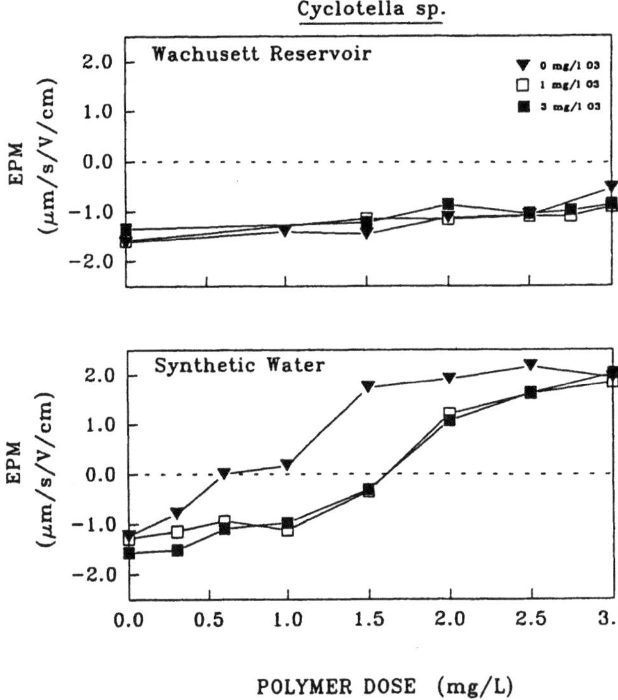

Fig. 3. Effect of preozonation on the surface charge of *Cyclotella sp..* Top figure for *Cyclotella* in Wachusett Reservoir water, bottom figure for *Cyclotella* in a synthetic water (Cyclotella at 2.1×10^4 cells/ml; Synthetic Water: pH 7, 10^{-3} M NaHCO$_3$, 30 mg/l Ca as CaCO$_3$)

Figure 4 illustrates that ozone increses the rate of flocculation for *Scenedesmus*. It is clearly seen that ozone increases α. To summarize our data, the relative effect of ozone on alpha is presented. Some explanation of this approach is presented next.

Kinetic flocculation experiments with algae are difficult to perform because of dealing with living organisms. As pointed out above, the particle volume concentration (ϕ) may change upon ozonation affecting rates of particle flocculation. Therefore, absolute alpha values lack meaning. What is more important is the relative effect of ozone on alpha. For these reasons the relative effect of ozone on alpha, not absolute alpha values, are presented. Relative alpha, α_{rel}, is defined as follows:

$$\alpha_{\text{rel}} = \frac{\alpha_{(\text{test})}}{\alpha_{(\text{ref})}} \tag{4}$$

$\alpha_{(\text{test})}$ Alpha value for any test case

$\alpha_{(\text{ref})}$ Alpha value for no preozone or the reference case

4.2.1 Synthetic Water Experiments. Relative alpha values are summarized in Table 2. The data for *Chlorella*, which were run in duplicate, show that ozone does not have a positive effect on the rate of algae-particle flocculation. Ozone reduces its flocculation rate, particularly at 3 mg/l ozone.

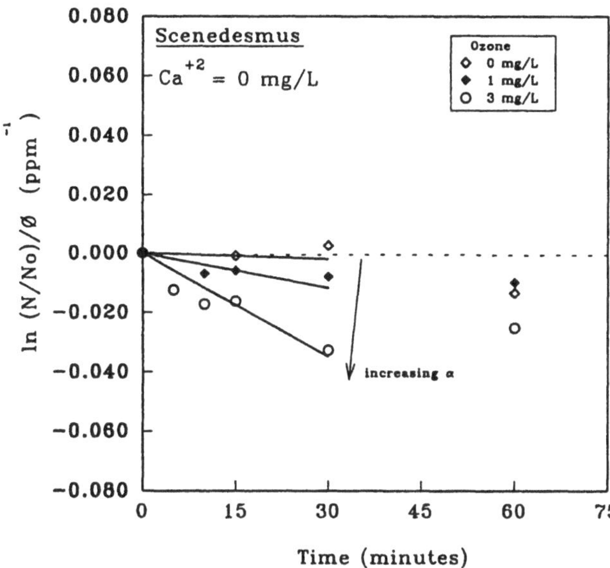

Fig. 4. Effect of preozonation on the flocculation kinetics of *Scenedesmus quadricauda*

Tab. 2. Relative alpha values for algae for experiments done in synthetic waters

Algae	Ca^{++} mg/l CaCO$_3$	Relative alpha		
		No O$_3$	1 mg/l O$_3$	3 mg/l O$_3$
Chlorella	30	1	0.4	< 0.01
Scenedesmus	0	1	5	18
	30	1	4	7
	50	1	19	6

Ozone increased the flocculation rate of *Scenedesmus*. Both Ca and ozone affect particle stability and thus flocculation kinetics. The data in Table 2 indicate that in the absence of Ca, a preozone dose of 3 mg/l has a greater impact on α_{rel} than 1 mg/l. On the other hand in the presence of 30 mg/l Ca, there is a slight increase in α_{rel} at 3 mg/l ozone compared to 1 mg/l. Finally, at the highest Ca concentration tested, ozone had a great effect at 1 mg/l (α_{rel} of 19), but increasing the ozone dose to 3 mg/l decreased α_{rel} to 6. This increase in α_{rel} at 1 mg/l and then a decrease at higher ozone may be due to interacting effects between ozone and calcium.

Singer (1989) and Chang and Singer (1991) reported α values before and after ozone addition for waters collected from seven locations around the United States. They found that whether ozone reduced particle stability (i.e., increased alpha) depended on both the ozone dose and the water hardness which they expressed in terms of the raw water hardness to TOC ratio. Specifically, they found increases in alpha for ozone doses of 0.4 to 0.8 $mg\,O_3/mg\,TOC$ and hardness to TOC ratios of at least 25 $mg\,CaCO_3/mg\,TOC$. Three of the supplies they tested should contain algae, at least on a seasonal basis: Los Angeles (Owens River Aqueduct), Monroe, Mi (Lake Erie), and Bay City, Mi (Saginaw Bay of Lake Huron). Singer (1989) found for all three of these supplies that ozone at low doses increased alpha while at higher ozone doses alpha decreased.

The experimental results for *Scenedesmus* in synthetic waters showed decreasing α_{rel} for Ca^{++} at 50 mg/l $CaCO_3$. Independent DOC measurements for these algae experiments indicated values of about 2 mg/l with little particulate carbon. This yields a hardness to DOC ratio of 25 and ozone doses of 0.5 to 1.5 $mg\,O_3/mg$ DOC.

4.2.2 Natural Water Experiments. A summary of α_{rel} values is presented in Table 3. Relative alpha values increase at a preozonation dose of 1 mg/l with a slightly greater effect on the flocculation rate of *Cyclotella* than *Synura* – α_{rel} values of 5 to 5.6 compared to 4.7. For all experiments at 3 mg/l ozone, lower α_{rel} values were obtained than at 1 mg/l ozone indicating overdosing with ozone may occur with regard to microflocculation. The experiments for *Cyclotella* and *Synura* in the natural waters were done at hardness to TOC ratios of 4 to 5 mg hardness per mg TOC and ozone doses of 0.3 to 1.2 $mg\,O_3/mg\,TOC$. The higher ozone dose reduced the α_{rel} value compared to the lower ozone dose. In summary, the results showed the effect of ozone on alpha depends on ozone dose, calcium or hardness, and algae type.

Tab. 3. Relative alpha values for algae for experiments done in natural waters

Algae	Supply/date	Relative alpha		
		No O_3	1 mg/l O_3	3 mg/l O_3
Cyclotella	Quabbin/July 1991	1	5.5	2.5
	Wachusett/April 1991	1	5.6	0.6
	Wachusett/May 1991	1	5.0	4.1
Synura	Wachusett/March 1991	1	4.7	1.3

4.3 Ozone Effects on Algae

This section presents material on the effects of ozone on 1) algae surface structure using scanning electron micrographs (SEM's), 2) particle size distributions (PSD's), and 3) extracellular organic matter (EOM).

Scanning electron micrographs (SEM's) were taken before and after ozonation to determine any effects on cell surface structure. SEM's are shown for two of the algae, *Scenedesmus* (green alga) in Fig. 5 and *Cyclotella* (diatom)

in Fig. 6. These micrographs demonstrate that extensive cell destruction (cell lysis) does not occur at an ozone dose of 3 mg/l. Similar results were obtained for *Chlorella* (green alga) and *Synura* (golden brown). Other SEM's showed resistance of cell breakup at ozone doses of 8 mg/l (i.e., some cells were broken, but most remained intact) with the diatom being most resistant to ozone. The SEM's showed evidence of cell surface alteration at all preozone doses (1, 3, and 8 mg/l). This is evident in both figures at a preozone dose of 3 mg/l. *Scenedesmus* cells are enclosed by a sheath (reticulate layer) and are characterized by two spines arising from the ends of the terminal cells. Fig. 5 shows for the control case of no ozone that the reticulate layer is intact and the spines are present. Following ozone, the reticulate layer is damaged and the spines have been broken off (they can be seen as separated pieces in the SEM). Likewise, Fig. 6 shows damage to the surface of *Cyclotella*.

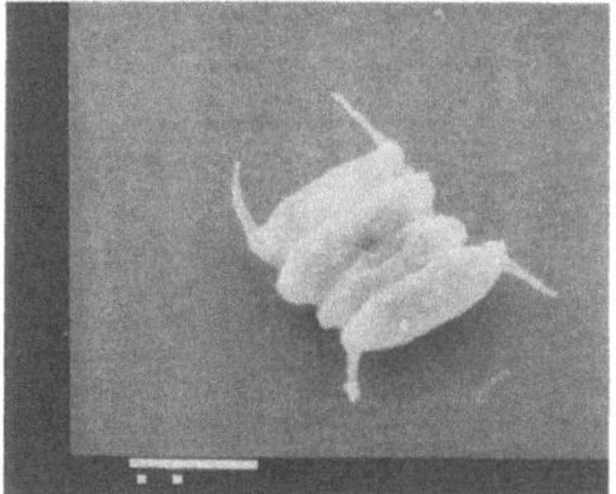
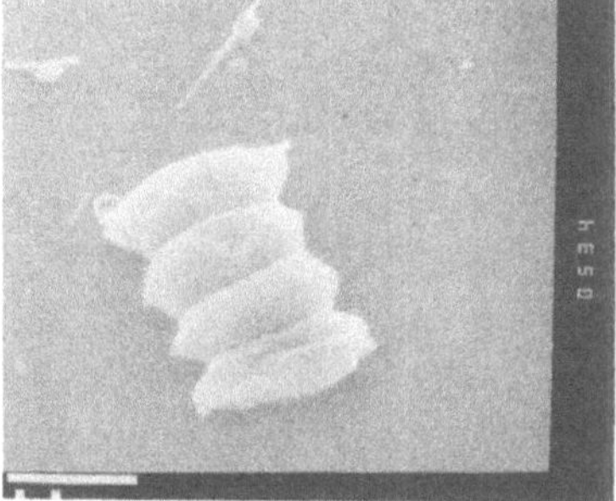

Fig. 5. Scanning electron micrographs for *Scenedesmus quadricauda*. Top figure, no ozone; bottom figure after 3 mg/l ozone (scale bar: both figures is 10 microns)

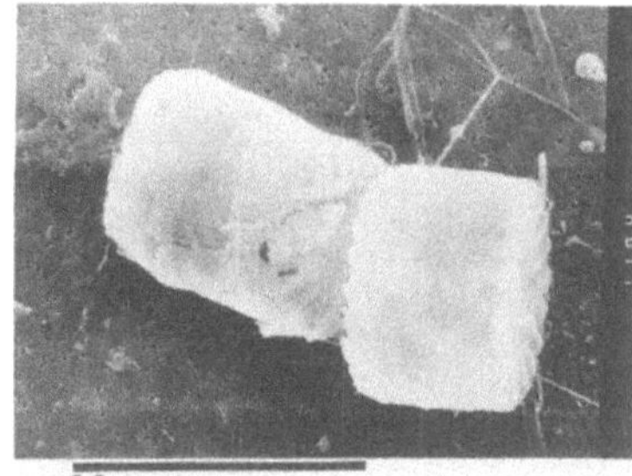

Fig. 6. Scanning electron micrographs for *Cyclotella sp.*. Top figure, no ozone; bottom figure after 3 mg/l ozone (scale bar: both figures is 10 microns)

Particle size distributions (PSD's) before and after ozone were made to determine any changes in the algae number and volume distributions. These distributions are presented in Figs. 7 and 8 for *Scenedesmus* and *Cyclotella*, respectively. *Scenedesmus* can colonize in groups of 4 or 8 cells. Colonies of 4 cells are shown in the SEM's (Fig. 5). The number distribution data for *Scenedesmus* (top of Fig. 7) show that before ozonation the cells have a size range between 7 and 22 μm with a mean of 11.4 μm. Ozone has a small effect on the particle (cell) number distribution. There is a slight decrease in the number of particles with a size greater than 12 μm and a concomitant increase in smaller particles although the total particle number concentration did not change significantly. The volume distribution data (bottom of Fig. 7) are instructive. They show a broad peak in cell volume for particle sizes of 12 to 18 μm with little effect at an ozone dose of 1 mg/l. At 3 mg/l ozone, there is a loss in cell volume of 22 percent. The mean size decreased from 11.4 μm to 10.6 μm. This agrees with the SEM data which showed evidence of alteration of cell surface structure and collapse of *Scenedesmus* cells.

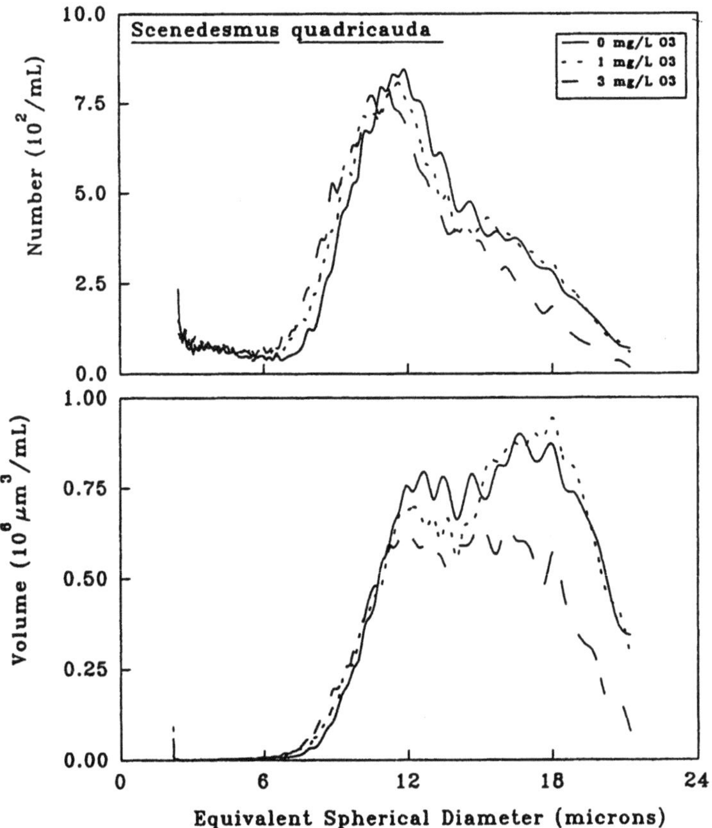

Fig. 7. Number and volume distributions for *Scenedesmus quadricauda* before and after ozonation

The particle number distribution data for *Cyclotella* in Fig. 8 show a shift to smaller size particles with ozone (see the particle number data in the 7 to 14 μm size range). It is more pronounced at 3 mg/l. The particle volume distribution

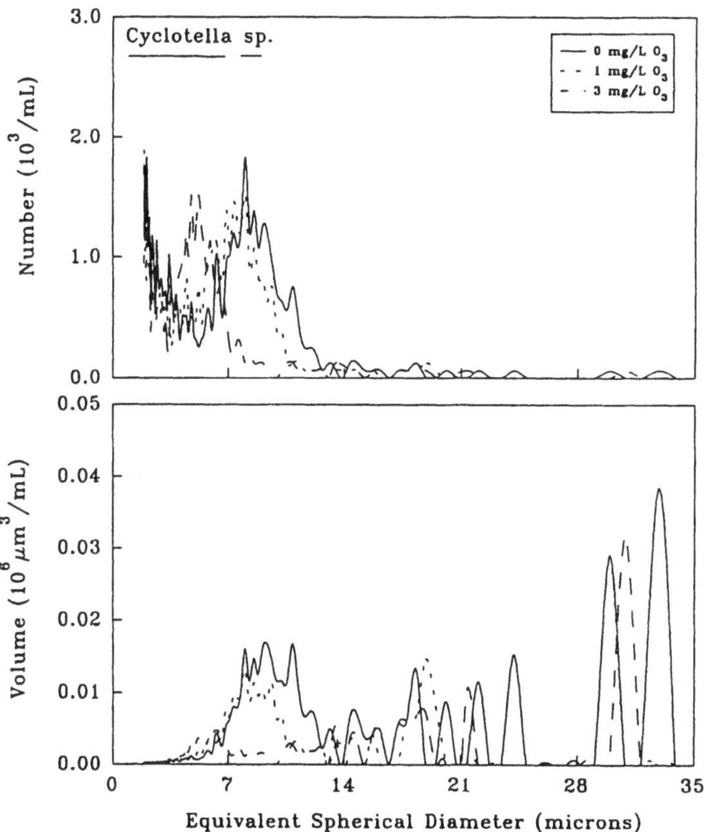

Fig. 8. Number and volume distributions for *Cyclotella sp.* before and after ozonation

data (bottom of Fig. 8) show a loss of cell volume with ozone which is attributed to cell shrinkage and not cell breakup (Fig. 6). The reduction in cell volumes for *Scenedesmus* and *Cyclotella* did not adversely affect flocculation kinetics of the cells (a decrease in particle volume will decrease flocculation kinetics if all other parameters are constant), on the other hand, ozone increased the rates of flocculation for three of the algae tested (Tables 2 and 3). This is discussed below.

The EOM production data are summarized in Table 4 and show that *Chlorella* produced a small amount compared to the other two algae. The diatom *Cyclotella* produced the greatest amount. The distributions of the EOM according to three molecular weight (MW) fractions are presented in Fig. 9 and defined as high MW as the EOM fraction greater than 30 Kdal (kilodalton), intermediate MW between 3 and 30 Kdal, and low MW less than 3 Kdal. No microflocculation effect was observed for *Chlorella* (Table 2). The data in Table 4 and Fig. 9 show that this algae produced very little EOM and that less than 50 percent of it was high MW. Ozone, even at 1 mg/l, converted most of *Chlorella's* EOM to low MW material.

Tab. 4. Production of extracellular organic matter (EOM) for different algae

Algae	Total volume processed l	Algae number #/ml	EOM produced μg DOC/10^6 cells
Chlorella	5	1×10^7	0.2
Scenedesmus	8	1×10^6	2.4
Cyclotella	4	5×10^5	7.2

Scenedesmus underwent microflocculation (Table 2). It produced 10 times more EOM (Table 4) than *Chlorella* and about 70 percent of the EOM prior to ozone was of a high MW. Ozone shifted the MW distribution to lower MW materials. After ozonation, about 40 to 50 percent of the EOM was of intermediate or high MW corresponding to approximately 1 μg DOC/10^6 cells. Sukenik et al. (1987) found that ozone damaged the reticulate layer and caused collapse of *Scenedesmus* cells. They also found that ozone aided alum coagulation (reduced the alum dose) of *Scenedesmus*. They proposed that ozone attacks the cell surface causing release of wall polysaccharides and leakage of cell proteins and nucleic acids. They proposed that ozone aided alum coagulation by altering the microalgal envelope (reticulate layer) of the cell surface causing a reduction in the colloid stability of the cells. Our results suggest that the intermediate and high MW EOM affect the stability and flocculation of *Scenedesmus*.

Cyclotella underwent microflocculation (Table 3). It produced 36 times more EOM (Table 4) than *Chlorella* and about 90 percent of the EOM prior to ozone was of a high MW. Ozone shifted the MW distribution to lower MW materials; however, after ozonation about 40 to 50 percent of the EOM was still present as intermediate or high MW material corresponding to approximately 3 μg DOC/10^6 cells.

5. Practical Aspects

Our coagulation and flocculation experiments were done for the following conditions: algae cell concentrations at 10^4 to 10^5 cells/ml in synthetic waters in the absence of other natural organic matter and mineral particles and in natural waters of high quality. The main findings are 1) ozone has little effect on reducing the polymer dose for effective coagulation, 2) ozone added to synthetic waters seemed to have an adverse effect on the EPM of the algae, which was not seen for experiments in natural waters; in either case, ozone tended to improve performance, 3) ozone, in the absence of a coagulant, increased the

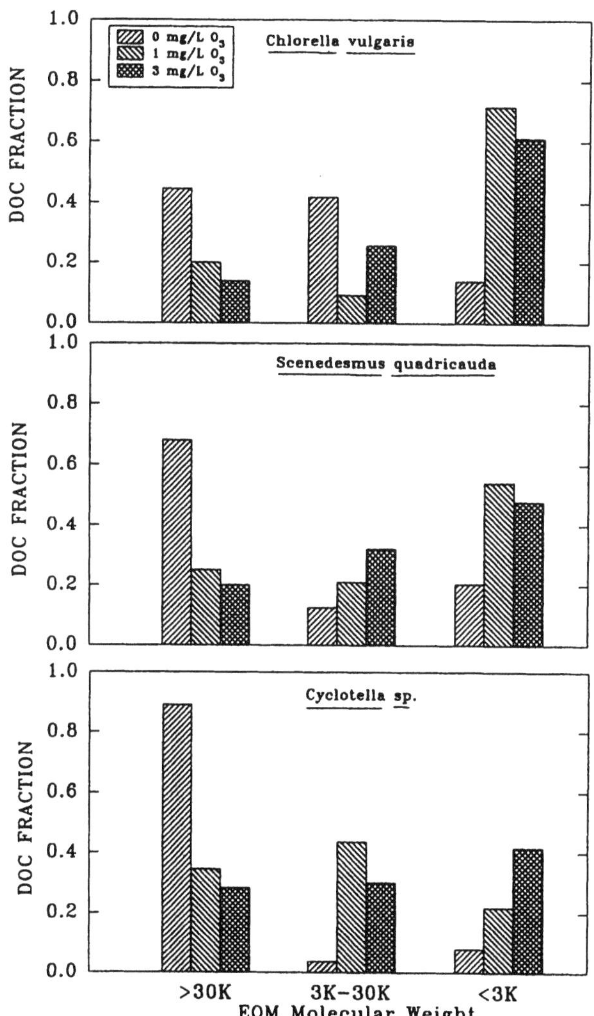

Fig. 9. Molecular weight distributions of extracellular organic matter before and after ozonation

flocculation rate (increased α_{rel}) for three of the algae tested (*Scenedesmus, Synura,* and *Cyclotella*), but not for *Chlorella*, and 4) ozone increased α_{rel} at a low preozonation dose, but in some cases at a higher preozonation dose there was evidence of overdosing. Whether ozone had a beneficial effect or not depended on the algae type as well as the concentration and molecular weight of the EOM released by the algae.

Ozone has little effect on the flocculation of *Chlorella*, but did increase the flocculation rates of *Scenedesmus, Synura,* and *Cyclotella*. Design engineers and water plant operators should be aware that ozone reactions with algae can affect flocculation rates. At absorbed ozone doses of 3 mg/l or less, cell lysis does not occur to any great extent indicating that EOM associated with the cell surfaces plays an important role in aiding flocculation of particles by interparticle flocculation. Data presented here showed that this material played a role in explaining the microflocculation of certain algae. Overdosing with ozone can occur affecting the flocculation rate. It was not studied here, but high ozone doses may release excess EOM or cause cell lysis and release of IOM which may increase the coagulant dose or adsorb and stabilize other particles.

Algae then should be considered when examining potential coagulating effects of ozone. Furthermore since phytoplankton or algae growth is seasonal, then ozone effects may also be seasonal. Engineers need to account for this in their observations at full-scale plants and in their planning of pilot plant studies.

The, SEM's showed that little algae cell destruction (lysis) occurs at ozone dosages up to 3 mg/l. This in contrary to popular opinion that ozone destroys algae. These data indicate that algae will be present (probably as dead cells) following preozonation, and would have to be removed in downstream water treatment processes.

Acknowledgments

The authors gratefully acknowledge the contribution and advice of D.A. Reckhow. Funding for this project was in part from CH2M Hill for a project being done for the Massachusetts Water Resources Authority and in part from the American Water Works Association Research Foundation.

Literature

Akiba, M., Gotch, K., Satoh, A.: The Influence of Algogenic Organic Matter on Coagulation under Pre-Chlorination. Wat. Supply *8* (1990) 103–109

Bernhardt, H., Hoyer, O., Lüsse, B., Schell, H.: Investigations on the Influence of Algal-Derived Organic Substances on Flocculation and Filtration. Proceedings from the Second National Conference on Drinking Water, Edmonton, Canada, 1986

Bernhardt, H., Clasen, J.: Flocculation of Micro-Organisms. J. Wat. SRT (Aqua) *40* (1991) 76–87

Chang, S.D., Singer, P.C.: The Impact of Ozonation on Particle Stability and the Removal of TOC and THM Precursors. J. AWWA *83* (3) (1991) 71–79

Dowbiggin, W.B., Singer, P.C.: Effects of Natural Organic Matter and Calcium on Ozone-Induced Particle Destabilization. J. AWWA *81* (6) (1989) 77–85

Farvardin, M.R., Collins, A.G.: Preozonation as an Aid in the Coagulation of Humic Substances – Optimum Preozonation Dose. Wat. Res. *23* (1989) 307–316

Grasso, D., Weber, W.J.Jr.: Ozone-Induced Particle Destabilization. J. AWWA *80* (8) (1988) 73–81

Hoyer, O., Bernhardt, H., Lüsse, B.: The Effect of Ozonation on the Impairment of Flocculation by Algogenic Organic Matter. Z. Wasser-Abwasser-Forsch. *20* (1987) 123–131

Jekel, M.R.: Particle Stability in the Presence of Preozonated Humic Acids. Wat. Supply *8* (1990) 79–85

Langlais, B., Reckhow, D.A., Brink, D.R. (eds.): Ozone in Water Treatment Application and Engineering. AWWARF Cooperative Report, Lewis Publ., Chelsea, Mich., 1991, pp. 190–213

Lüsse, B., Hoyer, O., Soeder, C.J.: Mass Cultivation of Planktonic Freshwater Algae for the Production of Extracellular Organic Matter (EOM). Z. Wasser-Abwasser-Forsch. *18* (1985) 67–75

Singer, P.C., Chang, S.D.: Impact of Ozone on the Removal of Particles, TOC, and THM Precursors. AWWARF Report, Denver 1989

Sukenik, J.M. et al.: Effect of Oxidants on Microalgal Flocculation. Wat. Res. *21* (1987) 533–539

James K. Edzwald
Kodak Professor of Environmental Engineering
Department of Civil and Environmental Engineering
Rensselaer Polytechnic Institute
Troy, NY 12180-3590
U.S.A.

Ashish Paralkar
Environmental Engineering Program
Department of Civil Engineering
University of Massachusetts
Amherst, MA 01003
U.S.A.

Improvements in Membrane Microfiltration Using Coagulation Pretreatment

M. R. Wiesner, S. Veerapaneni, and D. Brejchová

Introduction

An important consideration in water and wastewater treatment using low-pressure membrane filters is the potential for reductions in permeation rate due to the accumulation of colloidal and macromolecular materials on and within the membrane. One strategy for improving membrane performance has been proposed [1] in which colloids capable of entering membrane pores are coagulated to produce particles that are rejected by the membrane. Increasing the size of particles applied to membrane filtration units by coagulation may improve permeate flux by 1) reducing foulant penetration into pores, 2) forming a more porous cake on the membrane surface that may act as a dynamic membrane thereby protecting the membrane from long term fouling, and 3) decreasing the accumulation of foulants near the membrane due to particle size effects on particle transport. In this paper, a pilot study is reported in which metal coagulants were added to improve the permeate flux, fouling characteristics, and water quality produced by tubular ceramic microfiltration membranes with different nominal pore sizes.

Background

Reductions in the flux of permeate across a membrane may be categorized as being operationally reversible or irreversible. The irreversible loss in permeate flux due to accumulation of materials in and on the membrane is referred to as fouling. A reversible decrease in permeate flux is referred to as reversible fouling or sometimes as *"colmatage"* from the French word meaning clogging. The instantaneous flux of permeate across ultrafiltration (UF) and microfiltration (MF) membranes can be described using a modification of Darcy's law in which permeate flux, J, is assumed to be inversely proportional to a sum of resistances to the passage of permeate and proportional to the pressure drop across the membrane,

$$J = \Delta P/(R_m + R_f + R_c) \tag{1}$$

where R_m is the resistance to permeate flow associated with the clean membrane, R_f is the additional resistance due to membrane fouling, and R_c is resistance due to membrane colmatage. The colmatage term, R_c can be further divided into the resistance due to the formation of a cake (or gel layer) on the membrane that can be eliminated either hydrodynamically or chemically, the resistance of a concentration polarization layer, and resistance due to reversible obstruction of membrane pores. When a membrane is first placed into service, the fouling and colmatage terms in Eq. (1) are zero, and permeate flux is limited by the resistance of the membrane. The accumulation of materials on (and in) the membrane over time produces an increase in the relative importance of R_f and R_c and these terms may eventually control permeate flux. In the case where R_c is dominated by the formation of a cake of constant resistance per unit depth,

$$R_c = aL \tag{2}$$

where a is the specific resistance of the cake (resistance per unit depth), and L is the depth of the cake. In many cases a will increase over time as the cake compacts and as smaller colloidal materials are "filtered" by the cake. Cake growth at any time should be proportional to the rate at which materials are carried to the membrane surface. However, as the cake depth increases, shear-induced back-transport of deposited materials should occur at a greater rate. These assumptions may be expressed in the following differential equation describing the change in cake thickness with time:

$$\partial L/\partial t = k_1 J - k_2 L \tag{3}$$

The rate constant k_1 describes the transport of cake-forming materials to the membrane and should increase with increasing feed concentration. Theoretical consideration of particle trajectories in membrane elements indicates that k_1 should decrease with increasing particle diameter and increasing crossflow velocity due to the effects of inertial lift [2]. The constant k_2 indicates the relative importance of back-transport mechanisms in removing materials from the membrane and is also predicted to increase with increasing crossflow velocity. If it is assumed that the cake is composed of a fluidized layer of particles, a minimum in k_2 with respect to particle size is predicted for colloids near 0.1 μm in diameter based on the combined effects of Brownian diffusion and shear-induced diffusivity [3].

Under the assumption of a constant membrane resistance, constant pressure drop, colmatage resistance, R_c, dominated by cake growth, and a specific resistance of the cake that is invariant with time, the time derivative of Eq. (1) yields,

$$\partial J/\partial t = -Ja(k_1 J - k_2 L)/(R_m + aL) \tag{4}$$

During the early stages of cake formation, deposition of materials to the membrane may be much greater than loss by fluid shear ($k_1 J > k_2 L$) and initial cake resistance may be less than the resistance of the membrane ($R_m \geq aL$). Under these assumptions, the time rate of change of permeate flux during the early stages of filtration can be approximated as a pseudo-second order equation in J:

$$\partial J/\partial t \cong -k_1^* J^2 \qquad (5)$$

where $k_1^* = a k_1 / R_m$. It is important to note that a higher value for the specific resistance of the deposit, a, results in a higher value of k_1^*. Cakes composed of larger particles should produce a lower specific resistance [4]. Thus, an increase in particle size may reduce k_1^* both through effects on particle transport and cake morphology. The slope of a linear plot of $(1/J)$ versus time yields the cake formation rate constant, k_1^*. As cake formation proceeds, $aL \geq R_m$. For the case of active cake formation where $k_1 J > k_2 L$ and cake resistance controls permeate flux, Eqs. (1)–(4) reduce to,

$$\partial J/\partial t \cong -k_1^* (a/\Delta P) J^3 \qquad (6)$$

and the cake formation rate constant is derived from a least-squares fit to a plot of $(1/J^2)$ versus time. The trans-membranar pressure drop, ΔP, often increases as J decreases due to a decreasing back-pressure on the permeate side of the membrane. When ΔP is not constant over time, equations similar to (5) and (6) for second- and third-order flux decay can be applied to the specific permeate flux, $J^* = J/\Delta P$.

Expressions also can be derived assuming other mechanisms for membrane colmatage and fouling such as the development of a concentration-polarization layer, the blockage of membrane pores and the adsorption of materials with the membrane matrix. Some of these expressions are summarized in Table 1. The generic rate constant, K, in each of these expressions is an indication of the rate at which permeate flux declines. In all cases, K is directly proportional to a deposition, adsorption, or cake-formation rate constant (e.g. k_1 in Eq. 3).

Tab. 1. Kinetic expressions for different mechanisms of permeate flux reduction

Mode of fouling/colmatage	Permeate flux equation	Linearized form of equation
Cake formation		
membrane-limited flux	$J = J_0/(1 + J_0 K t)$	$(1/J) = (1/J_0) + Kt$
cake-limited flux	$J^2 = J_0^2/(1 + J_0^2 K t)$	$(1/J^2) = (1/J_0^2) + Kt$
Concentration-polarization	$J = J_{SS} + B \exp(-Kt)$	$\ln(J - J_{SS}) = \ln B - Kt$
Adsorptive pore fouling	$J = J_{SS} + B \exp(-Kt)$	$\ln(J - J_{SS}) = \ln B - Kt$
Pore blockage (initial)	$J = J_0 \exp(-Kt)$	$\ln(J/J_0) = -Kt$

Note that the units and interpretation of empirical constants K and B differ for each expression.
J_{SS} = permeate flux at steady state

Analysis of permeate flux data using the kinetic expressions in Table 1 allows for a comparison of the effect of operating conditions, raw water quality and membrane characteristics on the rate, extent, and nature of permeate flux decline. The values of rate constants evaluated from field data assuming a single mechanism of flux decline can be used to compare the treatability of various raw waters as well as the impact of various pretreatments on permeate flux. Identification of rate expressions that best describe permeate flux may suggest mechanisms of membrane colmatage and/or fouling. While different mechanisms sometimes lead to similar rate expressions, some of this ambiguity can be reduced by designing pilot studies to differentiate between reversible and irreversible reductions in permeate flux. However, in practice several rate expressions may be found to yield an acceptable fit to the data and it is likely that several fouling and colmatage mechanisms will play a role in reducing permeate flux.

A pilot study of surface water treatment using membrane microfiltration was performed to evaluate the effects of coagulation pretreatment on the performance of tubular ceramic membranes. Membrane performance was quantified with respect to the rate of permeate flux decline, the relative effects of fouling and colmatage, the asymptotic value for permeate flux achieved after extended operation, as well as the removal of colloidal material and organic carbon by the membrane.

Methods

Site. Pilot work was performed at the Houston water purification facility located east of central Houston near the Houston ship channel. This facility includes three independent potable water treatment plants, each rated at 23 625 m^3/hr. Plant 1 treats a blend of approximately 80 % Trinity River and 20 % San Jacinto River waters. Plant 3 treats only water from the San Jacinto River. Waters from the Trinity and San Jacinto Rivers are pumped to the Houston facility and are initially discharged into presedimentation basins. In this study the raw water feed to the membrane pilot was taken from the presedimentation basin at the inlet structure of either Plant 1 or Plant 3. Because Plant 1 treats predominantly Trinity River water, raw water taken from the presedimentation basin of Plant 1 in this study is referred to as Trinity River water. Water taken from the presedimentation basin at Plant 3 is referred to as San Jacinto River water.

Pilot Unit. The microfiltration system and membranes used in this work were provided by the Alcoa Separations Technology Division (Warrendale, PA). The system consisted of two skid mounted units. The first unit was a chemical pretreatment unit used to add coagulant, adjust pH, and mix raw water with recycled concentrate from the membrane. The pretreatment skid included a 189 liter tank with a mixer and a variable speed and stroke pump. Chemical feed from two 75-liter storage tanks was provided via variable speed and stroke pumps. The second skid mounted unit was a 19 liter/min membrane filtration unit capable of operating with either 1 membrane module or with 2 modules in parallel or in series. The filtration unit included two centrifugal pumps to provide system pressure and crossflow velocity. A gear pump filled a pressurized backflush reservoir. Pressure and flow sensors provided input to a process control and data acquisition system. In addition, feed, recycle, backflush, and permeate flow rates could be read directly from calibrated rotometers. Flow through the system was controlled by solenoid valves which, in conjunction with the process control hardware, allowed for automatic operation of the unit. The system could also be operated in a manual override mode.

Membranes. The pilot was operated using a single Alcoa P19-40 module, with a membrane area of 0.2 m^2. A P19-40 module consists of a single ceramic rod, 100 cm in length, with 19 tubular membrane channels or "elements" running through the length of the rod. The diameter of each tubular element is 0.4 cm. Bonded to the interior of these tubes is a sintered layer of α-alumina or zirconia that defines the nominal pore size of the membrane. Work using membranes with nominal pore sizes of 0.05 μm (zirconia), 0.2, and 0.8 μm (α-alumina) are reported in this paper. These membranes are reported to tolerate waters at pH values from 1 to 14 and temperatures as high as 140 °C [5]. They also exhibit good mechanical stability, and are able to withstand rigorous cleaning using detergents, acids, bases, steam, and high backflush velocities. While these characteristics are likely to confer a long lifetime to the membranes, the initial cost of these membranes is greater than that of polymeric membranes.

Pilot Operation. The unit was operated at a crossflow velocity of approximately 167 cm/s, which produced a pressure drop across the length of the module of approximately 14 kPa. The system pressure on the retentate side of the membrane was approximately 380 kPa. The pressure drop across the membrane ranged from 50 to 340 kPa depending on the period of operation, the pretreatment conditions, and the nominal membrane pore diameter.

Two operating modes were used in the pilot experiments; 1) short-duration experiments with operator-initiated backflushing every 20 minutes and, 2) longer duration runs of 8 to 150 hours at fixed recoveries with automatic backflushing every 96 seconds. Membranes were thoroughly cleaned before each experiment using a sequence of base and chlorine chemical washes. An intermediate wash with acid was found to be ineffective or detrimental in the restoration of membrane flux. The pilot was operated at recovery ratios (permeate flow/raw water feed) of 90 to 100%. For experiments of 80 minutes to 8 hours in duration, all of the retentate was recycled to the membrane feed stream, yielding an effective recovery of 100%. During longer runs of 32 to 150 hours, retentate was wasted at a constant rate selected to produce a recovery of either 90 or 95%.

Short-duration experiments were performed to evaluate the dynamics of membrane fouling and colmatage as a function of membrane pore size and pretreatment conditions and to evaluate the cost-optimal backflush frequency. The permeation behavior of membranes of varying nominal pore size was evaluated for the case when untreated raw water was introduced directly to the membrane. Over a period of approximately 20 minutes flux decline and pressure drop were monitored. After 20 minutes of operation, the membrane was back-flushed and a second period of flux decline was monitored. These cycles of flux decline and backflushing were repeated 3 to 5 times for each membrane tested. Similar experiments were performed using either alum or ferric sulfate to coagulate the raw water. Coagulation pretreatment of the raw water involved the injection of a metal coagulant into the raw water feed line followed by a period of mixing in the feed tank ($\sim$ 1 hour). Mixing was accomplished by pumping material from the bottom of the tank and returning this flow to the tank inlet. Coagulant dose was selected on the basis of jar tests performed on the raw water. Coagulant dose ranged from $10^{-3.7}$ to $10^{-3.4}$ M of either aluminum or iron.

Pilot runs of longer duration were performed to establish the permeation rates and permeate qualities to be expected from each membrane during normal operation with programmed backflushing. Treatability of the Trinity and San Jacinto River waters was evaluated with and without coagulation pretreatment. The effect of pH on coagulation pretreatment using alum and ferric sulfate was evaluated.

Water Quality Measurements. In all experiments, the turbidities of the raw, feed and permeate waters were measured, as well as the pH and absorbance of UV light at 254 nm. The distribution of particles greater than 0.4 μm was measured using an electrical sensing zone instrument (Coulter Multisizer) fitted with 19 and 50 μm apertures. Photon correlation spectroscopy was used to assess the relative mass concentration of colloids in the raw water that passed through a 0.45 μm membrane and to estimate the mass-average diameter of these colloids. In addition to these parameters, the concentrations of total and dissolved organic carbon (TOC and DOC respectively) in the raw, feed and permeate waters were measured.

Raw Water Quality During the Study Period. Mean values for raw water quality parameters observed during the study period are listed in Table 2. Plant records indicate that historically, the turbidity and concentration of TOC in the San Jacinto River water entering Plant 3 are higher than those measured in the Trinity River water at the inlet to Plant 1. Analysis of these waters using electrical sensing zone (ESZ) and photon correlation spectroscopy (PCS) also indicated that there are substantial differences between the sizes and concentrations of colloidal materials in these two waters. Particles in the San Jacinto River tend to be smaller than those in the Trinity River. The number-volume average diameters of particles in the Trinity and San Jacinto Rivers as determined by ESZ were 2.2 μm and 0.9 μm respectively. In addition, PCS analyses indicated that samples of San Jacinto River water filtered through a 0.45 μm membrane contained a significant mass of particles. These particles had an estimated mean diameter of 0.3 μm. Concentrations of both TOC and DOC in the San Jacinto River water were almost twice those observed in Trinity River water.

Tab. 2. Mean values for raw water quality parameters during the pilot study

Parameter	Trinity River water	San Jacinto River water
Turbidity (NTU)	25	70
Suspended solids (mg/l)	15	27
Particles $> 0.4~\mu$m (#/ml)	4×10^6	1×10^7
Average diameter (μm)	2.2	0.9
PCS peak below 0.5 μm	no	yes
TOC (mg/l)	8.2	14.1
DOC (0.45 μm membrane, mg/l)	6.3	11.5
pH	7.8	7.2
UV @ 254 nm (cm^{-1})	0.16	0.33
Hardness (mg/l CaCO$_3$)	110	67
Alkalinity (mg/l CaCO$_3$)	102	56

Experiments using Trinity River water were conducted between May and September 1990. Experiments using San Jacinto River water were conducted between February and August 1991

Results and Discussion

Kinetics of Permeate Flux Decline. Short-duration pilot experiments with Trinity River water were designed to evaluate the initial rate of permeate flux decline for different feed waters and membranes. The permeation behavior shown in Fig. 1 of coagulated and uncoagulated Trinity River water filtered on a 0.05 μm nominal pore-size membrane exhibits many of the characteristics also observed in experiments with the 0.2 μm and 0.8 μm membranes. In all of the pilot experiments a rapid drop in permeate flux was observed after introducing the feed water to the membrane. The largest and most rapid drop in permeate flux occurred when untreated raw water was filtered directly on the membrane. Backflushing the membrane restored a significant portion of the initial flux. However, the effectiveness of backflushing as a means of restoring flux decreased with time. The addition of either ferric sulfate or alum to the raw water slowed the decline in permeate flux resulting in higher fluxes for the same service time. Although alum tended to produce higher fluxes than did ferric sulfate during the initial cycles, flux curves for alum and ferric sulfate coagulation converged with time.

In most cases, the short-term decrease in permeate flux was found to be described best by the expression for third-order decay of permeate flux (Eq. 6) corresponding to a cake-limited resistance to the flow of permeate. A plot of the linearized form of the data is shown in Fig. 2 along with the linear equations of best-fit to the data determined by least squares. A change in the intercept of the linearized data is anticipated over time due to the long term trend of deceasing permeate flux after backflushing. However, the change in slope observed with

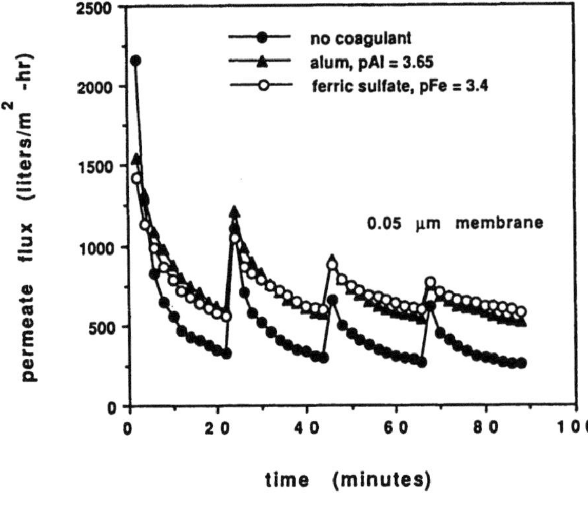

Fig. 1. Permeate flux as function of time with backflushing every 20 minutes

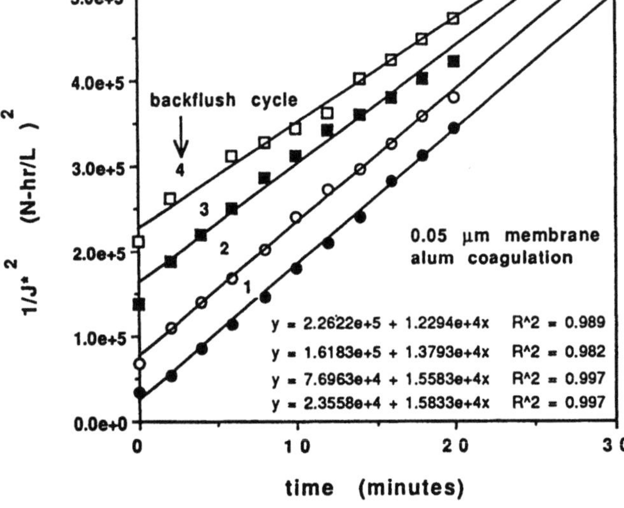

Fig. 2. Linearization of specific flux, $J^* = J/\Delta P$, assuming cake-limited permeation rate to be described by third-order kinetics in J^*

each backflush cycle is not predicted from the cake-limited model and may indicate that materials deposit within the membrane and/or that a portion of the cake is not removed by backflushing.

The rates of flux decline observed for each membrane and pretreatment combination are summarized in Fig. 3 using the third-order rate constants for cake deposition calculated from the data and averaged over 4 backflush cycles. The addition of alum or ferric sulfate to the raw water significantly reduced the rate of colmatage for each of the membranes tested. Alum appeared to be more effective than ferric sulfate in reducing the rate of colmatage; particularly for the 0.2 and 0.8 μm membranes. Rate constants for reversible flux decline obtained from experiments using coagulation pretreatment with the 0.05 μm membrane were significantly smaller than those obtained using the other two membranes. The observation that rate constants varied as a function of pore size, as well as the trend towards decreasing permeate flux after backflushing, are not predicted by the cake-resistance model and underscore the importance of additional mechanisms of colmatage and fouling.

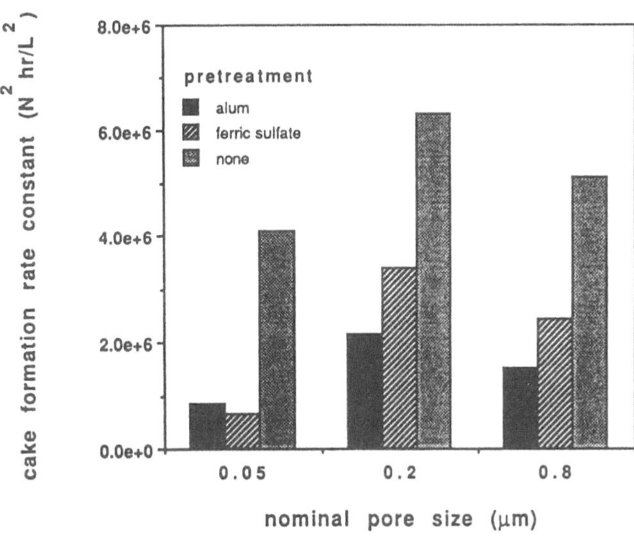

Fig. 3. Comparison of third-order rate constants for cake formation (K in Table 1) calculated from pilot data obtained using membranes with three different nominal pore sizes and pretreatments

The resistance, R_c, of the cake layer removed by backflushing was calculated in each experiment from the differences in permeate flux before and after each backflush. Values for R_c as well as values for R_m are listed in Table 3. The values for R_m reported in Table 3 were calculated from the pressure drops and permeation rates measured at the beginning of each experiment. With the exception of the 0.8 μm membrane, the values for the membrane resistances are greater than those reported by the membrane manufacturer. These higher values for R_m may indicate that chemical washing of the membrane before each experiment did not return the membrane to its initial permeability and that some fouling of the membrane had taken place. However, higher initial resistances may also reflect a very rapid colmatage of the membrane before or during the first measurement of permeate flux. On the average, R_c was lower when the raw water was coagulated with alum or ferric sulfate. Consistent with the assumptions made in deriving third-order kinetics for permeate flux decay, R_c was found to be greater than R_m for all of the experiments conducted. Differences in cake resistance affect the rate constants calculated for cake deposition. Thus, the smaller values for rate constants reported in Fig. 3 partially reflect a lower resistance to permeate flux by deposited materials.

The trend towards deceasing peaks in permeation rate after each backflushing was described using expressions for initial pore blockage and adsorptive fouling. In both cases, the rate constants were approximately equal 10^{-2} min^{-1} for different membrane pore sizes when the raw water was pretreated

Tab. 3. Membrane and cake resistance observed in the short-duration experiments $(g/cm^2 \cdot s)$ adjusted to $20°C$

	0.05 μm	0.2 μm	0.8 μm	Average
R_m	6×10^7	5×10^7	3×10^7	–
R_c				
no coagulant	2.4×10^8	2.6×10^8	2.7×10^8	2.6×10^8
ferric sulfate	9.0×10^7	2.9×10^8	1.7×10^8	1.8×10^8
alum	8.4×10^7	1.8×10^8	9.2×10^7	1.2×10^8

with alum or ferric sulfate. Without pretreatment, membranes with larger nominal pore sizes were irreversibly fouled at a slower rate than were membranes with smaller pore sizes.

Expressions for the kinetics of cake deposition and the associated rate constants calculated from pilot data can be used to estimate the optimal frequency for backflushing the membrane to remove deposited materials and reverse the effects of colmatage. A cost model for low-pressure membrane filtration [6] was used in conjunction with the expression for cake-limited permeate flux as a function of time to estimate the cost-optimal backflush frequency for these membranes. The values for rate constants calculated from the pilot data (Fig. 3) were used in these calculations. Results of one set of these calculations are shown in Fig. 4 for alum coagulation and a 0.05 μm membrane. Capital and operating costs vary as function of facility design capacity, the cost of concentrate disposal, membrane life, and many other factors. Without specifying these factors, the costs in Fig. 4 should be interpreted as relative rather than absolute values. Both capital and operating costs pass through minima for values of the backflush frequency near 100 seconds. Backflush frequencies near 100 seconds minimized costs for all of the membrane-coagulant combinations evaluated in the pilot study. Different results might be obtained using other membranes, other raw waters, and/or other pretreatments. Subsequent pilot experiments of longer duration were performed with an automatic backflush cycle every 96 seconds.

Extended Runs with Frequent Backflushing. Without coagulation, backflushing the membrane every 96 seconds did not increase permeate flux (Fig. 5). This indicates that backflushing was either ineffective in removing deposited materials, or that the deposition rate of materials was rapid in comparison with this backflush frequency. When the pilot was operated with automatic backflushing every 96 seconds, Trinity River water fouled the 0.05 μm membrane at approximately the same rate as that observed when the membrane was backflushed every 20 minutes. With backflushing every 96 seconds, the resulting curve of permeation rate versus time traces the lowest points of the 20-minute backflush curve. After 400 minutes of operation, permeate flux had decreased to approximately 200 $l/m^2 \cdot h$. Untreated San Jacinto River water appeared

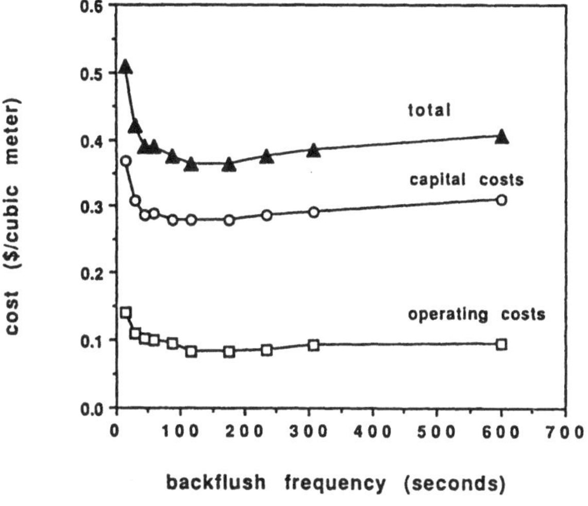

Fig. 4. Relative capital and operating costs as a function of backflush frequency calculated assuming cake-limited permeation. In this example, costs were calculated using the rate constant determined for a 0.05 μm membrane with alum coagulation

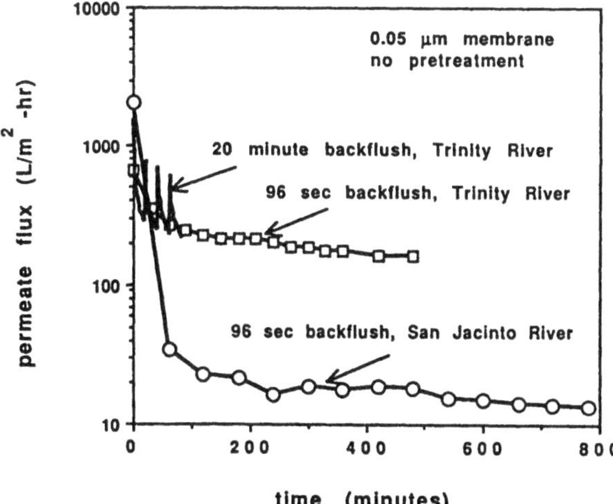

Fig. 5. Comparison of permeation rates through a 0.05 μm membrane for the San Jacinto and Trinity River waters without coagulation pretreatment. The effect of backflushing the membrane every 20 minutes is also compared with a 96-second backflushing frequency for Trinity River water

to foul the 0.05 μm membrane to a greater extent than did the Trinity River water. After 720 minutes of filtering San Jacinto River water, the permeate flux had decreased to 120 l/m$^2 \cdot$ h. A lower permeation rate for San Jacinto River water is anticipated based on the longer service time, smaller particle size and higher concentration of particles in this water compared with particles in the Trinity River water.

When these waters were coagulated with alum or ferric sulfate, frequent backflushing substantially enhanced permeate flux (Fig. 6). With backflushing every 96 seconds, the resulting curve of permeation rate versus time for the Trinity River water remains higher than a curve tracing out the maximum points of the corresponding 20-minute backflush curve. Coagulation pretreatment of these waters results in feed waters with very similar permeation characteristics. We speculate that the higher fluxes observed during the treatment of the two coagulated river waters are due to the transformation of smaller colloids and macromolecular materials to larger particles through particle aggregation, as well as adsorption, complexation, or precipitation of humic materials. This reduces the penetration of materials into membrane pores and thereby reduces fouling. In addition, coagulation increases the size of materials deposited on the membrane surface, resulting in a more porous cake. Larger particles should also experience a greater inertial lift and have greater shear-induced diffusivities; factors which in the former case may reduce particle deposition to the cake (k_1) and in the latter case may increase back-transport (k_2) from the membrane.

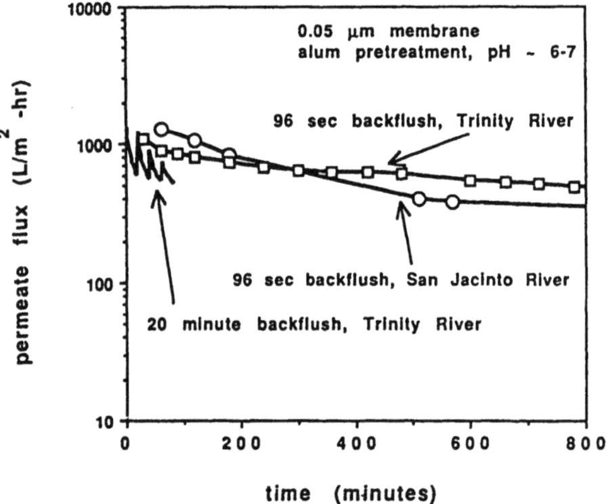

Fig. 6. Comparison of permeation rates through a 0.05 μm membrane for the San Jacinto and Trinity River waters with coagulation pretreatment using alum. The permeation rates obtained in treating these two river waters were similar when they were pretreated with coagulant. With coagulation, a 96-second backflushing frequency increased the permeation rate of Trinity River water over that observed with a 20 minute backflush frequency

Although coagulation with alum or ferric sulfate improved permeation rates, membrane performance was found to be sensitive to the conditions of pH and coagulant dose (Fig. 7). The highest permeation rates were obtained when pH was maintained near neutral values that minimize the solubility of aluminum or iron. The accumulation of aluminum or iron in the recycle loop of the pilot over time results in a decrease in pH. Without compensating for this drop in pH by adding base, permeate flux decreases, and the concentration of soluble aluminum or iron in the permeate increases. It is likely that some of the decreases in permeate flux observed during the first portions of these experiments were due to increasing concentrations of materials in the feed water over time as the unit approached a steady-state operation. (In fact, the unit never operates at steady-state due to continual changes in raw water quality.) The long-term, sustainable permeate flux to be expected in these applications should be estimated from pilot data collected after the unit has approached a steady-state with respect to the concentrations of materials in the feed stream that may cause fouling or colmatage of the membrane. During these experiments, the turbidity of the feed water often increased to over 1 000 NTU, the TOC increased to over 150 mg/l, and iron and aluminum concentrations increased by a factor of 10 or more in comparison with the coagulant dose.

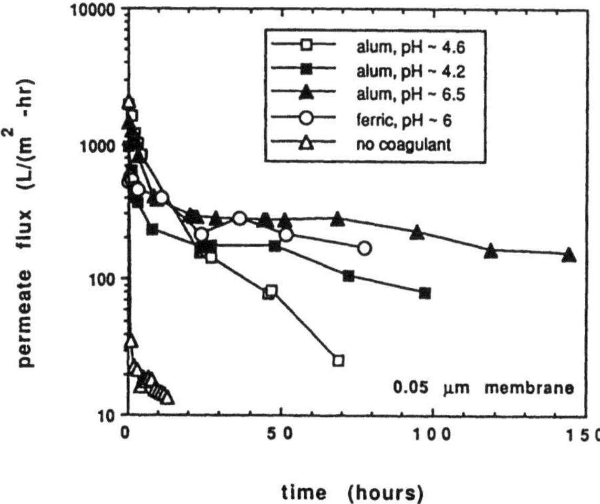

Fig. 7. Effect of coagulation pH on permeate flux. Although coagulation pretreatment improved permeate flux, higher permeation rates were obtained when pH was maintained between 6 and 7

For a constant permeation rate, the time required for the system to reach steady state, t_{ss}, with respect to materials rejected by the membrane can be estimated as,

$$t_{ss} \cong 3V_{sys}/Q(1 - Rr) \tag{7}$$

where Q is the flow of raw water to the membrane pilot, V_{sys} is the volume of the water within the membrane system (primarily the feed tank), R is the recovery of the system (permeate flow rate divided by raw water flow rate) and r is the rejection factor calculated as $(1 - C_p/C_f)$. C_p and C_f are the concentrations of the material of interest in the permeate and feed respectively. The time to "steady-state" in Eq. (7) is defined as the time required for C_f to be equal to 95 % of the concentration that would be measured if the system were operated for an infinite period of time. When pH was maintained near a value of 6.5 with alum coagulation, the concentration of aluminum rose from the initial dose of $10^{-3.6}$ M to approximately $10^{-2.9}$ M. In this instance, the average concentration of aluminum measured in the permeate during the experiment was $10^{-6.3}$ M. For these conditions and a recovery of 90 %, the time to steady-state in the pilot experiments summarized in Fig. 7 is estimated to be approximately 90 hours. In this experiment a permeate flux of 160 $1/m^2 \cdot h$ was achieved after 140 hours of operation.

In addition to improving permeate flux, coagulation pretreatment also significantly improved the quality of the membrane permeates. Average values for water quality parameters measured in the permeates of 0.05 μm membranes during longer pilot runs are listed in Table 4.

Tab. 4. Water quality of 0.05 μm membrane permeates during "steady state" operation

Parameter	No coagulant	Alum, pH 4.5	Alum, pH 6.5	Ferric sulfate*, pH 6
Turbidity (NTU)	1.29	0.18	0.44	0.47
TOC (mg/l)	8.6	5.1	4.1	3.1
p[Me]	–	2.1	5.9	5.1
Particles > 0.4 μm	1.6×10^6	5.1×10^4	2.5×10^6	4.5×10^5
p[Me][OH$^-$]^3	–	30.6	28.4	29.1

* Pilot may not have reached steady state

Turbidity and TOC data indicate improvements in permeate quality with coagulation pretreatment. The presence of significant numbers of particles greater than 0.4 μm in the permeate of the 0.05 μm membrane is attributed to the aggregation and/or precipitation after sampling of materials that may have passed through the membrane. Samples were diluted in electrolyte solution for analysis by the ESZ method. The electrolyte may have enhanced the aggregation of materials that were small enough to pass through the 0.05 μm membrane. Particle contamination on the permeate side of the membrane may have also contributed to particle counts in membrane permeates. However, when coagulation was used to treat the raw water, many of the particles present in membrane permeates may have been precipitates of the residual coagulant. The tendency for metal hydroxide precipitation in the permeate should increase with increasing pH and with increasing concentration of the metal in the permeate. A representation of the driving force for precipitation can be calculated from the metal concentration and pH measured in the permeate in a similar fashion to a solubility product. This quantity can then be normalized by the solubility product for the hypothetical solid phase. When the concentration of particles in the permeates of coagulated waters filtered through a 0.05 μm membrane is plotted as a function of this driving force a trend towards increasing particle number with increasing driving force is evident. Such a plot is shown in Fig. 8 where particle count, aluminum concentration, and pH data at different times during two pilot experiments are summarized. Points on Fig. 8 lying towards the right represent low aluminum concentrations but relatively high pH values. Points on the left of Fig. 8 represent high aluminum-low pH combinations in the permeate.

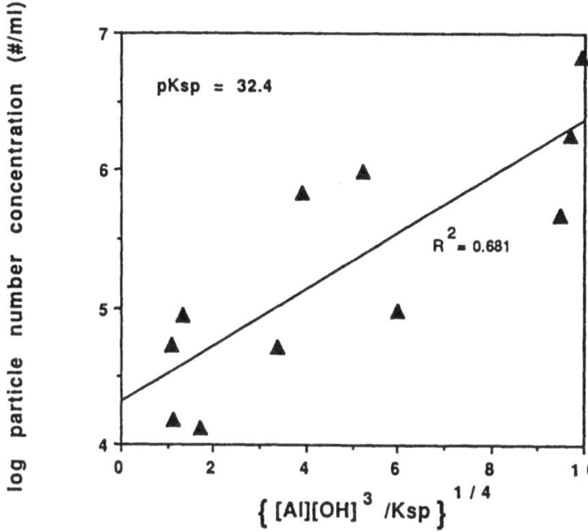

Fig. 8. Particle concentration in the permeate of 0.05 μm membrane as a function of the driving force for precipitation. Solubility products reported in the literature for the precipitation of aluminum hydroxide typically fall within the range labeled for gibbsite and the amorphous precipitate

Conclusions

Coagulation using alum or ferric sulfate significantly improved the flux and quality of permeates produced from turbid surface waters with moderate to high concentrations of TOC. Curve-fitting procedures suggest that reductions in permeate flux reversed by backflushing were due, in this pilot study, primarily to the formation of a cake of materials deposited on the membrane surface. Although the cakes formed when treating coagulated water presented less resistance to the flow of permeate, the cake resistance, with or without coagulation pretreatment, was greater than the resistance of the membrane. In addition to creating a more porous cake, coagulation of the raw water also reduced the rate of colmatage and improved the efficiency of backflushing as a means of maintaining permeate flux. In experiments with San Jacinto River water the highest "steady-state" permeate flux (160 1/m$^2 \cdot$ h) was obtained using a 0.05 μm membrane with alum coagulation at pH 6.5 and with frequent backflushing.

Although coagulation pretreatment was found to enhance most aspects of membrane performance, satisfactory performance was sensitive to the conditions of coagulant dose and pH. Near-neutral pH values that minimized the solubility of aluminum or iron produced the highest permeate fluxes and the lowest concentrations of TOC in the permeate. However, in comparison with lower pH values, turbidities and particle counts were higher when the pH for coagulation was maintained near neutral values (6–7).

Acknowledgements

The authors gratefully acknowledge the contributions of Daniel Schmelling, Faten Nazzal, Karen Pickering, and Shankararaman Chellam to this work. Jim Greenlee, Carl Henriques, and Aubrey LaFargue of the Houston Public Works Department provided for the installation of the pilot at the Houston Water Purification Facility and contributions of time and equipment by Ken Thomas and the Alcoa Separations Technology Division made this work possible. This work was supported by the Texas Advanced Technology Program under Grant No. 003604022.

References

[1] Wiesner, M.R., Clark, M.M., Mallevialle, J.: Membrane Filtration of Coagulated Suspensions. J. Envir. Engr., ASCE *115* (1989) 20–40
[2] Altena, F.W., Belfort, G.: Lateral Migration of Spherical Particles in Porous Flow Channels: Application to Membrane Filtration. Chemical Engineering Sci. *49* (1984) 343–355
[3] Wiesner, M.R., Chellam, S.: Mass Transport and Colloidal Fouling in Pressure-Driven Membrane Processes: Considerations in Potable Water Treatment. J. Am. Wat. Wks. Assoc. *84* (1992) 88–95
[4] Fane, A.G.: Ultrafiltration of Suspensions. Membrane Sci. *20* (1984) 249–259
[5] Hsieh, H.P.: Inorganic Membranes. In: New Membrane Materials and Processes for Separation, K.K. Sirkar and D.R. Lloyd (eds.), Vol. 84. American Institute of Chemical Engineers, New York 1988
[6] Pickering, K.D., Wiesner, M.: Cost Model for Low-Pressure Membrane Filtration. J. Envirn. Engr. Div., ASCE (in press)

Mark R. Wiesner, Srinivas Veerapaneni, and Drahomira Brejchová
Department of Environmental Science and Engineering
Rice University
P.O. Box 1892
Houston, Texas 77251
USA

Wastewater Treatment

Characterization of Wastewater:
The Effect of Chemical Precipitation
on the Wastewater Composition
and its Consequences
for Biological Denitrification

M. Henze and P. Harremoës

Abstract

Preprecipitation changes the wastewater composition by reducing the actual concentrations of the components. More significant are the changes between the various fractions of a given component. Preprecipitation increases the soluble fraction of COD and nitrogen, which affects the biological down-stream process for nitrogen removal. Remaining COD has a high denitrification rate but a low denitrification capacity, which might necessitate the addition of an external carbon source in order to fulfill the effluent criteria. The seeding with biomass present in raw or preprecipitated wastewater into the activated sludge increases the denitrification rates which are obtained in practice.

1. Introduction

Wastewater can be characterized in many ways. Today it can be described in such detail that it is important to consider the purpose in each separate case (Sollfrank and Gujer, 1991; Henze, 1992). In chemical precipitation, the following basic characterization into three fractions is appropriate:

- suspended
- colloidal
- soluble.

This fractionation is purely artificial and a compromise between complexity and practical use of the characterization. The division of organic matter into 3 fractions is in line with the present day modelling of biological wastewater treatment processes (Henze et al., 1987, and Sollfrank and Gujer, 1991). At the same time, such a division is needed to explain the observations of chemical precipitation, where significant parts of what classically is considered soluble organic matter (particle size $< 1\ \mu m$) is removed. It is important to realise that the composition of each of the three fractions varies considerably from fraction to fraction. By removing parts of these fractions by chemical preprecipitation, the composition of the wastewater after precipitation will be changed significantly, not only in absolute concentrations but also with respect to the ratios between the various substances and fractions. The changes thus created affect the biological processes downstream of the preprecipitation. The consequences with respect to design and operation of the biological nitrogen removal processes will be presented in this paper.

2. Composition of Raw Wastewater

Preprecipitation of wastewater removes mainly suspended matter, but significant removal of colloid material and some removal of soluble matter can also be encountered. The broad spectrum of precipitants available today presents an opportunity to control the degree of removal of particles and colloids. The concentration of various components in the three main fractions in wastewater is illustrated in Table 1. Suspended matter is the dominant component.

Tab. 1. Fractionated components in wastewater

Fraction	Component [g/m^3]			
	C_{COD}	C_{BOD}	C_{TN}	C_{TP}
Suspended	200	85	4	1
Colloidal	65	35	6	1
Soluble	85	40	20	6
Total	350	160	30	8

(Based on Henze, Kristensen and Strube, 1992,
and Ødegaard, 1992)

Figure 1 shows an example of a detailed domestic wastewater analysis for organic matter.

In domestic wastewater, about one third of the nitrogen is organic and thus linked to the COD fractions given in Fig. 1. As seen in Fig. 2, most of the nitrogen is found as ammonia, which passes through every pretreatment process.

3. Parameters Affected
by Chemical Precipitation

The composition of the wastewater after primary sedimentation or preprecipitation will vary according to the efficiency of the pretreatment process. Thus the relative amounts of the various fractions will change. As an example, Fig. 3 illustrates biological sludge production, FSP, after the various pretreatment methods given in Fig. 1 and 2 were used on the wastewater. The calculations

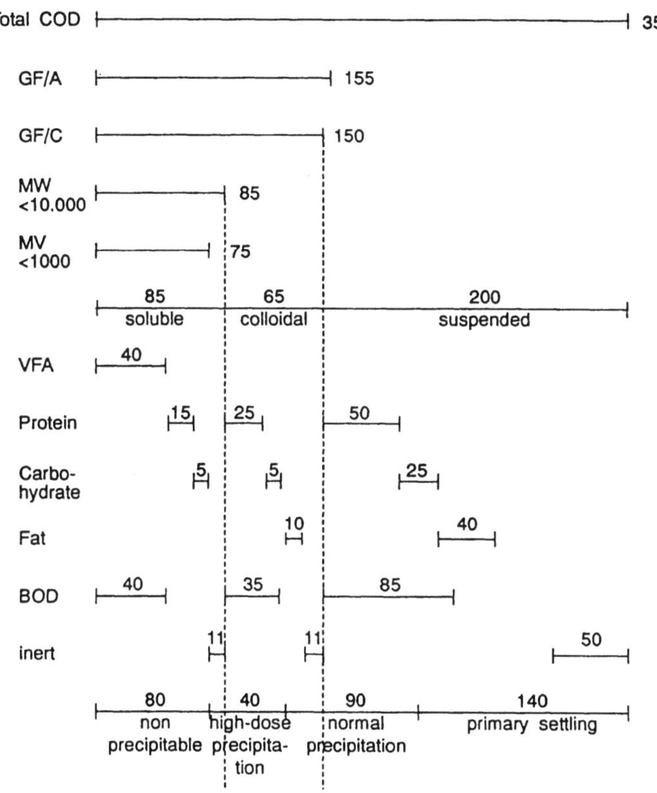

Fig. 1. Detailed fractionation of organic matter in domestic wastewater. All figures in $g\,COD/m^3$. Based on Henze, Kristensen and Strube (1992)

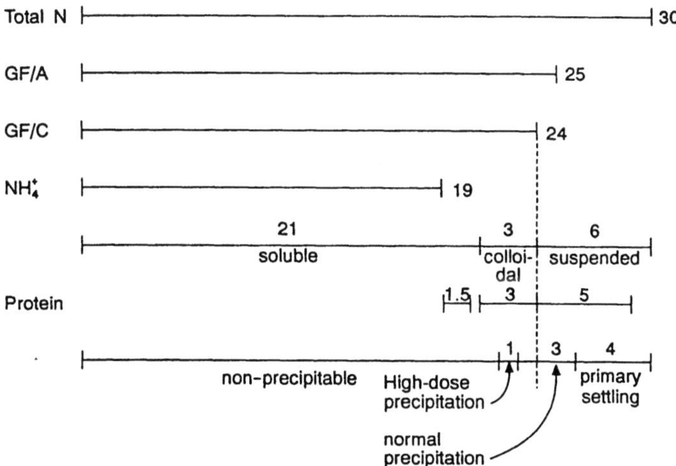

Fig. 2. Detailed nitrogen fractionation in domestic wastewater. All figures in $g\,N/m^3$. Based on Henze, Kristensen and Strube (1992)

are based on the sludge production found for nitrification-denitrification following preprecipitation, where the *actual* yield was 0.32 g sludge COD/g COD removed (Kristensen et al., 1992 a).

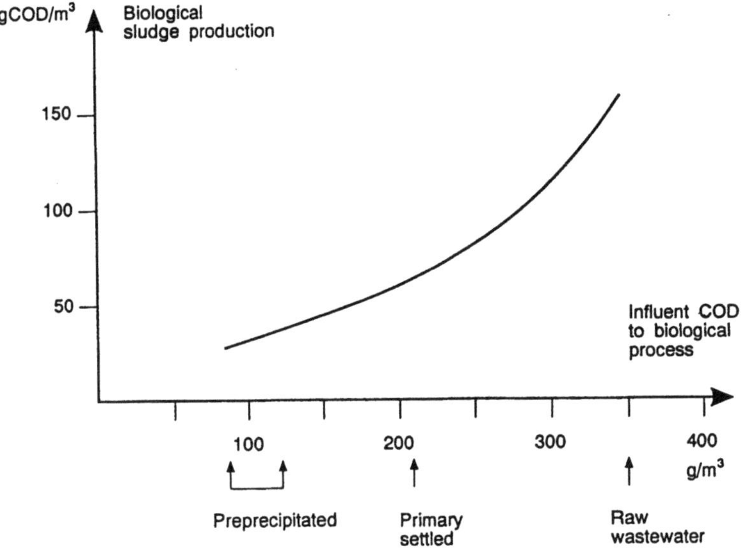

Fig. 3. Biological sludge production, F_{SP}, at 14 °C as a function of pretreatment (COD concentration in the influent to the biological process)

The nitrification tank volume, V_2, is inversely related to the sludge production in the biological process:

$$V_2 = (\Theta_x \cdot F_{SP})/X_2$$

where X_2 is the sludge concentration in the aeration tank and Θ_x is the aerobic sludge age needed for nitrification.

In Fig. 4 this is shown in more totality as the total sludge production (pretreatment and biological process) and the total tank volume for pretreatment and nitrification. The required aerobic sludge age has been assumed to be 10 days. The figure illustrates the range of treatment plant design possibilities with respect to total tank volume and total sludge production.

Primary settling affects the biological sludge production considerably and even more so does preprecipitation. Figure 4 illustrates the range which can be obtained depending on the efficiency of the preprecipitation, and it is seen that this range is rather narrow. This means that the efficiency of the preprecipitation only affects the biological sludge production marginally. The important issue is the total design ranges observed in Figs. 3 and 4. It is possible to select a process which requires very small tank volumes (and small area) but which results in high sludge production. Such a process might be the best solution when the land area available is limited. The overall economy is much more complicated and will not be discussed here.

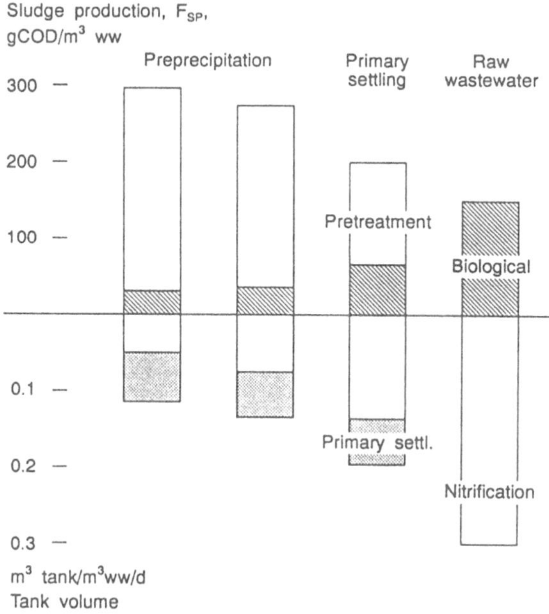

Fig. 4. Sludge production at 14 °C (from pretreatment and biological process) and tank volumes needed for pretreatment ($\Theta = 1.5\,\text{h}$) and nitrification

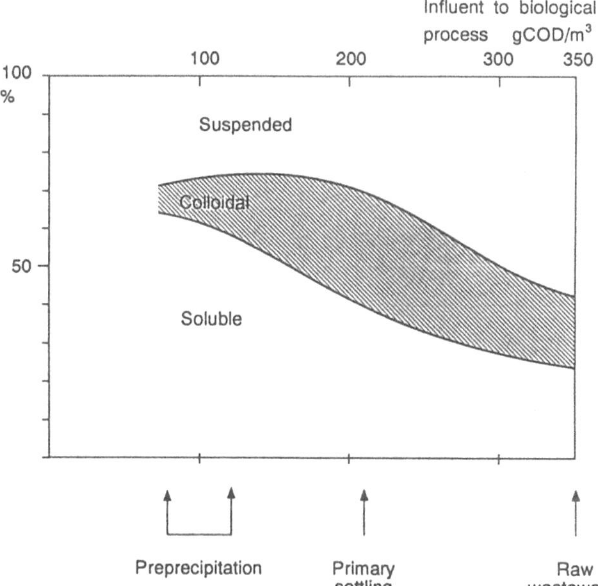

Fig. 5. COD distribution after pretreatment of domestic wastewater

The effect of the pretreatment on the detailed COD distribution is shown in Fig. 5. Chemical pretreatment results in a predominance of soluble and colloidal matter, which influences the denitrification rate, the denitrification capacity and the biomass activity considerably.

The effect of pretreatment on nitrogen and phosphorus in the influent to the biological process is seen in Figs. 6 and 7.

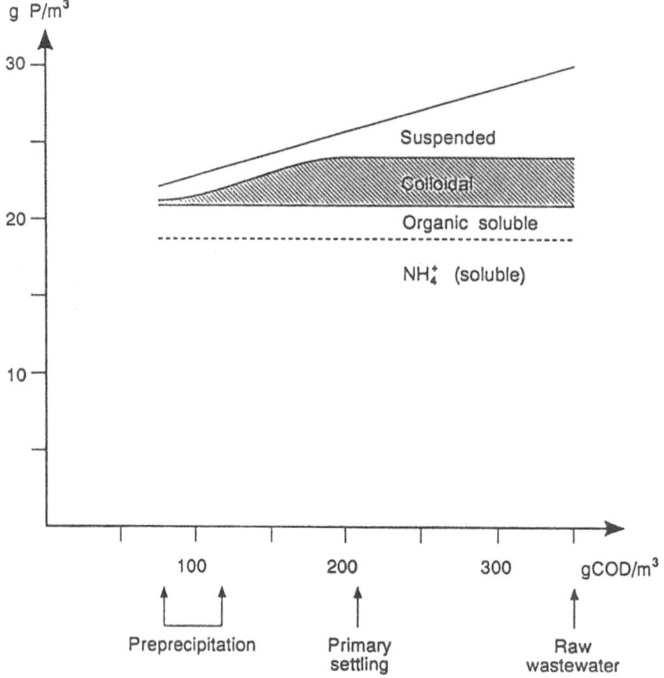

Fig. 6. Distribution of nitrogen compounds after pretreatment

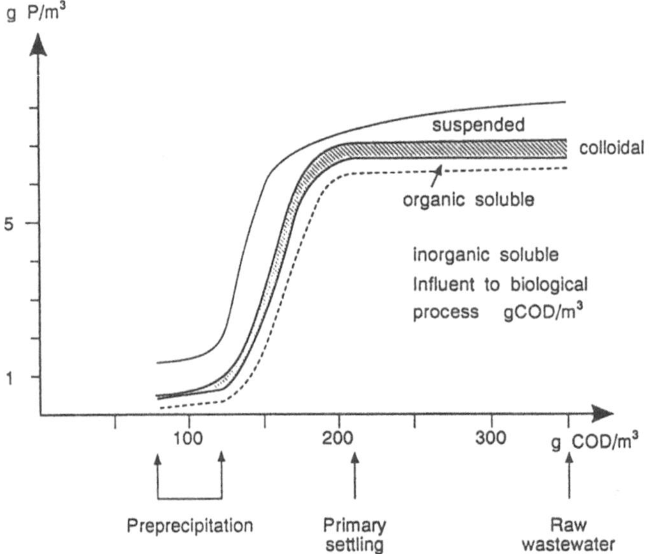

Fig. 7. Distribution of phosphorus compounds after pretreatment

4. Characterization of Wastewater for Biological Processes

The three fractions into which the wastewater was separated in Table 1 are closely related to the chemical treatment processes. At the same time, they influence the biological process. In order to explain the observations from biological processes, a more detailed fractionation is needed. Figure 8 shows a detailed fractionation which can be directly linked to the biological processes.

Suspended material can be regarded to consist of three fractions:

– biomass
– slowly degradable (hydrolysable) material
– inert matter.

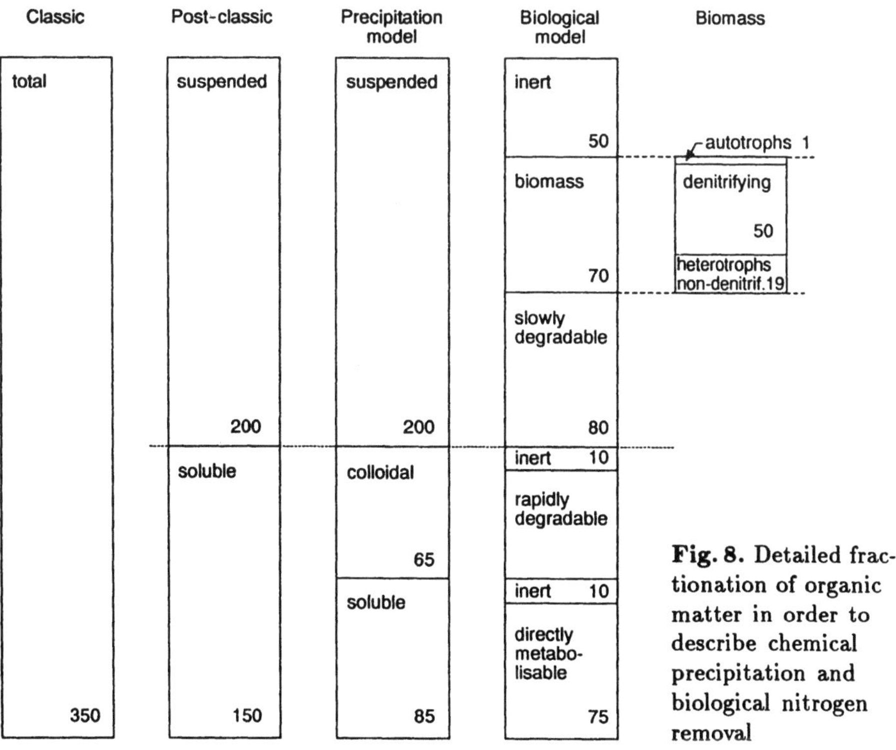

Fig. 8. Detailed fractionation of organic matter in order to describe chemical precipitation and biological nitrogen removal

The biomass in wastewater suspended solids plays an important role with respect to the development of the biomass in the biological process (Henze, 1989). Thus the removal of more or less suspended solids in the pretreatment will affect the biomass composition and the reaction rates in the biological process. The biomass can be subdivided into autotrophic, heterotrophic and denitrifying biomass, each of which influences the biological nitrogen removal process separately. The slowly degradable material has to be hydrolyzed before being metabolized. Removal of organic matter within a biomass of sufficient size and structure requires diffusion into the biomass. That applies to biofilms (Harremoës, 1978) and to flocs in activated sludge. This demands fractionation into two groups:

– diffusible substrate
– non-diffusible substrate.

The demarkation line between the two groups is not well defined, but estimated to be in the range 10^4–10^5 amu (average molecular weight), (Larsen, 1992). It requires molecular weights on the order of 10^3 amu ($5 \cdot 10^{-3}$ μm) to penetrate the cell membrane. Accordingly, further hydrolysis is required within the biomass. Today there is no specific analytical procedure suitable for an explicit fractionation.

The hydrolysis is the rate limiting step in the degradation of the non-diffusible fraction. Hydrolysis can take place by surface attached enzymes or by exogenic enzymes released into the bulk water. The latter case has proved valid for the hydrolysis of starch to smaller sized starch-monomers that can diffuse into the biofilm, where mineralization can take place after further hydrolysis to glucose. In fixed film reactors such a mechanism may prove essential to design, because poor hydrolysis may be a result of wash-out of the enzymes due to a short hydraulic residence time (Larsen and Harremoës, 1992)

The removal of slowly degradable suspended material in the biological process will affect the biological process. By removing part of the slowly degradable suspended material in the pretreatment, the oxygen consumption and sludge production in the biological process will be reduced. At the same time, the denitrification capacity will be reduced, see Table 2. The inert suspended matter ends up in the sludge without any change.

Tab. 2. Effect of pretreatment on nitrogen removal process. The removal of various fractions affects the biological process as indicated in the table (+ = increase; − = decrease; 0 = no effect)

Removal of fraction, pretreatment	Effect on					
	oxygen consumption	bio sludge production	total sludge production	denitri-fying biomass	denitri-fication rate	denitri-fication capacity
Biomass	−	−	+	−	−	0
Suspended degradable	−	−	+	0	+	−
Suspended inert	0	−	0	0	+	0
Colloidal degradable	−	−	+	0	−	−
Colloidal inert	0	−	0	0	+	0
Soluble degradable	−	−	+	0	−	−
Soluble inert	0	0	+	0	0	0

Colloidal material can be regarded to consist of two fractions:
- rapidly degradable material
- inert matter.

The definition of the colloidal fraction given in Fig. 1 includes material in the size interval 1 μm–MW 10 000. It is recognized that this might not reflect a concise chemical definition of colloidal matter, but for municipal wastewater it is a realistic approximation. Rapidly degradable organic matter needs to be hydrolysed before being metabolized by the microorganisms. The hydrolysis of this material is faster than that of the slowly degradable material, but is governed by the same basic phenomena. Again, for degradation by biomass the distinction between diffusible and non-diffusible organic matter may be essential. The hydrolysis is the rate limiting reaction and the potential risk of poor performance due to wash-out of the exogenic enzymes has to be evaluated. Inert colloidal matter can contribute more or less to sludge production, depending on the degree to which it is captured in the biological or chemical processes in the plant.

Soluble material can be divided into two fractions:
- directly metabolisable
- inert.

Directly metabolisable organic matter can be metabolized without any preceding hydrolysis step. The uptake of this fraction is very rapid, and a direct response on the oxygen utilisation rate and the denitrification rate is seen (Kristensen et al., 1992). The inert soluble material passes through the treatment plant without being changed, possibly with the exception of a minor amount that could be precipitated if a very efficient chemical precipitation is performed.

5. Wastewater Composition and Nitrogen Removal

The changes in wastewater composition due to preprecipitation influence the nitrogen removal process by the mechanisms shown in Table 2. Table 3 shows the denitrifying biomass in the influent to the nitrogen removal process. The denitrifying biomass in the process will be a mixture of biomass inoculum from the influent and biomass developed (grown up) in the process. An estimate of this is made based on a yield of denitrifyers, $Y_D = 0.32$ g COD/g soluble and colloidal COD in the influent (Kristensen et al., 1992 a).

The actual denitrification rate in a biological process depends mainly on the basic denitrification activity and the carbon source available. The basic denitrification rates are influenced by the fraction of denitrifying biomass which comes from the influent. This is shown in Fig. 9, which is based on data from Christensen and Harremoës, 1977, and Henze, 1989. It is seen that the biomass present in the raw wastewater has a high denitrification rate as compared to

Tab. 3. Denitrifying biomass in the influent to the biological process and generated in the process

Denitrifying biomass	Pretreatment			
	None	Primary	Precipitation normal	Precipitation high dose
From influent, g COD/m³	50	16	8	7
Generated, g COD/m³	41	41	24	15
Per cent from influent	55	28	25	32

that of the biomass produced in the treatment process. Why this is so is not known, but one explanation could be that the denitrifying organisms which grow in the treatment plant have a slower growth rate than those organisms found in the raw wastewater. Another explanation could be that the fraction of active biomass in a denitrifying treatment plant is smaller than that in the raw wastewater biomass. Whatever the explanation, the high denitrification rates found in raw wastewater biomass cannot be obtained in an activated sludge plant, because a significant fraction of the denitrifying biomass will always be generated within the plant.

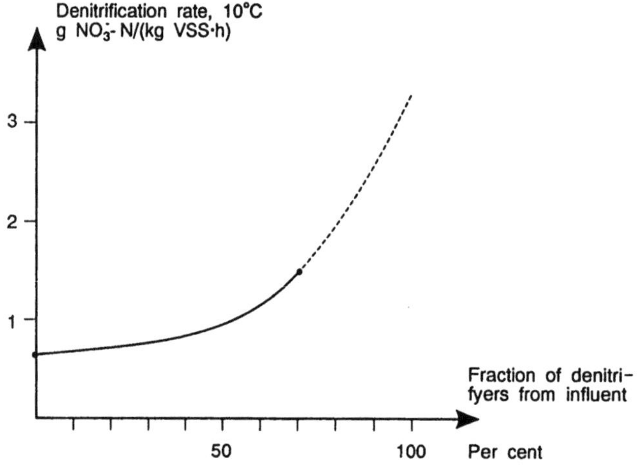

Fig. 9. Denitrification rates with organic matter from raw wastewater (soluble and colloidal)

The actual wastewater composition determines the denitrification rate which can be obtained and the denitrification capacity. The more organic matter that is removed from the wastewater by the pretreatment, the less denitrification capacity is left. The resulting rate depends on the amounts of the various organic fractions that are left over from the pretreatment. As seen from Fig. 5, the fractions with slow denitrifying rates are reduced relative to the fast rate fractions. The overall result with respect to the denitrification rates and nitrogen removal capacity is shown in Fig. 10. The capacity depends strongly upon

the effectiveness factor with respect to COD. The COD/N ratio will in practice
vary from 5–10 kg COD oxidized/kg N removed. A good COD household can
keep the COD/N ratio down around 6. In Fig. 10 a ratio of 6 has been used.
Observe that the COD/N ratio as defined here is smaller than the one often
used, based on influent COD and nitrogen. The inert part of the influent COD
will result in a higher COD/N influent ratio needed to get the same amount of
nitrogen removed. The denitrification rates and the capacities given in Fig. 10
can be used to calculate the retention time required for denitrification and the
additional carbon source required. This is shown in Fig. 11.

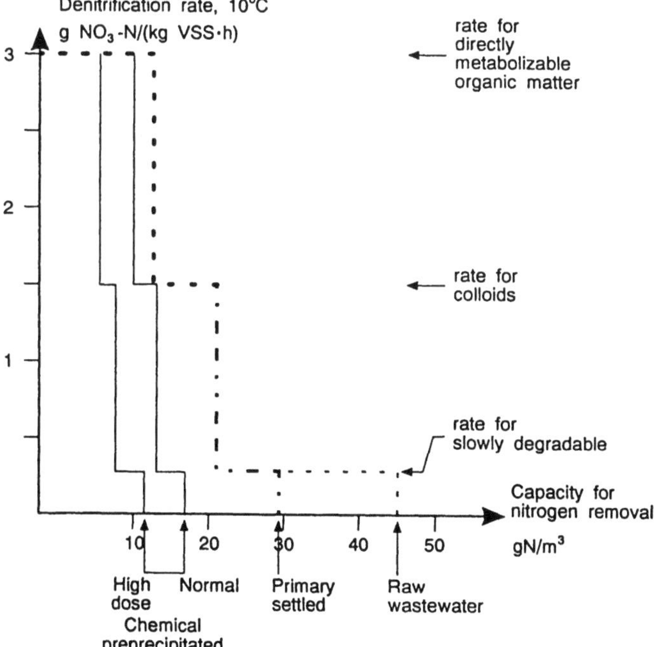

Fig. 10. Denitrification rates and capacities for the three organic fractions considered. Based on wastewater composition as given in Fig. 8

When the denitrification capacity for a given wastewater is too small, it is
necessary to add an additional carbon source. This can be hydrolysed sludge,
acetic acid, industrial waste products, etc. One does not need to wait until
all available carbon in the wastewater has been used for denitrification. It
is possible to add an additional carbon source to the denitrification process
in such a way that the high rate is maintained during the removal of all the
nitrogen. This will save denitrification volume but might have other effects that
are economically counterproductive, for example, increased sludge production.
It is beyond the scope of this paper to go into these details, but they are
important for the overall analysis of a wastewater treatment process needed
before the final decision is made concerning plant type.

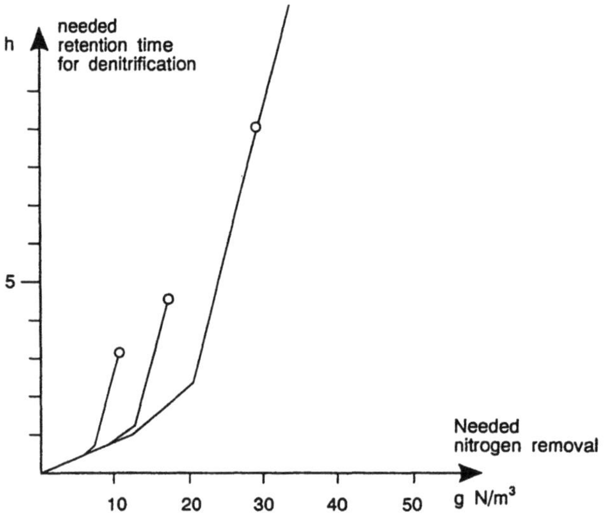

Fig. 11. Required retention time for a denitrification process at 10 °C as a function of the degree to which the nitrogen removal capacity has to be used. Sludge concentration assumed to be 4 kg VSS/m³

6. Conclusion

A detailed characterization of wastewater makes it possible to predict effects of pretreatment on the downstream nitrogen removal processes. The sludge production of the biological process and the tank volumes needed for nitrification and denitrification are significantly influenced by the pretreatment.

The pretreatment increases the soluble fraction in the wastewater. This results in higher denitrification rates of the organic matter in the wastewater. This is partly counterbalanced by the reduced denitrifying activity of the wastewater caused by a smaller inoculum of denitrifying biomass from the influent to the biological process.

The denitrification capacity is strongly reduced by preprecipitation, and this means that an additional carbon source is needed in many cases in order to obtain the required removal of nitrogen.

List of Symbols

C_{BOD}	Total BOD concentration
C_{COD}	Total COD concentration
C_{TN}	Total nitrogen concentration
C_P	Total phosphorus concentration
F_{SP}	Sludge production
X_2	Sludge concentration in aeration tank
Y	Yield coefficient
Θ	Hydraulic retention time
Θ_x	Solids retention time

Acknowledgements

Some of the results reported here are from the EUREKA project: HYPRO. The partners in this project are the Department of Environmental Engineering, Technical University of Denmark, I. Krüger Systems, Kemira Kemi, the Water Quality Institute, Denmark and the University of Trondheim, Norway.

7. Literature

Christensen, G.H., Henze, M., Harremoës, P.: Biological Denitrification of Sewage: A Literature Review. Progress in Water Technology *8* (4/5) (1977) 509–555

Harremoës, P.: Biofilm Kinetics. In: Water Pollution Microbiology, Vol. 2, R. Mitchell (ed.). Wiley & Sons, New York 1978, pp. 82–109

Henze, M.: The Influence of Raw Wastewater Biomass on Activated Sludge Oxygen Respiration Rates and Denitrification Rates. Wat. Sci. Tech. *21* (6/7) (1989) 603–609

Henze, M.: Characterization of Wastewater for Modelling of Activated Sludge Processes. Water Sci. Tech. *25* (6) (1992)

Henze, M., Kristensen, G.H., Strube, R.: Determination of Organic Matter and Nitrogen in Wastewater. Paper to be submitted to Water Research (1992)

Kristensen, G.H., Jørgensen, P.E., Henze, M.: Characterization of Functional Groups and Substrate in Activated Sludge and Wastewater by AUR, NUR and OUR. Wat. Sci. Tech. *25* (6) (1992)

Kristensen, G.H., Jørgensen, P.E., Strube, R., Henze, M.: Combined Preprecipitation, Biological Sludge Hydrolysis and Nitrogen Reduction – a Pilot Demonstration of Integrated Nutrient Removal. Wat. Sci. Tech. *26* (1992 a)

Larsen, T.A.: Degradation of Colloidal Organic Matter in Biofilm Reactors. Ph.D. Dissertation. Department of Environmental Engineering, Technical University of Denmark, 1992

Larsen, T.A., Harremoës, P.: Degradation of Colloidal Organic Matter in Biofilm Reactors. Paper. Department of Environmental Engineering, Technical University of Denmark, 1992

Sollfrank, U., Gujer, W.: Characterization of Domestic Wastewater for Mathematical Modelling of the Activated Sludge Process. Wat. Sci. Tech. *23* (4–6) (1991) 1057–1066

Ødegaard, H.: Norwegian Experiences with Chemical Treatment of Raw Wastewater. Presented at Conference on Management of Wastewater in Coastal Areas, Montpellier, France, March 31–April 2, 1992

Mogens Henze and Poul Harremoës
Department of Environmental Engineering
Technical University of Denmark
Building 115
DK-2800 Lyngby
Denmark

Pre-precipitation Followed by Biological Denitrification Supported by Addition of Biological or Thermal/Chemical Hydrolysis Products

G.H. Kristensen and P.E. Jørgensen

Abstract

Pilot scale experiences are reported for the process combination of pre-precipitation followed by biological nitrogen removal. Biological denitrification was supported by addition of hydrolysis products from hydrolysis of pre-precipitated primary sludge. Pre-precipitation was performed with PAX and with JKL, respectively. The biological nitrogen removal was performed based on the Bio-Denitro system. Hydrolysis products from biological sludge hydrolysis and from thermal/chemical sludge hydrolysis were applied in the experiments.

Both precipitation chemicals resulted in efficient reductions in COD, phosphorus, and suspended solids. JKL was more efficient for precipitation of phosphorus than PAX. Reduction in nitrogen was low for both chemicals. Due to the very different levels to which COD and nitrogen were reduced, the ratio of COD to nitrogen was reduced from 10–12 in the raw wastewater to 4.5–5.0 in the pre-precipitated wastewater.

Nitrogen reduction in the Bio-Denitro plant was studied with pre-precipitated wastewater alone and with pre-precipitated wastewater with addition of biological hydrolysate and various doses of thermal/chemical hydrolysate, respectively. Mean nitrogen reductions of 63 % and 67 % were found for the two chemicals with an influent of chemically precipitated wastewater alone. When biological hydrolysate was added, the mean nitrogen reduction was increased to 76 %; with the addition of thermal/chemical hydrolysate, the mean reduction was 77 %.

Introduction

The main result of chemical pretreatment as compared to conventional primary settling is the increased removal of suspended organics and total phosphorus. Suspended solids in the raw wastewater are coagulated and flocculated by adding chemicals, thereby enabling a subsequent removal by sedimentation. Dissolved and suspended phosphorus is removed by precipitation, coagulation, flocculation, and sedimentation. Due to the very efficient primary treatment of the wastewater it is possible to establish a compact secondary treatment stage (Karlsson and Smith, 1990).

The fact that chemical pretreatment is highly efficient for removal of organics but inefficient for removal of nitrogen poses certain questions when nitrogen removal in the secondary treatment is in question. Efficient denitrification in the secondary treatment stage demands a sufficient ratio of COD : N. One possibility for supporting and accelerating biological denitrification in such process combinations is to expose the primary sludge to a hydrolysis process. In the hydrolysis process, a portion of the particulate organics is solubilized. After separation from the sludge phase, the soluble organics can be introduced into the denitrification process. The hydrolysis process may be biological, thermal, chemical, or a combination of these.

The process configuration including pre-precipitation, hydrolysis, and biological nitrification and denitrification is being intensively studied in a Scandinavian project called HYPRO. The basic ideas in the HYPRO concept are described by Henze and Harremoës (1990).

In the present paper, hydrolysis products from biological and thermal/chemical hydrolysis are compared in lab-scale and pilot-scale investigations. Furthermore, pilot scale experiences with pre-precipitation with a pre-polymerized aluminium compound and a ferric salt are presented.

Experimental Materials and Methods

Experimental Programme

The pilot plant investigation covered several phases with different operations of the pilot plant system. This paper reports on the results with four process configurations, phases A–D, as illustrated in Fig. 1.

Phase A. Chemical pretreatment with PAX. Biological nitrogen reduction in chemically pretreated wastewater without adding hydrolysate.

Phase B. Chemical pretreatment with PAX. Biological nitrogen reduction supported by adding biologically hydrolyzed carbon source (liquid hydrolysate phase was separated from sludge phase by centrifugation). Simultaneous precipitation with PAX.

Phase C. Chemical pretreatment with JKL. Biological nitrogen reduction in chemically pretreated wastewater without adding hydrolysate.

Phase D. Chemical pretreatment with JKL. Biological nitrogen reduction supported by adding thermally/chemically hydrolyzed carbon source (liquid hydrolysate phase was separated from sludge phase by centrifugation).

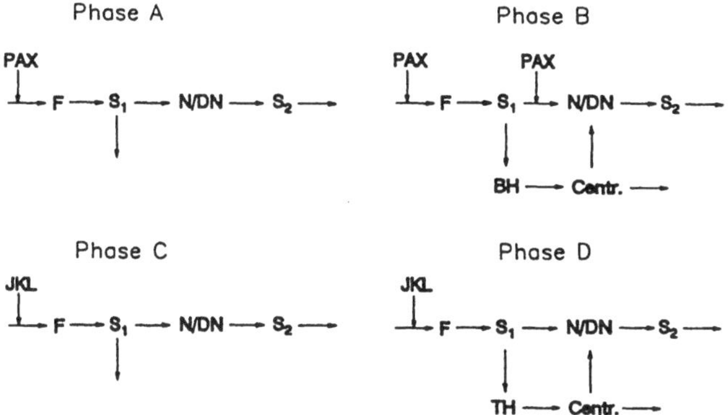

Fig. 1. Basic flow schemes for the four experimental phases. F = flocculation; S_1 = primary sedimentation; BH = biological sludge hydrolysis; TH = thermal/chemical sludge hydrolysis; Centr. = centrifugation; N/DN = nitrification and denitrification, S_2 = secondary sedimentation

Pilot Plant System

The pilot scale system consisting of pre-precipitation, biological sludge hydrolysis, and biological nitrogen removal is shown in Fig. 2. The figure illustrates the set-up for phase B. Data for the pilot plant units are given in Table 1.

Raw wastewater (2 m³/h) from the Lundtofte Treatment Plant north of Copenhagen was treated using pre-precipitation. Pre-polymerized aluminium salt, PAX, or ferric chloride sulphate, JKL, were used as coagulants. Flocculation took place in four tanks in series at decreasing G-values. Hydraulic retention time in each tank was 10 minutes. The flocculated suspended solids were separated from the wastewater in a primary settling tank.

Part of the chemically pretreated wastewater (150–200 l/h) was pumped into the activated sludge plant consisting of two biological reactors and a secondary clarifier. In experimental phases A and B, the activated sludge plant was operated for nitrogen removal based on the Bio-Denitro principle, as described by Bundgaard and Nielsen (1989). In phases C and D, the system was changed and operated according to the TRIO-system in which the mixed liquor passes through an aerobic tank before entering the clarifier. In the activated sludge system, phosphorus removal could be intensified by simultaneous precipitation.

The chemically precipitated primary sludge could be taken from the bottom of the clarifier to an anaerobic biological sludge hydrolysis stage. The hydrolyzed primary sludge was separated by centrifugation in a decanter centrifuge into a liquid phase, hydrolysate, and a solid phase. The hydrolysate could be dosed to the activated sludge plant to support biological denitrification.

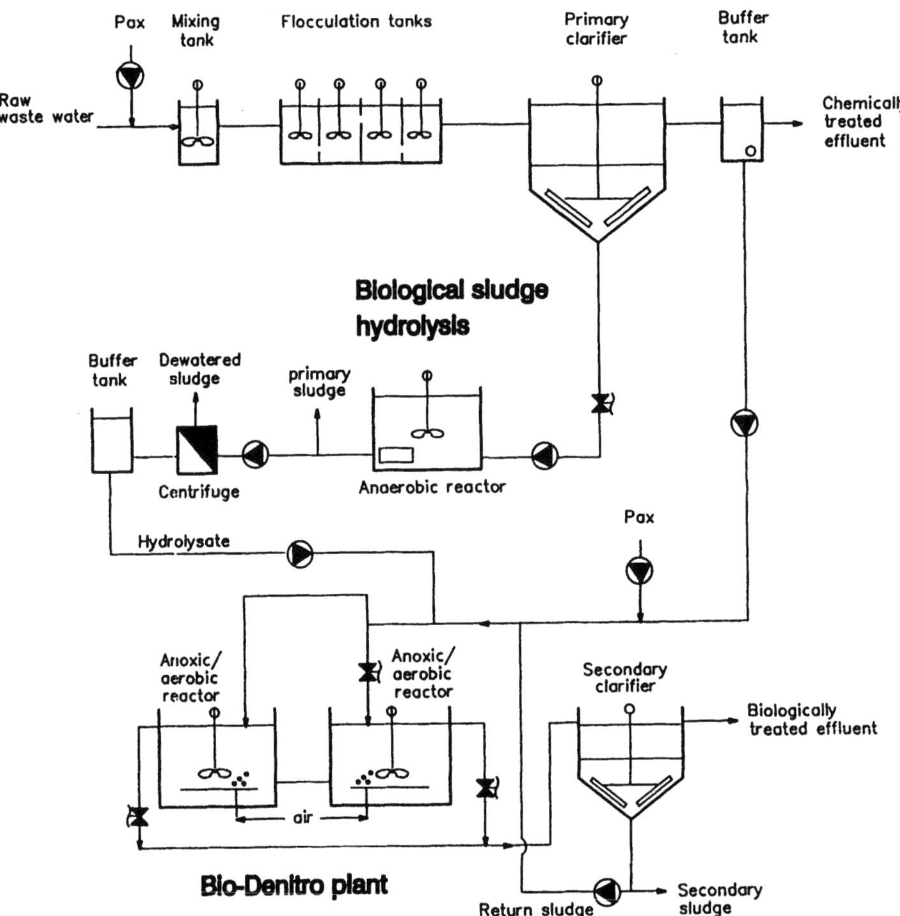

Fig. 2. Pilot plant system including pre-precipitation, biological sludge hydrolysis and biological nitrogen removal combined with simultaneous precipitation. Experimental phase B

Tab. 1. Data for the pilot plant units

Primary sedimentation		*Biological sludge hydrolysis*	
Primary clarifier	3.1 m^2	Anaerobic reactor	975 l
Pre-precipitation plant		*Bio-Denitro plant*	
Flocculation tanks	4 × 300 l	Reaction tanks	2 × 825 l
Primary clarifier	3.1 m^2	Secondary clarifier	0.8 m^2

Chemical Pretreatment

In experimental phases A and B, a pre-polymerized aluminium salt, PAX 61 (KEMIRA AB, Sweden) was used. Some advantages of PAX over traditional metal salt coagulants are less alkalinity consumption and a wider working pH range (Ødegaard et al., 1990). The basicity (OH : Al ratio) of PAX 61 was 1.7. In phase A, PAX was added to the wastewater at a dose of 0.45 mole Al/m^3 (from 08.00 to 24.00) and 0.30 mole Al/m^3 (from 24.00 to 08.00). In phase B, the PAX dose was 0.3 mole Al/m^3 throughout the day. For simultaneous precipitation, a PAX dose of 0.2 mole Al/m^3 was added.

In experimental phases C and D, JKL (KEMIRA AB, Sweden) was used as a coagulant for the chemical pretreatment. The JKL dose was 0.3 mole Fe/m^3.

Sludge Hydrolysis

Biological sludge hydrolysis was carried out in the pilot unit in a closed, totally mixed reactor. In phase B, the hydrolysis was carried out at a temperature of 30 °C and a hydraulic retention time of 2 days.

In phase D, hydrolysate from thermal/chemical treatment was dosed to support and accelerate the biological nitrogen removal.

Thermal/chemical hydrolysate was obtained from a full scale (60–120 m^3 sludge per day) thermal/chemical hydrolysis plant, the Tammerfors sewage treatment plant, 300 km north of Helsinki, Finland. The thermal/chemical hydrolysis plant was fed continuously with chemically precipitated primary sludge, and was operated with a sludge retention time of 30–35 minutes at a temperature of 150 °C. In order to increase the degree of COD solubilization, pH was lowered to approx. 2 during the thermal treatment by adding sulphuric acid. The thermally/chemically hydrolyzed sludge was then neutralized by adding $Ca(OH)_2$ to pH 8, and centrifuged to separate the liquid phase (hydrolysate) from the solid phase. Before shipment to Denmark, acid was added to the batch which was actually used, in order to avoid biological activity. The full scale thermal/chemical hydrolysis plant is described in more detail by Smith and Göransson (1991).

Activated Sludge Treatment

Main operational data of the four phases of the activated sludge plant are given in Table 2.

The temperature was kept constant in the activated sludge plant, all experimental work carried out at 12–13 °C. pH was not controlled, but resulting pH according to actual performance and wastewater was accepted. In phases with pre-precipitation with PAX, the pH remained rather constant in the range of 7.3–7.6. In phases with pre-precipitation with JKL, the pH showed a slight decrease to the range of 7.0–7.3 due to an increase in alkalinity consumption. In the experiments, the thermal/chemical hydrolysate was dosed in the acid form, thus resulting in a further decrease in pH.

Tab. 2. Operational data for the activated sludge plant

	Phase A	Phase B	Phase C	Phase D
Hydr. retention time [h]	8–9	12	9	10
Cycle duration [h]	4, 3, 2	3	3	3
Aerobic/anoxic [%]	60/40	60/40	60/40	60/40
Sludge age [d]	40–50	29–35	36–44	26–30
Aerobic sludge age [d]	25–30	17–20	22–26	16–18
MLSS [g/l]	3–4	4–5	4–5	4–5
Temperature [°C]	12–13	12–13	12–13	12–13
pH	7.3–7.6	7.3–7.6	7.0–7.3	6.8–7.1

Sampling and Analyses

2–3 times per week, composite samples (24 hours) of raw wastewater and of chemically pretreated wastewater were taken. COD, Kj-N, P_{tot}, PO_4-P, SS and VSS were analyzed. Once a week total samples were furthermore analyzed for BOD and filtered samples were analyzed for COD, BOD, Kj-N, and P_{tot}. Influent and effluent to the biological treatment were also characterized using 24-hour composite samples taken 2–3 times per week. The samples were analyzed for the same parameters as mentioned above, plus NH_3-N and NO_{2+3}-N. Grab samples were taken 2–3 times per week from the applied batches of hydrolysates and analyzed for COD, $COD_{filt.}$, Kj-N, Kj-$N_{filt.}$, NH_3-N, P_{tot}, SS and VSS. Periodically VFA was determined in the hydrolysates. Suspended solids and volatile suspended solids in the activated sludge reactors were analyzed 2 times per week and nitrogen, phosphorus, and COD contents in the sludge were determined once per week.

All analyses were performed according to Danish National Standard Procedures.

Results and Discussion

Pre-precipitation

Figure 3 shows concentrations of the main parameters: COD, P_{tot}, Kj-N, and SS in raw wastewater and in chemically pretreated wastewater. Table 3 shows mean values for the phases A + B (PAX precipitation) and C + D (JKL precipitation) together with the observed removal efficiencies with chemical pre-precipitation.

The data show that both coagulants are very efficient in removing COD, phosphorus, and suspended solids.

As can be seen from Table 3, the two coagulants seem equally efficient in removing suspended solids to a level around 30–35 mg/l in the actual system. Removal of soluble COD is low, just below 20 %, for both coagulants.

Removal of phosphorus is high for both coagulants, but JKL turns out to be more efficient due to the higher precipitation efficiency.

Figure 3 also shows that both coagulants have a low nitrogen removal efficiency. Removal efficiencies were typically in the range of 15–20 %. The observed differences in removal efficiency between PAX, 15 %, and JKL, 21 %, are probably caused by a higher concentration of influent suspended nitrogen in phases C and D.

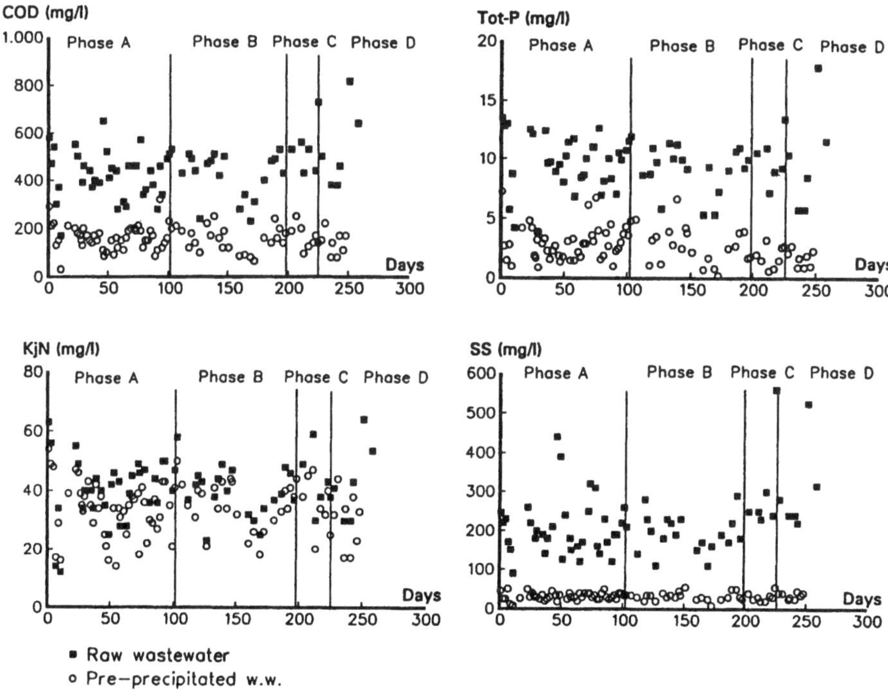

Fig. 3. COD, P_{tot}, Kj-N, and SS following chemical pretreatment

Since carbon must be available for biological denitrification, the ratio of carbon to nitrogen is important when the secondary treatment includes biological nitrogen removal. Table 3 shows how the different pretreatment efficiencies with regard to organics and nitrogen are reflected in the COD/Kj-N ratio. A mean ratio in the range 11–12 was found for the raw wastewater, while for the chemically pretreated wastewater, the ratio had decreased to around 4.6–4.8. As demonstrated below, the carbon to nitrogen ratio was significantly higher in the hydrolysates added to support the biological denitrification, thus increasing the COD/Kj-N ratio in the influent to the activated sludge plant.

The COD to Kj-N ratios in raw wastewater and in chemically pretreated wastewater are shown in Fig. 4. From the figure and from the data in Table 3 it can be seen that no obvious differences between the two coagulants can be observed.

Tab. 3. Mean values for raw wastewater and chemically pretreated wastewater

	Phase A + B			Phase C + D		
	Raw wastewater	Pretreated wastewater		Raw wastewater	Pretreated wastewater	
	mg/l	mg/l	% rem.	mg/l	mg/l	% rem.
COD	422	159	62	526	155	71
COD (filtr.)	139	114	18	134	108	19
P_{tot}	9.4	2.9	69	9.9	1.9	81
P_{tot} (filtr.)	6.1	2.1	66	5.5	0.8	85
Kj-N	40	34	15	43	34	21
Kj-N (filtr.)	30	31	–	–	–	–
SS	200	32	84	294	34	88
VSS	165	25	85	227	23	90
COD/Kj-N	10.9	4.8		12.4	4.6	

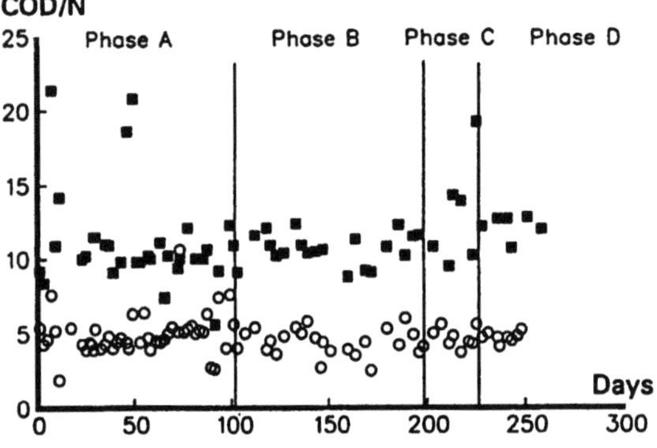

Fig. 4. COD to Kj-N ratio for raw wastewater and for chemically pretreated wastewater

Characterization of Hydrolysates

An important aspect of the present investigation was to compare hydrolysate produced by biological and thermal/chemical sludge hydrolysis.

The degree of COD solubilization in general is much lower when using biological hydrolysis than thermal/chemical hydrolysis. Thus the COD yield (ratio of COD in filtered sample of hydrolyzed sludge to COD in total sample) using biological hydrolysis typically varies between 3 and 15 % depending on sludge retention time (1–5 days) and temperature (15–30 °C) (Jørgensen, 1990, Kristensen et al., 1992). In comparison, the COD yield obtained by thermal/chemical sludge hydrolysis using typical operating conditions (retention time: 0.5–1.0 h, temperature: 120–160 °C and pH < 2) is in the range of 25–50 % (Smith and Göransson, 1991).

The composition of the hydrolysates used in the present study are shown in Table 4.

Tab. 4. Composition of biological and thermal/chemical hydrolysate

| | Biological hydrolysis | | Thermal/chem. hydrolysis | |
	Sludge mg/l	Hydrolysate mg/l	Sludge mg/l	Hydrolysate mg/l
SS	20 000	200	20 000	20
VSS		140		20
COD		2 500		6 900
VFA-COD		1 500		1 400
Kj-N		140		670
P_{tot}		12		14

It can be seen from the table that filtered COD in biological hydrolysate is only one third of that in the thermal hydrolysate, whereas sludge concentrations in the hydrolyzed sludges are approximately the same. Thus the yield of biological hydrolysis was only one third that of thermal/chemical hydrolysis.

The characteristics of the two types of hydrolysate are, however, very different.

From Table 4 it appears that the COD/N ratio of biological hydrolysate is 18 as compared to 10 for thermal/chemical hydrolysate. This considerable difference could be explained by a larger degree of nitrogen mobilization in the thermal/chemical process as compared to the biological process. A supplementary explanation could be differences in the nitrogen content of the two sludges.

From Table 4 it can also be seen that the phosphorus content is negligible in both types of hydrolysate. For the biological hydrolysate this is due to the fact that metal bound phosphorus in the chemically precipitated primary sludge is not mobilized during the biological processes. During the process of thermal/chemical hydrolysis nearly all iron bound phosphorus was solubilized. However, because of the neutralization of the hydrolyzed sludge with $Ca(OH)_2$, phosphorus is re-precipitated, and thus the phosphorus content of the thermal/chemical hydrolysate decreases.

Furthermore it appears from Table 4 that the percentage of volatile fatty acid COD (VFA-COD) in the filtered sample is 67 % of the total COD for biological hydrolysate and 20 % for thermal/chemical hydrolysate. This difference is due to the fermentation processes in biological hydrolysis, see Eastman and Ferguson (1981).

In this study the two types of hydrolysate were also characterized by measuring oxygen uptake rates (OUR) and nitrate uptake rates (NUR) of activated sludge in batch experiments after adding hydrolysate. The procedures used for these characterizations are described in Kristensen et al. (1991).

Figure 5 shows OUR rates after adding biological hydrolysate and thermal/chemical hydrolysate. A reference measurement after addition of a known easily degradable carbon source (acetate) is also shown. The amount of COD added was 60 mg/l in each experiment, and the temperature during the measurements was 22 °C.

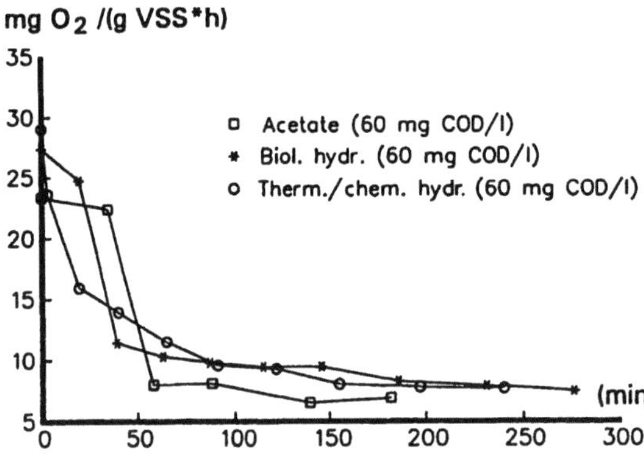

Fig. 5. OUR rates for activated sludge at 22 °C after addition of acetate and biological and thermal/chemical hydrolysate

As can be seen from Fig. 5, the initial OUR is higher in experiments where biological and thermal/chemical hydrolysate were added than in experiments where acetate was added. Such high initial OUR is often observed with carbon sources of a complex nature like hydrolysates or wastewater.

In experiments with thermal/chemical hydrolysate, there is a decrease in OUR only a short time after the hydrolysate was added. This indicates that only a small part of the added COD is easily degradable. With biological hydrolysate, the high OUR is observed for a considerably longer time than with thermal hydrolysate, although not as long as in experiments with acetate. This indicates that a relatively large part of the added COD is in an easily degradable form. These observations are in agreement with the VFA analysis (Table 4), which shows that 20 % and 67 % of COD, thermal/chemical and biological hydrolysate, respectively, is in the form of easily degradable volatile fatty acids.

Figure 5 also seems to show that there is a gradual decrease in OUR in experiments with biological hydrolysate and especially with thermal/chemical hydrolysate; this is in contrast to the experiments with acetate. This indicates that the hydrolysates are complex carbon sources, which contain a range of organic compounds that are more or less easily degradable.

Figure 6 and Fig. 7 show denitrification rates after addition of biological hydrolysate and thermal/chemical hydrolysate, respectively.

Figure 6 shows that the initial denitrification rate is approximately the same for all four doses of biological hydrolysate. The initial denitrification rate can be calculated to be 5–6 mg N/(g VSS · h) at 20 °C. This is comparable to denitrification rates obtained in corresponding batch experiments using acetate as the sole carbon source.

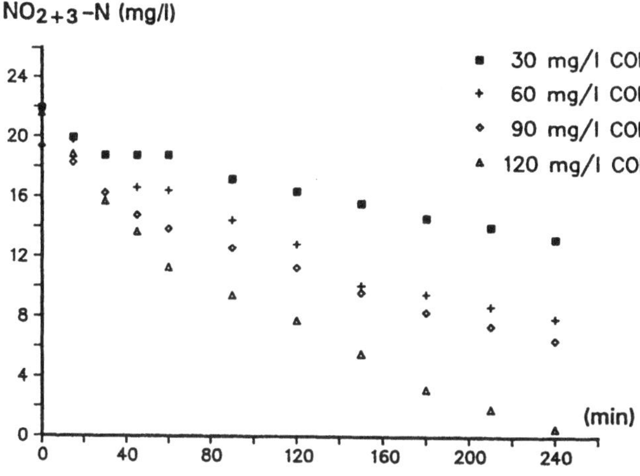

Fig. 6. Denitrification rates for activated sludge after addition of biological hydrolysate in various COD doses

Figure 7 seems to indicate that denitrification rates in experiments with thermal/chemical hydrolysate are considerably lower than those in experiments with acetate. Denitrification rates can be calculated to be 7.3–7.6 mg N/(g VSS · h) at 21 °C for acetate and 3.4–4.8 mg N/(g VSS · h) at 21 °C for thermal/chemical hydrolysate. COD consumptions of around 5–7 g COD per g N denitrified were found for both types of hydrolysate in the experiments.

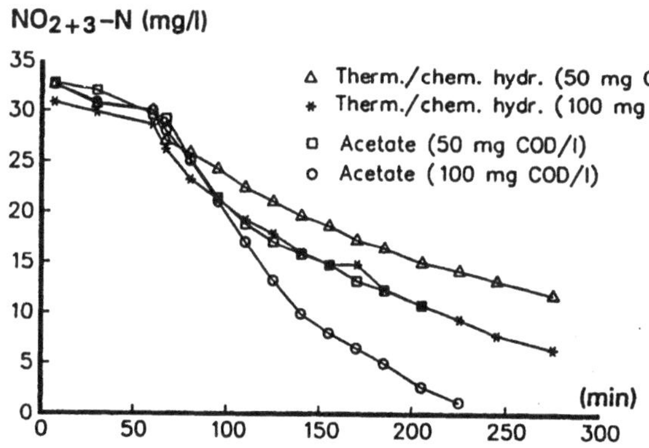

Fig. 7. Denitrification rates for activated sludge after addition of acetate and thermal/chemical hydrolysate in two different COD doses

Activated Sludge Nitrogen Removal

Figure 8 shows total-N in the influent and effluent to the activated sludge plant for the four phases. Mean values are given in Table 5.

The nitrogen removal data in phases A and C (without addition of hydrolysate) indicate that by far most of the nitrogen removal in all phases was based on organics from the chemically pretreated wastewater. For the pretreated wastewater alone, total nitrogen could be reduced from 37–39 mg/l in the influent to around 13 mg/l in the effluent. Total nitrogen in the effluent consisted mainly of nitrate-nitrogen, since effluent Kj-N typically was in the range of 1.5–3 mg/l.

The improvement in nitrogen removal by adding hydrolysis products in phases B and D can be easily recognized in the figure.

In phase B, the effluent total-N was improved to a mean value of 8.7 mg/l. In this phase, the contribution of the hydrolysate to the pretreated influent was typically 25–30 mg/l COD and 1.5 mg/l nitrogen. The hydrolysate dose which was actually applied corresponds to the amount of COD produced by biological sludge hydrolysis of the total primary sludge. Biological sludge hydrolysis can be seen as a way to improve nitrogen removal by some 10–20 %, depending on wastewater composition. However, a more important perspective is to use the process as a way of improving process stability and flexibility by producing and storing easily degradable carbon, which can be dosed when needed in the denitrification process.

In phase D the thermal/chemical hydrolysate was added in various COD doses. Based on influent, approx. 50 mg/l COD was dosed throughout most of the period. At the beginning of the phase and at the end of the phase, short periods were tested with doses approx. 25 mg/l and 75 mg/l, respectively.

On average, the effluent N_{tot} was improved from 13 mg/l in the reference period, phase C, to 8.5 mg/l in phase D. At increased COD dosages, only a moderate improvement in effluent quality could be observed. However, nitrogen loading seemed to be a parameter of great significance for the effluent quality during the actual relatively short retention time. Thus increased attention must be paid to optimizing dosage according to actual loading so that we may profit fully from such increased carbon dosing.

With regard to the usual COD yield from thermal/chemical hydrolysis, the actually dosed 50 mg/l of COD corresponds to the thermal/chemical hydrolysis of half the amount of primary sludge production. Thus thermal/chemical sludge hydrolysis provides a potential for producing enough carbon to account for a significant improvement in nitrate removal.

When dosing thermal/chemical hydrolysate it is important to keep in mind that most of the carbon needs to be hydrolyzed by the activated sludge before it can be used for denitrification, as shown by the lab-scale characterization. In order to optimize utilization and to minimize waste of COD by aerobic degradation, the thermal/chemical hydrolysate was dosed in the first third (30 min.) of each denitrification phase (each lasting 90 min.) to allow for hydrolysis of most of the carbon during the denitrification phase.

Another thing to keep in mind when dosing hydrolysates or other carbon sources to support denitrification is the possibility of controlling the activated sludge population dynamics. Addition of the carbon sources should be made in a way that aims at suppressing filamentous bulking. In phase B biological hydrolysate was added in the return sludge line, thereby obtaining an anoxic selector effect.

In Table 5 the nitrogen which is removed is quantified into the portion removed by incorporation into the biomass and the portion removed by denitrification. The portion removed by incorporation into the biomass is estimated from the sludge yield and observed nitrogen content in the activated sludge (typically 7–8 % based on SS). As can be seen, only a minor portion, 3.5–4.5 mg/l, is removed by biomass incorporation. This is due to the low sludge yield obtained with pretreated wastewater, as reported in Kristensen et al. (1992).

The ratios of removed COD to denitrified nitrogen are shown in Table 5. The table gives the calculated ratio under the assumption that influent COD is removed anoxically. Mean values for the different phases are found in the range of 4.8–7.6 g COD/g N. As can be seen from the table, phase D resulted in a high COD consumption ratio indicating the possibility of optimizing process performance and improving the effluent quality.

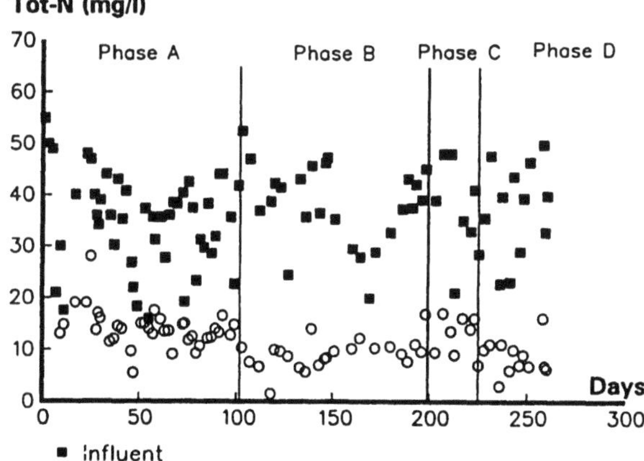

Fig. 8. Influent and effluent total nitrogen to activated sludge plant

Tab. 5. Mean values for nitrogen and phosphorus removal in activated sludge plant

	Phase A	Phase B	Phase C	Phase D
N_{tot} inf. [mg/l]	37	38	39	36
N_{tot} eff. [mg/l]	13	8.7	13	8.5
Removal eff. [%]	63	77	67	76
N in biomass [mg/l]	3.5	4.2	4.5	4.6
N denitrified [mg/l]	20.5	25.1	21.5	22.9
$\Delta COD_{anox}/N_{denit}$	5.2	4.8	6.4	7.6
P_{tot} inf. [mg/l]	2.9	3.4	1.9	2.2
P_{tot} eff. [mg/l]	1.9	0.5	0.6	0.2

Activated Sludge Phosphorus Removal

Figure 9 shows total-P in the influent and effluent of the activated sludge plant. Mean values for the four phases are given in Table 5.

When both coagulants were used for pre-precipitation, it was possible to obtain a very good effluent quality. From the data it can be seen that JKL (phases C and D) resulted in better pretreatment efficiency for phosphorus removal than the applied PAX (phases A and B). Phases C and D demonstrate how the effluent phosphorus concentration was kept very low without simultaneous precipitation due to the efficient pre-precipitation with JKL. Phase B demonstrates that the effluent phosphorus concentration could be reduced to around 0.5 mg/l by supporting the pre-precipitation with simultaneous precipitation with PAX.

In the experimental phases A, C and D, phosphorus removal in the activated sludge plant was limited to incorporation into the biomass due to growth and a slight precipitation caused by precipitation chemicals carried over from the pre-precipitation. In these phases typically 1–2 mg/l of phosphorus were removed.

When dealing with process combinations including pre-precipitation and biological treatment, it is important to keep in mind that the pretreatment should not be too efficient, since a certain amount of phosphorus is needed for the bacterial growth.

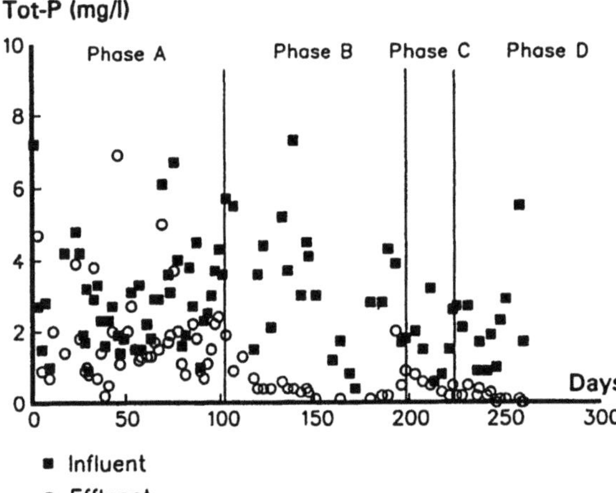

Fig. 9. Influent and effluent total-P in activated sludge plant

Conclusion

Chemical pretreatment with pre-polymerized aluminium salt, PAX 61, and JKL, respectively, was very efficient in removing COD, phosphorus, and suspended solids. The two coagulants were equally efficient in removing suspended solids, while JKL was more efficient in removing phosphorus. The removal efficiency of nitrogen was low for both coagulants, which caused a significant decrease in the COD : Kj-N ratio from 10–12 in the raw wastewater to 4.5–5.0 in the pretreated wastewater.

Biological hydrolysate COD consisted mainly of easily degradable COD, resulting in high denitrification rates similar to values obtained with acetate.

Thermal/chemical hydrolysate COD contained only a small fraction, typically less than 20 %, of easily degradable COD. Most of the COD needed to be hydrolyzed by the activated sludge before it could be used for denitrification. The denitrification rates found for the thermal/chemical hydrolysate were approx. half the values found for acetate.

By adding biological hydrolysate, 25 mg/l COD, and thermal/chemical hydrolysate, 50 mg/l COD, to the denitrification process, it was possible to improve the effluent nitrogen from a mean value of 13 mg/l, when operated with chemically pretreated wastewater alone, to a mean value of 8.5–8.7 mg/l. By optimizing process performance, the thermal/chemical hydrolysate has the potential to improve the effluent quality even further.

Both hydrolysis processes should be considered as valuable tools to improve effluent quality, and to increase process stability and flexibility through production and storage of more or less easily degradable organics to be dosed when needed in the denitrification process.

Acknowledgements

The study was carried out as part of the EUREKA project, HYPRO. The HYPRO project includes the following research institutions and companies: Dept. of Env. Eng., Technical University, Denmark; Norwegian Hydrotechnical Laboratory, Norway; Kemira Kemi AB, Sweden; I. Krüger Systems AS, Denmark; Water Quality Institute, Denmark.

References

Bundgaard, E., Nielsen F.M.: Stability of Effluents from Biological Nutrient Removal Plants – Danish Long-Term Operating and Optimization Experiments. Presented at the WPCF 62nd. Annual Conference, San Francisco, 1989

Eastman, J.A., Ferguson, J.: Solubilization of Particulate Organic Carbon During the Acid Phase of Anaerobic Digestion. J. WPCF *53* (3) (1981) 352–366

Henze, M., Harremoës, P.: Chemical-Biological Nutrient Removal – The HYPRO Concept. In: Chemical Water and Wastewater Treatment, H.H. Hahn and R. Klute (eds.). Springer, Berlin Heidelberg New York 1990, pp. 499–510

Jørgensen, P.E.: Biological Hydrolysis of Sludge from Primary Precipitation. In: Chemical Water and Wastewater Treatment, H.H. Hahn and R. Klute (eds.). Springer, Berlin Heidelberg New York 1990, pp. 511–520

Karlsson, I., Smith, G.: Pre-precipitation Facilitates Nitrogen Removal Without Tank Expansion. Wat. Sci. Tech. *23* (1990) 811–818

Kristensen, G.H., Jørgensen, P.E., Henze, M.: Characterization of Functional Microbial Groups and Substrate in Activated Sludge and Wastewater by AUR, NUR and OUR. IAWPRC Conf. on Interactions of Wastewater, Biomass and Reactor Configurations in Biological Treatment Plants. Copenhagen, August 21–23, 1991 (in press)

Kristensen, G.H., Jørgensen, P.E., Strube, R., Henze M.: Combined Pre-precipitation, Biological Sludge Hydrolysis and Nitrogen Reduction – a Pilot Demonstration of Integrated Nutrient Removal. To be presented at the IAWPRC 16th Biennial Int. Conf. Washington, DC, May 24–30, 1992

Smith, G., Göransson, J.: Generation of an Effective Internal Carbon Source for Denitrification Through Thermal Hydrolysis of Pre-precipitated Sludge. Proceedings from the 6th IAWPRC Conf. on Design and Operation of Large Wastewater Treatment Plants. Prague, Czechoslovakia, Aug. 26–30, 1991
Ødegaard, H., Fettig, J., Ratnaweera, H.: Coagulation with Prepolymerized Metal Salts. In: Chemical Water and Wastewater Treatment, H.H. Hahn and R. Klute (eds.). Springer, Berlin Heidelberg New York 1990, pp. 189–220

G. Holm Kristensen and P. Elberg Jørgensen
Water Quality Institute
Dept. of Wastewater and Process Technology
Agern Alle 11
DK-2970 Hørsholm
Denmark

Use of Internal Carbon from Sludge Hydrolysis in Biological Wastewater Treatment

I. Karlsson, J. Göransson, and K. Rindel

Abstract

Chemical pretreatment changes the composition of wastewater by reducing the content of particulate organic matter. The resulting lower sludge production facilitates the nitrification. The pretreatment reduces the amount of carbon available for denitrification. To increase the denitrification capacity, the precipitated sludge can be hydrolysed into a readily degradable form and used as a electron donor.

Two types of hydrolysis are investigated in the HYPRO project. At Tampere WWTP a thermal hydrolysis process is implemented, where sludge is hydrolysed at 150 °C and low pH. Tuelsø WWTP applies biological hydrolysis at 20–25 °C with no pH adjustment.

The hydrolysed sludge is separated into two fractions: One with particulate organic matter to be digested, the other containing a solution of organic and inorganic substances to be used for denitrification. The heavy metals liberated during thermal hydrolysis can be separated and the precipitant recovered and recycled. The carbon source produced from the hydrolysis increases denitrification rate and total denitrification capacity. The sludge dewatering characteristics were improved for thermally hydrolysed sludge and somewhat but not critically reduced by biological hydrolysis.

Introduction

Eutrophication along the coast lines and in the lakes in the nordic countries has resulted in increased demands for removing nutrients from wastewater. Hypro is a joint Nordic (Eureka) research project aimed at developing an advanced process for nutrient removal. The Hypro concept is based on three elements:

- Chemical pretreatment of the wastewater for removal of organic matter and phosphorus.
- Hydrolysis of pre-precipitated organics for production of a readily degradable carbon source for biological nitrogen removal.
- Biological nitrification and denitrification. Nitrification is accelerated by the pre-removal of suspended organic matter. Denitrification is accelerated by hydrolysed organic matter.

The removal effect of different metal precipitants was studied during the first part of the Hypro project. At the same time an analysis of different hydrolysis concepts was carried out, from which thermal and biological hydrolysis was selected for full scale investigation.

An essential part of the project during this stage was an in-depth characterization of all involved process streams, including raw and pretreated wastewater, activated sludge, hydrolysed sludge and hydrolyate. On the basis of this activity a new characterisation concept was established to allow for a better modelling of the biological processes.

Full scale demonstration of the Hypro concept was initiated in the late autumn 1991. In Denmark with pre-precipitation and biological hydrolysis, in Finland with pre-precipitation and thermal chemical hydrolysis.

Full Scale Experiments

Thermal Hydrolysis

Tampere Wastewater Treatment Plant. The Tampere sewage plant Viniikanlahden treats a flow of about $65\,000$ m^3 per day. A high industrial discharge comes from the textile industry. The demands for the effluent are BOD$_7$ = 15 mg O$_2$/l and P$_{tot}$ = 0.8 mg/l. The water temperature in winter is favourable for nitrification, since normally it is not lower than 10 °C.

The process sequence is coarse screening, aerated grit chamber and primary settling tanks (surface load ~ 0.5 m/h). Biological treatment takes place in 6 separate trains with separate sludge systems. Aeration average retention time is 4.7 hours. Secondary settling surface load is 0.6 m per hour.

Primary and secondary sludge are thickened separately and stabilised anaerobically. Reject waters are returned to the sewage influent water in the course of 24 hours.

Two of the six trains in the sewage works are built for nitrogen removal, with a pre-denitrification volume of 730 m^3 and an aerobic volume of $1\,400$ m^3. The recirculation of activated sludge was $2 \times Q$ and return sludge pumping $1 \times Q$. The flow to the test lines can be varied.

Pre-precipitation was introduced to the sewage plant in order to relieve the biological process and to secure nitrification. A ferric salt (~ 14 g Fe^{3+}/m^3) was dosed from a perforated plastic tube across the inlet channel to the primary tanks. Strong turbulence was created by a constriction at the dosing point. The removal efficiency can be seen in Table 1.

The effluent phosphorous was 0.4 mg/l, the BOD$_7$ 10 mg/l and the COD 40 mg/l. A remarkable result is the high removal of dissolved BOD, and the comparatively low removal of suspended solids. Characterisation of the sewage water into different COD fractions is shown in Table 2.

The low molecular fractions, VFA, are normally not influenced by chemical treatment at all and the difference between influent and primary effluent fractions must be a combined effect caused by the SS activity in the raw sewage, the

Tab. 1. Average results of daily composite samples
(September to December 1991)

	Influent mg/l	Pre-precipitation mg/l	Removal %
$BOD_{7_{tot}}$	195	58	70
$BOD_{7_{sol}}$	55	14	74
COD_{tot}	434	166	62
COD_{sol}	122	58	53
SS	205	87	58

Tab. 2. Filtered COD fractions (February 18, 1992)

	Influent mg/l	Pre-precipitated sewage mg/l
COD	520	250
COD_F (GF/A)	220	85
COD_F (0.45μ)	180	65
COD_F (0.22μ)	160	60
VFA-COD	29	0
KJN	46	46
KJNF (GF/A)	35	39
KJNF (0.45μ)	35	38
KJNF (0.22μ)	34	38

reject water and chemical precipitation. The aerobic conditions in the sewage
plant from the influent sampling point to the primary settling tank last about
one hour. In the primary settling tank the retention time is about 6 hours. The
oxygen in the primary effluent is zero.

A respiration test with oxygen and nitrate was used on raw sewage and pre-
precipitated water to characterise the water into different biologically degradable fractions (Jørgensen, 1992) (Table 3).

Tab. 3. Results from respiration tests

	Directly metabolised		Rapidly degradable	
	COD [mg/l]	% of COD	COD [mg/l]	% of COD
Raw sewage	51	13	49	13
Pre-precipitated	0	0	29	16

The oxygen and nitrate respiration of the waters where no activated sludge
was added was two to three times higher than in the tests where activated sludge

was added. The results of the respiration tests are in good agreement with the COD analysis of the different fractions. No tests were done with presettled sewage.

Full Scale Experiments. Two trains were used from September to December with pre-precipitated sewage water and pre-denitrification. The nitrification over the period was good, with a small increase in ammonia at the end due to lower water temperature, 18 °C falling to 12 °C. Aerobic retention time was 3.7 h, and anoxic retention time 1.9 h. MLSS was 3 000 g/m^3, MLVSS was 2 100 g. The aerobic sludge age was about 5 days.

Tab. 4. Reference period

	Influent g/m^3	Pre-precipitated g/m^3	Effluent g/m^3
N_{tot}	35.6	31.6	19.0
NH_4-N	22.4	23.8	3.7
NO_3-N		1.5	14.9
P_{tot}	7.1	3.4	0.4
P_{sol}		0.8	0.1

The assimilated nitrogen in the biological sludge was 5.4 g/m^3, and the nitrification rate 1.9 g N/kg MLSS · h. The total nitrogen removal over the biostep was 12.6 g and denitrified 7.2 g, which gives a denitrification rate of 1.3 g N/kg MLSS · h. Both lines gave the same result.

To increase the biological nitrogen removal, primary sludge was hydrolysed and dosed to the anoxic zone. The hydrolysed process consisted of thermal treatment at 150 °C in an acid environment. The retention time was 30–60 minutes. The pH was maintained below 4 by adding sulphuric acid.

Dissolved organic matter was separated after lime precipitation and centrifugation. The yield of dissolved COD from total COD was 15–35 % containing 6 000–15 000 g COD per litre. The yield is related to the hydrolysis pH. Analysis of the hydrolysate is shown in Table 5.

Tab. 5. Analysis of the thermal hydrolysate

SS	20 mg/l	VFACOD	1 426 mg/l
VSS	20 mg/l	Kj-N	670 mg/l
COD	6 950 mg/l	P_{tot}	14 mg/l
COD$_F$	6 900 mg/l		

The COD : N ratio was 10 : 1 and COD : P ratio 500 : 1. Dosing 60 g COD gives an additional 6 g N and 0.1 g P per m^3. In a sewage process with anaerobic or aerobic sludge stabilisation, the organic bound nitrogen will be returned to the treatment process as ammonia anyhow.

The results of the respiration test with oxygen and nitrate are shown in the following figures (Jørgensen, 1992).

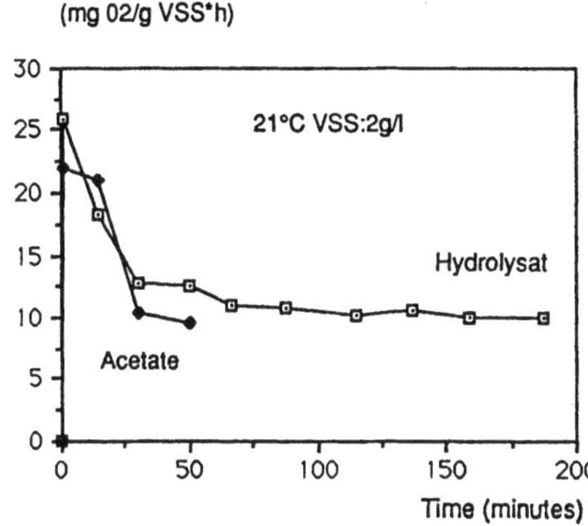

Fig. 1. Oxygen respiration with thermal hydrolysate and acetate

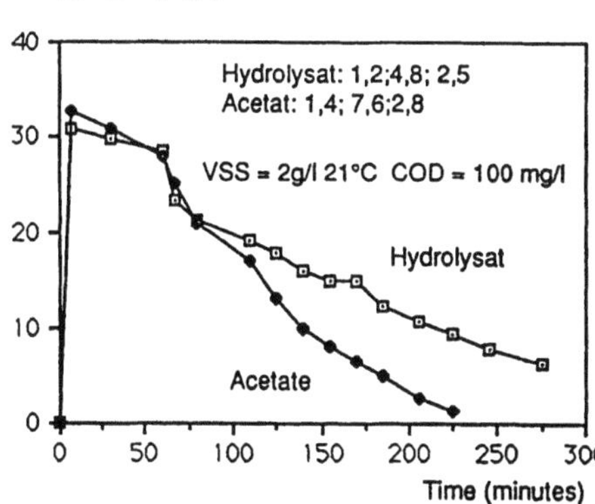

Fig. 2. Nitrate respiration with thermal hydrolysate and acetate

Based on the respiration tests, the 100 g hydrolysate COD can be characterised as 17 mg/l directly metabolisable COD, 33 mg/l rapidly degradable COD, and the final 50 mg/l COD as slower than rapidly degradable COD but faster than slowly degradable COD.

Laboratory tests at the Tampere sewage plant 10 °C showed similar results, Fig. 3. The hydrolysate was dosed to one of the lines from December to March. The temperature of the water was 8.8–12.0 °C. The sludge age was increased to about 9 days and the anoxic and aerobic retention times were 3.2 and 6.2 h, respectively. MLSS was $3\,500$ g/m^3 and MLVSS $2\,500$ g/m^3. The dosage of hydrolysate, with occasional hydrolysis maintenance problems, was estimated to be about 60 g COD/m^3.

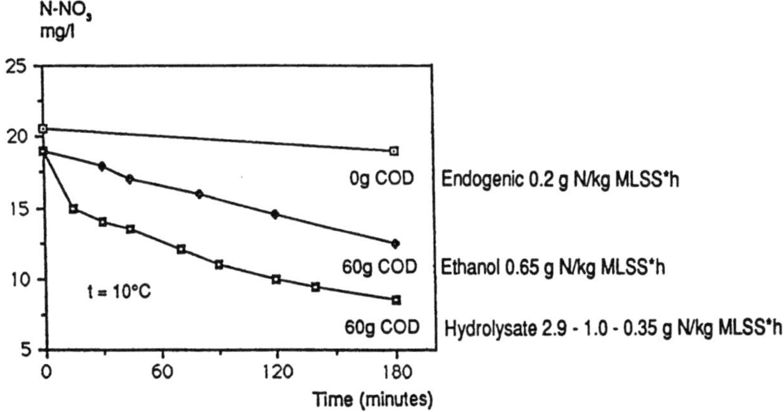

Fig. 3. Nitrate respiration with thermal hydrolysate and ethanol

Tab. 6. Test period with thermal hydrolysate

	Influent test line mg/l	Influent reference line mg/l	Effluent test line mg/l	Effluent reference line mg/l
N_{tot}	35 + 6	35	21	20
NO_3-N	1.0	1.0	14	16
NH_4-N			5.5	3.5

No increase in P_{tot} or COD was found in the effluent.

Assimilated nitrogen in the bio-sludge was 5.0 g/m³ in the reference line and 6.7 g/m³ in the test line water. The nitrification rates for the reference line and for the test line was 1.2 g N/kg MLSS · h. The denitrification rates were 0.9 and 1.2 g N/kg MLSS · h, respectively. The lab tests indicate a higher capacity of the carbon source and the process was later changed to a longer retention time in the anoxic zone.

The dewatering characteristics of the hydrolysed sludge were drastically improved. The CST value was very low at values below pH 4. Different dewatering devices are compared in Table 7.

Tab. 7. Results from hydrolysed sludge dewatering

	Directly TS %	Lime addition TS %	After first separation and lime addition TS %
Centrifuge	30–33	30	28
Chamber filter press	50	50	
Belt press	20–30		

Almost all of the directly separated sludge is organic if pH is about 2. The inorganic and about 25 % of the organic part is solubilised. Metals and phosphorous can be separated by precipitation. The process offers the possibility to recycle the precipitant.

Distribution of organic and inorganic matter with thermal treatment

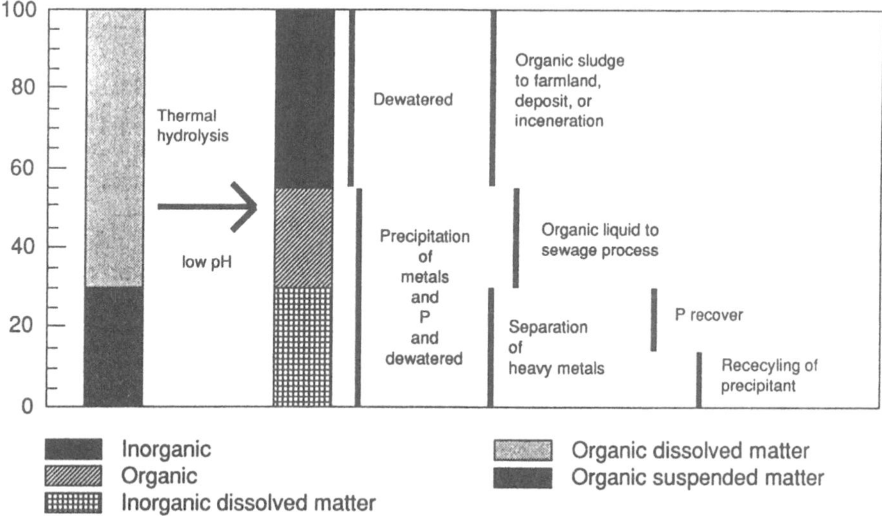

Fig. 4. Distrubution of organic and inorganic matter with thermal treatment

From hydrolysis of pre-precipitated sludge from the sewage plant in Västerås the following results were found for the separation of heavy metals.

Tab. 8. Heavy metals separation of hydrolysed sludge

	Total in Sludge μg/l	pH > 2 Hydrolysed decantate μg/l	pH 9 Precipitated decantate μg/l
Zn	46 000	58 000	1 100
Cu	16 000	1 600	80
Pb	4 500	4 400	1
Cr	2 200	2 100	60
Ni	4 000	4 400	790
Co	600	720	100
Cd	38	86	2
Hg	0.5	0.5	< 0.1

Biological Hydrolysis

Tuelsoe Wastewater Treatment Plant. The biological sludge hydrolysis was tested at the Tuelsoe municipal treatment plant. The plant is treating a wastewater flow of about $3\,000$ m^3 per day, mainly of domestic origin. Effluent demands are 10 ppm BOD, 8 ppm total N and 1 ppm total P. The plant was extended in 1990 for complete nutrient removal with primary sedimentation, biological treatment by the alternating Bio-Denitro process and simultaneous precipitation of phosphor. Primary sludge is stabilized anaerobically with recirculation of reject water to the primary treatment.

Full Scale Experiments. The full scale experiments with biological hydrolysis at the Tuelsoe plant are the direct continuation of laboratory- and pilotscale investigations carried out since 1989 (Kristensen et al., 1992). The experimental work was initiated in the autumn of 1991 after completion of site works. The experiments are divided into three main phases, of which phase 1 has been completed and phase 2 is in progress.

1. Preliminary testing of precipitation chemicals and conditions for hydrolysis.
2. Operation with pre-precipitation, biological hydrolysis of precipitated sludge and application of hydrolysate for denitrification.
3. Standard operation of the plant for reference.

During the first part of phase 2 the hydrolysed sludge is separated by a centrifuge. The dewatered sludge is pumped to anaerobic treatment. Nitrogen removal is optimised through a timer controlled dosing of the produced hydrolysate. At a later stage the hydrolysate addition will be controlled by on-line measurements.

During the second part of phase 2 the hydrolysed sludge will be recycled to primary treatment. This procedure ensures an efficient transfer of soluble organic matter to the wastewater without mechanical dewatering, but at the same time reduces the possibilities for control of hydrolysate addition.

Figure 5 shows a flow sheet for the Tuelsoe plant after the extension for Hypro operation. Key figures related to the design and operation of the Tuelsoe plant are shown in Table 9. The plant is at present operated at about 40 % of design loading.

Tab. 9. Average operational data for Tuelsoe WWTP

Influent flow, Q	$3\,000$ m^3/d
Temperature of water: winter	5 °C
summer	12 °C
Pre-precipitation: surface load	0.5 m/h
ferric-salt dosing	15 g Fe/m^3
Biological treatment: anoxic volume	$3\,000$ m^3
aerobic volume	$3\,000$ m^3
return sludge	0.5 $\times Q$

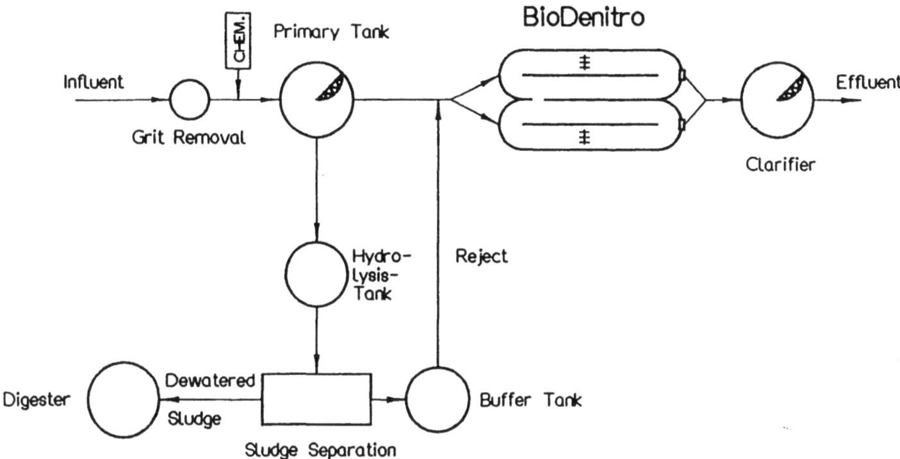

Fig. 5. Flow sheet for Tuelsoe WWTP after Hypro extension

Chemical Pre-precipitation. The effect of chemical pre-precipitation on wastewater characteristics and plant performance is shown in Table 10. The results presented are an average from 26 days of operation without addition of hydrolysate. Influent flow was $3\,700$ m^3/d and chemical addition 15 g Fe/m^3.

Tab. 10. Average results from 21 composite samples (August to September 1991)

	Influent mg/l	After pre-precipitation mg/l	Reduction %	Effluent Bio-Denitro mg/l
BOD$_5$	180	80	58	5
COD	480	220	57	30
COD$_{sol}$	120	100	22	20
SS	290	100	64	10
N$_{tot}$	–	–	–	11
Kj-N	38	31	25	1.5
NH$_4$-N	24	24	–	0.3
NO$_3$-N	–	–	–	10
P$_{tot}$	8	4	47	0.9
P$_{ortho}$	5	2	58	0.5
COD : N	13	7.7		

It can be seen from Table 10 that effluent demands for nitrogen are exceeded when pre-precipitation is applied without hydrolysate addition. Total nitrogen during the period was found in the range of 6–17 ppm.

Biological Hydrolysis. The biological hydrolysis of pre-precipitated sludge is carried out at 20–25 °C and a detention time of 2–3 days. Influent sludge solids is controlled in the range of 2.5–4 %, and process conditions results in a pH of 5.5–6 in the hydrolysis reactor.

Fermentation products are mainly volatile fatty acids with a typical yield of 9–12 % based on total COD in the sludge. The composition of hydrolysate after separation by centrifuge is shown in Table 10 with data from March 1992.

Tab. 11. Typical composition of hydrolysate after separation (March 1992)

SS	600 ppm	COD_{VFA}	1 500 ppm
VSS	400 ppm	Kj-N	100 ppm
COD	2 300 ppm	P_{tot}	12 ppm
COD_{sol}	1 900 ppm	COD : N	23

The high concentration of volatile fatty acids with acetic acid as the main component makes the biological hydrolysate an ideal carbon source for denitrification. More than 65 % of total COD is readily biodegradable and additionally 15 % is rapidly hydrolysed during the denitrification process. The applicability of the hydrolysate for denitrification is further favoured by the high COD to nitrogen ratio.

When mechanical separation is applied on the hydrolysed sludge a fraction of soluble COD is lost with the dewatered sludge. Dewatering by centrifuge results typically in a net yield of 75 % based on totally produced soluble COD.

Biological Nitrogen Removal. The nitrogen removal to be expected by addition of hydrolysate can be calculated from a massbalance based on the results from hydrolysis and pre-precipitation experiments. The calculation is based on the conditions stated below:

Influent flow	3 000	m^3/d
SS-reduction after pre-precipitation	70	%
Yield of soluble COD from hydrolysis	12	%
Net yield after dewatering	75	%

With a net production of 125 kg/d of soluble COD an additional nitrogen removal of 24 kg N/d by denitrification can be expected from the implementation of biological hydrolysis. After correction for nitrogen contained in the hydrolysate the reduction corresponds to an effluent concentration of 5–6 mg N/l, well below the effluent demand of 8 mg N/l. The calculations also indicate the possibility for a substantial reduction (15–25 %) of necessary biological tank volume due to the increased denitrification rate obtained with hydrolysate.

Conclusions

Chemical pretreatment of sewage water increases the possibilities for nitrification by lowering the biological sludge production.

The denitrification capacity of the water is decreased after pre-precipitation.

By hydrolysing the removed organic matter it is possible to produce a readily degradable carbon source to accelerate the denitrification process.

Hydrolysate from biological hydrolysis can be applied either after mechanical dewatering or directly by recycling to the primary treatment. An additional removal of 5–6 ppm of nitrogen is expected based on massbalance calculations.

The sludge dewatering characteristics after thermal-acid treatment are drastically improved compared to those of primary sludge.

It is possible to separate heavy metals from the sludge.

Recycling of the precipitant is also possible.

Acknowledgments

Some of results reported in this paper are from the Eureka project Hypro. The partners in this project are

Technical University of Denmark	Denmark
I Krüger Systems	Denmark
Water Quality Institute	Denmark
University of Trondheim	Norway
Kemira Kemwater	Finland
Kemira Kemwater	Sweden

References

Jørgensen, P.E.: Karakterisering af termisk hydrolysat og spildevand fra Tammerfors. Ilt- og nitratrespirations forsøg. WQI, Hørsholm 1992 (in Danish)

Kristensen, G.H., Jørgensen, P.E., Strube, R., Henze, M.: Combined Pre-precipitation, Biological Sludge Hydrolysis and Nitrogen Reduction – a Pilot Demonstration of Integrated Nutrient Removal. Wat. Sci. Tech. 5–6 (1992) 1057–1066

Ingemar Karlsson and Jonas Göransson　　Kim Rindel
Kemira Kemi AB　　　　　　　　　　　　I. Krüger Systems AS
Kemwater　　　　　　　　　　　　　　　Gladsaxevej 363
Kungstensgatan 38　　　　　　　　　　　DK-2860 Soeborg
S-113 59 Stockholm　　　　　　　　　　 Denmark
Sweden

Evaluation of Pre-precipitation in a Wastewater Treatment System for Extended Nutrient Removal

B. Andersson, H. Aspegren, U. Nyberg, J. la Cour Jansen, and H. Ødegaard

Abstract

In order to meet future effluent standards of 10 mg BOD_7/l, 0.3 mg P/l and 8 mg N/l, investigations were made on a full scale at the Klagshamn wastewater treatment plant in Malmö, Sweden. The primary goal was to find an upgrading concept that could meet the nitrogen removal standard within existing basin volumes.

Considering the wastewater characteristics, pre-precipitation followed by nitrification and post-denitrification in a single sludge system with addition of methanol as carbon source for denitrification was the upgrading concept that could meet the effluent nitrogen removal standard without new basin extensions. The biological process can be characterized as a high-rate process since the removal of particulate material by pre-precipitation improved conditions for nitrification and since methanol addition resulted in a low sludge yield and high denitrification rate. High phosphorus removal was achieved by pre-precipitation with an effluent concentration of less than 1 mg P/l. To meet the very stringent phosphorus removal standard, however, a filtration plant with possibilities for contact precipitation will be installed.

The additional costs for the extended nitrogen removal based on this upgrading concept was estimated to be only one-tenth of the costs required for a concept based on a pre-denitrification system without primary clarification.

Introduction

Requirements to remove organic material and phosphorus in municipal wastewaters were introduced in Sweden during the 1960s and 1970s. Wastewater treatment plants were built in most Swedish communities, and since the beginning of the 1980s almost 100 % of the urban areas are connected to wastewater treatment plants for biological and chemical treatment. The design of the plants was based on general design procedures supported by the Swedish EPA. The original concept with a high-loaded activated sludge process followed by post-precipitation has gradually been replaced by pre-precipitation, and the post-precipitation plants have in many cases become obsolete.

Based on international agreements in order to decrease the nutrient discharges to the North Sea and the Baltic, a new Swedish policy was proposed in 1988. According to this, extended nitrogen removal has to be introduced in wastewater treatment plants located near the coast in southern Sweden. Generally, an effluent level of less than 15 mg N/l has to be met, but the plants located in the so-called sensitive areas have to comply with effluent standards between 8 and 12 mg N/l. Increased phosphorus removal with effluent levels of 0.3 mg P/l is proposed for plants larger than 10 000 p.e.

Upgrading existing wastewater treatment plants for the new standards requires at least some consideration. The plants are designed for extensive BOD and phosphorus removal and the process schemes proposed are mainly based on pre-precipitation. Removal of BOD in an activated sludge process and removal of phosphorus by chemical precipitation was in most cases rather easy to implement. Introducing extended nitrogen removal may not be as easy since the wastewater quality plays a more decisive role for the result.

The Klagshamn wastewater treatment plant in Malmö is one of the plants that has to be upgraded for extended nutrient removal. The more stringent requirements are likely to be imposed, i.e. future discharge requirements of 10 mg BOD_7/l, 8 mg N/l and 0.3 mg P/l can be anticipated. Full scale tests have been performed since 1987 in order to find the most appropriate upgrading technology for the plant, with the existing basin volumes and the wastewater quality taken into consideration.

The Klagshamn Wastewater Treatment Plant

The Klagshamn wastewater treatment plant is designed for 90 000 population equivalents. The annual average wastewater flow to the plant is 15 400 m^3/d corresponding to 640 m^3/h. The diurnal dry weather flow variations have been equalized by utilizing the free volumes of the sewer system ahead of the plant. This gives an almost constant flow to the plant during dry weather conditions. However, the maximum flow during rains can reach up to 3 000 m^3/h.

The existing plant configuration includes primary and secondary treatment. A flow sheet is shown in Fig. 1. The flow to the plant is split into two parallel trains that can be operated separately. Screening, grit removal, and primary clarification are followed by biological treatment in an activated sludge process. Based on the average flow to the plant, the hydraulic detention time is about 7 hours in the aeration basin.

The sludge treatment includes raw sludge thickening, digestion, digested sludge thickening and dewatering by centrifuges. The supernatant from the sludge treatment is recirculated directly back to the inlet pumping station.

Figure 1 shows how the plant is run today, with pre-precipitation (ferric chloride added to the grit chamber) and post-denitrification (with external carbon source added to the denitrification unit).

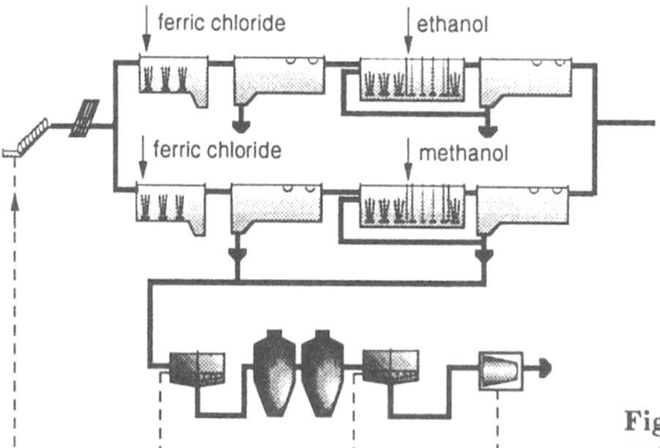

Fig. 1. The Klagshamn wastewater treatment plant

Process Selection

When a wastewater treatment plant is to be upgraded for enhanced nitrogen
removal, the basic concept is generally an activated sludge process designed
for pre-denitrification. A certain minimum carbon to nitrogen ratio in the influent wastewater to the biological step is then necessary in order to attain
the required degree of denitrification. The wastewater characteristics at the
Klagshamn plant are shown in Table 1. The temperature variations are usually
between 8 and 16 °C and the wastewater alkalinity is about 5–7 meqv/l.

Tab. 1. Wastewater characteristics at Klagshamn. The figures are averages over
the last year. Samples from the influent wastewater include the recycled water
from the sludge treatment

Parameter	SS	BOD$_7$		P$_{tot}$		N$_{tot}$		BOD$_7$/N$_{tot}$	
		tot	sol	tot	sol	tot	sol	tot	sol
	mg/l	mg/l	mg/l	mg/l	mg/l	mg/l	mg/l		
Sampling points:									
Influent wastewater	210	140	28	7.7	4.8	33	25	4.1	0.8
Effluent from primary clarifiers	62	68	24	6.3	4.8	29	25	2.3	0.8
Effluent from pre-precipitation	40	39	17	2.9	1.5	28	25	1.4	0.6

It can be anticipated that the carbon to nitrogen ratio in the primary
effluent (2.3 g BOD$_7$/g N) will not be sufficient for satisfactory nitrogen removal.
The carbon to nitrogen ratio can be increased at the Klagshamn plant by the
application of at least two different strategies, see Fig. 2.

The first strategy is based on a process configuration where the total amount
of organic material in the influent wastewater is utilized by excluding the primary clarification. For phosphorus removal, simultaneous precipitation can be

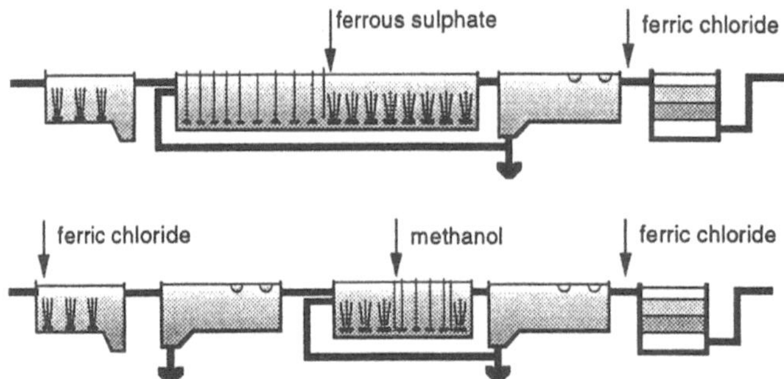

Fig. 2. Upgrading alternatives for the Klagshamn wastewater treatment plant

used in this case. The other strategy is based on pre-precipitation and the addition of an external carbon source in a post-denitrification process. Pre-precipitation will remove most of the organic material ahead of the biological process and conditions for pre-denitrification are unfavorable. On the other hand, conditions for maintaining nitrification are improved and phosphorus removal is accomplished at the same time. The addition of an external carbon source to a post-denitrification step can be expected to give a lower sludge yield and a higher utilization rate in comparison to the other alternative.

In order to comply with the requirements for phosphorus removal, a filtration plant with possible contact precipitation has to be installed. If the total amount of phosphorus was to be removed by precipitation in a tertiary filtration plant, a high coagulant dosage would be required because of the high wastewater alkalinity. This would have resulted in a very high load of suspended solids on the filter. By using pre-precipitation, most of the phosphorus is removed before the final filter step and alkalinity is also reduced. This will result in a much lower load on the filter in this case. Considering the wastewater characteristics, biological phosphorus removal does not seem feasible.

The two basic process alternatives were tested on a full scale using the existing basin volumes. In the first alternative, the primary clarifiers were converted to anoxic volumes in order to accomplish a higher utilization of the organic material in the influent wastewater, and the biological step was operated as a single sludge pre-denitrification process. In the second alternative, pre-precipitation was introduced and the biological step was operated as a single sludge post-denitrification process with methanol as external carbon source for denitrification.

The results from the tests showed that both alternatives could meet the proposed effluent standard of 8 mg N/l. The results are more thoroughly discussed in [1], [2] and [3]. However, the pre-denitrification system without primary clarification was very close to its limits with respect to both nitrification and denitrification capacity and the basin volumes have to be increased in order to be certain to comply with the effluent standard. Preliminary design considerations using the German ATV-Arbeitsblatt A131 [4] design procedures indicated that the necessary volume extensions would have been in the range of 14 000 and 17 000 m³ [1]. The upgrading concept based on pre-precipitation with the addition of methanol could meet the effluent standards within the existing basin volumes. This alternative was consequently selected for upgrading the Klagshamn plant.

Evaluation of the Pre-precipitation Concept

The system with pre-precipitation followed by single sludge post-denitrification with methanol as carbon source has been operated for about one year in one of the flow trains at the Klagshamn plant. The other flow train was converted to the same operating mode in March 1992 with the one exception that ethanol is used instead of methanol, see Fig. 1.

The physical outline of the aeration basins is shown in Fig. 3. Each of the two aeration basins is divided into 8 cells and 2 mixing chambers, all separated by wooden walls. The aeration system is of the flexible rubber membrane diffusor type that allows each cell to be operated either aerated or non-aerated. Agitation in these cells is achieved by mechanical mixers. The dissolved oxygen concentration in the aerated cells is maintained at a setpoint value of 2 mg O_2/l by an advanced control system.

Fig. 3. Physical outline of the aeration basins at the Klagshamn wastewater treatment plant

In the basic operating mode, the first four cells are kept aerobic, the next three cells anoxic and the last cell aerobic.

Ferric chloride is added at the inlet of the aerated grit removal basins. The coagulant dosage is kept constant at about 10 g Fe/m^3 on the average. There is also a system for handling methanol and ethanol. The dosing point is fixed at the mixing chamber just ahead of the 5th cell in the flow trains. The dosage is constant over time and adjustments are based on BOD equivalents. The average dosage has been 50–60 g BOD_7/m^3.

Pre-precipitation has a significant impact on both the primary and the secondary treatment in this upgrading concept. In the following sections, some of these effects are discussed.

Primary Treatment

The influent wastewater to the plant exhibits significant variations in concentration of particulate material on a day-to-day basis (see Fig. 4, upper curve). This is not mainly due to variations in external load but rather to an effect of the operational conditions within the plant and in the sewer system ahead of the plant.

As the sewer system is used for flow equalization during dry weather conditions, the water velocity in the sewers is often low and sedimentation of particulate material occurs. Much of this particulate material is flushed into the plant when the flow rate is increased. Supernatant from the sludge treatment containing high concentrations of particulate material is recirculated back to the influent pumping station. This also contributes to the varying concentration of particulate material entering the primary treatment.

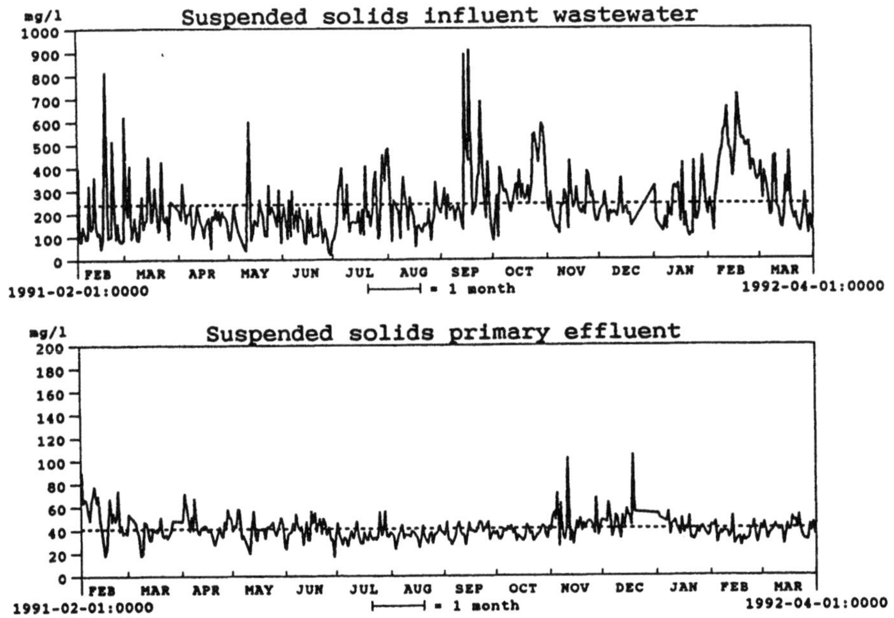

Fig. 4. Influent and effluent concentration of suspended solids in the primary treatment step with pre-precipitation

The variations in particulate material are effectively reduced by pre-precipitation, and a stable wastewater quality entering the biological treatment is reached, see Fig. 4, lower curve. The concentration of suspended solids in the primary effluent after precipitation normally varies between 30 and 50 mg SS/l with an average of 40 mg/l. The average overflow rate on the primary clarifiers is 0.6 m/h. The decrease in removal of particulate material at the end of 1991 was caused by increased hydraulic variations. During tests with constant flow rate in the parallel flow train, all the peak flow during rains was distributed to the pre-precipitation flow train. The hydraulic load on the primary clarifiers could be up to 6 m/h during this period.

Pre-precipitation tests were performed earlier with different dosages of ferric chloride [5]. The removal of phosphorus and suspended solids at different dosages is shown in Fig. 5. The removal of suspended solids did not seem to be affected by changes in the dosage, but the phosphorus removal was increased by increasing Fe-dosage. In order to reach 0.5 mg P/l in the effluent from the plant, a dosage of 260 g/m^3 ferric chloride (about 30 g Fe/m^3) was required at that time.

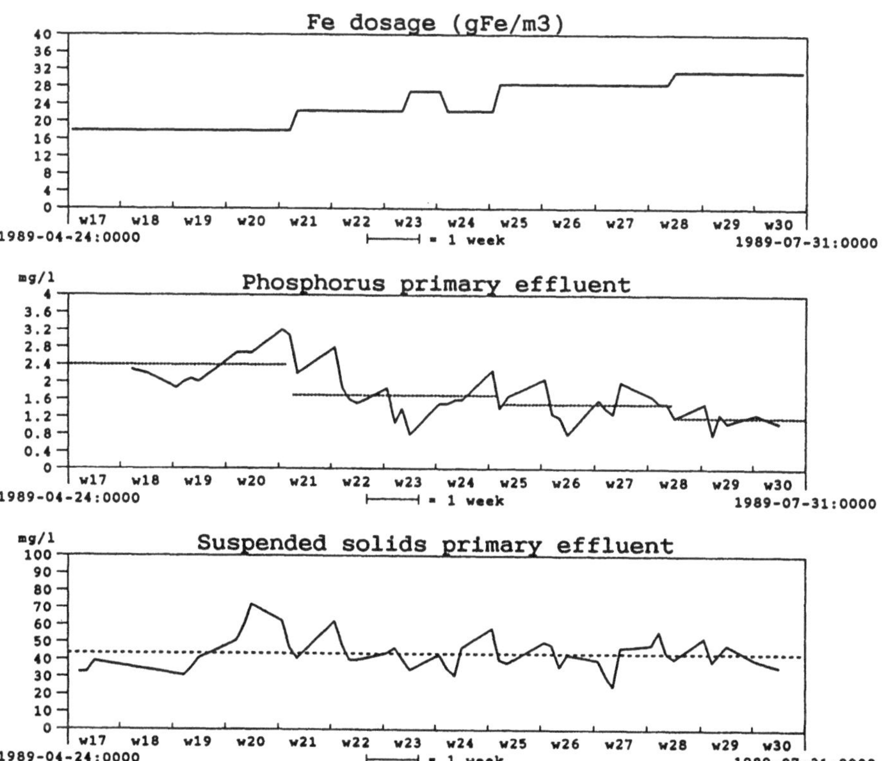

Fig. 5. Removal of phosphorus and suspended solids in the pre-precipitation step at various dosages of ferric chloride

In a pre-precipitation/post-denitrification system, the phosphorus concentration in the aeration basin must not be too low as this can cause rate limitations to denitrification. This has been observed in laboratory tests for denitrification rate measurements. In these tests, nitrate is added in excess. A higher rate was determined if phosphorus also was added to the test vessel. Therefore, it is important not to add too much coagulant for the pre-precipitation. Preferably, about 1 mg P/l as soluble phosphate should remain after pre-precipitation.

By limiting the dosage to 10 g Fe/m^3, enough phosphorus was available for the biological nitrogen removal in the full scale plant. The dosage is equivalent to 1.2 mol Fe/mol soluble P in the influent wastewater. The concentration of total phosphorus was decreased from about 8 mg P/l in the influent wastewater to about 3 mg P/l after pre-precipitation. Soluble total phosphorus concentration was reduced from about 5 to about 1.5 mg P/l. The concentration of phosphorus in the influent and in the effluent from the pre-precipitation step is shown in Fig. 6.

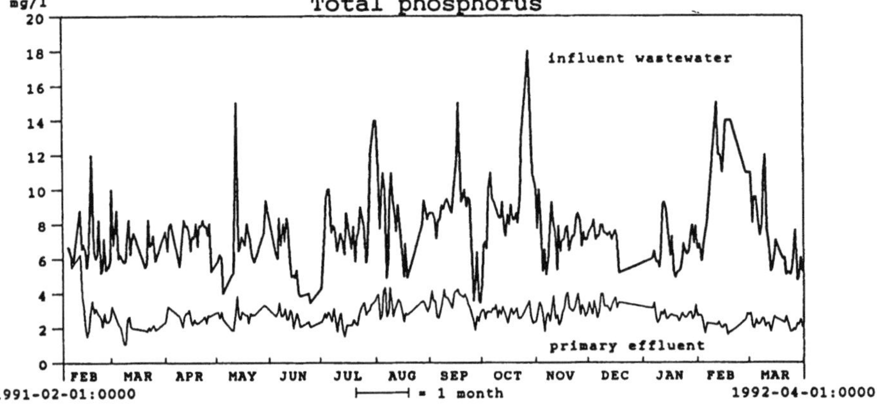

Fig. 6. Influent and effluent concentration of total phosphorus in the primary treatment step with pre-precipitation

The inorganic fraction in primary effluent wastewater changed because of the dosage of ferric chloride. The ratio between the volatile suspended solids and the total suspended solids in the primary effluent decreased from about 0.83 to 0.74 when pre-precipitation was introduced. Analyses of unfiltered samples from the primary effluent showed an increase in ferric ion concentration from about 1 g Fe/m^3 before to about 3 g Fe/m^3 after the start of iron addition.

Secondary Treatment

The results concerning nitrogen removal in the flow train with methanol addition are shown in Fig. 7. The dosage of ferric chloride was started in the beginning of February and the dosage of methanol in the beginning of April.

The period between May and October can be characterized as a steady state period. In October, the flow was deliberately increased by about 50%. The effluent total nitrogen concentration was on the average 3–4 mg N/l cor-

responding to an overall nitrogen removal of about 87 %. The methanol dosage
was about 60 g per m³. A decrease of the methanol dosage in July to about
50 g/m3 caused the effluent nitrogen concentration to increase to about 6 mg
N/l. More data from the steady state period are shown in Table 2.

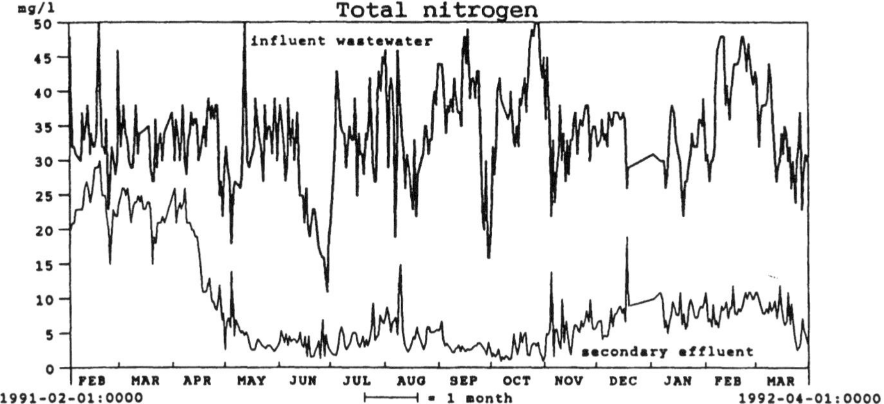

Fig. 7. Influent and effluent nitrogen concentration during full scale operation with
pre-precipitation and denitrification with addition of methanol

Tab. 2. Average results from full scale operation with addition of methanol. Data
from a steady-state period during May to October, 1991. The dosage of methanol
was 50–60 g BOD_7/m^3

Parameter	SS	BOD_7		P_{tot}		N_{tot}	NH_4	NO_x	Fe
		tot	sol	tot	sol				
	mg/l	mg/l	mg/l	mg/l	mg/l	mg/l	mg/l	mg/l	
Sampling points:									
Influent wastewater	220	145	26	8.1	4.9	34	25	0.2	–
Effluent from pre-precipitation	38	36	17	3.0	1.7	27	25	0.2	3.1
Effluent from biological treatment	7	< 5	< 5	1.4	1.1	3.8	0.2	2.7	0.4

The flow increase in October resulted in break-through of ammonia from
the biological step in November. Aeration was then started in one of the anoxic
cells and the ammonia concentration could again be brought down to low levels.
The nitrate concentration, however, increased to about 6 mg N/l and the total
nitrogen concentration increased to about 8 mg N/l.

The influent and effluent concentration of total phosphorus are shown in
Fig. 8. A further removal of phosphorus took place over the secondary step.
The effluent concentration was normally below 1 mg P/l and about 1.6 mg P/l
was removed in the secondary step. In August and September there was an

increase in the sludge water recirculation due to a decrease in the dry solids concentration after the raw sludge thickener. The sludge dewatering plant was not operated at the capacity it should have been. The phosphorus concentration in the primary and secondary effluent increased during this period and this is included in the averages in Table 2. The sludge dewatering operation was changed at the end of September.

Particulate bound phosphorus and iron is entrapped in the sludge, but the figures in Fig. 8 and in Table 2 also indicate that about 0.5 to 1 mg of phosphate phosphorus was needed for biomass growth. The entrapment of the highly inorganic suspended solids leaving the primary step decreases the MLVSS/MLSS ratio of the sludge from about 0.84 in a system without pre-precipitation to about 0.74 for the system with pre-precipitation.

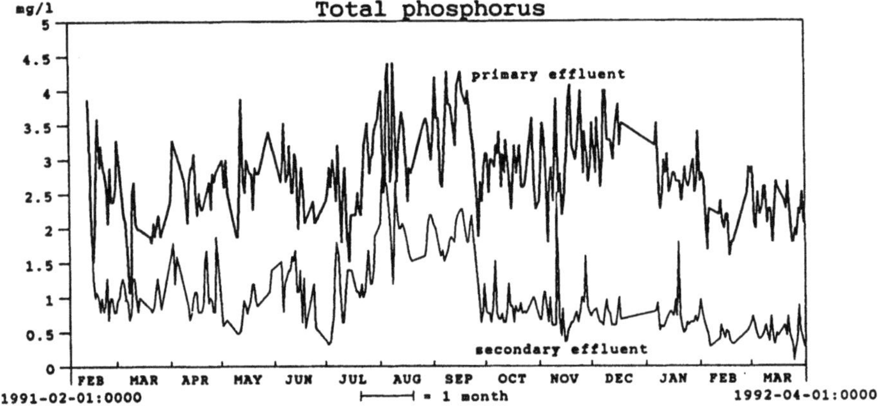

Fig. 8. Influent and effluent phosphorus concentrations in the secondary treatment step

Maintaining nitrification during winter conditions is one of the most important factors which make it possible to control the overall nitrogen removal. Normally, an aerobic sludge age of about 15 days is sufficient for complete nitrification at temperatures down to 7–8 °C. The removal of suspended solids by pre-precipitation reduces the sludge yield considerably. The sludge yield at the Klagshamn plant has been estimated to be around 0.4–0.6 g SS/g BOD$_7$. The corresponding figure for the pre-denitrification system without primary clarifiers was around 1.2–1.4 g SS/g BOD$_7$. It should be noted that simultaneous precipitation was not used during the tests with this system. The difference in sludge yield means that under similar loading and operational conditions, the aeration volume for sustaining the same sludge age must be about two to three times larger in the pre-denitrification system without primary clarification than the system with pre-precipitation/post-denitrification.

The ammonia uptake rate (AUR) [6] was repeatedly determined in laboratory tests. The AUR over the test period is shown in Fig. 9. The rates have been converted to a temperature of 10 °C. For the pre-denitrification system without primary clarification, the rate was around 2 mg N/g VSS · h. For the pre-precipitation system the rate is around 4–5 mg N/g VSS · h. This difference in AUR between the two systems reflects the difference in sludge production in the biological step. The break-through of ammonia in November due to loss of suspended solids during peak flows is also clearly seen in the figure. The rate suddenly drops and as the ammonia concentration again decreases in the effluent water, the rate increases as well. It can also be seen in Fig. 9 that the AUR increases in February when pre-precipitation is started and decreases as the methanol dosage is started. This again shows the effect of an increased or decreased sludge production.

The denitrification capacity for a biological nitrogen removal system is limited either by the stoichiometric relationship between carbon and nitrogen, by the nitrate concentration in the anoxic step or by the relationship between the denitrification rate and the detention time in the anoxic parts of the basin. In a concept based on addition of a readily biodegradable external carbon source, such as methanol, the plant should preferably be operated such that the limiting factor for denitrification is the stoichiometric relationship between carbon and nitrate.

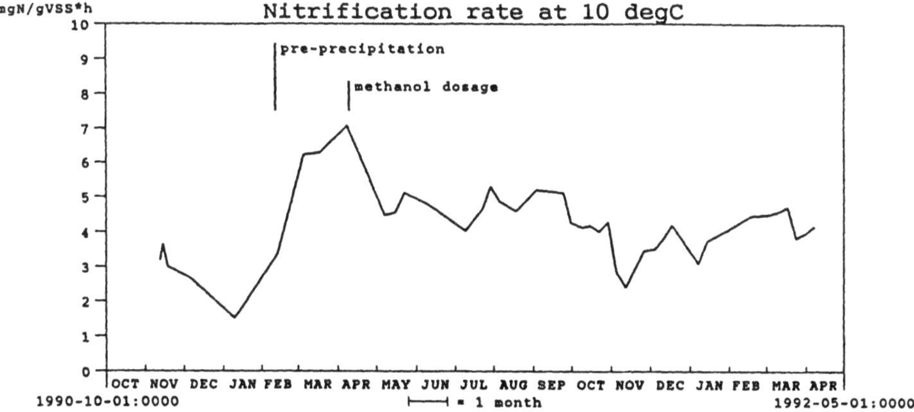

Fig. 9. Ammonia uptake rate (AUR) measurements in the sludge from the pre-precipitation/post-denitrification flow train

The ratio between BOD$_7$ and denitrified nitrogen was estimated to be about 6–8 g BOD$_7$/g N for the pre-denitrification system without primary clarification. The corresponding figure for the post-denitrification system with pre-precipitation was around 5–6 g BOD$_7$/g N. This shows that the carbon was better utilized in the post-denitrification system.

Methanol as carbon source requires an adaptation period before full denitrification is achieved. This can be seen in Fig. 10, which shows the nitrate uptake rate (NUR) [6] from the start of the methanol dosage. The denitrification capacity increases continuously during the adaptation period. Later, the denitrification rate was stabilized between 3 and 4 mg NO$_3$-N/g VSS · h.

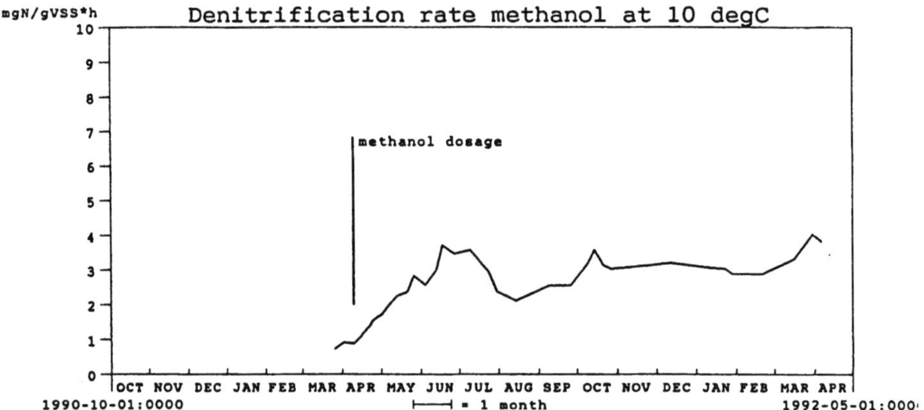

Fig. 10. Nitrate uptake rate (NUR) measurements in the sludge from post-denitrification with methanol as carbon source

The use of methanol as carbon source for denitrification results in a highly select microbial system consisting mainly of Hyphomicrobium spp [2]. This group of bacteria is very specialized and can, for example, not utilize acetate for growth.

Sludge Production

During the test program, it was not possible to accurately determine the total sludge production from the two different system configurations, because the sludges could not be separated in the sludge treatment process. A rough estimate of the total sludge production is presented in Fig. 11.

The sludge produced from the primary treatment with pre-precipitation is composed of the suspended solids removal from the influent wastewater and the sludge due to precipitation by addition of the coagulant. With an addition of 10 g Fe/m^3, the sludge production due to precipitation amounts to about 30 g SS/m^3. With the removal of about 170 g/m^3 of suspended solids from the influent wastewater, the sludge production from the pre-precipitation is about 200 g/m^3. By including the sludge production of about 40 g/m^3 from the biological treatment, the total sludge production from this system is calculated to be about 240 g SS/m^3.

The sludge production from the pre-denitrification system without primary clarification is composed of the sludge production due to simultaneous precipitation and the biological sludge production. Assuming a sludge yield of 1.4 g SS/g BOD$_7$, sludge production is almost 200 g SS/m^3. Using the same conversion between precipitation and sludge production as in the concept with

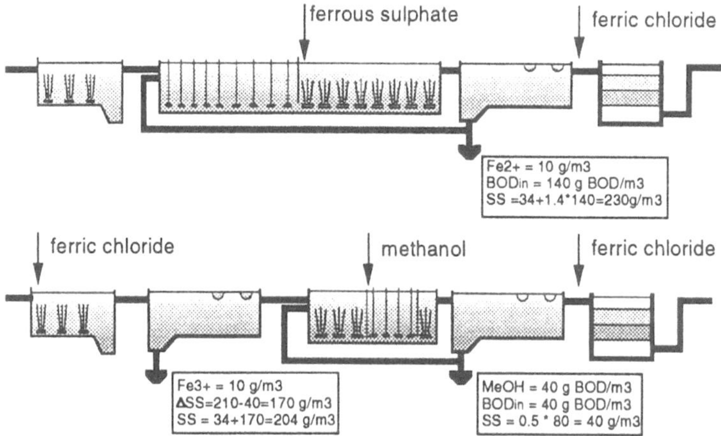

Fig. 11. Calculated sludge production from two different upgrading alternatives at the Klagshamn plant

pre-precipitation, the total sludge production from the pre-denitrification system amounts to about 230 g/m^3.

The rough calculation indicates that the total sludge production from the two systems is fairly the same. Most important, however, is the difference between the two systems concerning the sludge production from the biological treatment, the sludge yield. This amounted to 230 g/m^3 in the pre-denitrification system without primary clarification and to about 40 g/m^3 from the pre-precipitation/post-denitrification system. This means that the total sludge mass produced in the pre-denitrification system was almost six times greater than the mass produced in the other system. If the mixed liquor suspended solids in the aeration basins is kept at the same level, the basin volume needed in order to reach the same sludge age would have to be about six times greater in the pre-denitrification system without primary clarification.

Costs for Nitrogen Removal

The additional cost for extended nitrogen removal for the concept based on pre-precipitation/post-denitrification with methanol addition constitutes only that for methanol. The tests were performed on a full scale and most of the investments made within the upgrading test program can be used for the proceeding operation. In the pre-denitrification concept, new basin volumes would have to be built. The investment costs for the required basin extensions has been estimated to 40–50 MSEK, which means an annual cost of 4–5 MSEK. Addition of 40–50 g/m^3 of methanol, enough to reach 8 mg N/l in the effluent, will give annual costs of about 0.5 MSEK.

Ferrous sulphate instead of ferric chloride could be used for phosphorus removal in the pre-denitrification alternative. The cost for precipitation chemicals would be higher in the pre-precipitation alternative since ferrous sulphate is less expensive than ferric chloride. On the other hand, the energy costs for the air supply would be much lower for the pre-precipitation alternative. This will roughly balance the increased coagulant costs.

Thus the annual additional costs for the extended nitrogen removal at the Klagshamn wastewater treatment plant based on a concept with pre-precipitation/post-denitrification and external carbon addition is only about one-tenth of the costs required for a concept based on pre-denitrification with new extra basin volumes.

Conclusions

Full scale tests at Klagshamn wastewater treatment plant have shown that upgrading for extended nitrogen removal based on a concept with pre-precipitation and addition of an external carbon source in a single sludge post-denitrification process could meet the future proposed nitrogen effluent standards within existing basin volumes.

Using pre-precipitation with ferric chloride, the large variations in suspended solids concentration in the influent wastewater to the plant could be decreased, resulting in a very stable wastewater quality entering the biological process. Thus, conditions for maintaining nitrification were improved and phosphorus was removed at the same time.

Methanol addition for denitrification resulted in low sludge yield and high denitrification rates.

To meet the very stringent phosphorus effluent standard, a filtration plant with possible contact precipitation will be installed. By using pre-precipitation, most of the phosphorus is removed before the filters, resulting in a much lower load on the filters.

A pre-denitrification process, utilizing the total amount of organic carbon in the raw wastewater by excluding primary clarification, would require extensions with new basin volumes. Thus, the additional cost for the extended nitrogen removal was much lower with the process selected.

References

[1] Andersson, B., Nyberg, U., Aspegren, H., Jansen, J., la Cour: Extended Nutrient Removal at the Klagshamn Wastewater Treatment Plant in Malmö. Paper presented at the EWPCA conference Advanced Sewage Purification to the Protection of the North and Baltic Sea in Lübeck/Travemünde, November 7–8, 1991

[2] Nyberg, U., Aspegren, H., Andersson, B., Jansen, J., la Cour, Villadsen, I.: Full Scale Application of Nitrogen Removal with Methanol as Carbon Source. Paper presented at the IAWPRC 16th Biennial Conference in Washington D.C. in May 1992

[3] Nyberg, U., Andersson, B., Aspegren, H.: Nitrogen Removal in an Activated Sludge System Without Primary Clarification. Vatten/Journal of the Water Management and Research *47* (4) (1991) 278–285

[4] ATV-Arbeitsblatt A 131: Bemessung von einstufigen Belebungsanlagen ab 5 000 Einwohnerwerten. Ges. zur Förderung der Abwassertechnik, St. Augustin 1991

[5] Andersson, B., Aspegren, H.: Evaluation of Pre-precipitation at the Klagshamn Wastewater Treatment Plant. Internal Report, Malmö Water & Sewage Works, 1989

[6] Kristensen, G.H., Jørgensen, P.E., Henze, M.: Characterization of Functional Biomass Groups and Substrate in Activated Sludge and Wastewater by AUR, NUR and OUR. Pre-print to IAWPRC Conference on Interactions of Wastewater, Biomass and Reactor Configurations in Biological Treatment Plants. Copenhagen, August 21–23, 1991

Bengt Andersson
Henrik Aspegren
Malmö Water & Sewage Works
S-205 80 Malmö
Sweden

Jes la Cour Jansen
Water Quality Institute
Agern Allé 11
DK-2970 Hörsholm
Denmark

Ulf Nyberg
Department of Environmental Engineering
Lund Institute of Technology
S-221 00 Lund
Sweden

Hallvard Ødegaard
Div. of Hydraulic and Sanitary Engineering
Norwegian Institute of Technology
N-7034 Trondheim – NTH
Norway

High Rate Biofilters –
Simultaneous Phosphorus Precipitation
and Nitrogen Removal

R.F. Gonçalves, F. Sammut, and F. Rogalla

Abstract

Simultaneous precipitation of phosphorus was carried out on an industrial scale nitrifying/denitrifying fixed film reactor. The behaviour of the upflow biological aerated filter during the addition of sodium aluminate as precipitating agent was assessed. In addition to the phosphorus removal rate, particular attention is paid to the possible impacts on denitrification and nitrification rates of the filter, headloss development and particle retention. The effective dose of precipitant required per unit of phosphorus removed is obtained based on these practical trials.

The trial showed that residuals close to 2 mg PO_4-P/l are obtained with doses between 100 g and 150 g of reagent/m^3 settled wastewater containing 10 to 12 mg/l PO_4-P. From an influent concentration of 500 mg/l COD, 65 mg/l TKN and 15 mg total P, treated effluent with 70 mg/l COD, 2 mg NH_4-N/l, < 15 mg NO_3-N/l and 2 mg PO_4-P/l was obtained. Total hydraulic retention time was under 3 hours with a recycle of 200 % and at a temperature of 15 °C.

No inhibition of nitrification was observed, and eliminated loads reached 0.9 kg NH_4-N/m^3 aerated zone/d at 15 °C. Addition of the precipitating agent does not reduce the denitrification rate significantly, colloidal organic matter seems to remain available to the denitrifying bacteria. When the influent carbon concentration is sufficient, eliminated loads of 1 kg NO_3-N/m^3 anoxic zone/day are achieved.

Introduction

Increases in the quantities of phosphorus discharged to rivers and lakes, estuaries or enclosed seas such as the North, Baltic or Adriatic Seas, may disturb the ecological balance of these environments. In order to prevent this potential problem, the European Community directive of March 1991 imposes stringent limits on the concentrations of nitrogen and phosphorus in wastewater discharges. In order to conform to this new European directive, extensive refurbishment of existing treatment works, in addition to the construction of new works, will be required.

Nitrification and denitrification can be achieved in activated sludge plants by installing alternate anoxic and aerobic zones [1]. Total nutrient removal may

be achieved by combining nitrification/denitrification with phosphorus removal
using physico-chemical precipitation, biological treatment or a combination of
both [2].

Physico-chemical precipitation is an effective means of reducing phosphorus
levels, ensuring effluent P-levels below 1 mg/l. This technique relies upon the
addition of salts of aluminium, iron or calcium directly upstream of the primary
treatment (preprecipitation), within or immediately after the aeration tanks
(simultaneous precipitation) or after the clarifier (tertiary precipitation). To
achieve concentrations below 1 mg P/l, however, filtration is required [3].

To reduce land area requirements and environmental impacts of wastewater
treatment works, Biocarbone aerated filtration, a process which has been in use
for over 10 years, ensures aerobic degradation of pollutants and physical retention of particulate matter in a single reactor [4]. When coupled with lamellar
settling for primary treatment, this results in very compact treatment works
with no need for secondary clarification [5]. Close to 100 of these works have
already been built, proving the effectiveness of these processes for carbon and
nitrogen removal.

A few full scale installations of this type included phosphate removal by pre-
precipitation in the primary settler [6]. This solution can lead to a significant
consumption of reagents and the production of large quantities of sludge. In
addition pre-precipitation reduces the carbonaceous load of the water prior to
a possible denitrification stage. An innovative approach has been developed to
overcome this dilemma [7].

Biological phosphorus removal in a fixed film reactor has been shown to be a
very promising technique [8]. Upflow aerated filters functioning under alternate
aerobic and anaerobic conditions have been shown to carry out 80 % phosphorus
removal without chemicals, in addition to achieving complete nitrification.

The aims of this study were to explore the possibilities of simultaneous
precipitation on an industrial scale nitrifying/denitrifying Biostyr aerated filter
using an alkaline reagent. The potential advantages of such a process would be:

- retention of the precipitated phosphorous on the Biostyr itself, eliminating
  the need for a clarifier;
- no reduction in alkalinity of the wastewater, carbonates therefore remaining
  available for nitrification;
- lower sludge production due to precipitation being performed on settled water with a reduced solids content. This will also optimize the dosage of precipitant;
- physical retention of precipitated phosphorus and carbon on the filter, the
  latter therefore remaining available for denitrification.

In addition to the phosphorus removal rate, denitrification and nitrification
rates of the filter, headloss development and particle retention of the filter were
also studied.

Materials and Methods

Existing Installation

A schematic of the wastewater treatment works where the trials were carried out is shown in Fig. 1.

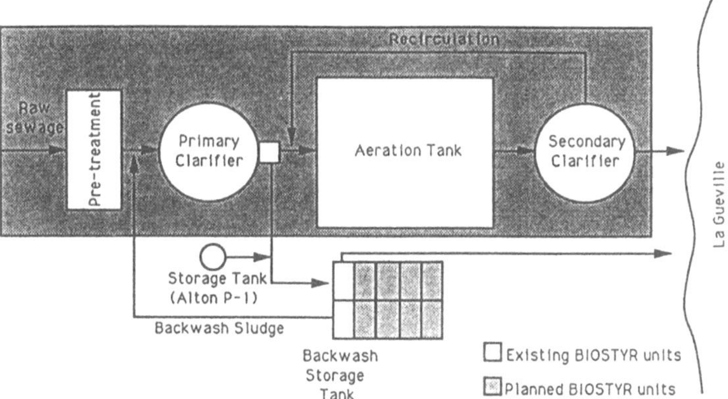

Fig. 1. Schematic of existing and planned installation

The installation receives wastewater from the town of Rambouillet near Paris. The treatment works is currently operating at the limit of its capacity (30 000 p.e.). In order to meet future needs and to conform to new European regulations, a refurbishment of the works is expected to be completed in 1993. The characteristics of the existing works and the future works are presented in Table 1.

Tab. 1. Characteristics of existing and future works

	Existing works	Future works
Population [p.e.]	30 000	40 000
Average daily flow [m³/d]	6 150	8 000
Peak hourly flow [m³/h]	450	450
COD load [kg/d]	3 600	6 000
Treatment	Activated sludge (high load)	BAFs (nitrification/denitrification) Phosphorus removal (physico/chemical)

The existing works consists of a preliminary treatment (screening, grit removal, grease removal), a primary sedimentation tank, biological treatment using high load activated sludge followed by a secondary sedimentation tank.

Refurbishment of the existing works using the currently installed technology would have been impracticable due to space restrictions. Thus the option chosen was to replace the existing installation with compact components such as lamellar settlers and aerated filters. The future works is to include a lamellar settler, followed by ten Biostyr units (submerged filters), each with a surface area of 22 m^2. Two of these units have already been built, and one has been used as an industrial scale pilot plant for these trials. The second unit is being used temporarily as a backwash water reservoir.

The future works should ensure a treated wastewater which conforms to the standards outlined in Table 2.

Tab. 2. Water quality standards for the future works

Parameter	Daily mean	2 hour mean
COD [mg/l]	90	120
BOD [mg/l]	15	20
TSS [mg/l]	20	20
TKN [mg/l]	10	15
N_{tot} [mg/l]	20	25
P_{tot} [mg/l]	2	2
P_{tot} removal [%]	80	80

The Biostyr Process

The biofiltration process chosen for this installation is an upflow process which uses a floating polystyrene media for biomass attachment and particle retention. The floating media allows aeration within the filter, which results in the creation of an aerated zone above an anoxic zone, allowing both nitrification and denitrification to be carried out on a single reactor [9].

As the first stage in the process, denitrification lowers the carbonaceous load of the water prior to entering the nitrifying stage, an essential requirement for the development of autotrophic nitrifiers. This also ensures that water entering the denitrification stage has the highest carbon content possible. A certain percentage of the nitrified water exiting the reactor is recirculated back to the denitrifying stage to achieve the required nitrate removal.

The filtering media is not fluidised because filtration is carried out in the direction of the media compression. This feature allows very low suspended solids residuals. A periodic counter-current backwash removes the accumulated excess biomass and the retained solids. The backwash gives priority to the highly loaded part of the bed in contact with the incoming solids, the light-weight beads significantly facilitating this operation.

Experimental Design

The industrial scale demonstration Biostyr has a surface area of 22 m^2, and the filter has a height of 3 m. The submerged floating media has a grain size between 3 and 4 mm. The anoxic zone has a height of 1 m, above which is the aerated zone with a height of 2 m. The wastewater flow through the filter is approximately 1 000 m^3/d. The trials lasted three months, from October through December 1991. The temperature throughout this period varied from 19 °C to 10 °C.

The wastewater treated by the reactor comes from a combined urban sewer and has undergone preliminary treatment and primary settling. The characteristics of this settled wastewater during the trial period are shown in Table 3.

Tab. 3. Mean concentration of settled wastewater

Parameter	Mean concentration mg/l
COD total	480
COD soluble	220
Suspended solids	170
NH_4-N	50
TKN	62
P_{tot}	12

Chemical precipitation was carried out using a sodium aluminate-based reagent called Alton P-1 [Na(Al(OH)$_4$)]. This product consists of an alkaline mineral reagent, with a concentration of 3.5 moles Al/litre and a density of 1.35, obtained from Nalco France SA.

Two metallic tanks with a total capacity of 10 m^3 of reagent were installed on site. The reagent was continuously dosed at the influent pipework of the aerated filter. An industrial dosing pump was used to inject the precipitant a significant distance upstream of the reactor in order to ensure a homogeneous mix. Figure 2 is a schematic view of the pilot installation showing the point of injection of the precipitant.

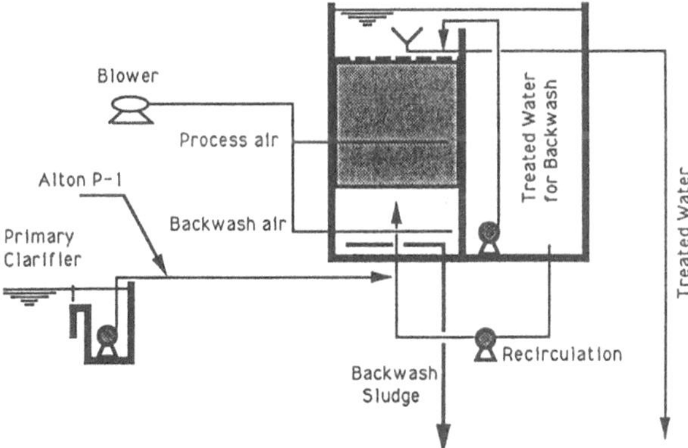

Fig. 2. Schematic of aerated filter unit used during trials

The trials were separated into three distinct periods, depending on the applied dosage of reagent, as outlined in Table 4 below.

Tab. 4. Applied Alton P-1 dosages

Dose of Alton P-1	Week number
100–150 g/m^3	1–5
150–200 g/m^3	6–9
> 200 g/m^3	10–12

The performance of the aerated filter throughout this period was monitored using the following sampling techniques:

- 24 hour average samples at the entry and exit of the filter (BOD, COD, TSS, TKN, NH_4-N, NO_3-N, PO_4-P). Analysis for total phosphorus was introduced in the tenth week.
- Sampling from the 5 sampling points along the height of the filter (COD, TSS, NH_4-N, NO_3-N, PO_4-P).

The effects of physico-chemical phosphorus removal on the performances of the aerated filter were evaluated by comparing the results with those obtained during the same period in 1989–1990.

Results and Discussion

Phosphorus Removal

The influent and effluent concentrations of PO_4-P before and after the addition of Alton P-1 are shown in Fig. 3. It can be seen from this figure that effluent PO_4-P concentrations of around 2 to 3 mg/l are consistently obtained. Peaks in the concentration of PO_4-P at the exit of the filter are due to problems with dosing of the Alton, as a result of the dosing pipe being blocked by a white precipitate. The performance of this process was monitored by measuring only the PO_4-P concentrations, whilst measurements of total-P were begun at the end of the third month.

The trial showed that 70 to 80 % removal of PO_4-P is obtained with doses between 100 g and 150 g Alton P-1/m^3 settled wastewater containing 10 to 12 mg/l PO_4-P, which results in an effluent PO_4-P concentration around 2 mg/l (Fig. 4). With a dosage of more than 200 g Alton P-1/m^3 settled wastewater, PO_4-P removal efficiencies of over 90 % are achieved, with a maximum of 97 % PO_4-P removal (Fig. 5).

As the dose of reagent increases, the molar ratios, β (moles Al/mole PO_4-P eliminated), increase proportionally as follows (Fig. 6):

- dose 100 g/m^3 $(1 < \beta < 1.5)$
- dose 150 g/m^3 $(1.5 < \beta < 2)$
- dose 200 g/m^3 $(2.5 < \beta < 3)$.

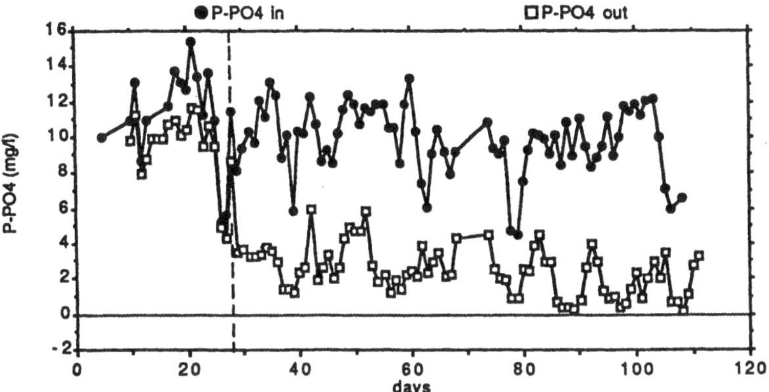

Fig. 3. Variations in PO$_4$-P concentrations before and after addition of reagent

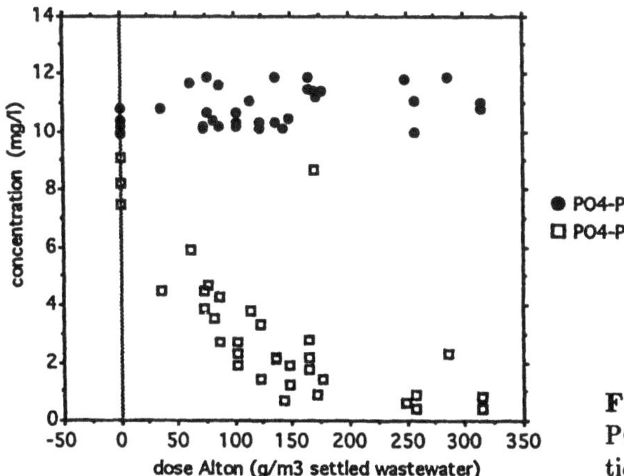

Fig. 4. Influent and effluent PO$_4$-P concentrations as a function of Alton dosage

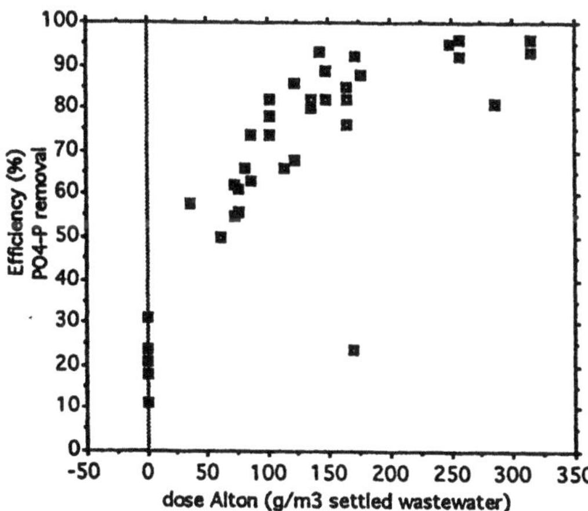

Fig. 5. Removal efficiency as a function of Alton dosage

In order to obtain very low phosphorus residuals in the effluent, (< 1 mg/l PO_4-P), a significant dose of over 250 g reagent/m^3 settled water would be required, with β being between 3 and 3.5 (Fig. 6).

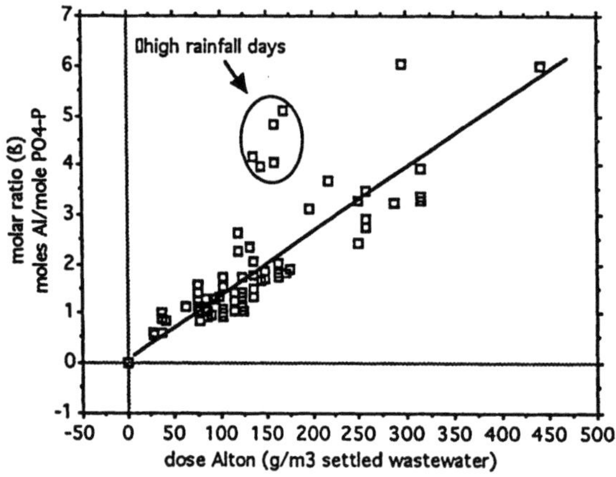

Fig. 6. Effect of increasing the Alton dosage on the β value

Measurements of total P at the exit of the filter were carried out during a period when the reagent dose was over 200 g/m^3. These show a difference of approximately 4 mg/l between effluent PO_4-P and total P concentrations for the same samples (Fig. 7). This indicates that a proportion of the flocs of precipitated PO_4-P are not retained within the filter. The flocs resulting from the precipitation of PO_4-P by Alton P1 are thought to be very fine [10] and may be too small to be effectively retained.

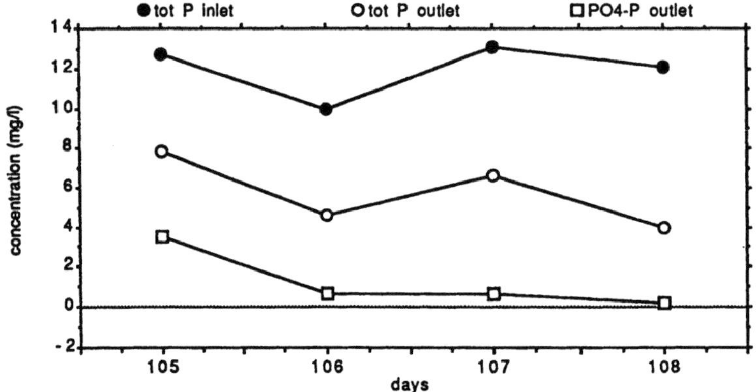

Fig. 7. Inlet and outlet total P compared to outlet PO_4-P concentration with very high dosages ($>$ 200 g Alton P1)

This phenomenon would account for the observation that increasing doses of Alton P-1 above 150 g/m^3 settled wastewater result in a proportional increase in effluent suspended solids concentration (Fig. 8). The effluent suspended solids would therefore contain flocs of precipitated PO_4-P, which is then measured as total-P. Addition of a polyelectrolyte may increase the size of the flocs, leading to a better retention within the filter.

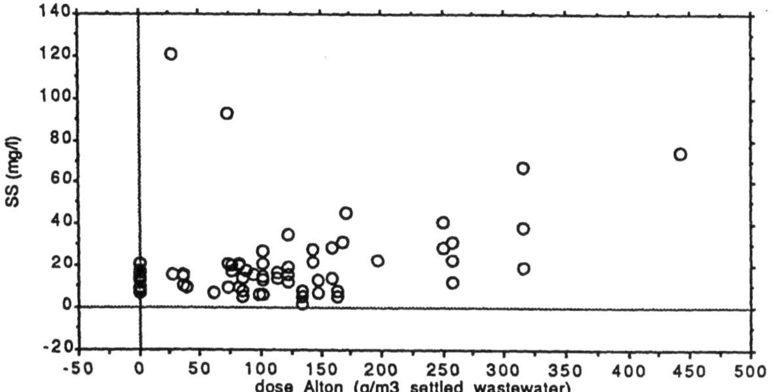

Fig. 8. Suspended solids concentration in the filter outlet as a function of Alton dosage

It is clear from these results that high removal efficiencies for PO_4-P are possible following the addition of the reagent. The applied dosages required to obtain these very high removal efficiencies, however, result in an increase in effluent suspended solids concentration. In addition, it is questionable whether the addition of the reagent at these dosages is economically viable.

Pre-precipitation of PO_4-P can be combined with simultaneous precipitation on the aerated filter using dosages of 100–150 g reagent/m^3, in order to obtain PO_4-P residuals of < 1 mg/l.

Nitrification

The performances of the aerated filter during the trial period and the period September to December 1990 can be compared using Fig. 9. The operational conditions during these two periods were similar.

The addition of Alton P1 does not appear to have any effect on nitrification rate, which would agree with the findings of Kraft and Seyfried (1990). No inhibition of nitrification was observed during the 3 month trial period, and a maximum nitrification rate of 0.9 kg NH_4-N/m^3 aerated zone/d at 15 °C, with applied loads of 6 kg COD/m^3/d and a retention time of 3 hours, was found consistently. Table 5 presents a summary of the performances of the filter during the two periods stated above.

A comparison can also be made using Fig. 10, which shows the variation in NH_4-N concentration within the filter during the 2 periods. The initial drop in NH_4-N concentration between the first two sampling points is due to recirculation of part of the nitrified wastewater. Although NH_4-N concentration is

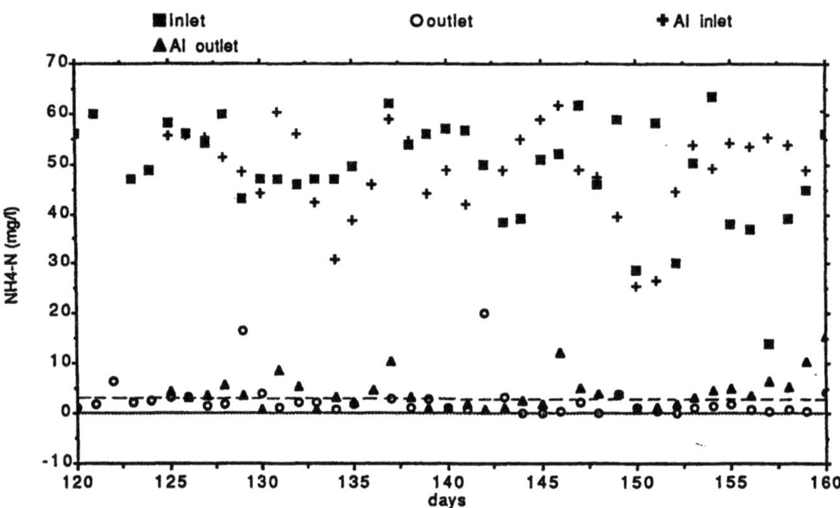

Fig. 9. Nitrification performance of the filter during the two different periods (Al = with Alton P1)

Tab. 5. Nitrification performances of the filter

	Without Alton P-1 (1990)	With Alton P-1 (1991)
NH_4-N applied load [kg/m^3/d]	0.85–1.0	0.9–1.0
max NH_4-N eliminated load [kg/m^3/d]	0.9	0.9
Elimination efficiency [%]	90–95	90–95
T [°C]	15	15
NH_4-N exit [mg/l]	2	2

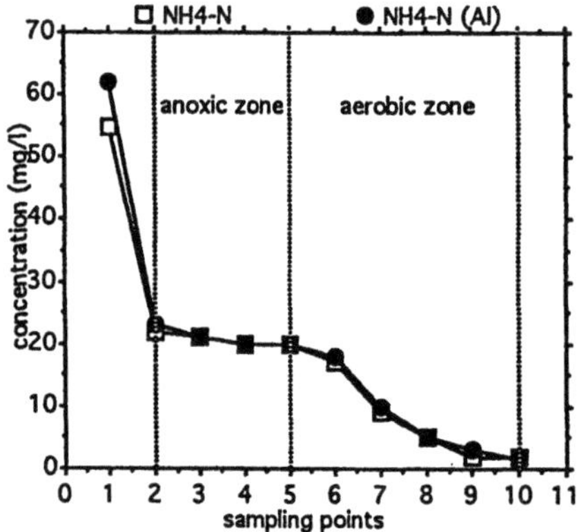

Fig. 10. Variation in NH_4-N concentration within the filter for the two periods (Al = with Alton P1)

high in the filter influent, complete nitrification is achieved in the aerobic zone with identical kinetics. The applied load of NH_4-N is 0.8 kg NH_4-N/m^3 aerated zone/d, whilst the applied load of COD is of the order of 5 kg/total m^3/d at 15 °C.

Denitrification

Denitrification rate in the Biostyr is directly related to the recirculation rate of the wastewater in the anoxic zone, in addition to the availability of a carbon source in the settled wastewater. The requirements for complete denitrification are a ratio of 3–5 mg BOD/mg NO_3-N denitrified (11). Trials on nitrifying/ denitrifying biological aerated filters have confirmed a ratio of 6 mg COD/mg NO_x-N eliminated [12].

Due to the characteristics of the wastewater at Rambouillet, this carbon to nitrogen ratio is not always assured. With a recirculation rate between 2 and 3 times the influent water flow, the effluent has a concentration of between 10 and 15 mg NO_3-N/l (Fig. 11).

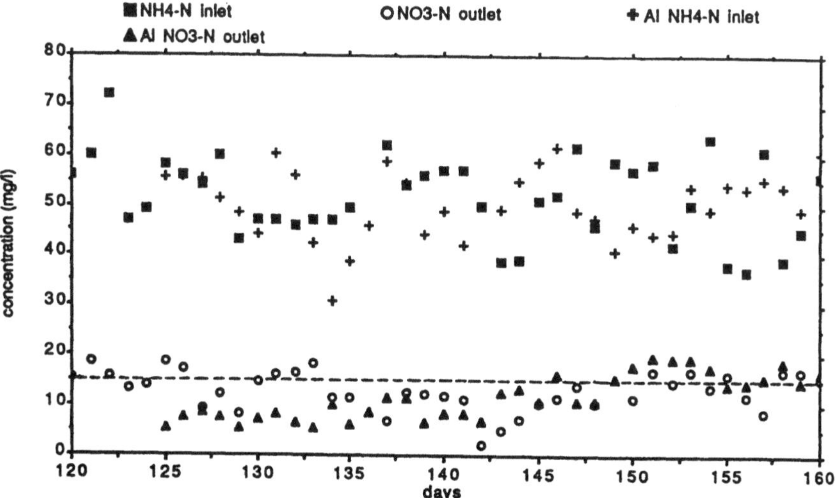

Fig. 11. Denitrification performance of the filter during the two different periods (Al = with Alton P1)

A possible reduction in the availability of carbon as a result of precipitation of colloidal organic matter due to the addition of Alton P1 was one of the main parameters to be verified during these trials. The addition of Alton P1, however, does not reduce the denitrification rate significantly, and colloidal organic matter seems to remain available to the denitrifying bacteria. A summary of the performances during the two periods is given in Table 6.

Tab. 6. Denitrification performances of the filter

	Without Alton P-1 (1990)	With Alton P-1 (1991)
NH_3-N eliminated load [kg/m^3/d]	0.8–1	0.8–1.1
T [°C]	15	15
NH_3-N exit [mg/l]	15	15

Figure 12 shows the reduction of NO_3-N along the height of the filter. A slight reduction in the rate of denitrification is noticed during the addition of Alton P1, but it should be noted that this figure is based on only two examples. In both cases denitrification is practically complete.

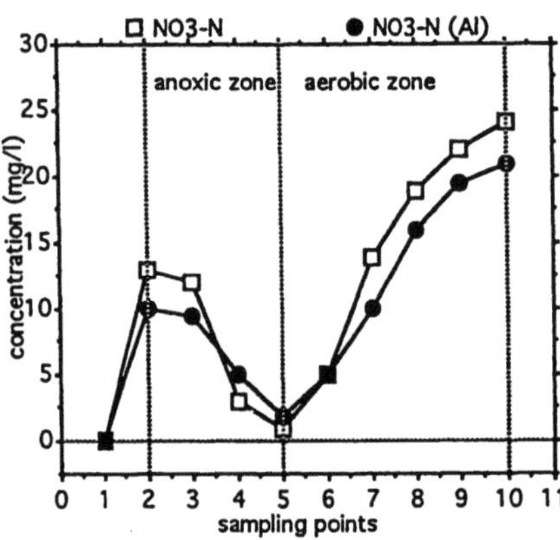

Fig. 12. Variation in NO_3-N concentration within the filter for the two periods (Al = with Alton P1)

Filter Headloss Development and Sludge Production

Operation of the aerated filter is divided into a filtration phase, followed by a backwash phase, during which the filter is rinsed in order to remove retained particles and the excess biomass. During the backwash phase, water is added in a counter-current manner whilst air is added co-currently to the operational water flow. Backwash sludge is collected by pipework from the bottom of the filter. Backwash is carried out every 24 hours.

As the amount of retained particles and biomass increases, so does the headloss through the filter. Headloss is monitored both across and within the filter. Normal headloss after backwash is about 50 cm, whereas the maximum headloss is limited to about 2 m. Between successive backwashes 2 or 3 short flushes are carried out in order to disperse the particles accumulated on the filter surface.

It was expected that the addition of Alton P1 would lead to a higher rate of headloss development and shorter cycles, due to the increase in particulate matter as a result of the formation of precipitated flocs. The effect of the addition of the precipitant was assessed by comparing headloss development during the trials with Alton P1 and after the trials were completed. Examples of the headloss development during these two periods is shown in Fig. 13. From this figure, the following observations are made:

– Contrary to possible expectations, the total increase of headloss is similar during and after the addition of Alton P1.

– Headloss development is of a more linear nature during the addition of Alton P1.

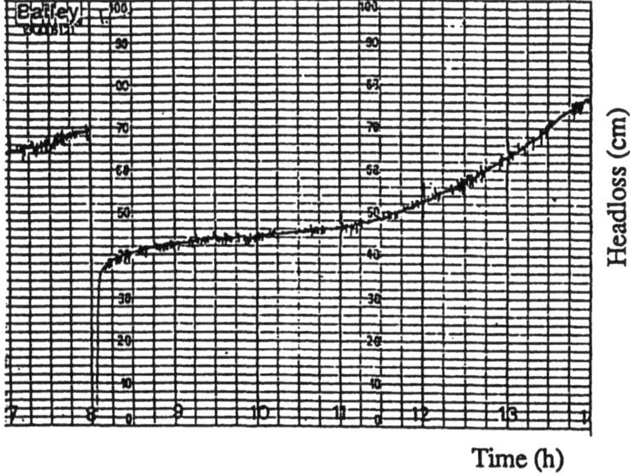

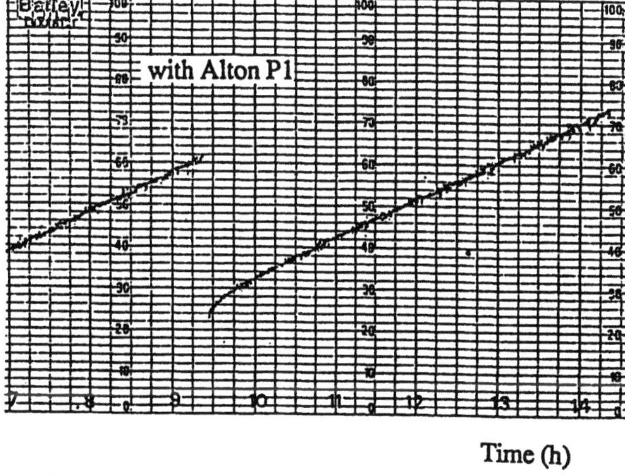

Fig. 13. Headloss development during the two periods

Both of these observations are probably due to a more even distribution of retained particles within the filter as a result of the addition of Alton P-1. This could be related to the observation that flocs resulting from the precipitation of PO_4-P by Alton P1 may be very fine, as stated above.

With regard to the production of sludge, the manufacturers of the reagent state that excess sludge production due to the addition of the reagent is low, of the order of 10 %. Sludge production was assessed by measuring suspended solids concentration in the backwash water, and relating this value to the eliminated COD. As can be seen from Fig. 14, there is an apparent increase in sludge production as a function of increasing reagent dosage. The average solids obtained are twice as high as during the reference period, where only 0.4 kg TSS/kg COD removed was measured, corresponding to usual values [13].

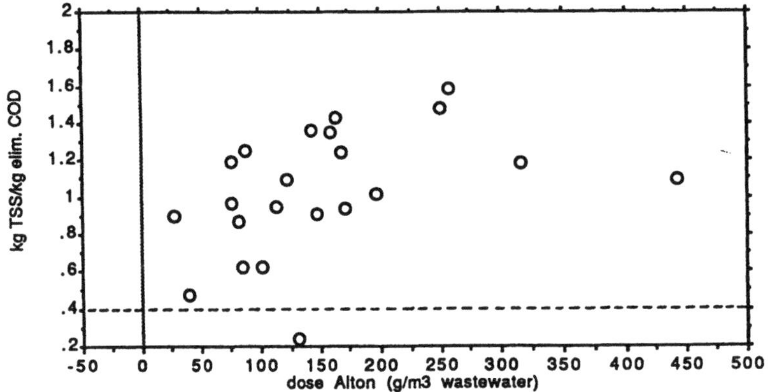

Fig. 14. Assessment of sludge production as a function of reagent dosage
(- - - = Biological sludge production without Alton P1)

Pre-precipitation and Nitrogen Removal

In parallel with trials for simultaneous precipitation, pre-precipitation trials were also carried out at a full scale installation. The first Biostyr plant went on line at the end of 1990 in St. Jean d'Illac, close to Bordeaux. The plant, designed for 2000 m³/d, is fully enclosed in a building of 20 by 30 m. The reagent used for these trials was PAX 10 (aluminium polychloride), obtained from Kemrhône.

The trials were carried out over a period of three and a half months, from October to mid-January. The reagent was added at the entry of the lamellar settler, ahead of the five aerated filter units of 16 m² each. When the concentration is 10 mg PO_4-P/l in the influent to the lamellar settler, treated effluent with a PO_4-P concentration of under 2 mg/l is obtained. Fig. 15 shows the increase in efficiency with higher dosages. This is achieved with a dosage of reagent of between 250 and 300 g/m³, or a molar ratio, β, of 1.5 to 1.75 moles Al/mole P [14].

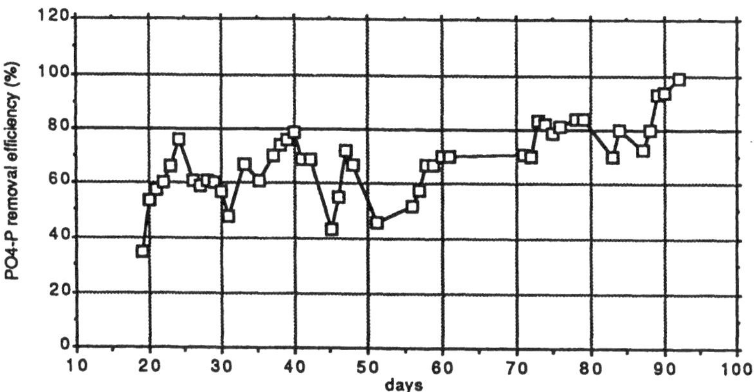

Fig. 15. Increase in removal efficiency with increasing dosage of PAX 10

Conclusions

These preliminary trials have shown that simultaneous phosphorus, nitrogen and carbon removal on a aerated filter is feasible. The Biostyr process allows the removal of carbonaceous matter, suspended solids, ammonia, nitrate and phosphorus on a single compact unit with a short retention time below 3 hours. With an influent concentration of 500 mg/l COD, 65 mg/l TKN and 15 mg total P, treated effluent with 70 mg/l COD, 2 mg NH_4-N/l, < 15 mg NO_3-N/l and 2 mg PO_4-P/l was obtained.

The trial showed that 70 to 80 % removal of PO_4-P is obtained with doses between 100 g and 150 g Alton P-1/m^3 settled wastewater containing 10 to 12 mg/l PO_4-P. With a dosage of more than 200 g Alton P-1/m^3 settled wastewater, PO_4-P removal efficiencies of over 90 % are achieved, with, however, an increase in the suspended solids content of the effluent.

No inhibition of nitrification was observed during the 3 month trial period. It would also appear that the addition of the reagent does not reduce the denitrification rate significantly, and that colloidal organic matter seems to remain available to the denitrifying bacteria.

The total headloss development is similar during and after the addition of coagulant, although headloss development would appear to be of a more linear nature during the addition of the precipitant. There is an apparent increase in sludge production as a function of increasing reagent dosage.

Trials are extended over an additional three month period in order to assess the long term effects of addition of the reagent on the performances of the filter and the possibility of a better retention of the flocs on the filter at high reagent dosages.

References

[1] Cooper, P.F., Collinson, B., Green, M.K.: Recent Advances in Sewage Effluent Denitrification, Part II. Wat. Poll. Contr. *76* (1977) 389–401

[2] Bundgaard, E., Pedersen, J.: Full Scale Experience with Biological and Chemical Phosphorus Removal. In: Chemical Water and Wastewater Treatment, H.H. Hahn and R. Klute (eds.). Springer, Berlin Heidelberg New York 1990, pp. 443–459

[3] Boller, M.A.: Chemical Optimization of Tertiary Contact Filters. J. Env. Eng. *110* (1984) 263

[4] Gilles,P.: Industrial Scale Applications of Fixed Biomass on the Mediterranean Seaboard. Design, Operating Results. Wat. Sci. Tech. *22* (1/2) (1990) 281–292

[5] Rogalla, F., Payraudeau, M., Bacquet, G., Bourbigot, M.M., Sibony, J., Gilles, P.: Nitrification and Phosphorus Precipitation with Biological Aerated Filters. J. Wat. Poll. Contr. *62* (2) (1990) 169–176

[6] Kantardieff, A.: Le premier filtre canadien pour le traitement des eaux usées du Canton d'Orford (Québec). Sciences et Techniques de l'Eau, *22* (1) (1989) 73–82

[7] Henze, M., Harremoës, P.: Chemical-Biological Nutrient Removal - The HYPRO Concept. In: Chemical Water and Wastewater Treatment, H.H. Hahn and R. Klute (eds.). Springer, Berlin Heidelberg New York 1990, pp. 499–510

[8] Franci Gonçalves, R., Rogalla, F.: Biological Phosphorus Removal in Fixed Film Reactors. AGHTM/IAWPRC Conference on Wastewater Management in Coastal Areas, Montpellier 1992, pp. 209–218

[9] Rogalla, F., Bourbigot, MM.: New Developments in Complete Nitrogen Removal with Biological Aerated Filters. Wat. Sci. Tech. *22* (1/2) (1990) 273–280

[10] Kraft, A, Seyfried, C.F.: Ammonia and Phosphate Elimination by Biologically Intensified Flocculation Filtration Process. In: Chemical Water and Wastewater Treatment, H.H. Hahn and R. Klute (eds.). Springer, Berlin Heidelberg New York 1990, pp. 471–481

[11] Barnard, J.L.: Biological Denitrification. Wat. Poll. Contr. *72* (6) (1973) 705–720

[12] Jimenez, B., Capdeville, B., Roques, H., Faup, G.M.: Design Considerations for Nitrification and Denitrification Process Using Two Fixed Bed Reactors in Series. Wat. Sci. Tech. *19* (1987) 139

[13] Pujol, R., Canler, J.P., Iwema, A.: Biological Aerated Filters: an Attractive and Alternative Biological Process. Water Quality International, IAWPRC Sixteenth Biennial Conference and Exhibition, Washington DC, 24–30 May 1992

[14] Sammut,F., Rogalla, F., Franci Gonçalves, R.: Phosphorus and Nitrogen Removal on Industrial Scale Upflow Biofilters. Aqua Enviro Technology Transfer Conference, Nutrient Removal from Wastewaters, Leeds, UK, September 1992

Ricardo Franci Gonçalves, François Sammut, and Frank Rogalla
Anjou Recherche
Research Centre Compagnie Générale des Eaux/OTV
Chemin de la Digue
F-78600 Maisons-Laffitte
France

Denitrification in Submerged Filters of Nitrified Wastewater and Chemical Pre-precipitated Wastewater

S.-E. Jepsen, K. Laursen, J. la Cour Jansen, and P. Harremoës

Abstract

To meet an effluent standard of 10–15 mg/l total nitrogen, as required in the EEC directive for ecologically sensitive areas, biological denitrification is the most cost-effective solution.

A compact solution for treating wastewater to meet this effluent standard is denitrification in submerged filters. Post-denitrification with an external carbon source as well as nitrification/denitrification of chemically pre-precipitated wastewater in biofilters are solutions to a compact treatment design problem. A major part of the phosphorus and suspended solids is removed during the pretreatment, while nitrogen and dissolved carbon are removed in the biofilm reactors.

Results from pilot plant investigations of post-denitrification and nitrification/denitrification after pre-precipitation show that both processes are feasible. Pilot plant experiments were performed in two different submerged biological filters, one down-flow reactor (Biocarbone) and one up-flow reactor (Biostyr). The post-denitrification filters were operated at loading rates up to more than 4 kg NO_3-N/m$^3 \cdot$ d at 10–17 °C with residual nitrate less than 5 mg N/l and external carbon in excess. The hydraulic load on the filters was up to 5 m/h. In the nitrification/denitrification operating mode, the nitrate loading rate was lower due to a less easily degradable carbon source, while the hydraulic load went up to 20 m/h.

The wastewater composition in the two experimental series has a major impact on the treatment efficiency. The wastewater used for the post-denitrification was nitrified wastewater with acetate as external carbon source. The wastewater used for the nitrification/denitrification experiments was chemically pretreated. The main difference was the amount of incoming easily degradable carbon source.

The head loss development in the two types of filter systems differs. In the down-flow filter, the head loss is stable until the surface clogs and then the terminal head loss is reached quickly. In the up-flow filter the head loss development is close to linear at constant load.

Introduction

In many countries, nitrogen removal from wastewater will be required in the future. The EEC wastewater directive gives a standard of 10 mg/l Total-N for large plants and 15 mg/l for small plants for discharge to ecologically sensitive waters, to be fulfilled before 1999. In Denmark, the national standard requires 8 mg/l for all treatment plants above 5 000 PE.

To meet such requirements, it is necessary to erect new plants or to extend existing wastewater treatment plants with nitrogen removal. Biological denitrification has proved to be the most cost-effective solution for nitrogen reduction. Different solutions are possible for this process. Today, activated sludge systems dominate but, during the last few years biofilm systems have had a revival. The technical basis for this development is a change of the former trickling filter to submerged filters where the processes are well controlled. This paper evaluates the performance of two different types of submerged filters for denitrification.

One of the major advantages of biofilm systems compared to activated sludge systems is the possibility of very compact design. This is important in cold climates and where space is limited in large cities or at existing plants. A large amount of experience exists with BOD-removal and nitrification in submerged biological filters with primary settled wastewater as well as pre-precipitated wastewater (Rogalla and Bourbigot (1990), Jimenez et al. (1987), Dillon and Thomas (1990)). In addition, some plants have been operated with denitrification, but only recently, plants have been built to meet effluent standards below 10 mg/l Total nitrogen (Rogalla et al. (1992)).

In submerged filters biological reaction and filtration can be performed in the same reactor. Backwash of the filters is needed in order to remove excess sludge and to keep the head loss at a proper level. Since filtration is influenced by the production of new biomass as well as by the incoming particles, submerged filters are favoured when the load of incoming particles is low.

Denitrification as a supplement to an existing nitrifying plant results in a very low load of incoming particles. Furthermore, the external carbon source can be selected so that sludge production is minimized.

A denitrifying submerged filter can also be advantageous in a recirculation system with a submerged filter as the nitrifying step. In such a system, pre-precipitation of the incoming wastewater is a good choice since it removes the phosphorus and reduces the particle load on the filter. Care should be taken in such cases to ensure enough degradable carbon for the denitrification after precipitation. The utilization of the incoming carbon will be high. If the content of degradable carbon in the influent water is not sufficient, it is possible to add an external carbon source and thereby lower the effluent nitrate.

Pilot Plants and Operation Scheme

Two different treatment systems were used in this investigation, a downflow filter with expanded slate as carrier material for the biofilm (Biocarbone) and an upflow filter with polystyrene pellets as carrier material (Biostyr). In both systems, the filter serves as a biological reactor where the nitrogen is removed and as a filtration unit which retains the suspended solids. Consequently, a secondary clarifier is not necessary.

The pilot plants were operated as pre-denitrification units as well as post-denitrification units as shown in Fig. 1. The main difference between the two operating modes is that in the recirculation mode, carbon is limiting while carbon is dosed in excess in the post-denitrification system. Therefore, the capacities are not directly comparable.

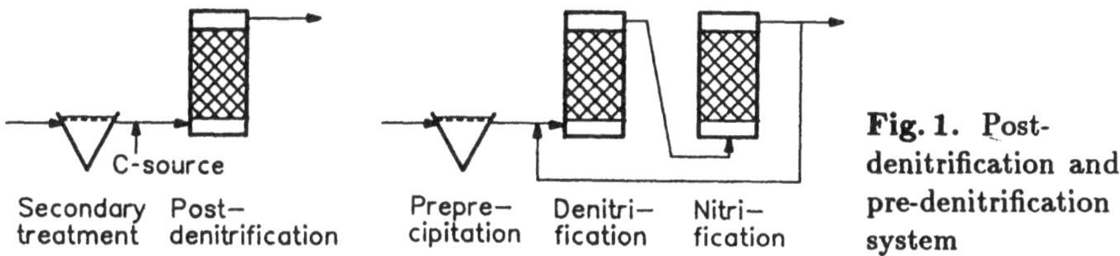

Fig. 1. Post-denitrification and pre-denitrification system

In the post-denitrification biofilter, wastewater from a nitrifying biofilter is introduced in the denitrifying biofilter together with an external carbon source.

In the recirculation treatment system, the wastewater is recirculated to denitrification from a nitrification biological filter or from a nitrifying zone in the same filter. The pre-precipitated wastewater is introduced in the denitrification filter where the incoming carbon source is used for nitrate reduction. The optimal recirculation ratio depends mainly on the necessary effluent standard, provided that sufficient carbon is available in the incoming wastewater or added.

In the post-denitrification operating mode, the composition of the influent wastewater is very stable. By dosing external carbon in excess it is possible to find the maximum capacity of the system under the most advantageous circumstances for bacterial growth.

Material and Methods

Pilot Plants

Two different pilot plants were used in this research. Differences and similarities were evaluated by operating the two plants in parallel. The physical data of the two systems are given below in Table 1. The size of the two treatment systems was identical, resulting in about the same treatment capacity. The two reactor types, Biostyr and Biocarbone, are developed by a French company, OTV, and further information about the systems is available in, for example, Rogalla et al. (1992) and Rogalla and Sibony (1992).

The Biostyr pilot plant has a diameter of 0.3 m. Polystyrene pellets are used as the carrier material and the height of this material is 2 m. To retain the pellets in the reactor, a cover was made which had nozzles for the passage of water. As illustrated in Fig. 2, the flow direction in a Biostyr treatment plant is up. In a nitrification/denitrification treatment plant, this makes it possible to establish nitrification in the top layer and denitrification in the bottom layer. In the course of time, the filter material is clogged by incoming particles and biomass. When the head loss reaches a pre-determined level, the filter is backwashed. The backwash level can be varied between 0.7 and 2.0 m of water pressure. During a backwash, air and water are flushed through the filter, air co-currently, but water counter-currently to the operational water flow.

Tab. 1. Physical data on the Biocarbone and Biostyr pilot plants

	Biocarbone	Biostyr
Flow direction	Down-flow	Upflow
Diameter	0.3 m	0.3 m
Height of carrier material	2.0 m	2.0 m
Carrier material	Expanded slate	Polystyrene pellets
Average size	4–6 mm	2–5 mm
Terminal head loss	0.7 m water	0.7–2.0 m water
Feed and recirculation pumps	Monopumps	Monopumps
Feed pumps operation range	60–350 l/h 0.84–4.95 m/h	60–350 l/h 0.84–4.95 m/h
Recirculation pumps	200–1200 l/h 2.8–17 m/h	200–1200 l/h 2.8–17 m/h

The dimensions of the Biocarbone pilot plant are also shown in Table 1. Expanded slate is used as carrier material. A flow scheme is given in Fig. 2. The recirculation pump was used only during the period of operation as a re-circulation plant. The flow direction in a Biocarbone treatment plant is down. After some time of operation, the filter material is clogged by incoming parti-cles, biomass production, and nitrogen bubbles. When the head loss reaches a pre-determined level the filter is backwashed. The head loss, where backwash is initiated, is usually around 0.7 m of water pressure. During a backwash, air and water are flushed through the filter counter-currently to the operational flow.

Backwash

The backwash procedures were fixed during the experiments. The filter run time was not optimized in this investigation. Short back flushes are a possibility for extending the filter run time significantly.

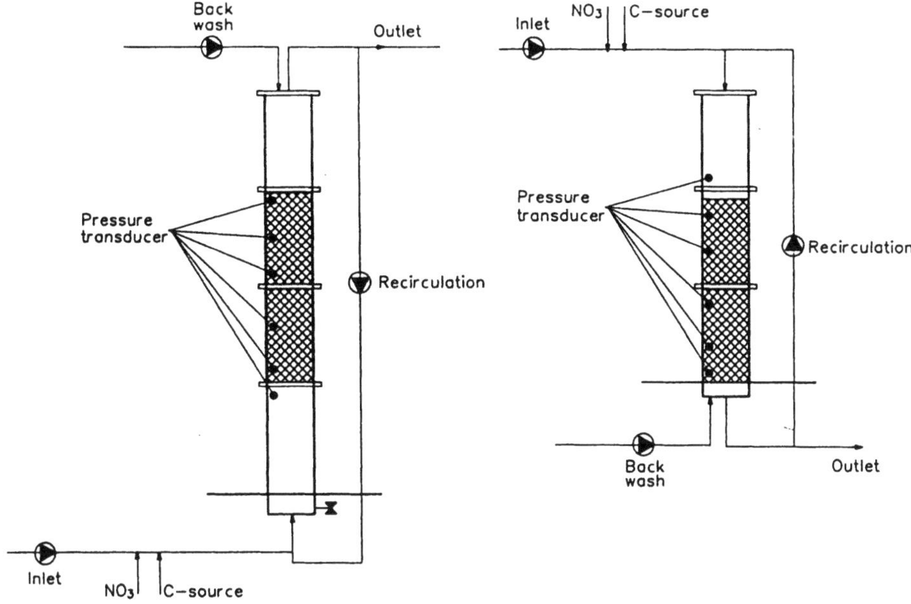

Fig. 2. Biostyr and Biocarbone pilot plants

Carbon Source

In the post-denitrification experiments, an external carbon source must be dosed. Acetate was selected in order to establish a broad bacterial population.

Wastewater Composition

In both the post-denitrification and recirculation experiments, raw wastewater from Lundtofte Wastewater Treatment Plant was used. Lundtofte is located in the northern part of Copenhagen, Denmark and receives mainly domestic wastewater. Its size is about 120,000 PE. The wastewater composition is given in Table 2 and is typical of Danish wastewater.

In the post-denitrification experiments, nitrified wastewater from a rotating disc, nitrifying pilot plant was used for feeding the pilot plants. At high loadings, the capacity of the rotating disc was insufficient and tap water and nitrate were dosed. Table 3 shows the average composition of the wastewater. Nitrate was varied in the given range and the high concentrations were used to find the maximum capacity of the pilot plants. The external carbon source, acetate, was generally dosed in excess, at a C/N ratio of more than 6.

In the recirculation experiments, chemical pre-precipitated wastewater was used. The raw wastewater was precipitated with iron salt (JKL from Kemira Kemwater). Instead of a nitrifying reactor in the recirculating system, nitrate was dosed to the wastewater. The resulting wastewater characteristics are given in Table 3.

Tab. 2. Mean values for raw wastewater
from Lundtofte Wastewater Treatment
Plant (from Kristensen et al. (1992))

	Raw wastewater mg/l
COD	446
COD (filtr.)	142
BOD	173
BOD (filtr.)	67
P_{tot}	10.1
P_{tot} (filtr.)	6.8
Kj-N	42
Kj-N (filtr.)	34
SS	216
VSS	179

Tab. 3. Wastewater characteristics

		Wastewater composition	
		Post-denitrification	Recirculation
COD	[mg/l]	33–700	75–352
COD (filtr.)	[mg/l]		124–260
PO_4-P, avg.	[mg/l]	1.6	5.0
pH, avg.	[-]	7.3	7.3
Temperature	[°C]	10–17	11–17
NO_3-N	[mg/l]	1–140	1–52
NO_4-N, avg.	[mg/l]	30	30
Flow	[l/h]	100–350	100–300
Recirculation	[l/h]		200–1000

Analytical Procedures

Samples were taken every day from the inlet to the pilot plants and from the
effluent from each pilot plant. A magnetic valve, opening 3 seconds every 15
minutes connected to a refrigerated container, was installed at each sampling
point.

Grab samples were used in intensive campaigns and for evaluating the effect
of shock loads on treatment efficiency.

Suspended solids and volatile suspended solids were analysed according to
Danish standards. Nitrate was analysed on a Technicon auto-analyser. Quick
analysing equipment from Dr. Lange was used for the daily operation.

Pressure Transducers

Six pressure measuring points in each pilot plant made it possible to follow the head loss development at different places in the filter media. All six measuring points were connected to the same pressure transducer (Danfoss). Magnetic valves (Danfoss) at each point were connected with a control and data logging system which made it possible to open each valve 10 seconds each minute and record the pressure at the measuring point. Usually a 10 minute average was stored. When all the measurements are made by the same transducer, the relative differences are due to differences in head loss, and not to the pressure transducers. The pressure transducer signal is converted into cm water pressure so that the results can be directly controlled by visual observations.

Results

Biostyr – Upflow

The Biostyr pilot plant returned very quickly to stable operation after changes in the operating conditions. Figure 3 shows the effluent nitrate content at different nitrate loadings when the pilot plant was operated as post-denitrification with acetate as carbon source as well as when it was operated as recirculation system under carbon limitation.

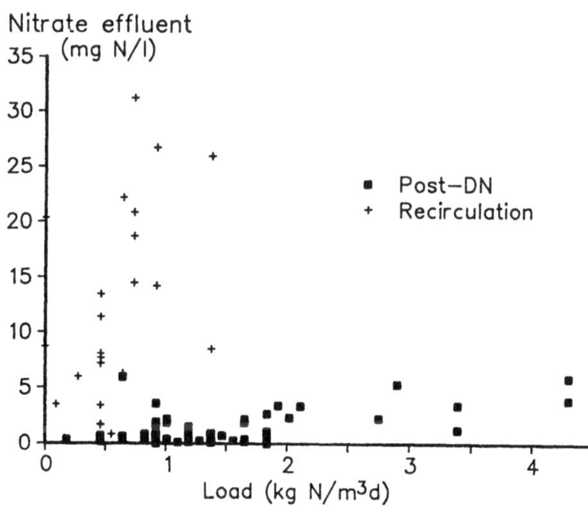

Fig. 3. Effluent quality at different loading rates for Biostyr pilot plant when operated as post-denitrification reactor (C/N ratio $\sim$ 6) and recirculation system (C/N ratio $\sim$ 3.5)

In the post-denitrification experiments, acetate was dosed in excess, while in the recirculation system, carbon was limiting growth. This difference explains the high effluent values in the recirculation period. When the incoming organics are used as carbon source, a large variation can be expected due to the differences in the influent concentrations. The consumed C/N ratio in the post-denitrification system was around 6 while the ratio was 3.5 for recirculation where carbon was the limiting substrate. The connection between C/N ratio applied and effluent nitrate in the recirculation system is illustrated in Fig. 4.

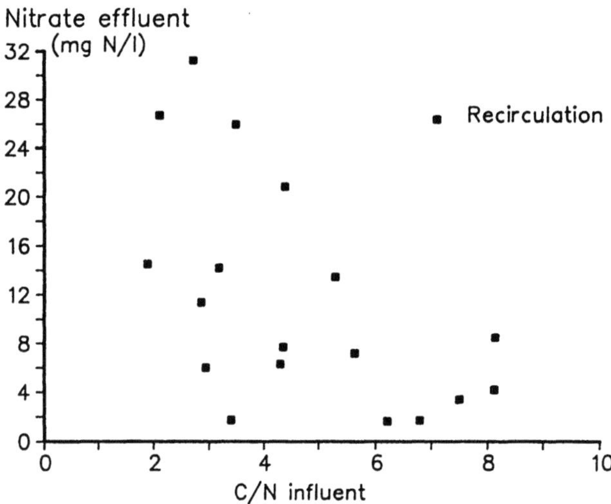

Fig. 4. Effluent nitrate as function of C/N ratio in the influent when Biostyr is operated as recirculation system. The large variations are due to variations in inlet concentrations of nitrate

In the recirculation system, the incoming and effluent carbon was characterized by the oxygen utilization rate (OUR); the method is described by Kristensen et al. (1991). The two major fractions in the incoming wastewater were easily degradable carbon and easily hydrolyzed carbon. In the effluent wastewater all easily degradable carbon was utilized, while around 30 % of the easily hydrolysed carbon passed through the filter.

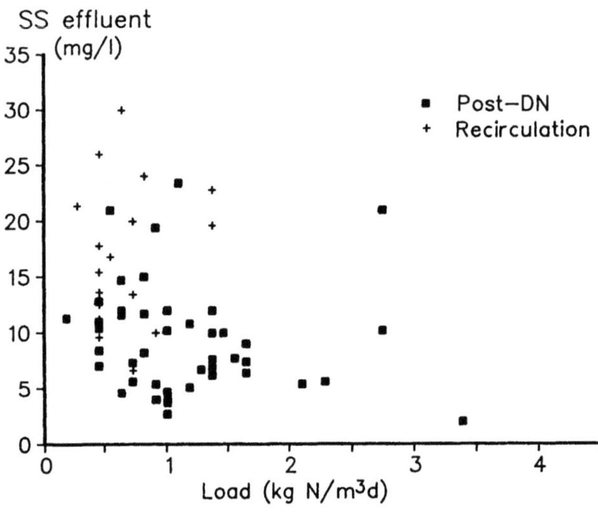

Fig. 5. Effluent suspended solids at different nitrate loading rates in the Biostyr pilot plant

Figure 5 shows the effluent suspended solids in the two operating modes. It can be seen that no significant change occurs as a function of the nitrate load. When operated as a recirculation system, the hydraulic load may be higher and this results in higher suspended solids in the effluent. In the recirculation mode, the denitrifying filter will always be followed by a nitrifying filter zone where solids removal will occur.

When backwashing, the head loss is lowered to between 5 and 30 cm water pressure depending on the hydraulic loading. Figure 6 shows the typical head loss between two backwashes. The increase in head loss is close to linear, whereby the slope of the curve is determined by the applied load of nitrate and the hydraulic loading. A major part of the head loss build-up is situated in the lower 50 cm of the filter.

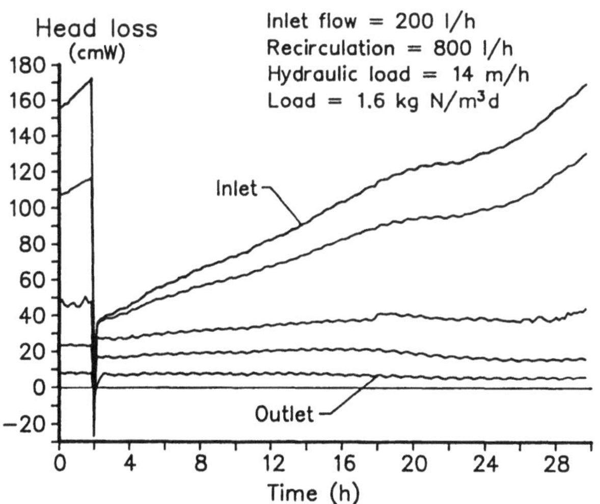

Fig. 6. Biostyr, head loss development between two backwashes at six different places in the pilot plant, situated at 0, 5, 15, 67, 117, and 192 cm from the inlet layer of the filter

The Biostyr pilot plant operated as a post-denitrification reactor was exposed to shock loads with stoichiometric amounts of nitrate and acetate. The week before this experiment was conducted, the filter was operated at stable low load, 1 kg NO_3-N/m3. Figure 7 shows the results from this experiment. The capacity for absorbing shock loads was high, since the load could be increased 5 times with only a minor deterioration of the nitrate effluent.

Biocarbone – Downflow

The results from the Biocarbone pilot plant are very similar to those of the Biostyr plant. Response was very rapid after changes in the operation conditions. Figure 8 shows the effluent nitrate content at different nitrate loadings when the plant was operated as post-denitrification with acetate as external carbon source as well as when it was operated as recirculation system under carbon limitation.

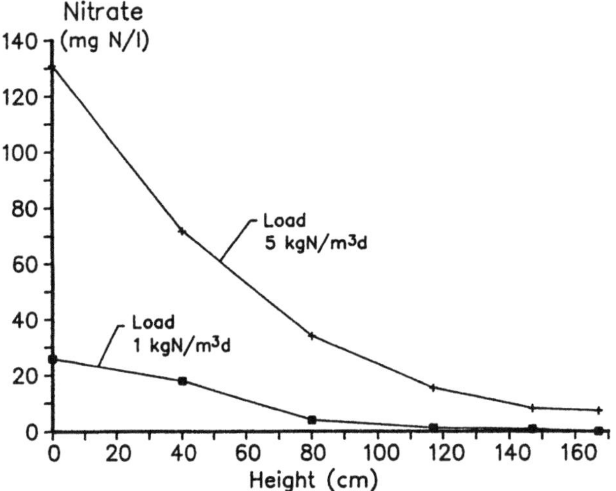

Fig. 7. Biostyr: Profile through the filter when a shock load was dosed

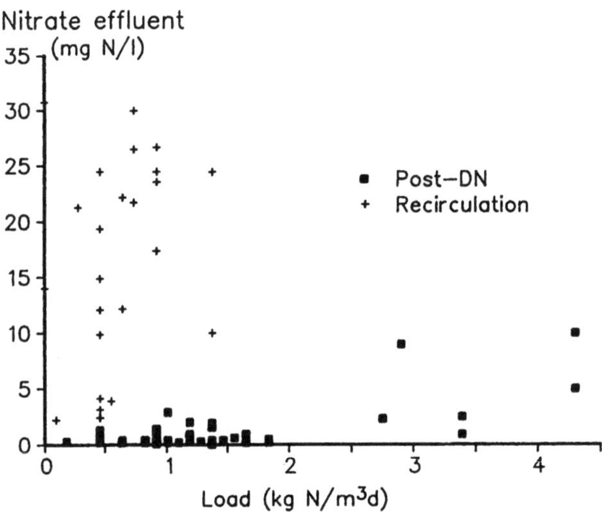

Fig. 8. Effluent quality at different loading rates for the Bio-carbone pilot plant when operated as post-denitrification (C/N ratio ∼ 6) reactor and as a recirculation system (C/N ratio ∼ 3.5)

In the post-denitrification experiments, acetate was dosed in excess, while carbon limited the growth in the recirculation system. This difference explains the high effluent values in the recirculation period. When the incoming carbon is used as carbon source, a large variation can be expected due to the differences in the influent concentrations. The consumed C/N ratio in the post-denitrification system was around 6 while the ratio was 3.5 in the recirculation experiments when carbon was the limiting substrate. The connection between influent C/N ratio and effluent nitrate in the recirculation system is illustrated in Fig. 9.

The two major fractions in the incoming wastewater, as characterized by the oxygen utilization rate (OUR), were easily degradable carbon and easily hydrolyzed carbon. In the effluent wastewater, all easily degradable carbon was utilized, while around 35 % of the easily hydrolysed carbon passed through the

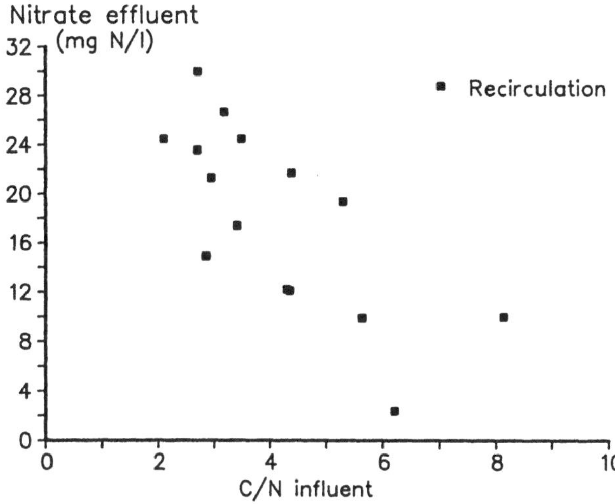

Fig. 9. Effluent nitrate concentrations as a function of influent C/N ratio in Biocarbone operated as a recirculation system. The large variations are due to variations in inlet concentrations of nitrate

filter. In the Biocarbone pilot plant, the utilization of carbon was about 5 % less than in the Biostyr pilot plant, and the effluent concentrations of nitrate were slightly higher, for the same loading rate and C/N ratio.

In Fig. 10, the suspended solids in the effluent is illustrated as a function of the applied load. No significant differences can be seen in suspended solids with increased nitrate load. When operated as a recirculation system, the hydraulic load may be higher and this results in higher suspended solids in the effluent. As for the Biostyr, it should be noted that the anoxic filter would be followed by an aerobic reactor.

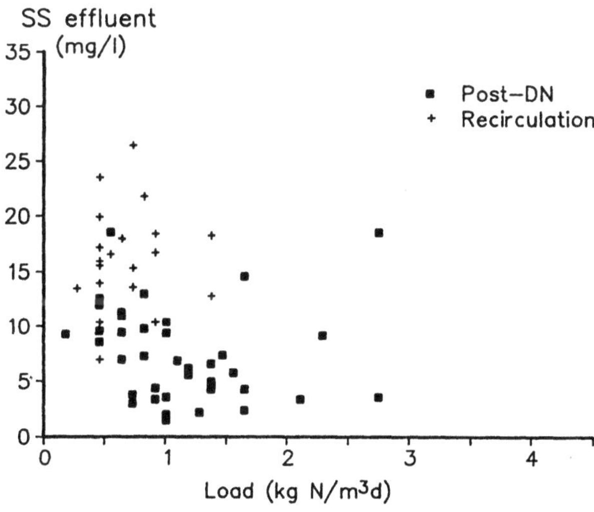

Fig. 10. Suspended solids in the effluent from the Biocarbone pilot plant in the two different operating modes with different nitrate loads

At the Biocarbone pilot plant the head loss is different from the linear headloss development at the Biostyr plant. Figure 11 shows a typical head loss curve. It can be seen that the head loss is very stable at the six measuring points in the reactor over a long period, but once the head loss starts to rise, it reaches terminal head loss very quickly. No significant change occurs in head loss in the depth of the filter.

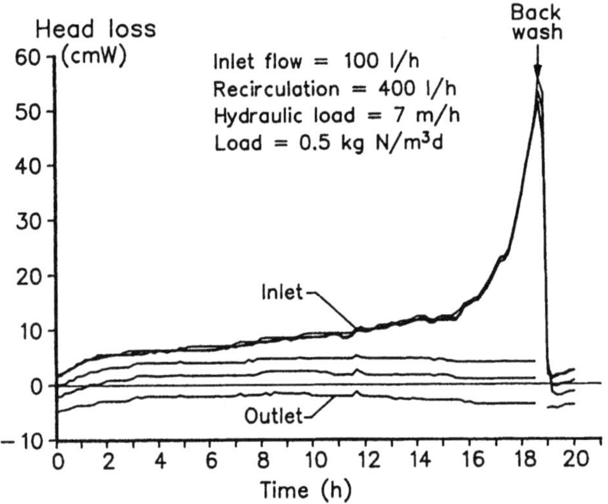

Fig. 11. Biocarbone, head loss development between two backwashes at six different places in the pilot plant, situated at 0, 5, 15, 65, 115, and 190 cm from the top of the carrier material

Shock loads of both nitrate and acetate were introduced in the Biocarbone filter. At extreme loads, more than 3–4 kg $NO_3N/m^3 \cdot d$, a breakthrough of nitrate occurred.

Discussion

Head Loss Build-Up

The development of head loss is caused partly by incoming suspended solids, and partly by biological growth and the nitrogen bubbles that are produced. Earlier investigations (Harremoës et al. (1980)) have tried to identify the influence of the three aforementioned phenomena. A rapid increase in head loss just after backwash was found. This was suggested to be due to the formation of nitrogen bubbles. In this investigation, such initial head loss in the first hours after backwash did not occur, see Figs. 6 and 11.

The head loss can be caused both by incoming suspended solids and by biomass growth. The influence of the two phenomena in the upflow and the downflow system may differ. In the Biostyr system, the head loss is close to linear with time, while the head loss in the Biocarbone system develops slowly until accumulation of solids at the top clogs the surface and prevents the release of bubbles. This causes head loss to reach the terminal value very quickly.

In the upflow system, the main part of the head loss is situated in the first half meter of the filter, see Fig. 6. At normal loading rates, most of the nitrate will be removed in the bottom section and thereby cause biomass growth and head loss development there. Furthermore, the filtration of suspended solids will be most efficient in the inlet layer.

A layer of solids develops on the top layer of the Biocarbone carrier material. After a while this layer makes it impossible for nitrogen gas, which is trapped inside the filter, to escape from the filter. This is illustrated in Fig. 11. A large part of the filter space will be occupied by the gas. At a typical post-denitrification load of 3 kg NO_3-$N/m^3 \cdot d$, the time between two backwashes is around 12 hours with hydraulic load at 5 m/h and 40 mg SS/l in the inlet. With a yield factor of 1.17 g VSS/g NO_3-N from activated sludge systems (McClintock et al. (1988)), the production of biomass in the filter is 250 g. The removal of suspended solids was about 20 mg SS/l. With a flow rate of 300 l/h this results in 75 g SS feed into the reactor. It is assumed that nitrogen for biological growth comes from ammonia, which means that all the removed nitrate is converted into nitrogen gas, N_2. The total production in one filter run will be about 175 l nitrogen gas. Before nitrogen forms bubbles, the liquid has to be saturated or super-saturated. At saturation, 6 % of the nitrogen can be removed by the effluent water.

Even though biomass production is three times as high as the influent suspended solids, visual observations indicate that suspended solids accumulate on the surface of the filter. This is expected to play a major role in the head loss build-up. When the surface layer is formed, the void volume below will rapidly be filled with gas. This phenomenon is especially connected with the post-denitrification mode, since operation of a recirculation system secures removal of a significant part of the nitrogen dissolved in the water. Figures 6 and 11 show typical filter run times for the two systems. Long term experience with the systems operated in parallel (same hydraulic loading and loading of suspended solids and nitrate) shows that the Biocarbone has a shorter run time than the Biostyr, due to the clogging of the surface layer in the Biocarbone.

Capacity

When operated as post-denitrification, both the upflow and the downflow systems are very efficient. When carbon source is dosed in excess it is possible to remove more than 4 kg NO_3-$N/m^3 \cdot d$. Here acetate was used as the external carbon source, but similar denitrification results are usually obtained with methanol after a lag phase (Andersson et al. (1991)). When operated below the maximum capacity, dynamic experiments have shown that extra capacity is available. Effluent quality below 5 mg NO_3-N/l is stable until the maximum capacity is reached.

If sufficient amounts of easily degradable carbon are present in the influent or if external carbon is dosed, the capacity of the recirculation system is comparable to that of the post-denitrification system.

The consumption of carbon source is nearly identical in the two systems. If the carbon in the influent drops, a very quick response in effluent nitrate occurs. The availability of the incoming carbon or external carbon source is of major importance. In the recirculation system, it has been shown that all easily degradable carbon is used for denitrification. The easily hydrolyzed carbon, which requires an exoenzyme for degradation, is not completely used. Approximately 65–70 % will be used for denitrification. This is expected to be due to the very short hydraulic retention time. If operated as a post-denitrification system, the carbon source must be easily degradable and dosed carefully.

Comparison Between Post-Denitrification and Recirculation

The major disadvantage of post-denitrification is the need to dose external carbon, and thereby the running costs will be higher.

The removal rate of the post-denitrification step can be very high if an easily degradable carbon source is selected. In the recirculation systems where the denitrification can be based on the incoming carbon, the removal rate may be lower due to the lower degradability of the carbon. Addition of external carbon in such systems will increase the removal rate.

The major problem of post-denitrification will be dosage of the right amount of carbon. This requires a reliable sensor, stable inlet concentrations or an aeration in the outlet to remove excess carbon.

The hydraulic load on a recirculation system may be higher than in post-denitrification. A normal recirculation ratio will be 2 to 4 depending on the effluent quality.

Conclusions

Both the Biostyr and the Biocarbone reactor are efficient solutions for denitrification. It has been shown that the filters can serve both as post-denitrification reactors and in nitrification/denitrification treatment plants.

The denitrification efficiency and capacity of the two reactor types are nearly identical. The major difference is the head loss build-up and duration of one filter run. Due to accumulation of solids on the surface of the filter in the downflow filter, the time of the filter run is shorter than in the upflow filter.

Submerged filters have their major application where space is limited or in cold climates. The advantage is that it is possible to keep a high biomass content in the reactor and to avoid the secondary clarifiers. A process which consists of chemical pre-precipitation followed by a filter will be the most space-saving solution. Wastewater composition, existing facilities, and price will determine whether a recirculation system or a post-denitrification system is the most favourable.

Acknowledgements

The study was carried out as part of the EUREKA project, SIMBIOSE. The SIMBIOSE project includes the following companies and research institutions: Générale des Eaux (Anjou Recherche/OTV), France; Danfoss, Denmark; EAWAG, Switzerland; the Technical University of Denmark, Denmark; and the Water Quality Institute, Denmark.

References

Andersson, B., Aspegren, H., Berg, L., Gustelius, A.: Efterdenitrifikation i sandfilter. Försök i ett DynaSand-filter. Denitrification in a DynaSand Filter, 1991, VATTEN 47:14-23. Lund

Dillon, G.R., Thomas, V.K.: A Pilot-Scale Evaluation of the "Biocarbone Process" for the Treatment of Settled Sewage and for Tertiary Nitrification of Secondary Effluent. Wat. Sci. Tech. *22* (1/2) (1990) 305–316

Harremoës, P., la Cour Jansen, J., Kristensen, G.H.: Practical Problems Related to Nitrogen Bubble Formation in Fixed Film Reactors. Prog. Wat. Tech. *12* (1980) 253–269

Jimenez, B., Capdeville, B., Roques, H., Faup, G.M.: Design Considerations for a Nitrification-Denitrification Process Using Two Fixed-Bed Reactors in Series. Wat. Sci. Tech. *19* (1987) 139–150

Kristensen, G.H., Jørgensen, P.E., Henze, M.: Characterization of Functional Microorganism Groups and Substrate in Activated Sludge and Wastewater by AUR, NUR and OUR. IAWPRC Conf. on: Interactions of Wastewater, Biomass and Reactor Configurations in Biological Treatment Plants. Copenhagen, 21–23 August, 1991

Kristensen, G.H., Jørgensen, P.E., Strube, R., Henze, M.: Combined Pre-precipitation, Biological Sludge Hydrolysis and Nitrogen Reduction – A Pilot Demonstration of Integrated Nutrient Removal. Wat. Sci. Tech. *26* (5/6) (1992) 1057–1066

McClintock, S.A., Sherrard, J.H., Novak, J.T., Randall, C.W.: Nitrate Versus Oxygen Respiration in the Activated Sludge Process. J. Wat. Poll. Contr. Fed. *60* (1988) 342–350

Rogalla, F., Bourbigot, M.-M.: New Developments in Complete Nitrogen Removal with Biological Aerated Filters. Wat. Sci. Tech. *22* (1/2) (1990) 273–280

Rogalla, F., Sibony, J.: Biocarbone Aerated Filters – Ten Years After: Past, Present, and Plenty of Potential. Wat. Sci. Tech. *26* (9/11) (1992) 2043–2048

Rogalla, F., Badard, M., Hansen, F., Dansholm, P.: Upscaling a Compact Nitrogen Removal Process. Wat. Sci. Tech. *26* (5/6) (1992) 1067–1076

Svend-Erik Jepsen
Karin Laursen
Jes la Cour Jansen
Water Quality Institute
11, Agern Allé
DK-2970 Hørsholm
Denmark

Poul Harremoës
Department of Environmental Engineering
Technical University of Denmark
Building 115
DK-2800 Lyngby
Denmark

Observed Synergistic Effects
of Aluminium and Iron Salts
in Nutrients Removal

M. D'Elia and A. Isolati

Foreword

The simultaneous removal of nutrients in a standard active sludge plant has always been the aim of all people attending to the operation of these plants, but the interaction of the particular mechanism and kinetics, both chemico-physical and biological, has always lead to only partial solutions, often causing undesired side effects such as "bulking" or "foaming". For example, the synthesis of nitrates and their presence in the anoxic area prevents the biological removal of phosphates, whereas the constantly aerobic or even anoxic conditions of the system cannot prevent the forming of filamentous protozoa anyway, which, in turn, results in "bulking". This problem becomes greater both because of the increasingly reduced limits admitted for the phosphorus and nitrogen content (up to 0.5 ppm P and 10 ppm NH_3), and because of the tendency towards an imbalance in the BOD/TKN ratio of the municipal sewage waters fed to the treatment plants. To date, however, the main concern was and still is phosphates removal, for which the solution most often adopted consists of the addition of chemical conditioning compounds to the water, whereas less attention has been paid to the removal of nitrogen.

A number of systems have been proposed for simultaneous removal of P and N, but all of them are based on often remarkable modifications of the plants, whereby aerobic or anaerobic sections (A/O, PHOSTRIP, PHORE-DOX, etc.) would be added to the existing processing cycle, with very high investment costs. The availability, therefore, of a simple and flexible technology to improve nitrification/denitrification processes, phosphates hydrolysis and co-precipitation should be greatly appreciated by treatment plant managers [1].

Chemical Treatment

The purpose of adding chemical conditioning agents is generally to agglomerate the material particles present in the sewage waters as a colloidal dispersion and to remove the orthophosphates by forming insoluble salts, which can be separated from the liquid phase through the mechanism of coagulation-flocculation. The chemical reaction by which insoluble orthophosphates are formed takes place just before the chemical-physical coagulation-flocculation process.

The compounds commonly used for this purpose are the inorganic salts of iron (bivalent and trivalent) and of aluminium (sulfate, chloride, polyhydroxide chloride and aluminate), the characteristics of such salts generally being different. The iron salts, for example, lead very readily to the formation of well developed flocs with a favourable sludge index, thus tending to sediment quickly.

The effect of sludge "weighing down" due to the presence of the iron ions can readily be seen from a typical pattern of the sludge index (SVI) [2] (Fig. 1); this figure shows the SVI data both when $FeCl_3$ is added (beginning and end) and in the absence of the ferric salt (central period). As you can see, in the presence of iron salts, SVI remains stable at around 100. This fact is very important since, with a "heavy" sludge available, the phenomena of dragging and rising towards the surface are much easier to control even under critical conditions of hydraulic load.

A negative side effect is the greater volume of sludge which is produced, which involves higher treatment and disposal costs. Lastly, the compounds which the iron salts can form with orthophosphates have a very low solubility, and, for this reason, are clearly separated from the liquid phase: Ks $FeSO_4$ = $0.8 \cdot 10^{-22}$ at 20 °C. On the other hand, the capacity of the iron salts for charge neutralization seems not to be particularly relevant, because these salts are poorly efficient as regards the removal of highly charged colloids and thus show contrary behaviour in comparison with the aluminium salts.

The flocs resulting from aluminium salts are lighter and slower to form than those resulting from iron salts, since the iron hydroxide is already formed at pH 3.2 while the aluminium hydroxide appears at pH values higher than 4.5 (with the maximum activity at pH values between 5.5 and 6.5). The advantage of these compounds is shown in a higher efficiency in the neutralization of the surface charges (tendency of Z potential to be 0) as indicated in Fig. 2 A, and, consequently, a higher efficiency in the removal of the orthophosphates and turbidity. Among the available aluminium salts, the polyaluminium hydroxide chloride (PAC) seems to be the most effective, since polynuclear ions with very high positive charge are also formed at pH 7–8 [3, 4] (Figs. 2 A, 2 B, 2 C).

The properties of aluminium and iron salts seem, therefore, to be complementary, with the Fe ions being the best sludge conditioners and the Al ions the most effective in charge neutralization and coagulation-flocculation processes. This paper will discuss the advantages that can be obtained by their synergic use, especially in existing and overloaded plants.

Experimental

Two series of full scale trials were carried out for several months using both aluminium and iron salts (Polyaluminium hydroxide chloride and ferric chloride) in two different waste water plants. The choice of PAC as the source for Al ions is due to its superior performance in comparison to other Al salts. The choice of ferric chloride as the source of Fe ions arises from the slower growth of nitrifying bacteria induced by ferrous salts (chloride and sulphate) [5]. The plants treat sewages from two towns situated on the Adriatic Coast, where coastal waters suffer severe eutrophication problems. Both plants are overloaded during the summer.

The first trial was carried out in Plant A where ferric chloride was used only for phosphorus removal in the so-called simultaneous precipitation system.

Plant A consists of fine grating, an aeration basin, a sedimentation sludge aerobic stabilizer, and sludge dewatering. The inlet flow rate varies from less than $2\,000\ \mathrm{m^3/d}$ in the winter up to $10\,000\ \mathrm{m^3/d}$ in the summer, with flow peaks over $15\,000\ \mathrm{m^3/d}$. Ferric chloride was dosed in the summer (June–September) at the inlet point of the aeration basin, at a concentration of 120 ppm. During the trial, 50 ppm of ferric chloride were dosed at the same injection point plus 30 ppm of PAC on the activated sludge recycling line (Fig. 3).

The second trial was carried out in Plant B, where PAC was used for phosphorus removal in pre-post precipitation. Plant B consists of coarse and fine grating, lifting, a sand-oil collector, primary sedimentation basin, aeration basin, secondary sedimentation basin, sludge aerobic stabilizer, and sludge dewatering. The inlet flow rate varies from less than $5\,000\ \mathrm{m^3/d}$ in the winter up to $10\,000\ \mathrm{m^3/d}$ in the summer, with flow peaks over $20\,000\ \mathrm{m^3/d}$. PAC was dosed in the summer (June–September) at the inlet points of the primary and secondary sedimentation basins at a concentration of 50 ppm.

During the trial, 50 ppm of ferric chloride were dosed at the inlet point of the aeration basin and 50 ppm of PAC in the activated sludge recycling line (Fig. 4). It was very difficult to manage the plant during flow peak periods and treatment with the PAC + $FeCl_3$ system was therefore planned for July and August to check its performance in the worst situation.

During the two trials, corresponding samples of inlet and outlet water were monitored over 24 hours with regard to the following parameters: pH, COD, orthophosphates, total phosphorus, ammonia, nitrites, nitrates, suspended solids.

Results

The data obtained during the two trials with the PAC-$FeCl_3$ system were compared with those collected by the plants' management staff during the period of the previous year.

For Plant A, the inlet flow rate was actually the same in both periods (Fig. 5) and even the behaviour of the SVI (Fig. 6) was quite similar (confirming the peculiar ability of iron salts to increase sludge sedimentability). As far as COD/phosphorus and nitrogen removal are concerned, the PAC + $FeCl_3$ system seems to be significantly more efficient than only $FeCl_3$ conditioning (Figs. 7, 8, 9). Furthermore, the depuration efficiency remains quite constant during the experimental period, showing a peculiar characteristic of the PAC + $FeCl_3$ system to dampen the effects of daily and weekly variations of inlet flow rate and organic matter content. The most unexpected result concerns nitrogen reduction. The PAC + FeCl$_3$ system improved the efficiency of the nitrification/denitrification processes by about 30 %.

A summary of the results obtained in Plant A is given in Fig. 10. As far as Plant B is concerned, the inlet flow rate was about 2 000 m^3/d higher during the trial than in the previous year, whereby in both periods the flow rate increased sharply in the middle of August (Figs. 12 and 13). Thus it can be said that the trial was carried out under the worst conditions. The SVI values are consistent with the flow rate trends (Fig. 11), namely, with a flow rate of 10 000 m^3/d, the SVI value increases sharply to 130–150 with FeCl$_3$-PAC treatment, while above 10 000 m^3/d, the SVI values remain constant. A summary of the results obtained in Plant B is given in Fig. 14.

Discussion

The full scale trials carried out in two different plants having different lay-outs, but both overloaded in flow rate and COD, showed that the PAC + FeCl$_3$ system may be a helpful alternative for improving plants' depurative efficiency. Regarding COD and phosphorus removal in the plant without primary sedimentation (Plant A), the PAC+FeCl$_3$ system seems to be much more efficient in simultaneous precipitation than FeCl$_3$ conditioning. The depurative efficiency, furthermore, does not suffer the daily or weekly variations of inlet flow rate and COD.

In Plant B, due to the presence of a primary clarifier, the effect of the PAC + FeCl$_3$ system on COD/P removal is obviously less evident. As far as nitrogen and suspended solids removal are concerned, the PAC + FeCl$_3$ system seems to be the best chemical treatment for both plants studied. Owing to the complexity of the mechanism and processes involved, it is difficult to interpret the phenomena. It is, however, possible to make some suppositions.

The Fe salts seem to act on the physical properties of the sludge, particularly on the settling behaviour, resulting in the ready formation of bulky and heavy flocs. The Al salts seem to improve the effectiveness of the chemical and chemico-physical processes, which may explain the improved reduction of the phosphates. It is known that nitrogen removal takes place through the biological oxidation of ammonia nitrogen to nitrates (nitrification) under aerobic conditions, and the subsequent biological reduction of nitrates to molecular nitrogen (denitrification) under anoxic conditions. The growth and maintenance of colonies of nitrifying bacteria (Nitrosomonas and Nitrobacter) are difficult, since these are influenced by limiting factors such as dissolved oxygen, temperature and pH. The growth of autotrophic bacteria (for nitrification) is also significantly slower than that of heterotrophic bacteria (for organic matter). Furthermore, it appears that populations of nitrifying bacteria mainly develop on the finest particles of sludge.

From this point of view, it is clear that a chemical conditioning agent which improves the effectiveness of suspended solids removal (such as the Al salts) leads to the enrichment of the sludge with nitrifying bacteria, which means increasing artificially, together with the effect of the $FeCl_3$ on the SVI, the so-called "sludge age". In addition, the increased exchange with the atmosphere can be mentioned, owing to the reduction of foams due to the removal of anionic surface active agents. All of the mechanisms involved in the $PAC+FeCl_3$ system thus take place in the presence of a sludge having an optimum SVI index, which prevents or limits the release and the upward motion of the finest and lightest flocs.

The overall result is a stabilized process with higher depurating effectiveness. The results obtained using aluminium polyhydroxide chloride (PAC) and ferric chloride simultaneously and separately show that the purification yields achieved are higher not only than those obtained using a single chemical conditioning agent, but also than the sum of the results of the single additives.

Acknowledgements

Thanks are due to the Municipalities of Cattolica and Cesenatico (Italy) and to the managers and employees of municipal sewage treatment plants for their collaboration during the trials.

Bibliography

[1] D'Elia, M., Isolati, A: Improved Process for Water Treatment. (Patent pending)
[2] Puzzarini, P.: Municipality of Cervia – Trial for Phosphate Removal. (Personal communication)
[3] Masschelein, W.J.: The Use of Polymerised Aluminium Flocculants. (Personal communication)
[4] Caffaro S.P.A. – PRODECO Water Treatment Department: The PRODEFLOC AC Line-Technical Report
[5] Schertenleib, R.: Experience de la dephosphatation a l'echelle d'installations pilotes A.R.P.E.A.

Mauro D'Elia and Antonio Isolati
PRODECO Water Treatment Department
Strada per Poasco
I-20 097 San Donato Milanese
Italy

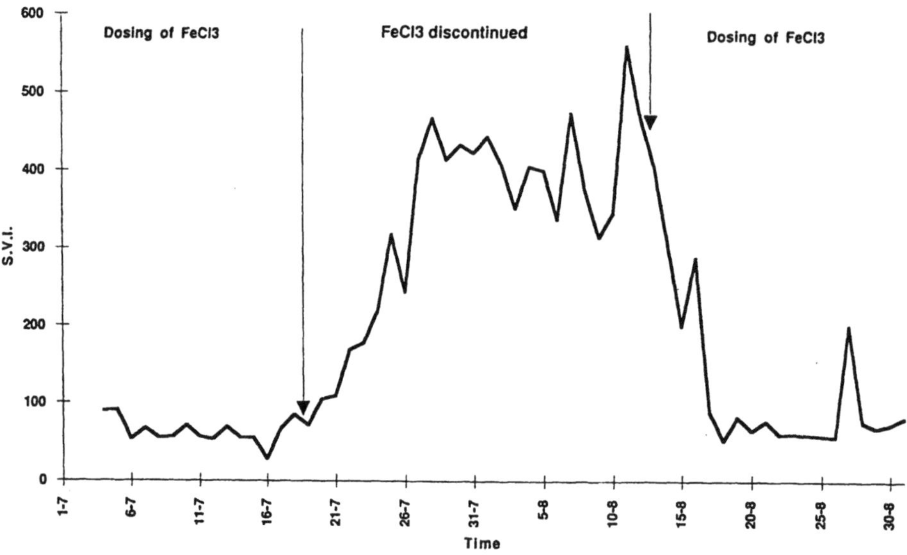

Fig. 1. S.V.I. with and without FeCL$_3$ dosing

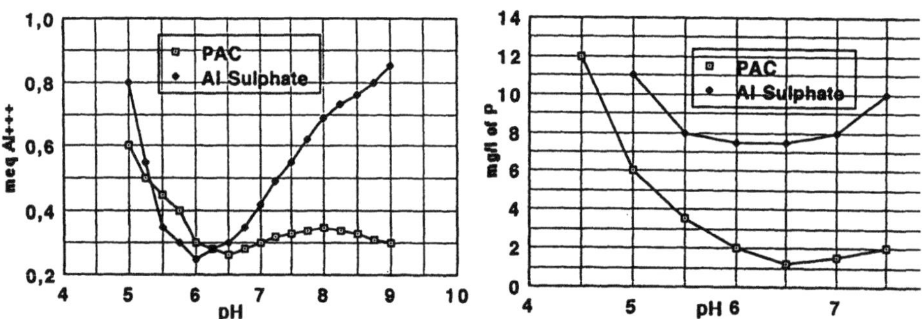

Fig. 2A. Dosage for Z potential → 0   **Fig. 2B.** Efficiency of phosphorus remov

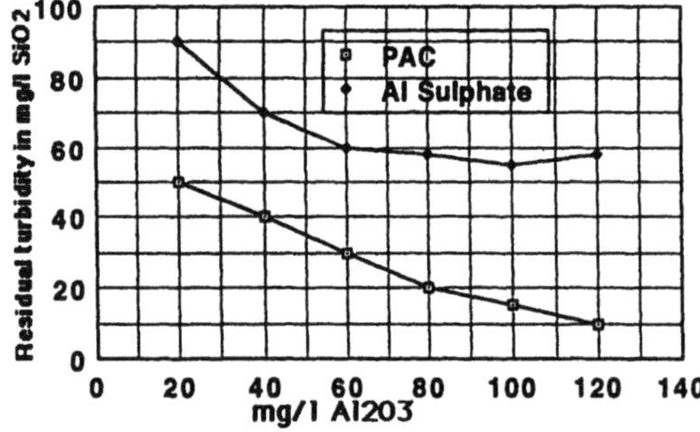

Fig. 2C. Efficiency of turbidity abatement

PLANT A

a) SIMULTANEOUS PRECIPITATION BEFORE TRIAL (1989)

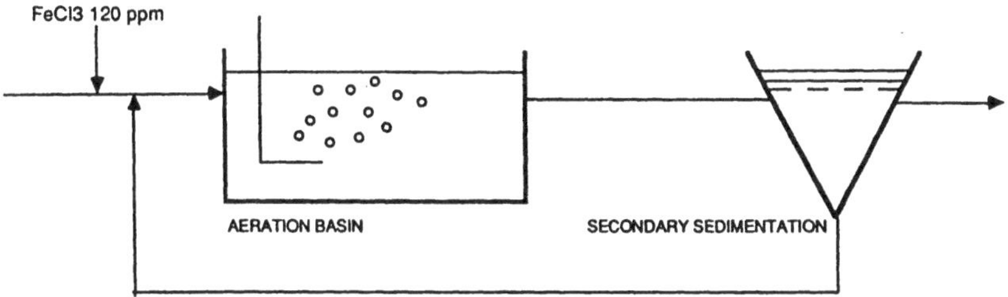

b) SIMULTANEOUS PRECIPITATION DURING TRIAL (1990)

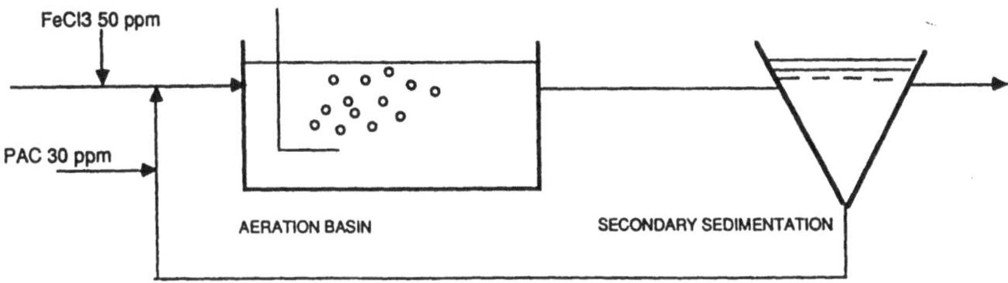

Fig. 3. Plant A lay-out

PLANT B

a) PRE - POST PRECIPITATION BEFORE TRIAL (1990)

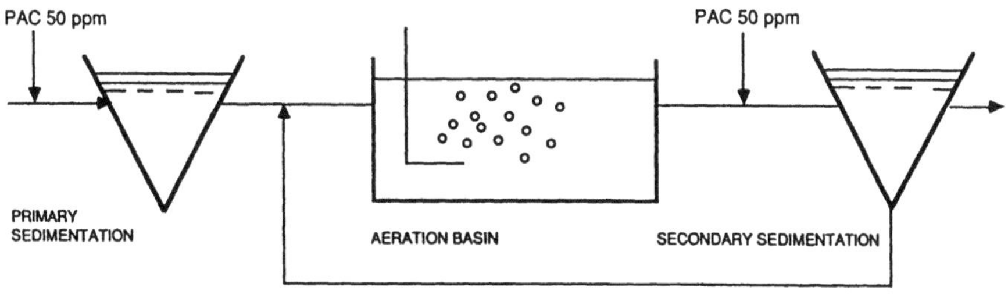

b) SIMULTANEOUS PRECIPITATION DURING TRIAL (1991)

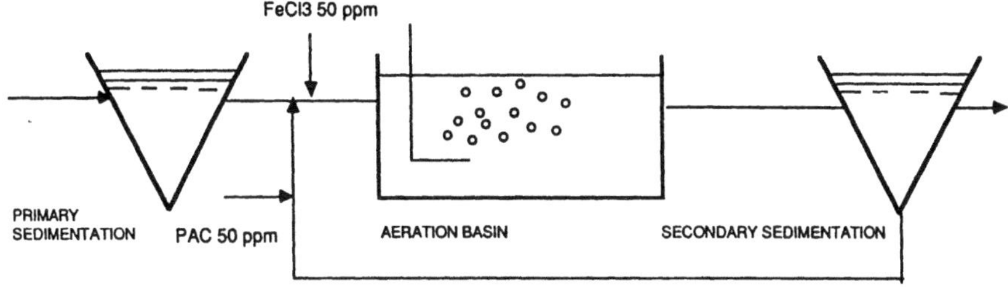

Fig. 4. Plant B lay-out

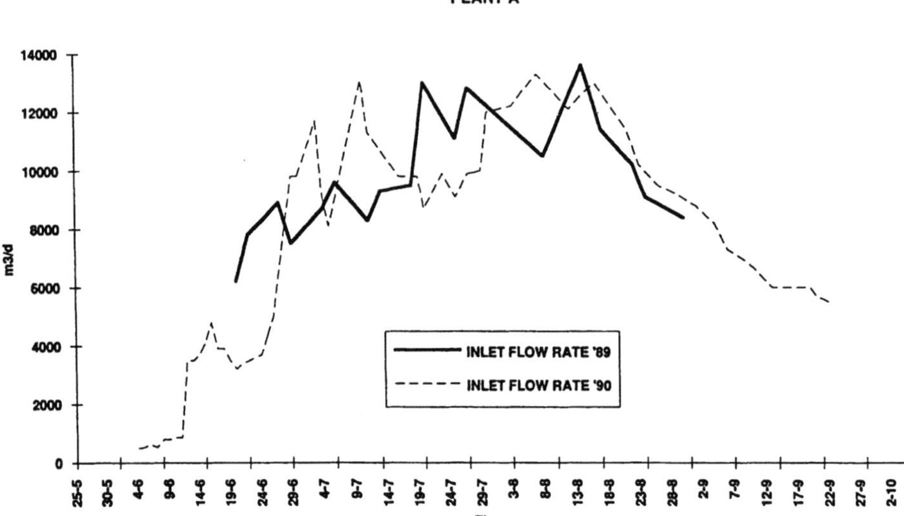

Fig. 5. Inlet flow rate

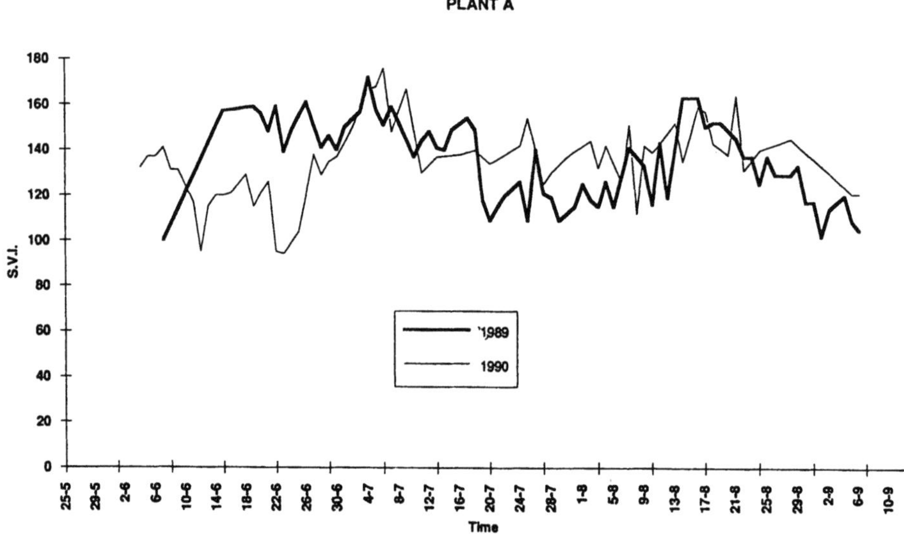

Fig. 6. S.V.I. during FeCl$_3$ conditioning (1989) and PAC + FeCl$_3$ trial (1990)

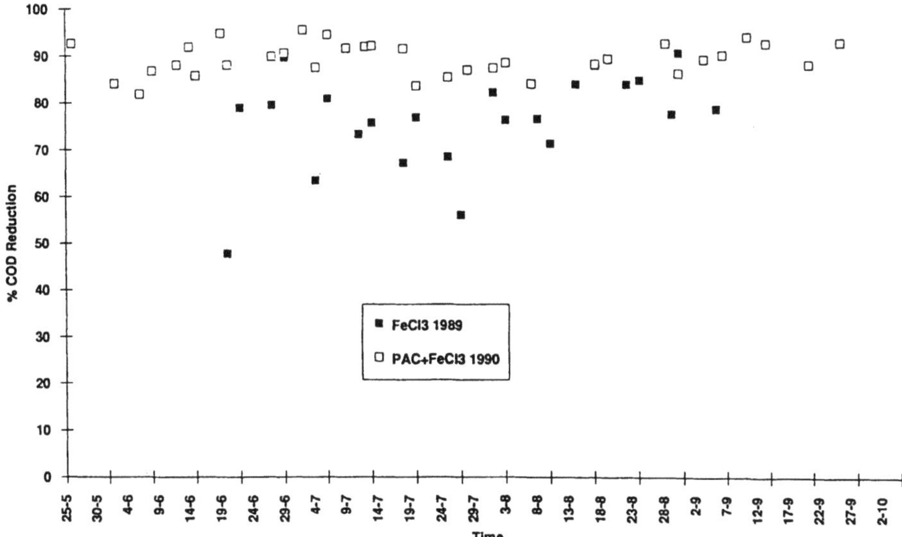

Fig. 7. Efficiency of COD reduction (Plant A)

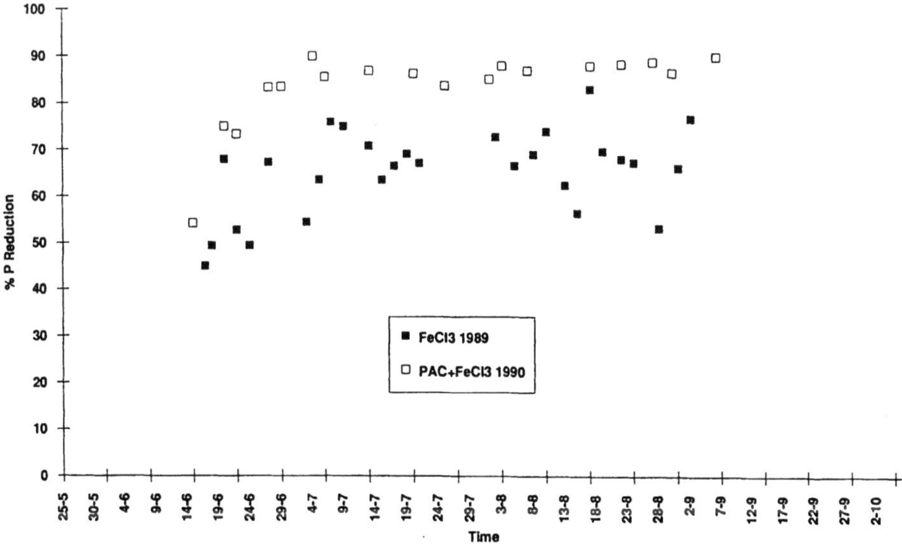

Fig. 8. Efficiency of phosphorus reduction (Plant A)

    M. D'Elia and A. Isolati

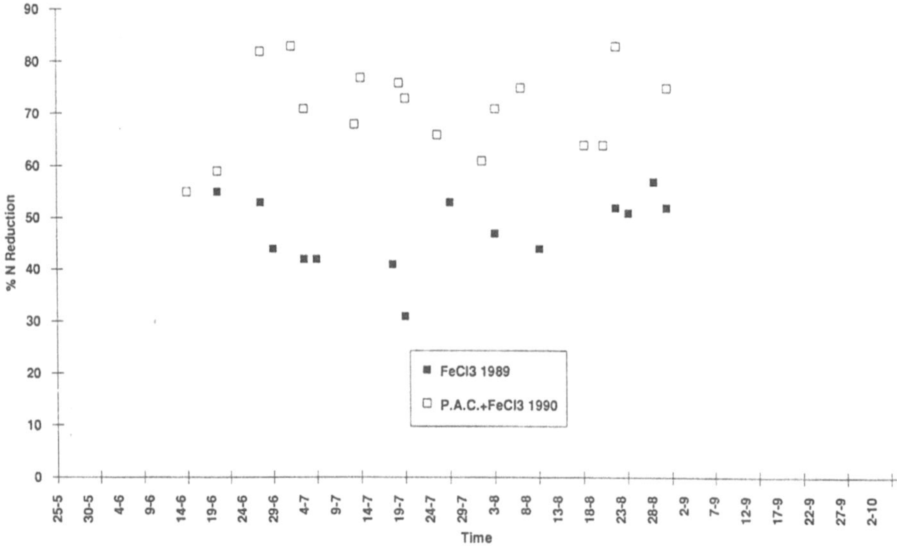

Fig. 9. Efficiency of nitrogen reduction (Plant A)

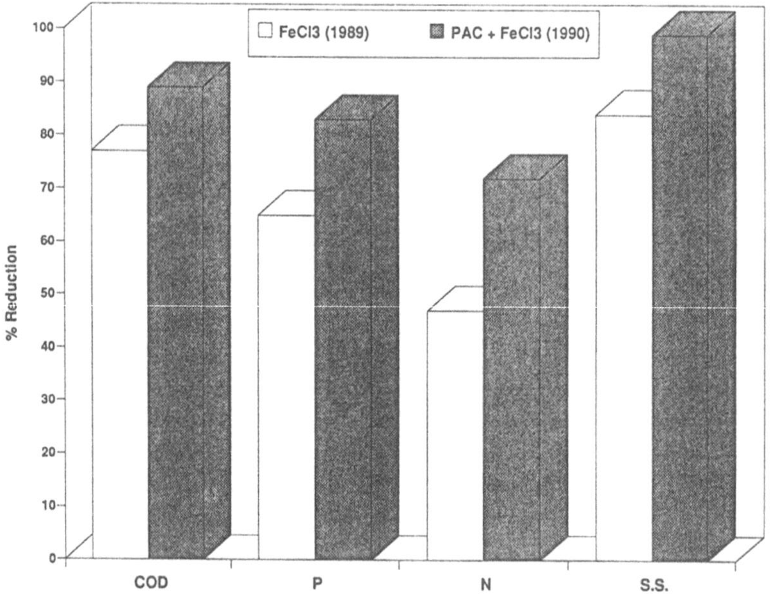

Fig. 10. Summary of the results (Plant A)

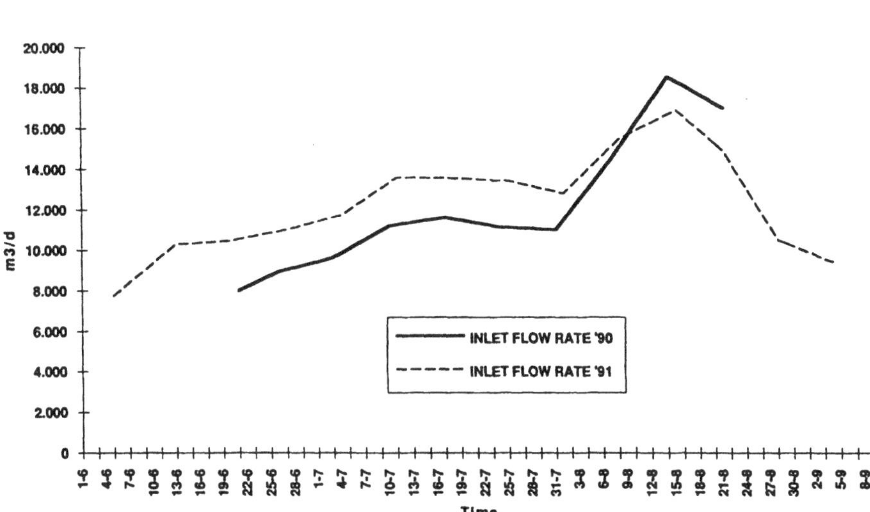

Fig. 11. Inlet flow rate

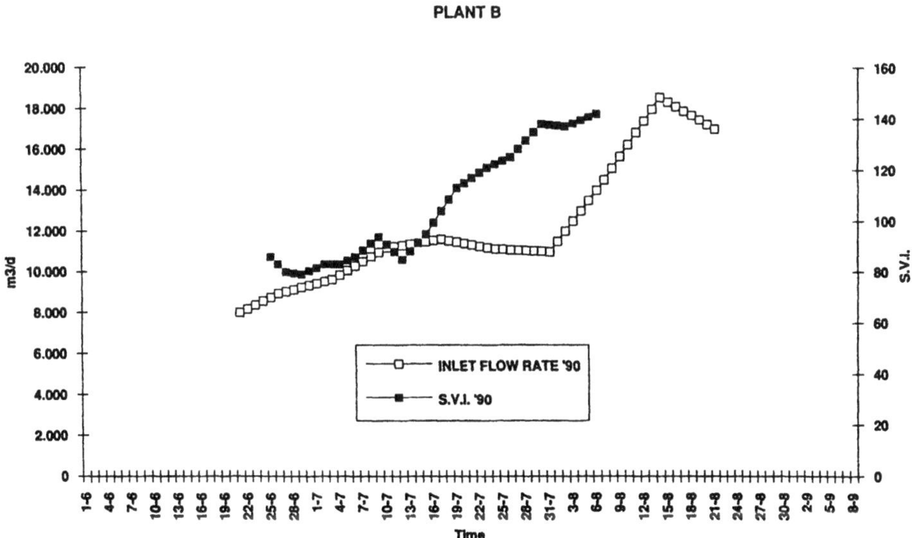

Fig. 12. S.V.I. versus the inlet flow rate for 1990

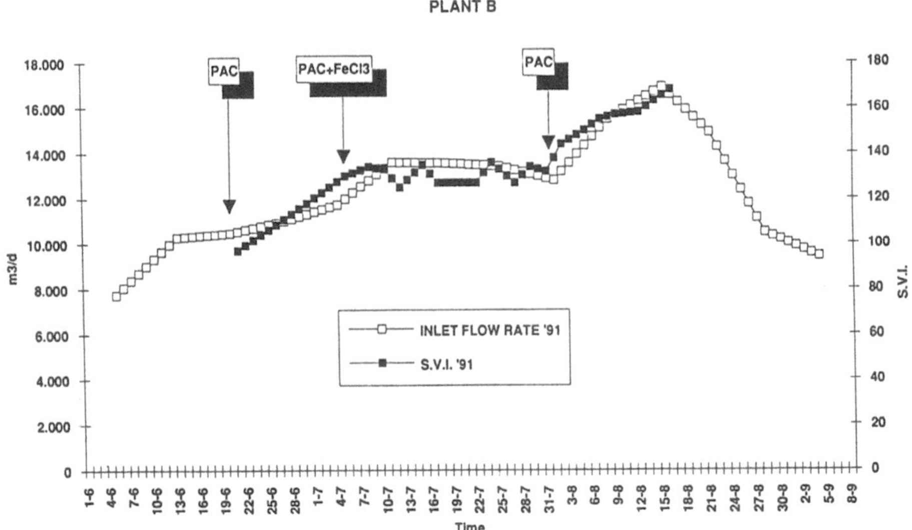

Fig. 13. S.V.I. versus the inlet flow rate for 1991

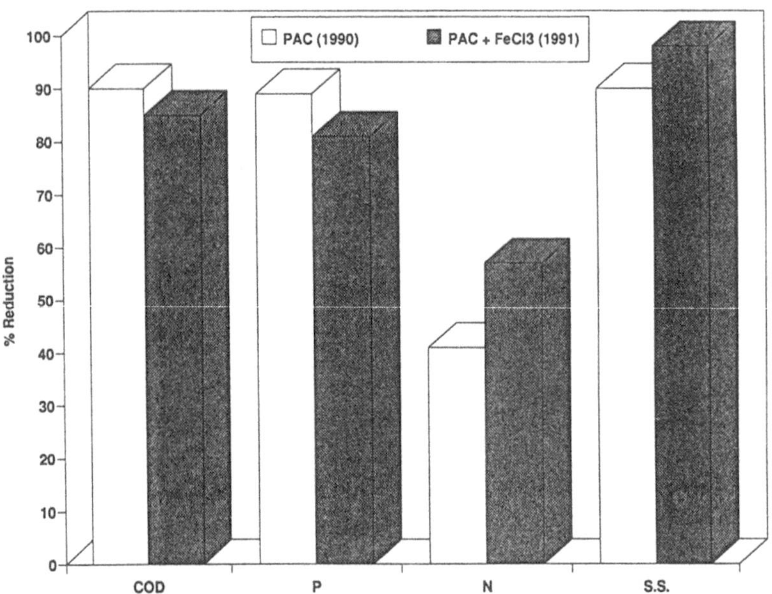

Fig. 14. Summary of the results (Plant B)

Retrofitting Conventional Primary Treatment Plants for Chemically Enhanced Primary Treatment in the USA

S.P. Morrissey and D.R.F. Harleman

Abstract

Chemically enhanced primary treatment (CEPT) has been in use for more than 100 years. However, it fell into disfavor in the 1930s when compared to biological treatment because the large concentrations of metal salts produced large quantities of sludge. In the USA, most of the emphasis on chemical treatment has been on phosphorus removal after biological treatment. However, Scandinavian and other European countries are focusing on phosphorus removal before biological treatment. Again, large concentrations of metal salts are needed to achieve low phosphorus effluents. With the technological advancements in polymer chemistry, it has become worthwhile to revisit chemically enhanced primary wastewater treatment with total suspended solids (TSS) and biochemical oxygen demand (BOD) removal as the focus. By replacing some of the metal salt with polymers, significant increases in TSS and BOD removals, compared to conventional primary settling, are possible. In addition, these higher removals are obtained at overflow rates twice that of conventional primary settling.

The paper discusses the regulatory requirements in the United States and the role CEPT has played for three municipalities in meeting these requirements. Finally, the paper discusses the overall perception of CEPT in the USA and emphasizes a number of points important in implementing this process in the USA.

Introduction

The proceedings of the four prior Gothenburg Symposia contain no examples from the USA of the use of chemical additives in municipal wastewater treatment in other than the conventional post-precipitation nutrient removal mode. The use of direct precipitation, a single stage, physical-chemical wastewater treatment process commonly used in Scandinavia, is virtually unknown in the USA. Pre-precipitation, also used in Europe, is a two stage process in which chemical precipitation augments sedimentation in the primary stage and reduces solids and BOD loading on the subsequent biological treatment stage. Pre-precipitation may be used to reduce the space, energy and cost of conventional primary-secondary treatment. Although pre-precipitation is employed to a limited extent in the USA, it has usually been implemented by treatment plant operators by retrofitting conventional primary sedimentation tanks for chemical additives. Engineering literature within the USA on pre-precipitation treatment processes is very limited. Consequently, state and federal environmental regulatory agencies and consulting engineering firms engaged in the design of treatment plants tend to ignore or downplay this innovative and evolving treatment technology.

In the USA, municipalities needing to build new wastewater treatment plants, or upgrade or expand existing facilities, generally retain the services of a consulting firm, on the basis of prior association or reputation to select the treatment technology and produce the plant design. The element of competitive bidding comes later at the construction stage in which there is little potential cost saving. This type of public works facilities acquisition procedure appears to be a disincentive to innovation. This is evidenced by the planned construction, between 1991–1999, of a completely new wastewater treatment facility for the metropolitan Boston area, with a capacity of 50 m^3/s, using conventional primary settling and activated sludge treatment technology. The writers have made the case [1] that use of pre-precipitation would result in large cost, space and time saving in the Boston Harbor cleanup plan.

Another aspect of disincentive to innovation in the USA is related to the enactment of the 1972 Federal Clean Water Act. This act has had many positive benefits and without question has resulted in significant water quality improvements in the nation's waterways. This was achieved by imposing uniform, technology based secondary treatment and effluent standards for all publicly-owned facilities regardless of the type, nature or size of the receiving waters.

Regulatory Background

The U.S. Environmental Protection Agency (EPA) defined the uniform treatment and effluent standards of the Clean Water Act on the basis of what was achievable using conventional primary settling followed by activated sludge. Specifically, the law requires 85 % removal of TSS and BOD$_5$ as well as 30 day average effluents of 30 mg/l for TSS and BOD$_5$.

Soon after the 1972 act was passed, many municipalities argued to Congress and the EPA that secondary treatment was not universally necessary particularly for the protection of the coastal environment. They contended that high removal of BOD was of little benefit to the coastal ocean where treatment plant effluents are mixed and dispersed by tidal currents and aerated by large water surface areas. They also pointed out that long outfall pipes could terminate in coastal areas of significant depth and that these areas exhibit substantial tidal flushing action. Furthermore, multi-port diffusers could be attached to the outfall in order to reduce the concentration of treated effluents by more than a hundredfold through the process of jet mixing. A number of communities in

Southern California that were discharging primary effluent through ocean outfalls had accumulated evidence that demonstrated the scientific merit of their claims for exemption from secondary treatment requirement. Congress was persuaded and in the Clean Water Act of 1977 directed EPA to allow dischargers to test their case in the administrative process.

In response to the 1977 Congressional directive, the EPA published guidelines in 1979 by which municipalities could apply for waivers of the full secondary treatment requirement for discharges into coastal waters. These stated that applicants "will bear a particularly heavy burden in demonstrating to the EPA that such (less-than-secondary) treatment is sufficient to protect marine waters". Applicants were not allowed to compare the environmental impacts of less-than-secondary with secondary effluents through the same outfall, thus contradicting an accepted principle of environmental impact analysis. The possibility that incremental benefits of secondary treatment might be negligible, or unjustifiably costly in relation to other environmental or societal needs, was not factored into the decision process. Of the 208 applicants filed, only 25 % were approved and the majority of these were small treatment plants discharging less than 0.2 m^3/s. The remainder were withdrawn or denied. In many cases, the EPA's decision process took five or more years thus placing a number of large urban areas beyond the time frame of the federal cost sharing program that initially paid 75 % of the capital treatment plant costs.

There is little doubt that the uniform, technology based standards legislated by the Clean Water Act reduced USA research and innovation in municipal wastewater treatment technology. In addition, the legislated priority for full secondary treatment has made it difficult to deal with local needs that are in conflict with the federal. Two prime examples are Boston and San Diego. In Boston, it is generally recognized that the major cause of harbor pollution is the frequent wet weather overflows of untreated waste from the old inner city combined sewer system. This contamination will continue even after the expenditure of 6 billion dollars to construct a new secondary plant and a 15 km ocean outfall to comply with federal regulations. The San Diego case involves a conflict between ocean disposal of treated water and water reclamation and reuse. The San Diego case is discussed in a following section.

Chemically Enhanced Primary Treatment in the USA

Physical-chemical wastewater treatment has been in use for more than 100 years [2]. This process fell into disfavor in the 1930s when it was compared to biological treatment because large quantities of sludge were produced when large concentrations of metal salts were added. Today, chemical treatment in the USA is primarily thought of in terms of phosphorus control [3–6] with the focus on phosphorus removal after biological treatment [7]. Again, large concentrations of metal salts (> 200 mg/l) are needed to achieve low phosphorus effluents [7]. With the recent technological advancements in polymer chemistry, it has become worthwhile to revisit chemical wastewater treatment with total suspended solids (TSS) and biochemical oxygen demand (BOD) removal as the major focus of concern. Replacing some of the metal salt with polymers results in excellent TSS and BOD removal, less volume, and more manageable sludge [8, 1].

The focus of this paper is on pre-precipitation with low dose flocculants (< 50 mg/l) and polymer additives in primary settling tanks. In the USA, this is usually referred to as chemically enhanced primary treatment (CEPT). We are interested in calling attention to some successful examples of CEPT brought about by retrofitting existing facilities as a means of upgrading treatment performance and expanding plant capacity. While European experience in pre-precipitation is extensive, we believe that little attention has been given to the impact of CEPT in sustaining high removal efficiencies at overflow rates two to three times that of conventional primary settling.

Point Loma Treatment Plant, San Diego California

Point Loma was designed as a conventional primary plant. The average annual flow at Point Loma is 9 m^3/s and this plant services a population of approximately 2 million. In 1979, Point Loma requested a federal waiver from full secondary treatment, based on a 4 km ocean outfall discharging into 60 m of water, under the premise of meeting California Ocean Plan effluent standards. The State plan requires 75 % removal of suspended solids, a level intermediate between the 60 % possible with conventional primary treatment and 85 % removal required by secondary treatment. The State plan contains no standard for BOD removal for ocean discharges but does require that dissolved oxygen not be changed by more than 10 % below ambient levels.

In 1985, plant operators began retrofitting Point Loma for CEPT to meet the California Ocean Plan. The plant consists of bar screens, aerated grit chambers, an influent channel leading to rectangular primary sedimentation tanks, sludge digesters and the ocean outfall. At minimal capital cost, Point Loma began adding 35 mg/l of FeCl$_3$ before the grit chambers and 0.26 mg/l of an anionic polymer in the influent channels (Fig. 1). The average annual influent TSS and BOD$_5$ concentrations are 300 mg/l and 280 mg/l respectively. The annual average removals for 1989 were 80 % for TSS, 57 % for BOD, 42 % for fats, oils, and grease (FOG), and 75 % for total phosphorus (TP). This was obtained at an average surface overflow rate of 70 m/d as compared to the standard design of 30 m/d for conventional primary plants. During a 30 day test, Point Loma was able to maintain high removal efficiencies at surface overflow rates of 110 m/d. Figure 2 illustrates the significant increase in Point Loma's performance over typical conventional primary performance.

In 1987, EPA denied San Diego's waiver and a Clean Water Program for San Diego was established to (1) comply with full secondary treatment for wastewater prior to discharge to the ocean, (2) to plan for water reclamation

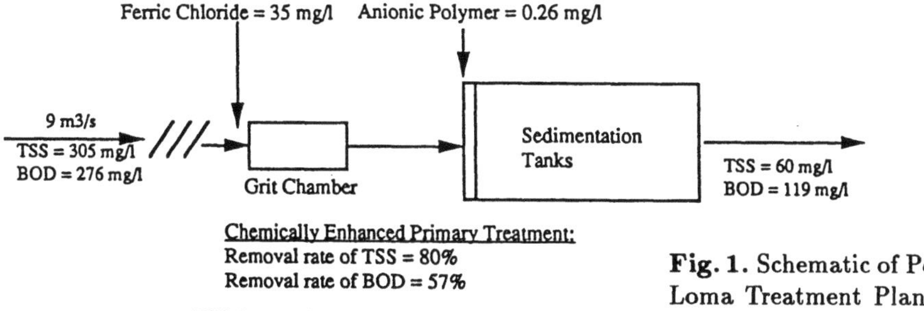

Fig. 1. Schematic of Point Loma Treatment Plant, San Diego California

by building tertiary treatment plants for a portion of the total sewage flow in order to lessen dependence upon imported water, and (3) provide a regional sludge management plan. The estimated cost of this facility plan is 2.5 billion dollars, to be paid by local rate payers.

After six years of CEPT and monitoring the San Diego coastal zone, scientists, including those from Scripps Institution of Oceanography, contended that effluent from the outfall does not harm the oceans and that any marginal benefit of secondary treatment to the ocean would be more than offset by the environmental impact of disposing of the larger amount of sludge generated by secondary treatment. In addition, strong claims were made that investment in small tertiary treatment plants for water reclamation and reuse should have a much higher priority for San Diego than full secondary treatment for water being discharged to the ocean.

In 1991, a federal judge with urging from environmental organizations required the City to complete a master plan which set reuse of reclaimed water as a higher priority than secondary treatment and required the Point Loma plant to implement a program to improve the effectiveness of CEPT. San Diego is currently in the process of fulfilling these requirements.

Point Loma's raw sludge production from CEPT can be calculated based on the stoichiometric Eq. (1). This equation has predicted their annual raw sludge production with less than 10 % error.

$$RS = 1.0\,TSS_{rem} + 1.42\,P_{rem} + 0.66\,FeCl_{3in} \tag{1}$$

RS	Raw sludge concentration [mg/l]
TSS_{rem}	Influent TSS minus effluent TSS [mg/l]
P_{rem}	Influent phosphorus minus effluent phosphorus [mg/l]
$FeCl_{3in}$	Concentration of metal salt ($FeCl_3$) added [mg/l]

This stoichiometric equation is obtained by assuming that the addition of $FeCl_3$ results first in the formation of $FePO_4$ and then any excess Fe^{3+} will react to form several species of ferric hydroxides which are simply represented as $Fe(OH)_3$. This assumption yields Eq. (2).

    S.P. Morrissey and D.R.F. Harleman

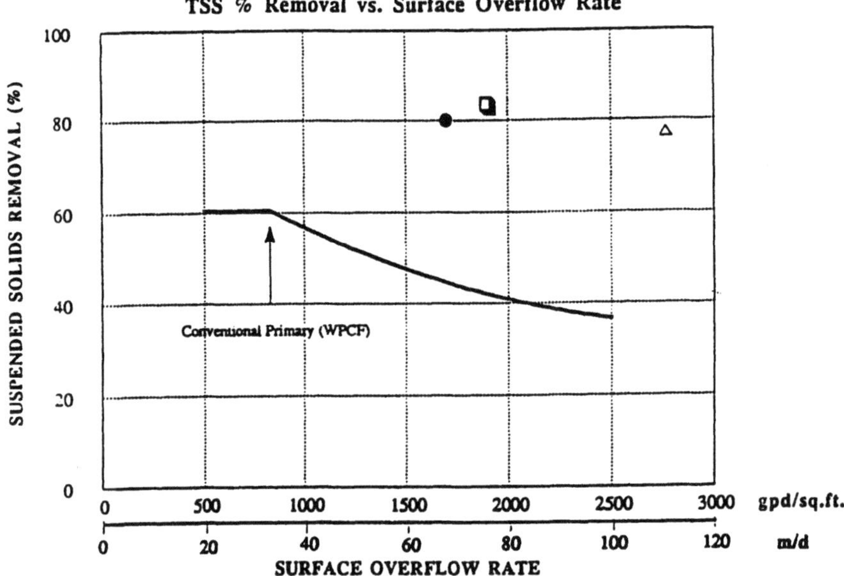

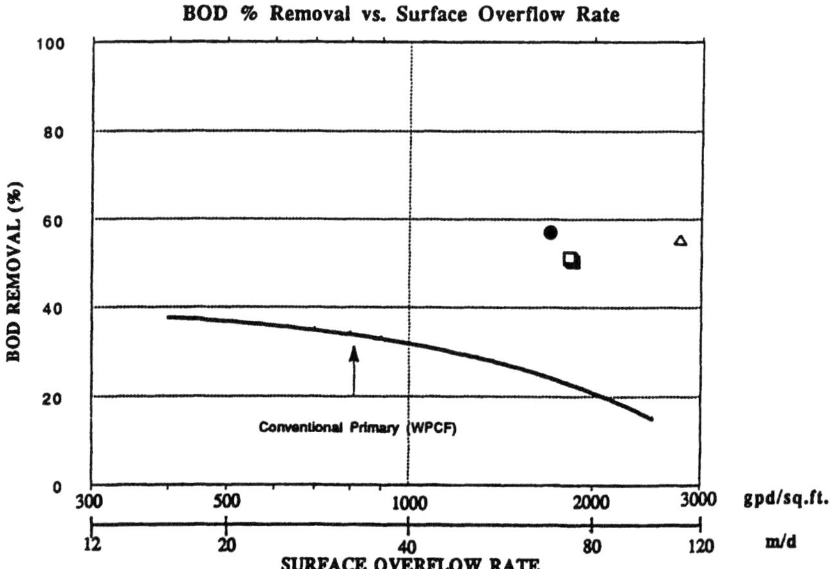

Fig. 2. Point Loma and Hyperion Chemically Enhanced Primary Treatment Plants

$$S = m\,FePO_4 + \frac{P_{rem}}{m\,P} + m\,Fe(OH)_3 \left[\frac{FeCl_{3in}\left(\frac{m\,Fe}{m\,FeCl_3}\right)}{m\,Fe} - \frac{P_{rem}}{m\,P} \right] \quad (2)$$

S   Chemical sludge concentration [mg/l]
m   Molecular weight

Combining terms and substituting the known molecular weights into Eq. (2) yields Eq. (3) which is represented in Eq. (1).

$$S = 1.42\,P_{rem} + 0.66\,FeCl_{3in} \quad (3)$$

Typical values for RS and S at Point Loma are 267 mg/l and 23 mg/l respectively. If Point Loma did not add chemicals, a typical value for RS would be 183 mg/l assuming 60 % TSS removal.

Hyperion Treatment Plant, City of Los Angeles California

Hyperion is the largest CEPT plant in the USA, servicing over 4 million people and currently treating an average of 16 m^3/s. The City of Los Angeles also operates three water reclamation facilities upstream of Hyperion that return their solids to the sewer lines transporting wastewater to Hyperion. All of the flow entering the plant receives chemical treatment and 50–60 % of the flow also receives biological secondary treatment using a conventional activated sludge process. The plant consists of bar screens, detritor type grit chambers, influent channels leading to conventional primary sedimentation tanks, aeration basins, secondary clarifiers, sludge digesters and an ocean outfall. Hyperion began using chemicals in 1985 and currently uses 20 mg/l of FeCl$_3$ added before the bar screens and 0.25 mg/l of a anionic polymer in the influent channels (Fig. 3). The annual average influent concentration for TSS and BOD are 270 mg/l and 300 mg/l respectively. Hyperion has an annual average removal of 83 % TSS, 52 % BOD and 80 % TP. This is obtained at an average surface overflow rate of 75 m/d. The success of CEPT at Hyperion in improving solids and BOD removal efficiencies since 1985 is shown in Fig. 4 [9]. Figure 2 illustrates the significant increase in Hyperion's performance over typical conventional primary performance.

Because of reduced BOD loading due to CEPT, installation of a fine air bubbling system in the aeration tanks and some modifications to the effluent weirs, Hyperion has been able to double the flow through their existing activated sludge system.

Hyperion has determined [10], that the increase in raw sludge production from CEPT, as compared to conventional primary treatment, is due to increased solids capture, chemical addition and colloidal capture. Equation (4) is used to predict raw sludge production at Hyperion. The addition of the colloidal capture (C_{rem}), a term not found at San Diego or other facilities using FeCl$_3$

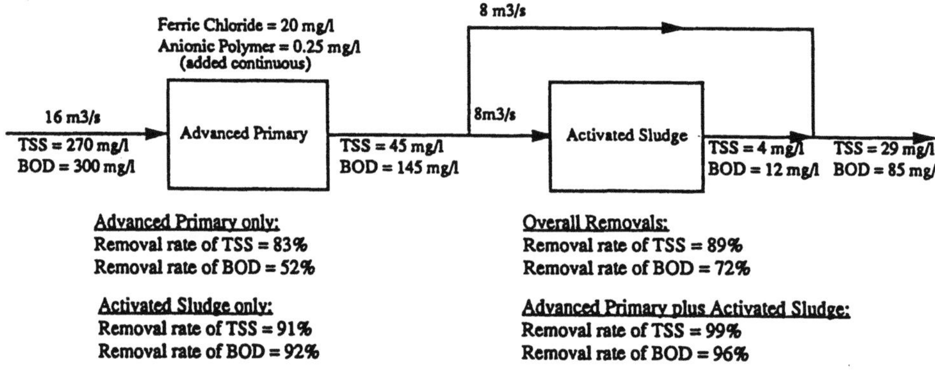

Fig. 3. Schematic of Hyperion Treatment Plant, Los Angeles California

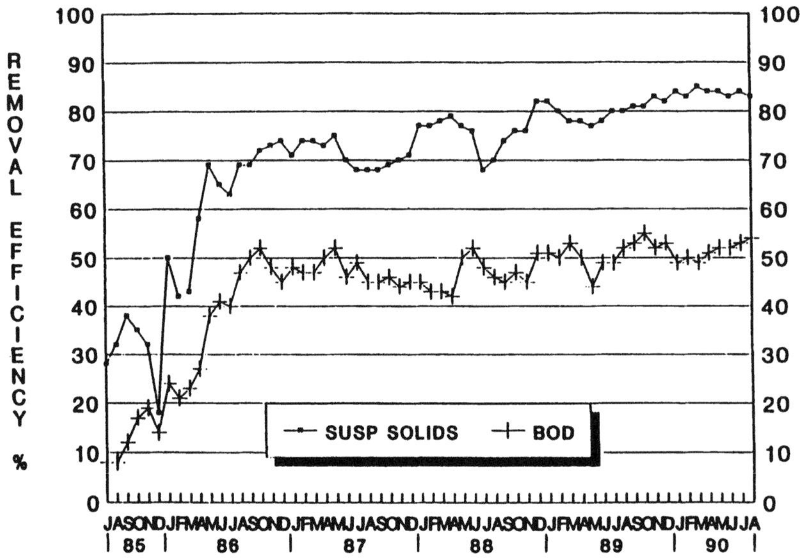

Fig. 4. Primary TSS and BOD removal efficiencies at Hyperion Treatment Plant

(Eq. (1)), may be due to the colloidal solids released into the sewer system from sludge produced at three upstream water reclamation facilities.

$$RS = 1.0\,TSS_{rem} + 1.42\,P_{rem} + 0.66\,FeCl_{3in} + 1.0\,C_{rem} \tag{4}$$

The following shows how sludge is generated due to chemical addition at Hyperion.

Conventional primary treatment

Sludge due to 60 % solids capture: 1.00

Chemically enhanced primary treatment

Additional sludge due to 83 % solids capture:	0.38
Sludge due to chemical addition:	0.11
Sludge due to colloidal capture:	<u>0.15</u>
Total sludge generated by CEPT:	1.64

Comparisons have been made [10] on total sludge production at Hyperion on the basis of full activated sludge secondary treatment with and without chemical addition in the primary stage. Before digestion, the sludge produced by CEPT plus secondary is about 10 % greater than with conventional primary plus secondary. After digestion, the difference is only 6 % due to recovery of the volatile content of the CEPT sludge.

The City of Los Angeles was denied its waiver in 1987 and the City has chosen to comply with full secondary treatment requirements and plans to complete the new facility by 1998. Their plan is to demolish the existing conventional activated sludge facility and construct a 24 m^3/s high purity oxygen activated sludge plant and increase the number of primary sedimentation tanks to maintain a lower surface overflow rate at a cost of 2 billion dollars.

Despite the successful seven year experience with CEPT, it is stated [10] that the new facility will operate without chemical addition as a conventional primary-secondary plant. This decision was based on an unsound economic analysis in which the benefits of CEPT in reducing the BOD loading and therefore the size of the activated sludge system were ignored.

South Essex Sewerage District (SESD), Salem Massachusetts

The SESD waiver was denied in 1986 and the district is currently under court order to comply with full secondary treatment requirements. M.I.T., SESD and others have recently completed a one-year CEPT study in an attempt to improve effluent quality at the existing conventional primary plant during the interim before the completion of a secondary treatment facility and to determine the impacts of CEPT on the design of the activated sludge plant.

SESD currently services 120 000 people and receives an annual average flow of 1.1 m^3/s. The annual average removals at the conventional primary treatment plant are 57 % and 24 % for TSS and BOD respectively [11]. The plant consists of aerated grit chambers, U-tubes which conveys the flow from the grit chambers to the influent channel leading to the primary sedimentation tanks, sludge belt-filter presses, sludge storage and an ocean outfall. The plant is configured such that one half of the plant can be operated completely independently from the other half (Fig. 5). This allows for a full scale comparison between conventional primary treatment and CEPT.

Bench Scale Tests for Optimization of CEPT. Because of the importance of BOD removal in determining the impact of CEPT on the future activated sludge plant, it was necessary to design a jar test protocol to evaluate various

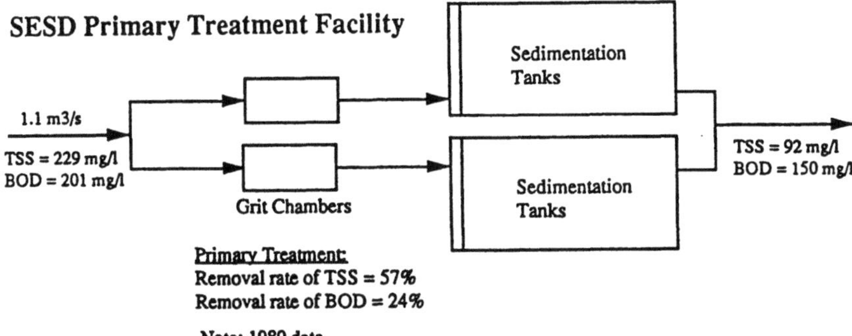

Fig. 5. Schematic of South Essex Sewerage District, Salem Massachusetts

screening devices, analytical parameters, chemical and dosages for coagulation and flocculation. Three screening devices were evaluated: a conventional one liter round jar with a Phipps & Bird gang stirrer, a two liter square jar, and a continuous flow reactor. All of the devices predicted COD removal equally well; however, the continuous flow reactor required one hour to perform one experiment while six round jar tests could be performed in less than half the time and were easier to use than the square jar. Therefore, the conventional jar testing device was chosen as the screening device.

A number of screening parameters were evaluated including visual results, turbidity, pH, orthophosphate, TSS, and COD. COD removal was the parameter eventually chosen because of its ability to be correlated with BOD removal and it required minimal analysis time.

Using the conventional jar testing device, the primary coagulant was chosen by varying the concentration from 0–100 mg/l and adding a small amount (0.5 mg/l) of an anionic polymer and then analyzing for COD removal. This process was done at various times throughout the day for several days in order to test various raw sewage influent conditions. As an example, Fig. 6 shows the averaged results from testing various primary coagulants at different COD influent concentrations. After an initial sharp increase in COD removal as the metal salt concentration is increased to a certain level, the curve flattens with very little improvement in COD removal as the metal salt concentration is increased beyond this level. Other factors such as cost, material handling and sludge production should also be used to determine which primary coagulant to use.

Full-Scale CEPT Tests at SESD. It was determined that a high molecular weight cationic polymer at a concentration of 0.2 mg/l would be necessary at SESD to strengthen the floc because of adverse mixing conditions caused by the U-tube connecting the grit chambers to the influent channel of the sedimentation tanks. There was only a slight improvement in COD removal when the cationic polymer was used as a coagulant aid in an effort to reduce metal salt concentration. A high molecular weight anionic polymer was used

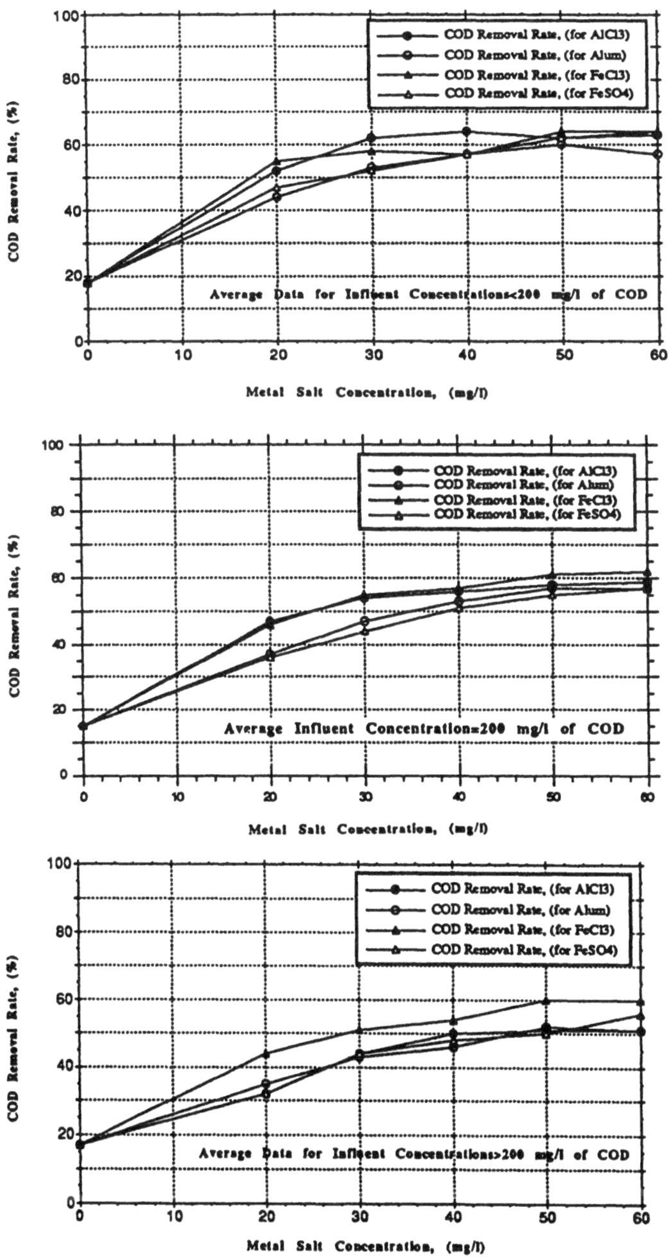

Fig. 6. Averaged data of COD removals versus metal salt concentration

at SESD to increase the settling velocity of the floc, although there was only slight improvement in COD removal, the floc settled three-to-four times faster with 0.5 mg/l of the anionic polymer.

Based on jar tests, it was predicted that 50% removal of BOD and COD would occur with the addition of 40 mg/l of alum, 0.2 mg/l of a cationic polymer, and 0.5 mg/l of an anionic polymer injected at the locations shown in Fig. 5.

When a conventional primary plant is retrofitted for chemically enhanced treatment, additional sludge is produced because of the increased removal efficiency. The effect of the increased sludge processing side-stream loading, such as filtrate from sludge filter presses, must be carefully considered in interpreting test results. This is especially important in the SESD plant, which is presently configured so as to introduce the liquid sludge processing side-stream downstream of the chemical addition points. The liquid sludge processing side-stream does not receive the benefits of chemical treatment. A complete analysis of the side-stream loading effect is most important in calculating BOD removal with CEPT and in determining BOD effluent loading to a future secondary treatment plant.

In order to quantify the effects of the side-stream loading at SESD, BOD concentrations and volumetric measurements were taken of the sludge belt-filter press side-stream. This allowed a comparison of conventional primary treatment and CEPT with and without the side-stream effect [12].

During *conventional primary treatment* operations, the effect of side-stream loading is small, but measurable. This is shown in Fig. 7 (a) using an average loading for the year 1990. With side-stream loading, the average BOD removal is 24%. Without side-stream loading, the BOD effluent decreases and the removal rate increases to 30%.

During the *half-plant chemical* addition test, all of the sludge processing side-stream entered the sedimentation tanks on the untreated side of the plant as shown on Fig. 7 (b). During this trial, the BOD removal rate averaged 49% on the chemically treated side while the apparent rate was 10% on the untreated side. Without the side-stream, the untreated side removal increases to 27%, which is close to the annual average for the untreated plant shown in Fig. 7 (a). The half-plant test clearly demonstrates the importance of correcting for the side-stream effect.

The *full plant test* is shown in Fig. 7 (c). The overall or apparent BOD removal rate of 32% increases to 44% when the side-stream is removed and the BOD effluent is correspondingly reduced. The fact that the corrected BOD removal rate of 44% is slightly less than the 49% observed in the half-plant test is undoubtedly due to the summer septic nature of the plant influent.

Even though this analysis clearly indicates the importance of the side-stream, no attempt was made by SESD to redirect the side-stream to the influent end of the plant so that it could benefit from the chemical addition.

Sludge production using alum as the primary coagulant was predicted using stoichiometric equation (5) with an error of less than 5%.

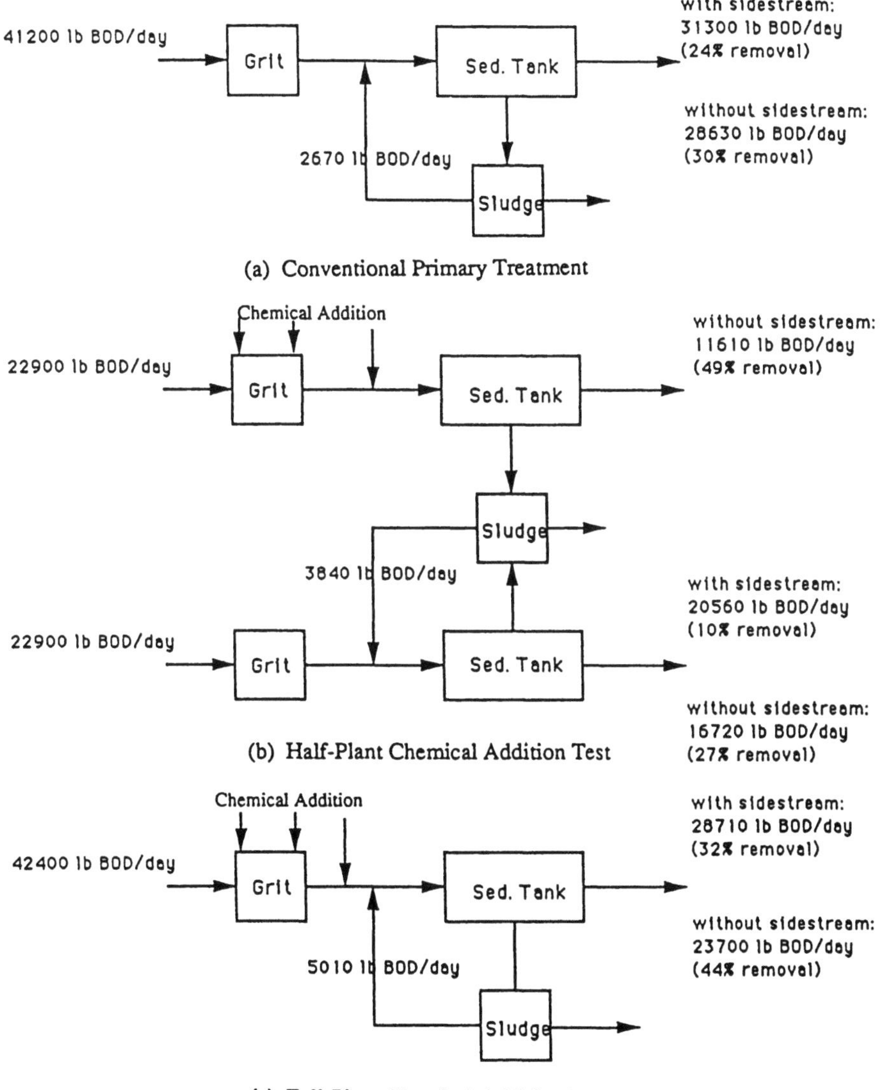

Fig. 7. Comparison of conventional and chemically enhanced primary treatment performance at SESD

$$RS = 1.0\,TSS_{rem} + 1.42\,P_{rem} + 0.26\,Alum_{in} \qquad (5)$$

A typical value for RS using CEPT is 142 mg/l and a typical value for RS using conventional primary treatment is 114 mg/l.

Currently, SESD has submitted the results from the study to the EPA for review. If the EPA feels that the increase in performance is warranted, SESD will be required to use CEPT before achieving full secondary treatment.

Other CEPT Plants in the USA

M.I.T., in conjunction with a National Research Council Committee study of wastewater management for coastal urban areas, has compiled operating data on a total of 19 plants using metal salt additives ($<$ 100 mg/l) in primary settling tanks [13]. Table 1 shows a summary of average influent/effluent characteristics as well as percent removals. Figure 8 shows a plot of annual average treatment efficiency as a function of surface overflow rate. These plots confirm the earlier observation that high removal efficiency is sustained at overflow rates at least twice the average for conventional primary treatment.

Tab. 1. Average influent/effluent concentrations and percent removal for CEPT plants

	TSS	BOD$_5$	TP	TN	NH$_4$-N	Raw sludge lb/lb TSS$_{rem}$
Influent/effluent [mg/l]	182/52	168/80	6/2	30/19	15/13	1.3–1.5
Percent removal	71	55	63	37	13	

Conclusions

We have attempted to convey operating data as well as an overall perception of CEPT in the USA and to emphasize a number of points that we believe are essential in bringing about the acceptance of CEPT in the USA.

1. In the past, phosphorus removal has been the primary objective of chemical addition requiring large concentrations of metal salts. With TSS and BOD removal as the primary objective, relatively low concentrations of metal salts can be used.

2. Synthetic polymers can be used to augment metal salt addition. First, as a coagulant aid to reduce the amount of metal salt added and to strengthen floc at plants with adverse mixing conditions. Second, as a flocculant to bridge coagulated particles and to increase the settling velocity which increases the surface overflow rate.

3. Due to the decrease in metal salt concentrations, sludge quantities are reduced; however, there is still an increase of approximately 45 % over conventional primary raw sludge production. Usually, 33 % of the increase is from increased solids capture and the remaining 12 % is from chemical addition. If an activated sludge system is added, the difference in the combined raw sludge produced from conventional primary treatment and activated sludge versus the CEPT and activated sludge is reduced to about 10 %. If both of the combined sludges are digested, the difference is less.

4. Side stream effects become increasingly important when chemicals are added to a conventional primary treatment plant. The difference should be quantified and the side-stream should be diverted to the head of the plant or treated separately.

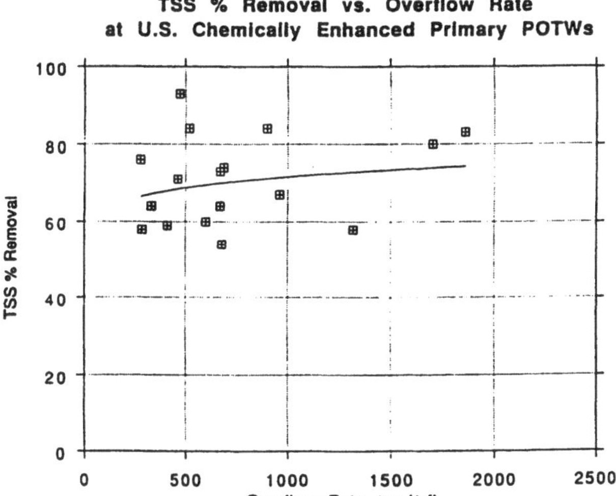

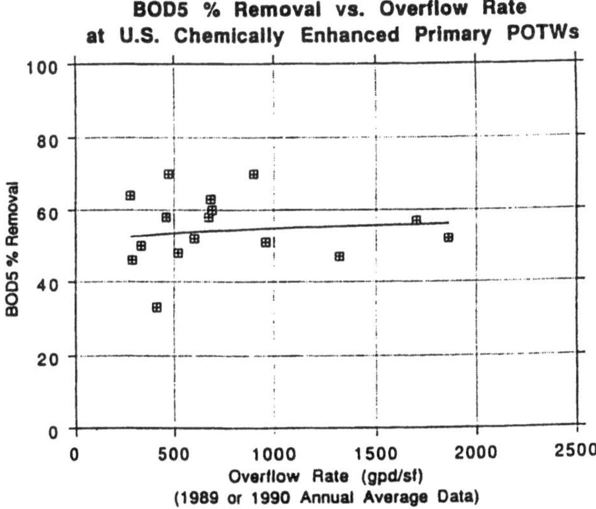

Fig. 8. Summary of annual average treatment efficiency as a function of surface overflow rate (1.0 gpd/ft^2 = 0.04 m/d)

5. CEPT is very effective in removing more than conventional pollutants such as TSS and BOD. It is also effective in removing FOG, nutrients, heavy metals and other priority pollutants [14].
6. CEPT is effective in reducing odors and VOC's [15].
7. CEPT can reduce the size and cost of any biological process that follows [16].

References

[1] Harleman, D.R.F., Morrissey, S.P., Murcott, S.: The Case for Using Chemically Enhanced Primary Treatment in a New Cleanup Plan for Boston Harbor. J. Boston Society of Civil Engineers Section/ASCE *6* (1) (1991) 69–84

[2] Culp, G.L.: Chemical Treatment of Raw Sewage. Water and Waste Engineering *4* (7) (1967) 61–63, *4* (10) (1967) 54–57

[3] Kreissl, J.F., Westrick, J.J.: Municipal Waste Treatment by Physical-Chemical Methods. In: Applications of New Concepts of Physical-Chemical Wastewater Treatment, W.W. Eckenfelder and L.K. Cecil (eds.). Pergamon Press, 1972

[4] Bowker, R.P.G., Stensel, H.D.: Phosphorus Removal from Wastewater. Noyes Data Corporation, 1990

[5] Farooq, S., Bari, A.: Physico-Chemical Treatment of Domestic Wastewater. Environmental Technology Letters *7* (1986) 87–98

[6] Weber, W.J. Jr.: Physiochemical Systems for Direct Wastewater Treatment. In: Applications of New Concepts of Physical-Chemical Wastewater Treatment, W.W. Eckenfelder and L.K. Cecil (eds.). Pergamon Press, 1972

[7] United States Environmental Protection Agency (USEPA): Process Design Manual for Phosphorus Removal. EPA-625/1-76-001a, 1976

[8] Fettig, J., Ratnaweera, H., Ødegaard, H.: Synthetic Organic Polymers as Primary Coagulants in Wastewater Treatment. Water Supply *8* Jonkoping (1990) 19–26

[9] Rad, H., Crosse, J.T.: Chemically Assisted Primary Treatment a Viable Alternative to Upgrading Overloaded Treatment Plants – The Hyperion Experience. Presented at Water Pollution Control Federation 63rd Annual Conference and Exposition, Washington, D.C., October 8–11, 1990

[10] Chaudhary, R., Shao, Y.J., Crosse, J., Soroushian, F.: Evaluation of Chemical Addition. Wat. Env. Tech. *3* (2) (1991) 66–71

[11] Morrissey, S.P., Harleman, D.R.F.: A Performance Analysis of the South Essex Sewerage District (SESD) Primary Treatment Plant. Massachusetts Institute of Technology, R.M. Parsons Laboratory Technical Note No. 28, February 1990

[12] Harleman, D.R.F., Morrissey, S.P.: Adjunct Report on 3-Month Full-Scale Chemically Enhanced Primary Treatment at the South Essex Sewerage District. Massachusetts Institute of Technology, R.M. Parsons Laboratory, November 1991

[13] Murcott, S.: Performance and Innovations in Wastewater Treatment. Massachusetts Institute of Technology, S.M. Thesis, February 1992

[14] Corcoran, E.F., Brown, M.S., Snedaker, S.C.: Water Quality Characteristics of a Southeast Florida Sewage Treatment and Bioeffects Laboratory. Florida Scientist *51* (1) (1988) 49–55

[15] Austin, T.: VOC's The New Effluent. ASCE Civil Engineering *62* (3) (1992) 42–45

[16] Karlsson, I.: Pre-precipitation for Improvement of Nitrogen Removal in Biological Wastewater Treatment. In: Pretreatment in Chemical Water and Wastewater Treatment, H.H. Hahn and R. Klute (eds.). Springer, Berlin Heidelberg New York 1988, pp. 261–272

Shawn P. Morrissey   Donald R.F. Harleman
Research Engineer   Ford Professor Emeritus
Massachusetts Institute of Technology   Massachusetts Institute of Technology
Ralph M. Parsons Laboratory   Ralph M. Parsons Laboratory
Rm 48–336B   Rm 48–317
Cambridge MA, 02139   Cambridge MA, 02139
USA   USA

Experiences with Chemical Treatment in Madrid

F. Morcillo and J.A. Sánchez

The Water in Madrid

The so-called Madrid Community includes Madrid City and the surrounding area, with an extension very close to $8\,000$ km^2, representing $1.6\,\%$ of Spain. The population concentrated in this area is as much as 4.9 million, which constitutes $13\,\%$ of the total Spanish population. About 1.8 million people ($37\,\%$) live in population nuclei outside Madrid Municipality.

We can separate two very well defined geographical regions in Madrid Community. The Northern Region is formed by the meridional versant of the Guadarrama Mountains (Sierra), part of the Somosierra and Gredos Mountains and the corresponding valleys. The Central-Southern Region, however, is constituted by a plain running in the NE-SO direction. The climate is continental, with low precipitation, although in the Northern Region the amount of rain and snow allow the exploitation of twelve reservoirs for water supply, with a total capacity of 920×10^6 m^3, and also the maintenance of large green areas.

This peculiar geographical distribution causes three kinds of problems concerning sewage purification in Madrid Community. In ecologically-sensitive areas, sewage is discharged into reservoirs, and consequently the treatment process is designed in order to guarantee and improve the effluent quality. In other areas, though, the trouble is produced by overloads of industrial origin in the Central-South Region, or by seasonal variations in the Northern Region. Finally, Madrid City area treats the sewage in seven conventional activated sludge plants. Figure 1 shows the location of sewage plants in Madrid Community and the type of treatment used in each one.

Canal de Isabel II is a public company responsible for the entire water cycle in Madrid Community, except, for instance, the sewage treatment works in Madrid Municipality. This company is the biggest in Spain for water distribution, with a staff of $2\,000$ people, assets of $13\,500$ million FF and a total cash flow of $1\,200$ million FF. The distribution net has a length of over $5\,000$ km, with flows varying between 1 and 2×10^6 m^3/day, with a global capacity of 41 m^3/s and a consumption of chemicals over $10\,000\times10^3$ kg/year.

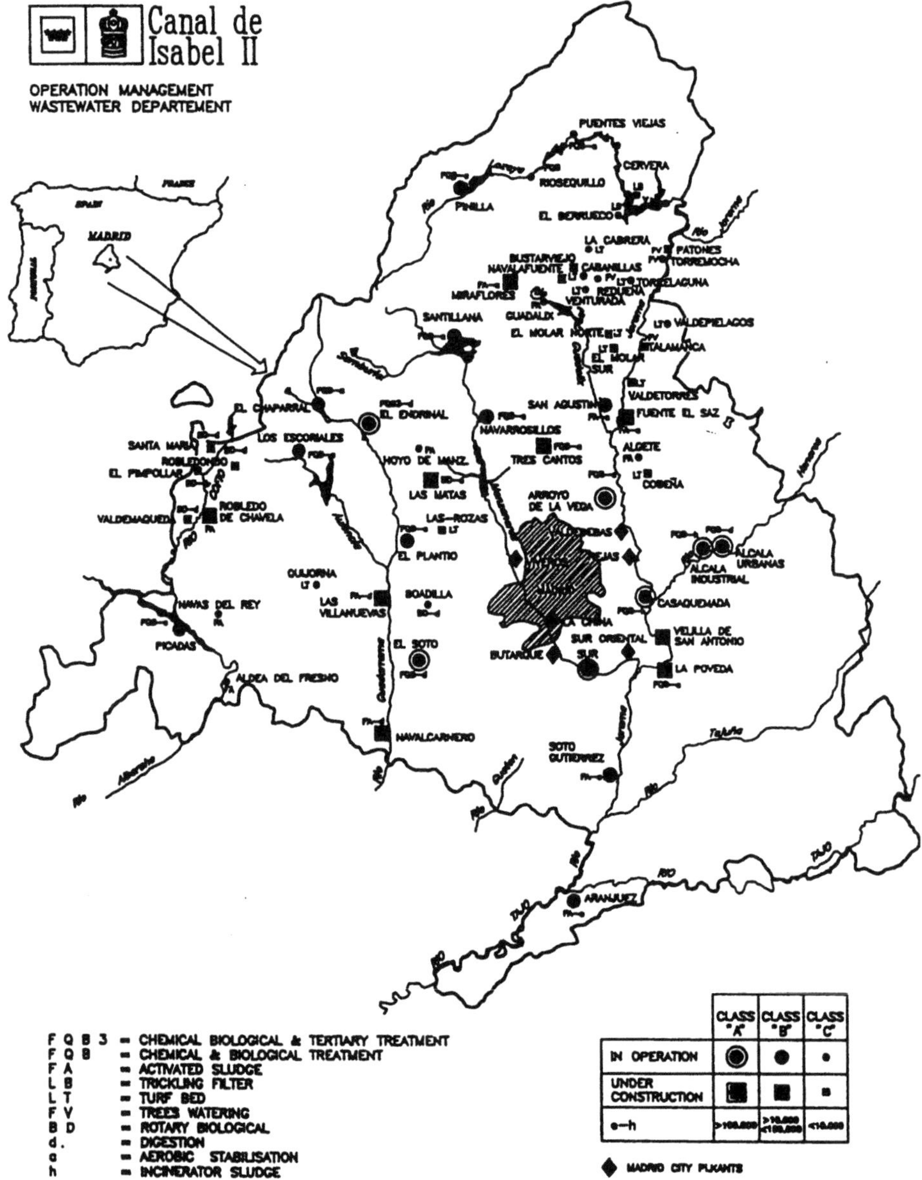

Fig. 1. Wastewater treatment plants in Madrid

Wastewater Purification

Before 1980, Madrid Municipality has some sewage plants where a small part of the wastewater produced was treated, and even this was not done in a very complete way. After the introduction of the PSIM (Plan de Saneamiento

Integral de Madrid) in 1983, seven activated sludge/anaerobic digestion plants were put into operation, with a total capacity of 14–16 m^3/s, to cover the needs of Madrid City, which has 63 % of the Madrid Community population. None of these plants incorporates chemical treatment in the water line.

The law regulating water supply and sewage purification in Madrid Community fixes 1985 as the starting point for an ambitious purification plan, known as PIAM (Plan Integral del Agua de Madrid). At present, 38 sewage works are in operation and treating the wastewater produced by 92 % of the population located outside Madrid City.

The average operating results are 24 mg/l for the BOD$_5$ and 22 mg/l suspended solids in the effluent. The treated flow is 187×10^6 m^3/year, which constitutes 34 % of the water supply in Madrid Community (about 550×10^6 m^3/year).

23 new sewage stations are now under construction. These will meet the needs of 400 000 pe, and will practically complete purification in Madrid Community.

Table 1 shows the average results of the analysis carried out on the influents and effluents for the different sewage plants that were active during 1991. In some of them, the exceptionally high values in the influent show a strong industrial activity.

Figure 2 shows the temporal change in the population served since the implementation of the PIAM. Since this time, the water supply has increased from 452×10^6 m^3/year in 1986 to 552×10^6 m^3/year in 1990. There are 10 stations discharging into water supply reservoirs which treat 20 % of the total flow purified in Canal de Isabel II's facilities, this is approx. 40×10^6 m^3/year (which treat 20 % of the total flow purified in Canal de Isabel II's facilities, which discharge their effluent into water supply reservoirs; this is approx. 40×10^6 m^3/year).

Of the 38 sewage treatment plants in operation, 10 of them are small installations, which incorporate low-cost technologies, such as green filters, turf beds and so on, and which represent about 1 % of the total flow. Of the remaining 28 plants, 20 of them have chemical treatment in the water line, and represent more than 80 % of the total flow.

Chemical Treatment

An approach to the wastewater purification problems in Madrid was given in the previous paragraphs. These problems can be classified into two big groups:

- *Type I*: Problems originating from the need to assure high quality in the effluent, usually because it will be discharged into reservoirs.
- *Type II*: Problems created by overloads in the influent. These overloads originate from industrial discharges in highly industrialised areas, or from seasonal population variations (summer period or weekends) at the Sierra, or even by daily peaks in bedroom towns close to Madrid City.

Tab. 1. Canal de Isabel II Plants (year 1991). Operation results (year averages)

Plant	Flow	Influent				Effluent			
		SS	BOD	COD	P	SS	BOD	COD	P
	m³/year	mg/l	mg/l	mg/l	mg/l	mg/l	mg/l	mg/l	mg/l
Cervera	5 300	44	42	106	4	6	6	28	1
El Berrueco	11 400	76	965	231	7	19	7	35	1
Quijorna	12 800	270	265	371	22	12	13	103	9
Venturada	21 300	150	126	231	4	12	19	43	2
Redueña	40 200	190	187	640	5	Trees watering			
Torremocha	48 100	214	113	283	4	Trees watering			
Cabanillas	74 000	261	178	338	6	22	31	64	2
Valdepielagos	75 700	760	325	645	8	44	66	122	3
Guadalix	350 700	278	261	458	3	22	24	49	1
Aldea del Fresno	411 600	142	238	8	12	13	45	3	
Soto Gutierrez	422 100	301	271	460		19	15	63	
Navas del Rey	554 800	208	218	328	9	17	5	57	3
San Agustin	686 600	510	514	843	4	14	14	41	1
Torrelaguna	689 200	561	460	965	3	70	140	293	1
Puentes Viejas	697 000	765	417	789	14	9	5	24	2
Algete	859 400	361	365	721	5	11	7	26	1
La Cabrera	881 300	207	153	298	2	28	28	60	1
Hoyo de Manzanares	883 800	120	163	255	5	15	15	32	3
Riosequillo	884 200	537	277	969	14	19	6	40	1
Boadilla	958 400	278	259	545	8	16	17	57	4
Pinilla	1 713 200	53	22	60	3	9	7	9	2
Picadas	1 877 400	272	224	406	8	10	11	44	2
Miraflores	1 940 300	127	111	217	1	26	18	45	1
Tres Cantos	2 698 100	466	433	740	4	17	21	66	1
Navarrosillos	3 349 600	463	570	925	9	18	21	55	2
El Plantio	3 590 900	599	631	1 035	15	11	17	45	1
La Poveda	3 706 000	455	834	1 611	13	44	70	180	2
Los Escoriales	3 804 100	210	363	535	8	10	15	36	1
Aranjuez	5 991 300	767	515	994	12	14	17	113	5
El Chaparral	6 235 500	91	120	185	2	6	8	20	1
Santillana	6 848 200	105	106	155	2	8	11	21	1
Alcala Urbana	8 443 400	205	301	564	3	14	19	86	1
El Endrinal	9 137 200	120	162	252	4	7	10	24	2
Arroya de la Vega	16 391 700	355	432	734	4	20	23	80	1
Alcala Industrial	19 475 700	201	248	476	6	21	27	97	4
Arroyo del Soto	20 082 700	301	310	605	14	25	20	80	2
Casaquemada	21 247 500	204	245	500	10	23	24	80	2
Madrid-Sur (parcial)	33 743 520	297	315	652		17	18	74	

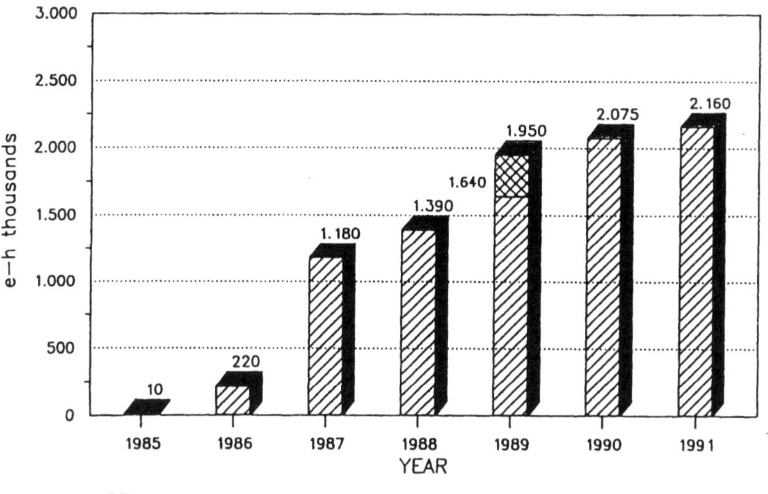

Fig. 2. Canal de Isabel II – Evolution of purification capacity

When the first sewage plants were put into operation in 1986, Canal de Isabel II carried out some full-scale trials with chemical treatment in two stations, each representing one type of problem. The selected plants were "Los Escoriales", with the effluent discharging into Valmayor Reservoir, and "Arroyo de la Vega", with exceptionally high overloads of industrial origin, which made correct treatment impossible.

Both full-scale trials were a success. In the first case, adequate phosphorus values in the effluent were achieved and, at the same time, stable BOD_5 and SS values were assured, whilst in the second case, continuous operation was made possible. Since then, chemical treatment is used with full reliance on the sewage plants run by Canal de Isabel II.

Experience at "Los Escoriales"

Figure 3 shows the treatment process implemented in this plant. The daily wastewater flow is about 12000 m³/day, which is pumped through an Archimedes screw into the pretreatment (screening, grit chambers and grease removal). The chemicals (solid AVR) are dosed at the Parshall flow measurement point. After sedimentation in two settling tanks, there are two aeration tanks with turbines and two clarifiers. The sludge is stabilised aerobically, thickened by gravity and dewatered in band filters.

Pre-precipitation was carried out with solid AVR, stored in a 30 m³ silo, from where it was diluted to a concentration between 5 % and 10 % and pumped into the effluent from the grit chambers.

Figure 4 shows the results of phosphorus removal over the 4-month period chosen for the trial. The influent phosphorus was in the range 10–16 mg/l.

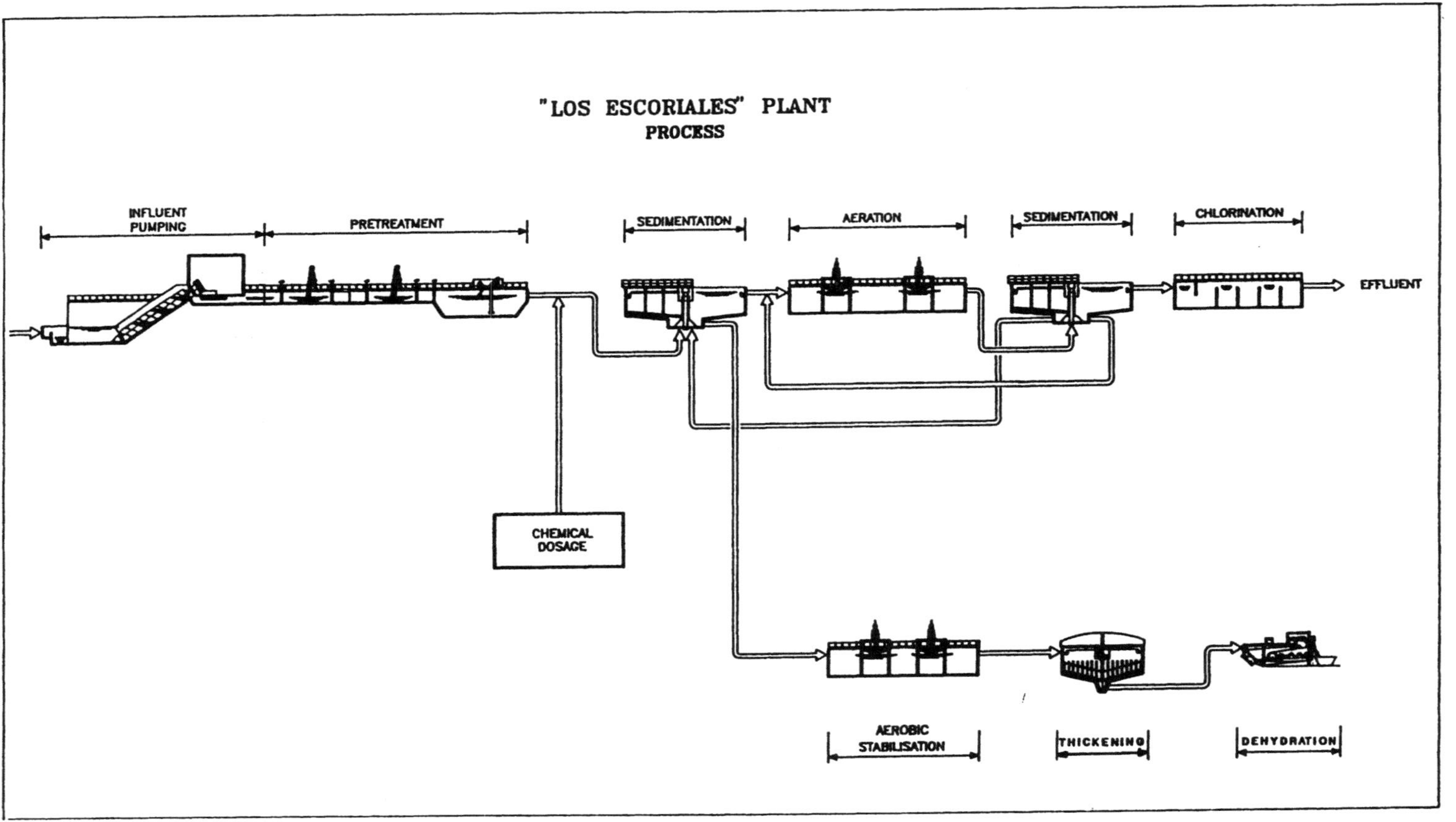

Fig. 3. "Los Escoriales" Plant. Process

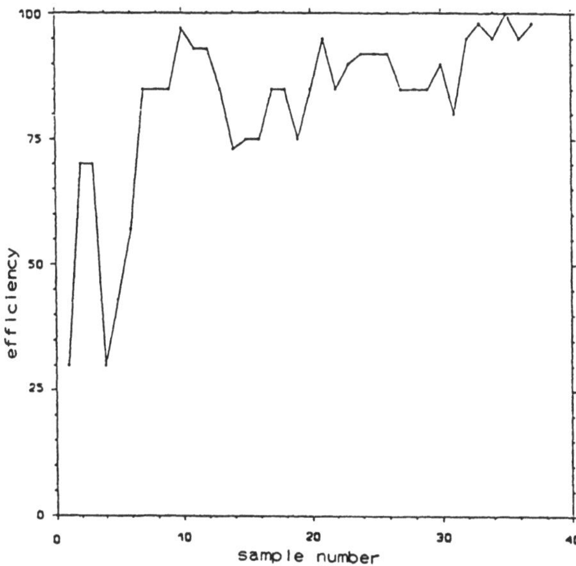

Fig. 4. "Los Escoriales" Plant. Total phosphorus removal efficiency

The main goal of this trial was phosphorus removal, which was achieved perfectly. Moreover, increased process stability was observed, with an effect on quality assurance and appreciably lower values in BOD_5 and suspended solids in the effluent.

Experience at "Arroya de la Vega"

Figure 5 shows the flow sheet for this plant. At the time the experiment was carried out, there were no mixing and flocculating chambers; these were built later as a consequence of the successful results achieved. The aeration chambers are 8 m deep, with medium-sized bubble diffusers. Primary sludge is thickened by gravity, and the excess sludge is thickened by flotation. The anaerobic digestion is followed by dewatering in band filters.

The experiment was also carried out using pre-precipitation with solid AVR, with different dosages during the day, in order to obtain a sensibly constant BOD_5 load over the biological stage. Figures 6 and 7 show the values for BOD_5 and SS in the influent, and the efficiencies achieved for both parameters in the primary settlers during the trial period.

The main goal in this case was to make the plant work continuously, something that was impossible without chemical treatment due to the fact that the extraordinarily high contamination in the influent exceeded the installed aeration availability. Here also the stability was increased and the control of the daily peaks became simpler.

On the other hand, it could be shown that the implementation of a pre-precipitation makes it possible, in a very simple way, to increase the capacity of any installation, and is a good weapon against peaks of any kind, both daily and seasonal.

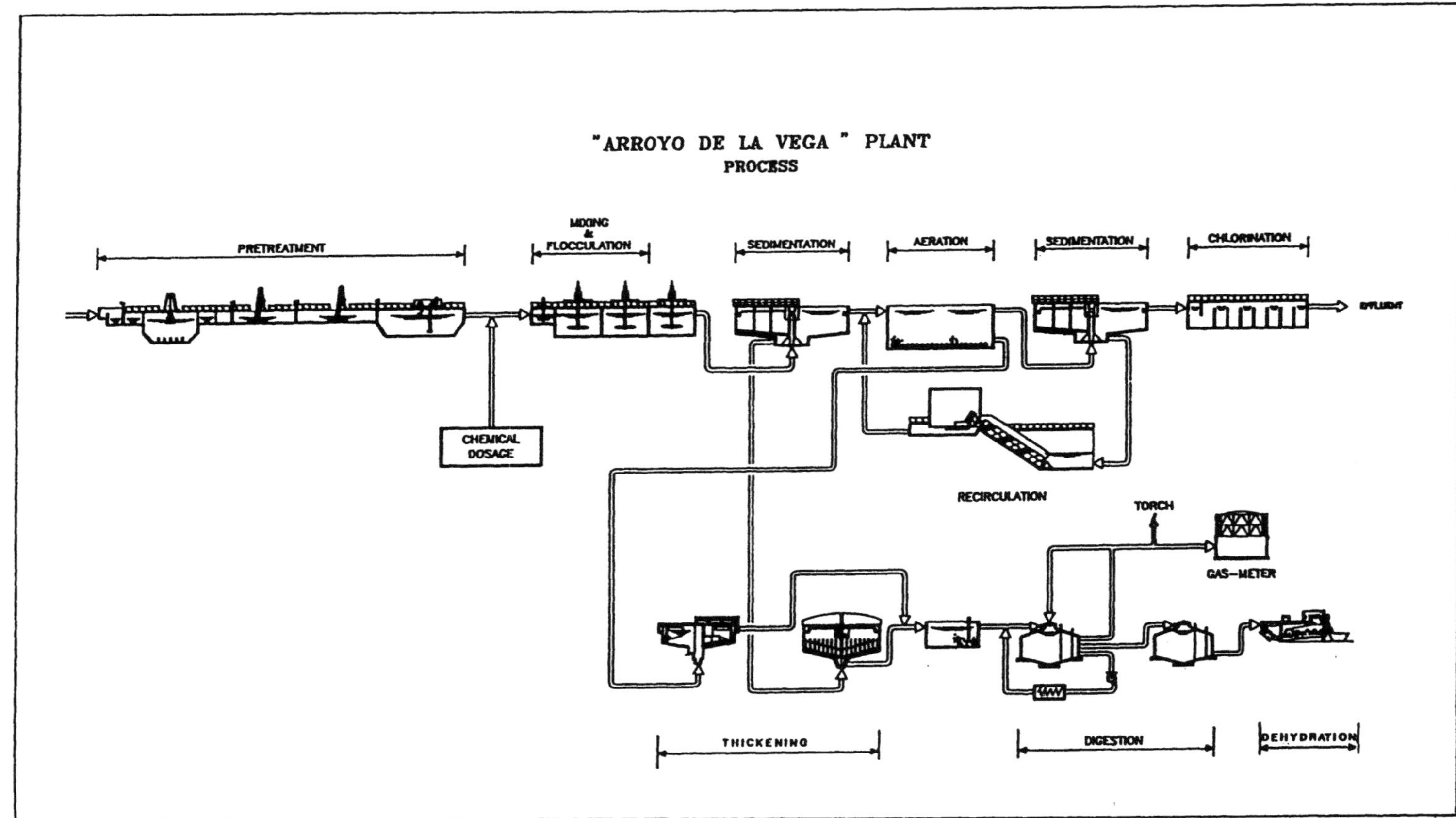

Fig. 5. "Arroyo de la Vega" Plant. Process

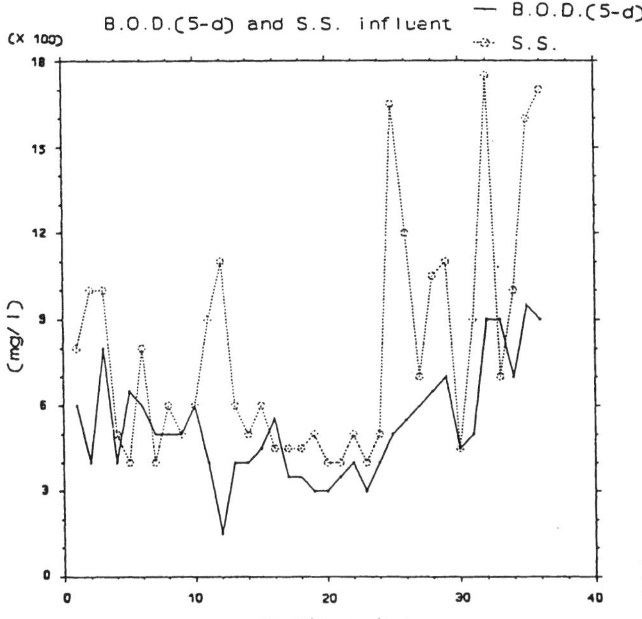

Fig. 6. "Arroyo de la Vega" Plant. Influent contamination

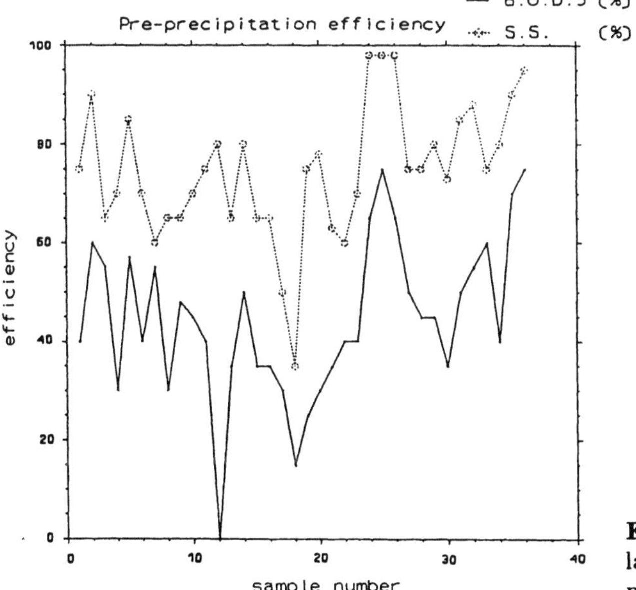

Fig. 7. "Arroyo de la Vega" Plant. Pre-precipitation efficiency

Present Situation

Chemical treatment has been widely adopted in the sewage plants run by Canal de Isabel II, so that 83 % of the purified flow receives some type of chemical treatment, and the necessary equipment has been installed permanently in 20 of these sewage plants.

The consumption of chemicals reaches $17\,000\times10^3$ kg/year, of which $6\,700\times10^3$ kg are lime.

In Figure 8, sewage plants with no chemical treatment, with chemical treatment for quality reasons and with chemical treatment for overload problems are distinguished in relation to the equivalent population. It can be observed from this figure that only 17 % of the discharges are not receiving any chemical treatment.

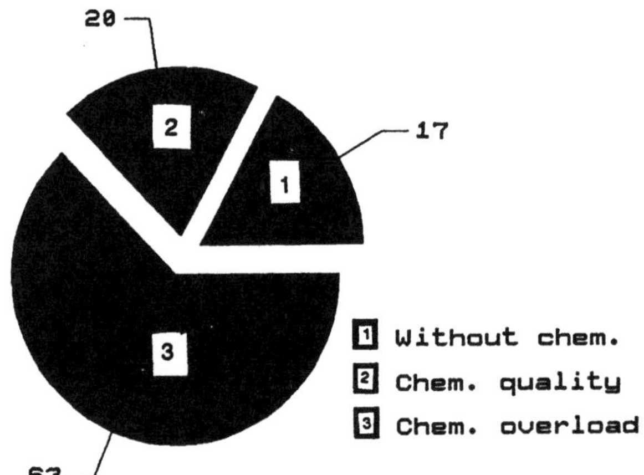

Fig. 8. Plants in Operation (pe). Chemical treatment implementation

24 % of the waste being treated chemically is Type I (quality requirements), and 76 % is Type II (overloads). Lime used in each case is 4 % and 96 %, respectively, and other chemicals are distributed 12 % and 88 %, respectively. These data are shown in Table 2. The average dosage for Type I is 33.6 mg/l.

Tab. 2. Chemical treatment plants

	Inventory	Type I*	Type II**
Flow	155 $\times10^6$ m^3/yr	24 %	76 %
Lime	6 700 $\times10^3$ kg/yr	4 %	96 %
Chemicals	10 200 $\times10^3$ kg/yr	12 %	88 %

*Type I: Chemicals due to quality
**Type II: Chemicals due to overload

This distribution in terms of percentage is maintained, in general, in the 23 works under construction, which will serve the waste for 400 000 pe. In the future, then, the criteria developed for chemical treatment by Canal de Isabel II will be maintained, with an increase in the consumption of chemicals on the order of 25 %.

The great volumes of sludge caused by the use of lime, together with the increased handling and transport cost of this sludge, are reasons for a gradual reduction in the use of this chemical. However, in sewage plants with significant industrial discharges, this reduction is not easily carried out. The substitution of lime with other chemicals must then be carefully evaluated in detailed trials with adequate alternative products.

Presently, the chemicals used by Canal de Isabel II in their sewage plants is as follows:

Solid products

- Lime
- AVR (7.2 % Al, 3.0 % Fe, both as sulphates)

Liquid products

- Ferric chloride
- Ferric sulphate
- Aluminium sulphate
- AVR (3.6 % Al, 1.5 % Fe, both as sulphates)
- PAC 18 (polyaluminium chloride)

Organic polymers

- Cationic
- Anionic
- Non-ionic

Conclusions

The PIAM (Madrid Integral Water Plan) has made it possible to provide practically the entire population of Madrid with some purification system, in a very short time and with an adjusted budget. It has, however, been very difficult to maintain a balance between planning and building accurately, and the effect on the environment by the reduction in the final terms is still uncertain. Due to the desire to implement the system as quickly as possible, this was unavoidable.

In this respect, the problems generated by the occasional failure to adapt the dimensioning to the real influent loads have found a simple solution by implementing chemical treatment units. In most of the cases, the necessary investment is below 12 % of the total cost. On the other hand, it is always possible to carry out full-scale trials with mobile equipment, and in this way to optimise future investments.

The experiences in Madrid with chemical treatment were always intended to achieve an effluent quality which suited the conditions set in advance or fixed later, when they were found to be convenient, with special emphasis on solving quality and/or overload problems, particularly:

- Nutrient removal when discharging into reservoirs
- Tertiary treatment and, consequently, improvement in quality
- Guaranteed results, leading to process stability
- Industrial overloads
- Daily peaks
- Overloads in the weekly or yearly cycle.

Generally speaking, it is difficult at first to predict which chemical must be used to achieve the desired result. Only full-scale trials, seriously implemented and pursued, will give a real idea of the appropriate chemical to use, and the way it should be dosed. It is a well known fact, for instance, that dosages predicted by laboratory tests are generally higher than those required in the actual plant.

Even though both liquid and solid chemicals are used presently, there is a trend toward the use of liquids, due to the ease of application and the simplicity of the equipment required.

In the products selected for their application in sewage treatment, important factors are fabrication quality, stability, reliability in the supplies and professionalism of the technical service.

Fernando Morcillo and José Sánchez
Canal de Isabel II
Departamento de Depuración
Santa Engracia, 125
28003 Madrid
Spain

Trade-Offs Between Physico-Chemical Lamella Separators and Aerated Biofiltration

F. Rogalla, G. Roudon, J. Sibony, and F. Blondeau

Abstract

The European Guideline for Wastewater Discharges distinguishes between three zones: less sensitive, normal, and sensitive. A tight schedule has been set to achieve certain effluent standards for each area, requiring extensive upgrading and construction of wastewater treatment plants throughout Europe. To achieve the objectives within reasonable investment and operational cost as well as with limited land use and reduced nuisances, high performance treatment systems have to be implemented.

Physico-chemical treatment allows a considerable removal of pollution in a rather simple and energy-efficient way. If coupled with lamella settling on parallel plates, space use is about one tenth of conventional primary treatment. To polish the effluent of the primary stage, Biocarbone aerated filters combine biodegradation with very high removal rates and retention of particles in one reactor, without additional clarification or filtration.

This paper shows the use of these three systems depending on the required treatment objective. The impact of the first stage on the performance of the bioreactor is studied at several sites where chemicals are used seasonally to remove peak loads. Solids removal and biodegradation efficiency can be balanced between the physico-chemical and biological treatment. Large scale examples of the compact technology and the operational flexibility is demonstrated, including plants exceeding capacities of 100 000 p.e. with very low environmental impacts that are located in downtown districts.

Introduction

In addition to stringent standards imposed in several regions of the US and many northern European countries, the European Community has adopted on March 18, 1991 a guideline that requires collection and treatment of wastewater for communities exceeding 2 000 population equivalents. As shown in Table 1, three different deadlines according to the size of the communities are given, as well as three different levels of effluent quality according to the sensitivity of the receiving water. It is up to the member-states to identify the respective zones of sensitivity before December 1993.

Tab. 1. Objectives of the European Guideline for Wastewater Treatment

Receiving water	Treatment	Removal* or quality** objective					Size p.e.	Date
		BOD	SS	COD	TN	TP		
Non-sensitive* [%]	Primary	20	50	–	–	–	< 100 000	2000
							< 15 000	2005
Normal* [%]	Secondary	70–90	90	75	–	–	> 15 000	2000
							< 15 000	2005
Sensitive** [mg/l]	Advanced	25	35	125	10	1	> 100 000	1998
		25	35	125	20	2	< 10 000	2005

* Effluent quality measured as 95 % compliance of daily composite samples
** Effluent quality measured as yearly average

It is clear from Table 1 that secondary treatment will be required for all plants larger than 100 000 population equivalent (p.e.), and only after justification with studies of environmental impact (for example for long and deep ocean outfalls) and for smaller plants is primary treatment sufficient. All receiving waters exposed to eutrophication or drinking water use should be regarded as sensitive and nutrient removal will be required.

The investment cost of wastewater treatment (without sewers) is estimated between 1 000 FF and 1 000 DM per p.e. treated depending on plant size and effluent quality. To comply with the European directive, under the assumption that 50 % of the population is already served according to the directive, close to 500 billion FF will have to be invested. In addition, a running cost of 100 FF/p.e. per year is expected. Thus great motivation exists to reduce running and investment cost. One way is to introduce compact technology that can be implemented close to population centers, reducing infrastructure cost and making it possible to reuse existing networks and outfalls.

In this text, the following technologies will be described as solutions to achieve the treatment goals of the European directive for non-sensitive zones:

- primary lamella settling to achieve solids removal above 50 %
- physico-chemical lamella settling to remove up to 90 % of SS and 70 % of BOD
- lamella settling with or without chemicals followed by Biocarbone aerated filters for complete secondary treatment.

All these options lead to reduced surface requirements of treatment plants, making it easy to integrate the units in urban areas and cover the plants for reduced nuisances [2]. More stringent effluent requirements can also be achieved using compact technologies [3, 4].

Lamella Settlers

Parallel plates make it possible to multiply settling surfaces in a given volume and thus reduce the necessary space for solids separation [5]. It has been widely applied in industrial [6] and drinking water treatment [7, 8] and was introduced in the last decade for wastewater primary settling [9]. Figure 1 shows the principle of a unitary lamella settler: water flows up through parallel plates, the particles are intercepted on the plates and slide downwards. Lamella settlers are designed according to surface loading theory [10], which states that parallel plates act as shallow basins that reduce the vertical distance a particle must fall and increase the surface area available for solids separation.

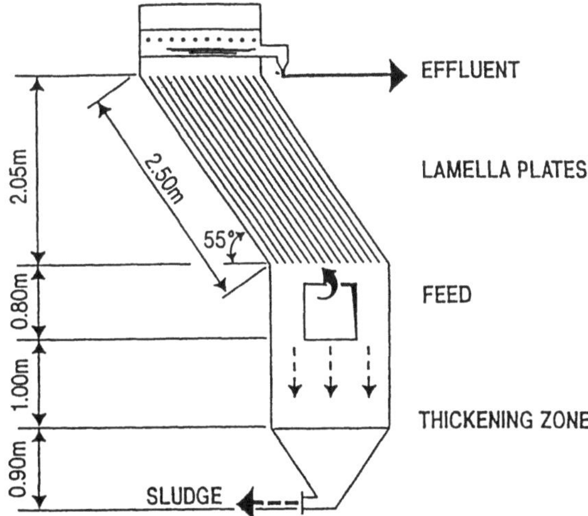

Fig. 1. Pilot unit of lamella settler

In practice, the plates are inclined between 50 and 60 ° towards the horizontal plane. The active surface area considered is the projected surface, multiplying the area of the lamella by the cosine of the inclination angle. The overflow rate in relation to this projected area is called Hazen or lamella velocity. For an angle of 55 ° and a lamella spacing of 10 cm, the ratio between the footprint of the settler and the projected surface is 1 : 11. The efficiency of lamella separators depends, as for conventional gravity settlers, on three factors which will be described below: flocculation technique, reagent dosing and overflow rate.

Flocculation Technique

The destabilisation of particles depends on a good dispersion of the reagent and an intense mixing. The initial rapid mixing should thus be rather long. An example of the influence of rapid mixing is given in Fig. 2: when rapid mix is increased from 1 to 6 minutes, residual suspended solids are reduced from 50 mg/l to below 10 mg/l. Once the colloids have been destabilised, a dense floc

with low water content and small drag should be created. This is best achieved
with a semi-rapid mix, purveying high local energy to aggregate particles and
preventing the formation of large feather-like flocs [9].

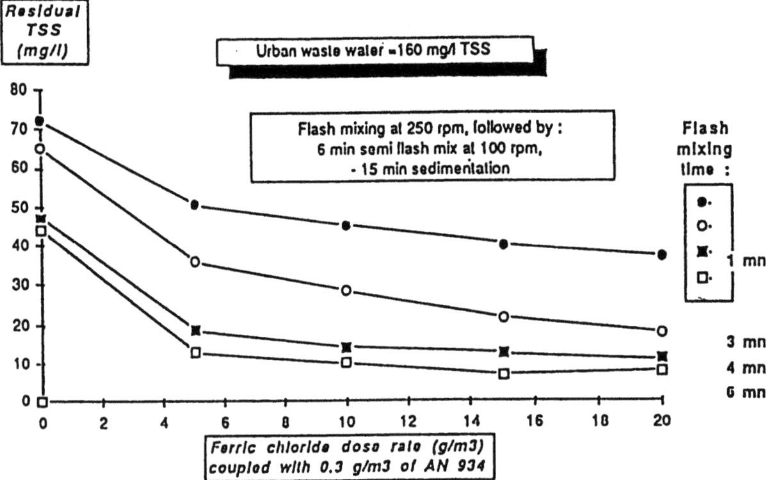

Fig. 2. Variation in residual TSS depending on different flash mixing times

Reagent Dosing and Raw Sewage Characteristics

The effluent quality of physico-chemical treatment depends on the influent
characteristics: the more settleable the solids or the destabilisable matter, the
higher the separation efficiency. The result of a jar test is given in Fig. 3 for
one wastewater with a relatively constant ratio of filtrable to total pollution:
the higher the influent solids, the higher the effluent residual, and the higher
the reagent dosage, the lower the effluent solids concentration. This leads to an
almost direct dependency of the removal efficiency on reagent dosage (Fig. 4).

Figures 3 and 4 were obtained at a full scale plant using physico-chemical
lamella settling. Samples were collected over several different hours to have
varying influent concentrations. Rapid mixing was carried out for 6 min., followed by 6 min. semi-rapid and 15 min. settling [9]. Whereas without chemical
and under static conditions, about 70 % of the solids can be separated, removal
efficiencies above 90 % can be achieved with higher dosages of $FeCl_3$, leading
to residuals below the EC-Guideline of 35 mg/l for secondary effluent.

The removal efficiency for COD and BOD is likewise dependent on the iron
dosage and influent characteristics. Figure 5 shows results obtained on the pilot
sketched in Fig. 1 [4]. Whereas without $FeCl_3$, a removal efficiency between 30
and 40 % is obtained for COD, which is the classic expectation of primary
settling, the removal efficiency can be increased to the minimum requirement
of the EC guideline for normal areas of above 70 % with higher reagent dosages.

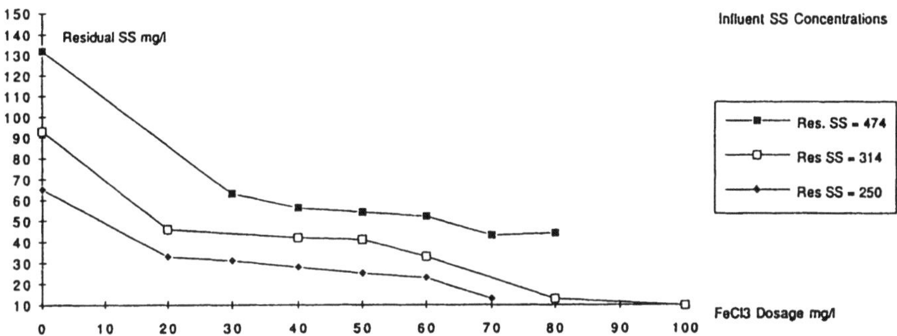

Fig. 3. Residual suspended solids in Porto Vecchio jar test

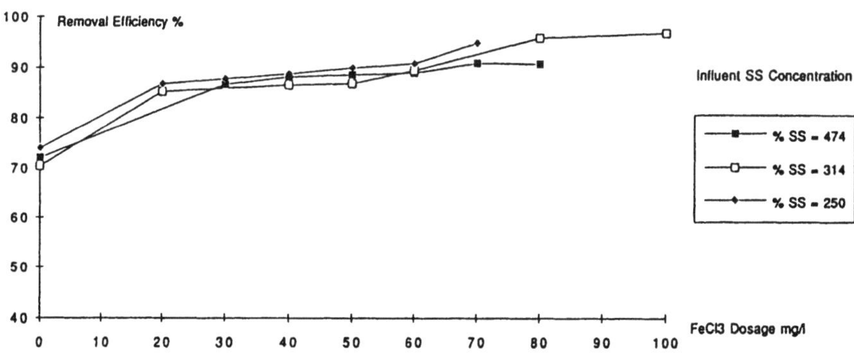

Fig. 4. Suspended solids removal efficiency at Porto-Vecchio

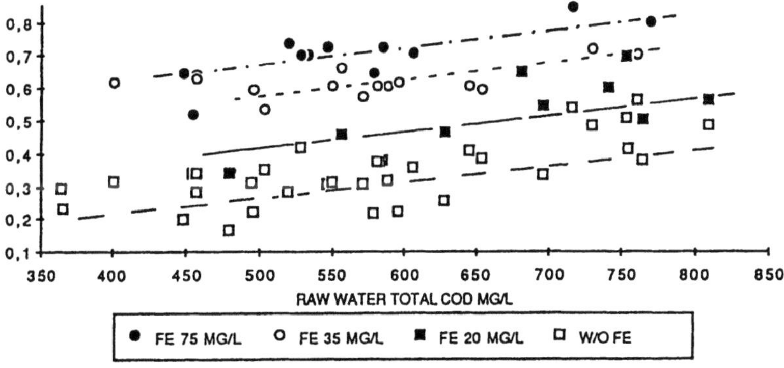

Fig. 5. COD removal efficiency

Overflow Rate

Hydraulic similarity is difficult to obtain between small scale tests and full scale results in order to simulate this parameter. By adding chemicals, much better removal efficiencies even at higher flow rates can be achieved in conventional settlers [11, 12]. When the reagent dose is sufficiently high, for example in the case of phosphorus precipitation, overflow rates up to 4 m/h can be tolerated [13]. The performance of full scale lamellae at the two plants described in Table 2 are given in Fig. 6.

Tab. 2. Design and operation at two full scale physico-chemical lamella plants

Plant location	Peak flow	Lamella spacing	Inclination	Reagent dosing		
				$FeCl_3$	Lime	Polymer
	m^3/h	cm	Angle °	mg/l	mg/l	mg/l
La Ciotat	750	10	55	250	150	0.3
Hyères	900	7.5	60	180	50	0.3

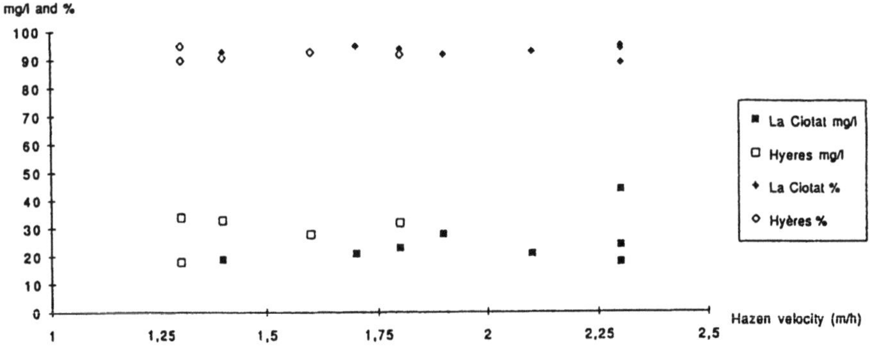

Fig. 6. Residual SS and removal efficiency in La Ciotat and Hyéres

The plants of La Ciotat and Hyères are located on the Mediterranean shore of France, and are designed for highly variable flow rates: units can be taken out of service during low-flow periods. The plants are designed so that even without chemicals a full primary efficiency can be obtained, and if one of the units is shut down, the other can still obtain the required effluent quality. The average overflow rates are thus far below the maximum allowable values. Figure 6 shows the removal efficiency and residual solids when all the peak flow is directed over one of the two lamella settlers. 90 % of the suspended solids are eliminated even at overflow rates above 20 m/h and residuals are below 30 mg/l.

Demonstration of Lamella Performance

The largest plants using lamella separation are located in Marseille (300 000 m^3/d) and Québec ($2 \times 200\,000$ m^3/d). In the former, the compactness of the settling technology made it possible to locate this plant treating about 1 million population equivalents in the center of the city beneath a stadium, even though the lamella is preceded by conventional settling to reduce coagulant dosage [14].

Before building these facilities, extensive tests were carried out on an intermediate scale. The pilot with a nominal capacity of 200 m^3/h has a cross-section of 5.55 m^2 and is 3.5 m high. Within this volume, 10 plates of 2.6 m^2 each are installed, resulting in a projected surface of 60 m^2 lamella, thus multiplying by ten the available settling surface. Sludge withdrawn from the lamella pilot averaged in concentration between 3.5 and 4.5 % DS.

Table 3 summarizes the pilot results from the Marseille and the Québec tests. Depending on the type of sewage and the reagent dosage used, around 80 % of the solids could be removed by chemically aided lamella settling. The detention time in the physico-chemical step is around 30 minutes, including flocculation. This step was later optimized to last less than the settling time [9]. If the reagent is adjusted accordingly, phosphorus can also be completely precipitated in the primary step [4, 13].

Tab. 3. Average results on a lamella settling pilot

Test site Sewer type	Québec Combined	Québec Combined	Marseille Separate	Marseille Combined
Suspended solids [mg/l]				
Influent	108	110	203	129
Effluent	20	18.6	44	42
Flocculant dosage [mg/l]				
Alum ($Al_2(SO_4)_3 14H_2O$)	75			
$FeCl_3$ (pure)		46	19	7
Detention time [min]				
Flocculator	19	13	22.75	19.5
Settler	11	10	17.5	15
Overflow rate [m/h]				
Total settler surface	18.9	20.9	12	14
Lamella projected area	1.75	1.9	1.1	1.3
Residual phosphorus	0.55 mg/l	0.5 mg/l	53 %	58 %

Besides precipitating solids and phosphorus, lamella settling can also be used to replace conventional settling. About 35 large wastewater treatment plants have been equipped with lamella settling [9]. A few results are summarized in Table 4 (adapted from 15). Most treatment plants using only physico-chemical treatment are designed to achieve a SS removal of more than 80 % (French level B, applied for most sea discharges). In such cases, BOD and COD removal normally averages 50 %. Higher removal is obtained by increasing chemical dosage, or by adding a compact biological stage, as described below.

Aerated Biofiltration

The Biocarbone Aerated Filter was developed at the end of the seventies [16] and is now used at almost 100 wastewater treatment plants worldwide [4]. By supplying oxygen to a filter medium, the granular material can be used as biomass carrier without losing its capacity to retain suspended solids. Because the fixed bacteria can be concentrated on the filter grain to around 15 g/l [18], independent of sludge settleability, very high removal rates can be achieved.

The Biocarbone is similar to conventional rapid sand filters, except that air is sparged into the lower portion of the filter bed, and a coarser filter medium is used. A schematic drawing of the filter is given in Fig. 7. The filter grain retains the suspended solids and provides surface for biofilm development. Fixed bacteria in the filter bed allow the combination of aerobic degradation of pollution and physical clarification in one reactor, without additional clarifiers or solids removal steps. The granular bed consists of expanded schist with grain size variable according to the desired treatment objective.

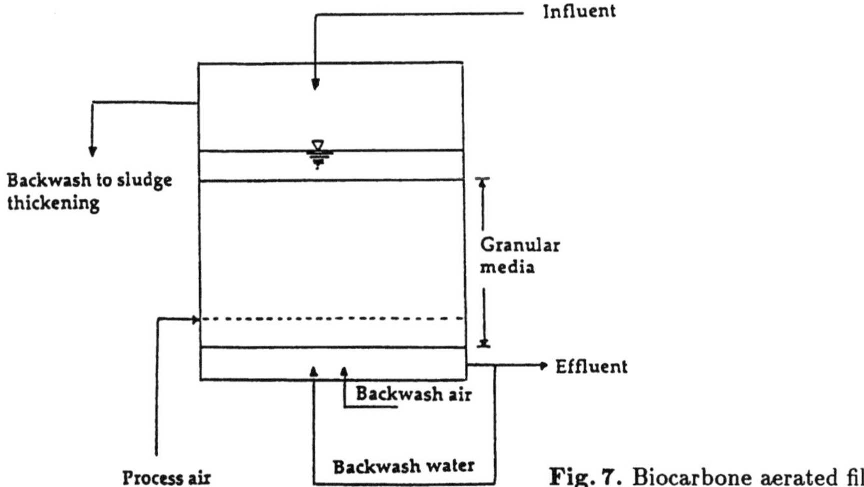

Fig. 7. Biocarbone aerated filter

The downflow submerged granular bed was designed for high oxygen transfer by countercurrent aeration. Sewage is fed into a mixing zone at the top of the filter bed. Any distribution device sensitive to clogging by influent solids is

Tab. 4. Average results of lamella plants

Treatment plant	Peak flow m^3/h	Lamella velocity m/h	Overflow rate m/h	TSS raw infl. mg/l	TSS lam. effl. mg/l	TSS removal %	BOD_5 raw infl. mg/l	BOD_5 lam. effl. mg/l	BOD_5 removal %	COD raw infl. mg/l	COD lam. effl. mg/l	COD removal %
Physico-chemical												
Toulon	1 900	–	–	269	71	74	200	106	47	592	276	53
Marseille*	26 640	–	–	72	30	58	200	111	45	500	268	46
La Ciotat	900	1.1	–	500	20	96	–	–	–	–	–	–
Aime La Plagne	520	–	–	175	30	83	200	80	60	500	200	60
Longueil	18 700	1.4	–	94	19	80	64	23	64	–	–	–
Physico-chemical + BAF												
Sanary-Bandol	1 200	–	–	492	140	72	331	198	40	641	264	59
Le Barcares	750	1.2	14	208	56	73	259	103	60	558	232	58
Orford (Canada)		1.1	10	147	20	86	–	–	–	–	–	–
Primary + BAF												
Le Touquet	1 000	–	4.5	275	124	55	286	168	41	–	–	–
Antibes	2 700	–	–	–	–	–	200	140	30	–	–	–
Monaco	1 800	–	–	310	125	60	–	–	–	520	300	42
Québec-East	15 625	1.55	–	138	62	55	131	85	35	–	–	–
Québec-West	13 125	1.82	–	138	54	61	106	64	40	–	–	–

* Marseille: Lamella settler comes after conventional primary
** Québec: Québec data is design data only

therefore unnecessary. Also, countercurrent backwashing first rinses the top of the filter bed, where the solids retention is greatest and the most intense rinsing is required. Backwashing is performed automatically according to headloss buildup or to time scale, and the washing sequences are controlled by microcomputer.

Aerated biofilters can be applied on primary settled wastewater or as an additional polishing step to upgrade existing plants. Besides the construction of around 100 plants worldwide [17], different independent pilot tests achieved the following performances:

a) Secondary treatment for BOD and SS removal on settled wastewater
   - Effluent total BOD less than 30 mg/l at soluble loadings up to 1.4 kg $SBOD/m^3$ d [19]
   - BOD removal efficiency above 90 % for loadings up to 4 kg BOD/m^3 d (JSWA)
   - Residual BOD values below 20 mg/l with loadings up to 4.1 kg BOD/m^3 d [20]
   - Residuals below 10 : 10 : 5 ($BOD/SS/NH_4$) at loadings up to 0.4 kg N/m^3 d [21].

Most Biocarbone plants in operation today have been designed for standard secondary treatment with effluent quality objectives between 15 and 30 mg/l of BOD and SS [25]. In those applications, settled water is fed into the aerated filter, and the media volume has an empty bed contact time of around one hour. In comparison with conventional activated sludge, space need is reduced by a factor of five through the absence of clarification.

b) Tertiary nitrification on pretreated clarified effluent
   - Greater than 90 % nitrification at loading rates up to 0.63 kg $Kj \cdot N/m^3$ d [20]
   - Maximum removal rates up to 0.74 kg $N\text{-}NH_4/m^3$ d at 20 °C [22]
   - Ammonia elimination rates of 0.4 kg N/m^3 d at 10 °C [23]
   - Simultaneous reduction of 95 % NH_4 and of 60 % BOD even below 10 °C [24]
   - Ammonia concentrations below 2 mg/l at loadings of 0.6 kg $Kj \cdot N/m^3$ d [21].

For average wastewater concentrations, these removal rates result in empty bed contact times below 2 hours. Because of the more stringent requirements for low residuals even during peak hours, the dynamic behavior of the plants is of primary importance [26]. Full scale results have confirmed the pilot data for tertiary nitrification [27]. Advanced nitrogen elimination to reach effluent values for sensitive areas of the EC guideline can also be achieved by adding an anoxic reactor, as practiced in drinking water denitrification [28] and wastewater nutrient removal [3, 29].

Enhanced Primary Treatment and Biological Polishing

Many ideas have been tried to enhance the efficiency of primary treatment in order to replace or reduce the size of the biological reactor [30]. Chemical treatment before activated sludge can upgrade treatment plants [31], or a combined biological and physical unit where lamella are used for anaerobic biofilm attachment achieves 75 % solids removal with detention times slightly more than one hour [32]. Similar results can be achieved with a high-load activated sludge plant, which results in a very compact solution if followed by an aerated biofilter [33].

The main idea of increasing efficiency of primary treatment is to reduce the load downstream. Therefore the bioreactor can be made smaller, less biological sludge is produced and less aeration energy is necessary. In activated sludge systems, energy requirements amount to about 1 kWh/kg BOD removed. With chemical treatment, about 0.5 mg Fe is needed per mg BOD removed. The tradeoff can easily be calculated by comparing electricity and reagent cost. Even in countries with low energy cost, the savings with chemical treatment is about 50 % [30].

Similarly, sludge production can be estimated for enhanced primary or biological treatment. Sludge production tends to be higher in quantity but less in volume with chemical treatment, and this sludge rich in organic matter can be degraded anaerobically to a mass having a final residue similar to that using biological treatment alone, yielding high biogas production. Primary sludge yield is about 1 kg SS/kg SS removed, and with chemicals this figure tends towards 1.5 [15]. Similarly, biological sludge production is about 0.8 kg SS/kg BOD. If BOD removal in the primary tank is enhanced from 30 % to 60 %, biological sludge production will be reduced by half.

To verify the reaction of a bioreactor to the two different primary effluents, pilot tests were carried out on two Biocarbone filters in parallel to compare the performance with and without chemical pre-precipitation [4]. Figure 8. shows the removal efficiency of the aerated filter as a function of the total COD loading. Since more solids are removed in the filter without chemical pretreatment, overall COD removal efficiency is higher.

A rather similar effluent quality (around 10 mg SS/l and 60 mg COD/l), however, was obtained with two different detention times: 1 h for the normal primary settling, and half an hour for enhanced primary lamella treatment. Elimination of filtered COD is comparable in both systems. As shown in Fig. 9, a consistent removal of around 60 % of filtered COD is obtained throughout a wide range of loading rates. Thus considerable savings in capital cost for the biological reactor are expected through physico-chemical pretreatment.

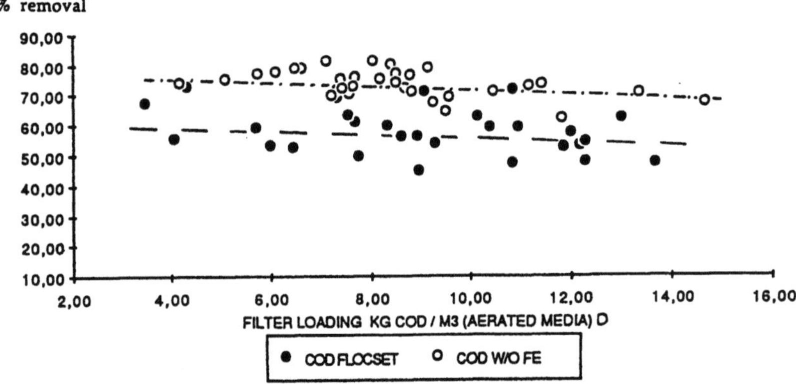

Fig. 8. Total COD removal efficiency of biocarbone

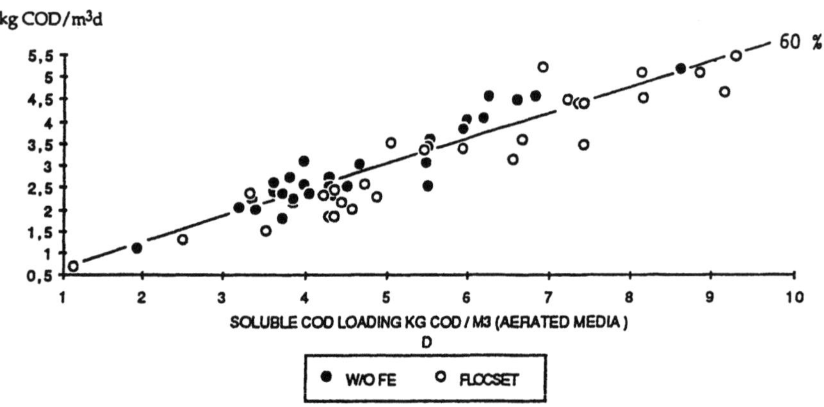

Fig. 9. Filtered COD removal rate

Full Scale Examples of Primary Lamella Settlers

Many full scale biofilters were examined on a full scale in a special study to evaluate performance [34]. Three Biocarbone units of similar size but with three different primary treatments were examined, the characteristics are summarized in Table 5. The first consisted of conventional primary settling in a circular tank, the second performed gravity settling on lamella, and the third had a lamella separator with physico-chemical enhancement. Upstream of primary treatment, screening on 20 mm is provided as well as an aerated grit chamber for sand and grease removal with a detention time of 30 minutes.

Tab. 5. Characteristics of primary stages at three biocarbone plants

Plant	Decazeville	Port Barcarés	Le Touquet
Dry weather peak flow $[m^3/h]$	580	550	994
Primary	Circular	Chemical + lamella	Lamella
Dimensions [m] number × (length × width)	Diameter 22	4 × (4 × 3.5)	2 × (8.5 × 4.3)
Total surface $[m^2]$	380	56	73
Projected lamella surface $[m^2]$	–	795	588
Volume $[m^3]$	580	246	205

The plant at Decazeville (100 km north of Toulouse) is designed to treat a load of 25 000 p.e. to secondary quality and was started up in 1985. The performance of the conventional circular settler was followed during a period where the loading averaged about half of the design. As shown in Table 6, influent concentrations are rather dilute and are eliminated by 50 % for SS and 35 % for COD. The long detention time due to underloading separates all the settleable matter, obtaining the maximum removal performance for a primary settler.

Tab. 6. Performance results of primary stages

Plant	Decazeville	Port Barcarés	Le Touquet
Test duration [d]	5	9	30
Flow $[m^3/d]$	2 005	8 363	3 238
Raw wastewater			
COD [mg/l]	244	558	5
SS [mg/l]	112	208	315
COD/BOD	2.5	2.2	2.4
Settled wastewater			
COD [mg/l]	157	230	354
SS [mg/l]	57	47	180
COD/BOD	2.6	2.3	2.0
Performance			
Settler velocity [m/h]	0.25	1.22	0.18
Detention time [min]	420	42	91
COD removal [%]	35	59	40
SS removal [%]	49	77	43

The plant at Port Barcarès (close to the Spanish border on the Mediterranean Sea) alleviates the loading to a lagoon. This tertiary treatment totals 16 ha. and its basic objective is disinfection. The plant was first started up in

1989 and is designed for a capacity of 40 000 p.e. During the tests in June and July, it was operating close to the design loading. With a rather low chemical dosage of 35 mg/l of $FeCl_3$ and 0.2 mg/l of anionic polymer, the removal of suspended solids is 75 % and COD removal is 60 %, as summarized in Table 6.

Le Touquet is situated on the channel coast in the northwest of France, and has a resident population of 10 000 inhabitants. The plant is designed for a fivefold load increase during the summer, and during the tests in July about two thirds of the capacity was treated. The primary capacity is much larger than the biological treatment in order to handle storm peaks. Thus, the overflow rate was a similar range as the above described conventional settler and comparable results were obtained with a COD and SS removal of 40 %.

Biocarbone Performance at Full Scale Plants

The above three primary treatments are all followed by Biocarbone aerated filters, which makes it possible to assess the reaction of the biological stage to different influents. The filter characteristics are summarized in Table 7. Unitary surfaces are limited to 43 m^3 in these plants, but larger units with about twice the size are installed at bigger plants [25]. Filter size and number of units are selected in order to have enough backup capacity when one filter is backwashed during the night flow. Media height is chosen in order to keep the filtration velocity below a certain value even at high organic loadings and depends thus on effluent characteristics.

Tab. 7. Dimensions of the biocarbone aerated filters

Plant	Decazeville	Port Barcarés	Le Touquet
Dry weather peak flow [m^3/h]	580	550	350
Number and dimensions [m]	3 × (8.7 × 4.9)	4 × (8.7 × 3.6)	6 × (8.7 × 3.8)
Total surface m^2	128	125	200
Media height [m]	1.75	2.1	1.7
Volume [m^3]	224	262	340
Design load [kg BOD/d]	1 200	1 200	2 000

The aim of the independent study was to verify under what practical loading conditions Biocarbone aerated filters could achieve secondary effluent quality. The number of filters was thus adjusted to the loading, leading to the use of one third of the units in Decazeville and two thirds in Le Touquet. Filter velocity averaged between less than 1 m/h up to close to 3 m/h, whereas average loading ranged between 5 and 8 kg COD/m^3 d. On the aerated biofilters, COD removal was 62 % for the most diluted wastewater and up to 75 % for the more concentrated sewage. A similar pattern was observed for SS, the filter capture ranging from 80 % on physico-chemical effluent to 90 % on a settled sewage.

Tab. 8. Results of biocarbone treatment

Plant	Decazeville	Port Barcarés	Le Touquet
Test duration [d]	5	9	30
Flow $[m^3/d]$	2 005	8 363	3 238
Filters in Service/Total	1/3	4/4	4/6
Settled Wastewater			
COD [mg/l]	157	230	354
SS [mg/l]	57	47	180
COD/BOD	2.6	2.3	2.0
Treated Effluent			
COD [mg/l]	59	74	88
SS [mg/l]	9	10	18
COD/BOD	2.8	2.7	2.9
TKN removal [%]	75	25	48
Operational Data			
Load $[kg\,COD/m^3\,d]$	4.8	7.6	6.5
Filter velocity [m/h]	0.65	2.8	1.0
Sludge yield [kg SS/kg COD]	0.35	0.44	0.4
Energy need [kWh/kg COD]	1.3	0.9	0.7

Depending on the loading, simultaneous nitrogen uptake and oxidation was obtained, with up to 75 % removal of ammonia and organic nitrogen at loadings below 5 kg COD/m^3 d. Complete oxidation and removal of nitrogen is possible with Biocarbone filters at lower loadings (35). Sludge production averaged 0.4 kg SS/kg COD removed, a value that most reports agree on [19–21]. Energy consumption depends on the loading factor and the possible regulation, it was highest for the plant with the lowest load, where blowers could not be regulated and nitrification occurred. About 80 % of the energy is needed for aeration, whereas 10 % is used for backwash and the rest is pumping energy.

Conclusion

Lamella settlers and Biocarbone aerated filters are compact wastewater treatment technology that reduce the space needed to 10^{-3} m^2/p.e. for primary and to 5×10^{-3} m^2/p.e. for secondary treatment. A typical wastewater treatment plant consisting of these two units, sludge treatment and odor removal, can accommodate between 30 and 50 p.e./m^2. Parallel plates can increase the settler overflow by a factor of 10, and chemical treatment in primary settlers can enhance SS removal up to more than 90 %. BOD removal higher than 70 % can be reached with chemical enhancement of primary treatment, achieving secondary effluent objectives for BOD concentrations below 100 mg/l in the raw sewage. Biological treatment stages can be reduced in proportion to the additional load removal of the first stage. Trade-offs can be calculated between investment cost, sludge production and energy consumption to give a flexible answer to the various effluent requirements of the European Guideline on wastewater discharges.

Acknowledgements

All of the results presented here were only obtained due to the tireless effort of many designers, builders and operators of pilot and full scale plants. We appreciate all the help and information. Special thanks in data collection to Christian Laplace, Nicolas de Longeaux, Jean-Yves Bontonou and Philippe Bonet. The full scale tests on the Biocarbone aerated filters were performed by Cemagref under the supervision of R. Pujol and financed by a common fund of the French river basin management agencies 'Agences de l'Eau' through the initiative of A. Iwema and R. Vidou.

References

[1] Commission of the European Communities: Directive regarding the treatment of urban effluents. Official Journal C 300/8, 1989

[2] Sibony, J., Rogalla, F.: Minimising Nuisances by Covering Compact Sewage Treatment Plants. Wat. Sci. Tech. *25* (1992) 1992

[3] Sauvegrain, P., Rogalla, F., Payraudeau, M., Sibony, J.: Reduced Hydraulic Detention Time for Nutrient Removal. Wat. Sci. Tech. *24* (10) (1991) 217–229

[4] Payraudeau, M., Bacquet, G., Rogalla, F., Bourbigot, M.M., Sibony, J., Gilles, P.: Nutrient Removal with Biological Aerated Filters. JWPCF *62* (2) (1990) 169–176

[5] Forsell, B., Hedström, B.: Lamella Sedimentation – a Compact Separation Technique. JWPCF *47* (1975) 834

[6] Kurbiel, J., Sapulak, A., Schade, H.: Turbulent Pipe Flow for Precipitating Electroplating Wastewater. Wat. Sci. Tech. *24* (7) (1991) 255–259

[7] Ignajtovic, L. : Efficient Tube Settling Tanks. Wat. Sci. Tech. *24* (7) (1991) 223–228

[8] Culp, G., Hansen, S., Richardson, G.: High Rate Sedimentation in Water Treatment Works. JAWWA *60* (6) (1968) 681

[9] Desbos, G., Laplace, C., Rogalla, F.: Extended Coagulation for Reagent and Space Savings with Wastewater Lamella Settling. In: Chemical Water and Wastewater Treatment, H.H. Hahn and R. Klute (eds.). Springer, Berlin Heidelberg New York 1990

[10] Hazen, A.: On Sedimentation. Transactions ASCE *53* (1904) 45

[11] Rad, H., Crosse, J.T.: Chemically Assisted Primary Treatment to Upgrade Overloaded Plants. 63rd WPCF Conf., Washington 1990

[12] Morissey, S.P., Harleman, D.R.F: Retrofitting Primary Treatment for Chemical Enhancement in the US. In: Chemical Water and Wastewater Treatment II, R. Klute and H.H. Hahn (eds.). Springer, Berlin Heidelberg New York 1992

[13] Sagberg, P., Soether, R., Baggerud Berge, A.: Increasing the Surface Load for Direct Precipitation. In: Chemical Water and Wastewater Treatment, H.H. Hahn and R. Klute (eds.). Springer, Berlin Heidelberg New York 1990

[14] Bourdon, F., Jarosz, J., Lavergne, G., Mallen, A.: Marseille Sewage Treatment Plant: 300 Days of Performance Demonstration Including Automatic Reagent Dosing. TSM l'Eau *84* (5) (1989) 311–322 (in French)

[15] Murcott, S.E., Harleman, D.R.F.: Performance and Innovation in Wastewater Treatment. Technical Note 36, R.M. Parsons Lab., MIT, Feb. 1992

[16] Leglise, J.P., Gilles, P., Moreaud, H.: A New Technology Development: the Biocarbone Aerated Filter. 53rd WPCF Conference, Las Vegas 1980

[17] Sibony, J., Rogalla, F.: Biocarbone Aerated Filters – Ten Years After: Past, Present and Plenty of Potential. Wat. Sci. Tech. *26* (9–11) (1992) 2043–2048

[18] Baccquet, G., Joret, J.C., Rogalla, F., Bourbigot, MM.: Biofilm Start-up and Control in Aerated Biofilters. Env. Tech. *12* (1991) 747–756

[19] Stensel, H.D., Brenner, R.C., Lee, K.M., Melcer, H., Rakness, K.: Biological Aerated Filter Evaluation. ASCE J. Env. Eng. *3* (6) (1988) 655–671

[20] Dillon G., Thomas,V.: Biocarbone Evaluation for Settled Sewage Treatment and Tertiary Nitrification. Wat. Sci. Tech. *22* (1/2) (1989) 305–316

[21] Smith, A.J., Hardy, P.J., Edwards, W.E., Woods, C.: Biocarbone Process Evaluation. Thames Water Annual Report, Project R 61, 1989

[22] Paffoni, C., Gousailles, M., Rogalla, F., Gilles, P.: Aerated Biofilters for Nitrification and Effluent Polishing. Wat. Sci. Tech. *22* (7/8) (1989)

[23] Fuchu, Y., Kimura, H., Tochikubo, E.: Advanced Sewage Treatment by BAF Process. 5th World Filtration Congress, Nice, June 1990

[24] Lilly, W., Bourn, G., Crabtree, H., Upton, J.: Production of High Quality Effluents Using the Biocarbone Process. JIWEM (1992)

[25] Gilles, P., Sibony, J.: Industrial Scale Applications of Fixed Biomass Reactors: Design and Operational Results. Wat. Sci. Tech. *22* (1/2) (1990) 281–292

[26] Rogalla, F., Roudon, G., Ravarini, P., Bourdon, F.: Continuous Follow-up of BAF with on Line Sensors. 5th IAWPRC ICA workshop, Kyoto, in R. Briggs (ed.), Adv. Wat. Pol. Ctrl. Pergamon, Oxford 1990

[27] Wheale, G., Williamson, S., Cooper-Smith, G.: Biocarbone for Advanced Sewage Treatment. IWEM Specialised Seminar, London, November 1991

[28] Ravarini, P., Coutelle, J., De Larminat, G., Rogalla, F.: Biological Nitrate and Ammonia Removal at Large Scale. JIWEM *4* (4) (1990) 319–329

[29] Gilles, P., Bouron, Y.: Nitrification and Denitrification with Fixed Bacteria. L'Eau, l'Industrie *93* (6) (1985) 53–57 (in French)

[30] Karlsson, I.: Chemical Treatment Alternative or Complement to Biological Treatment, Chemical Treatment. Schr. Reihe WaBoLu 62. Fischer, Stuttgart 1985

[31] Churchley, J., Lilley, W., Upton, J.: Operational Experiences of Chemical Addition for Enhanced Organics Removal. Severn Trent Water Report 0665 R

[32] Verhaegen, K., Van Rompu, K., Verstraete, W.: Pretreatment with Parallel Plate Separators. Tribune Eau *42* (540) (1989) 52–57

[33] Firk, W., Ghandehari, N.: Combining Highest-Loaded Activated Sludge and Biologically Intensified Filtration. Wat. Sci. Tech. *19* (1987) 981–992

[34] Pujol, R., Canler, J.P., Iwema, A.: Biological Aerated Filters, an Attractive Alternative. Wat. Sci. Tech. *26* (3/4) (1992) 693–702

[35] Rogalla, F., Payraudeau, M, Paffoni, C., Gilles, P.: Belüftete Filter zur weitergehenden Abwasserreinigung. GWF Wasser/Abwasser 131 (4) (1990) 178–185

Frank Rogalla
Anjou Recherche
Research Center Compagnie
Générale des Eaux - OTV
Chemin de la Digue
F-78600 Maisons Laffitte
France

Georges Roudon and
Jacques Sibony
OTV-Omnium de
Traitements et Valorisation
11, Avenue Dubonnet
F-92407 Courbevoie
France

François Blondeau
SPACE
International
Consulting
14–30, rue de Mantes
F-92411 Colombes
France

Recent Developments in Wastewater Treatment

Recent Developments in Industrial Wastewater Treatment: The Role of Physico-Chemical Processes in Metal Control

G. Tiravanti

1. Introduction

The industrial pollutants in wastewaters can be broadly categorized into organic and inorganic compounds. The most common method for treating organic pollutants is to destroy them by biological or thermal oxidation, converting them into carbon dioxide, water and other oxidation products. Inorganic pollutants cannot be destroyed, of course, and biological processes cannot be generally applied for their treatment.

Current physico-chemical methods for avoiding the adverse environmental impact of toxic inorganic contaminants involve either dispersion into the environment or conversion into a different physical form, which is generally a sludge.

Current control technologies for inorganic waste management, in general, and heavy metals, in particular, are indeed based on "end of pipe" technologies. These technologies in use, overall, are considered quite satisfying in the context of purging the wastewaters from metals prior to effluent discharge, but they generate a treatment residue, the sludge, which, in general, cannot be disposed of in a safe environmental manner [1, 2]. This is still recognized by the industrial generators of sludges who pay the expensive price of the clean-up operations of exhausted disposal sites, where leaching of metals impairs groundwater and soil quality.

This paper briefly describes the current physico-chemical processes for metal control and the new and safer management options based on "conservative technologies" for recovery and reuse of valuable by-products resulting from wastes and/or side liquid streams from manufacturing processes. Interest in developing these technologies is increasing due to the rising costs of raw metals coupled with the stringent need for their environmental control.

2. Current Physico-Chemical Technologies for Metal Control

Metals removal from wastewaters can be carried out using several processes, such as ion exchange, precipitation-coprecipitation [3] coupled to pre- or post-oxidation [4], reduction [4, 5] and concentration [6–8] of the species of interest.

All these processes are different in design, reagents, possibilities for recovering metals, etc. The chemical precipitation process including coprecipitation is the most utilized one because of its simplicity [3]. However the main disadvantages include a) large volumes of gelatinous solids with poor handling characteristics, b) production of large amounts of toxic sludges (after sludge dehydration) which must be disposed of and, in certain cases, c) poor quality of the treated water which is a function of the Ksp of the solid phases formed and of the water composition.

For metals like Al and Cr, the kinetics of precipitation is the slow stage and residual metal concentrations in solution are generally higher than those which can be predicted from thermodynamic equilibrium conditions. This phenomenon depends on the different velocities of the nucleation and of the crystal growth of the precipitates [9]. Under these conditions, coprecipitation of other substances can take place, which influences both the solid phase and the composition of the liquid phase [3].

The best operating pH value in treating wastewaters depends on the nature of the metal to be precipitated: it is generally in the range between 7 and 11. The most common reagent used is lime, due to its economy and availability.

In certain cases the velocity of sedimentation of the sludge is low, when the flocs are bulky and of colloidal nature. In these cases the process requires the addition of coagulation/flocculation aid polyelectrolytes in order to improve the settling and dewatering properties of the sludge.

The precipitation of Cd, Hg and Pb as hydroxides is generally achieved with residual concentrations in the treated water higher than those imposed by Italian law. These elements can be better precipitated as sulfides having lower solubilities than the corresponding hydroxides. Na_2S at pH 7 is generally used [10].

Interfering components in solution (forming ion complexes with metals to be precipitated) must be eliminated by suitable treatments. Cyanide ions are oxidized either by chlorination in alkaline media or by ozonation. Cr(VI) must first be reduced to Cr(III) while organo-metallic bonds must be broken by chlorination.

Summarizing, the following operations can be used in the precipitation process:

- pre-treatment of the feed water (stripping, ion exchange, chemical oxidation or reduction);
- addition of chemicals;
- coagulation and flocculation;
- solid/liquid separation (sedimentation, flotation, filtration, centrifugation, etc.).

Discontinuous plants generally operate either when the flow rates of the wastewater are below 200 m^3/d or when the composition of the feed water is highly variable so that frequent adjustments of the operating process parameters are needed. For higher flow rates, continuous operation plants are utilized; an equalization tank of suitable capacity is installed when the feed water composition is variable.

Ion exchange is a well established process, particularly for water softening. Its potential for metals pollution control is well recognized, but its applications quite limited. Three aspects of the process must be examined:

- the stoichiometry, which requires that every quantity of ions removed from the solution must be replaced by an equivalent quantity of other ionic species of the same sign;
- the concentration factors achievable in the regenerating solution can be as high as 10^4 times the concentration of the metal in the diluted solution;
- the selective nature of some ion exchange groups of the resins allows different ionic species to be separated during elution.

However, for many wastewaters, irreversible resin fouling and associated loss of exchange capacity may constitute severe problems.

The ion exchange applications for pollution control are either as tertiary (polishing) treatment of an effluent before discharge, or for direct recovery. Industrial examples include recovery of silver, where the recovered product is so valuable that the exhausted resin is directly fed to a furnace, and recovery of chromate in plating rinse waters [11].

3. Conservative Technologies

General opinion in this context involves several basic strategies to eliminate or reduce the masses of metal sludges now disposed into the environment:

- source avoidance and/or waste minimization;
- direct metals concentration after selective separation of liquid waste streams;
- waste exchange;
- selective extraction of metals for recycle, recovery and reuse.

The greatest potential for future applications is represented by the last option (recycle, recovery and reuse) up to now mainly limited to precious metals. Among the available technologies it is worth mentioning advanced precipitation, selective ion exchange, membrane processes (reverse osmosis, electrodialysis) evaporation and liquid extraction [3].

It is not possible to explore in this paper each of these treatments in detail; some comments are reported in the following and particularly on the results of research carried out by the Italian Water Research Institute (IRSA) in the field of selective metal recovery processes.

According to the strategies mentioned before, two main lines were followed. The first was to develop new precipitation processes characterized not only by attaining effluent metal concentrations down to the μg/l level for most heavy metals (Hg, Cu, Cd, Ag, Pb, etc.), but, at the same time, by ensuring recovery of metals as concentrated solutions after peculiar sludge post-treatments [12]. The second approach was to look at ion-exchange technology, applied to industrial segregated liquid streams [13].

3.1 The MEXICO Advanced Precipitation Process

The MEXICO (Metals Extraction by Xanthate Insolubilization and Chemical Oxidation) precipitation process is a typical example of the first approach. It utilizes some derivatives of annually renewable substances like starch or cellulose, as inexpensive raw materials to prepare starch or cellulose xanthates precipitating agents for metals.

The synthesis of the water soluble starch xanthate was reported previously [14] while new cellulose xanthate materials were prepared [15] and two cheap, stable and water insoluble precipitating agents were obtained: the sodium and magnesium cellulose xanthates, having chemical stability of more than one year.

It is worth noting that the MEXICO process can utilize the same unit operations in use for the traditional metal precipitation plants with soda or lime.

The MEXICO process requires a destructive oxidizing treatment of the sludge (metal xanthates) with NaClO. In this way, a concentrated solution of the metals to be recovered is obtained after filtration, and the detoxified solid residue can be safely disposed of. Metal recovery from the sludge can also be achieved by thermal treatments.

The process has been developed [16] to the stage of a demonstration continuous pilot plant of a capacity of 15 m^3/d treating, among others, real chloralkali wastewaters for mercury removal and recovery. Results show that Hg concentrations in the filtered effluent are in the range of 5–20 μg/l, with an average removal efficiency of 99.8%. The process needs small amounts of cationic polyelectrolytes (5 g/m^3) in order to increase solid/liquid separation.

The resulting sludge, after centrifugation, contains Hg in the range of about 20% (d.w.), which leads to a concentrated chlorocomplex Hg(II) solution of about 10 g/l after the oxidizing treatment with NaClO at acidic pHs. This solution can be directly recycled into the electrolytic cell or, as an alternative, it can be treated using an electrochemical cementation process, also developed at IRSA [17] to recover Hg in elemental form. This latter process is based on a rotating thin disc of aluminum as a sacrificial electrode in a CSTR reactor. Best operative conditions are reached at rotating speeds ranging between 100 and 150 rpm, where the elemental Hg recovery yield is around 95% and Al requirement for the operation on the order of 0.18 kg Al/kg Hg recovered [17].

The water insoluble cellulose xanthate can be utilized either under batch CSTR conditions or in column operations. In the latter case it works as a metal scavenger for the polishing of trace concentrations of metals in solution. As an example [15] the final product from a traditional ion exchange column of a plating wastewater, having a leakage concentration of 54 μg Cd/l, was treated in column operation with cellulose xanthate. After treatment Cd concentration was reduced below the maximum allowable limit set by Italian legislation (20 μg/l).

The technical-economical evaluation of the MEXICO process for Hg removal indicates that the total annual costs are comparable to other traditional precipitation processes such as the thiourea and sulfide processes for the treatment of chlor-alkali wastewaters. This evaluation does not take into account the benefits from the safe disposal of the treated sludge and the revenue for Hg recovery and reuse [16].

The Italian Ministry of University and Research is now funding a 5 million dollar project based on the MEXICO process for a demonstrative, skid-mounted $(150 \text{ m}^3/\text{d})$ plant to be used for the treatment of chemical, petrochemical and plating wastewater industries. The project will be operative in 1993 and it is believed that, in the mean time, this process will help in decreasing the generation of toxic sludges containing heavy metals.

3.2 The IERECROM Process for Selective Separation and Reuse of Cr(III) from Spent Tanning Baths

In connection with the strategy of the selective recovery of metals for reuse, there are many environmental problems related to the Italian tannery industry, which is, together with the United States and the former Soviet Union, the world's leading producer of leather.

It is known that tanning is among those activities most blamed for their negative environmental impact: 50 % of the incoming chemicals (essentially Cr(III) salts) are disposed of in the form of wastewaters and sludges from treatment operations. For example, in Italy about 40 million m^3 of wastewaters containing some 500 mg Cr/l and 120 000 tons (dry) of chromium containing sludges in the range of 1–4 % Cr(III) are discharged annually from the tannery industry; the final destination of these metals is generally the landfill [18]. A process has been developed in which the segregated streams from the tannery process are submitted to selective ion exchange. The process, called IERECROM (Ion Exchange Recovery of Chromium), makes it possible to separate and recover chromium from other interfering metals such as Fe and Al from spent tanning baths. The metal-free effluent is sent to the wastewater treatment plant [19].

The process is based on a carboxylic resin partially hydrolyzed in Na-form, which retains the three metals (Cr(III), Al(III), and Fe(III)) at pH 3 from the liquid phase during the exhaustion step, thus confirming the reasonably good affinity of the resin toward all these species.

After retention, it is not an easy task to regenerate metal ions from weak carboxylate resins using conventional chemicals. Quantitative elution of chromium and aluminum was obtained by regenerating the resin with hydrogen peroxide/sodium hydroxide solutions. The oxidation of Cr(III) to form anionic species, such as chromate, $CrO_4^=$, and/or dichromate, $Cr_2O_7^=$, allow these anions to be rejected, by the Donnan effect, from the resin-phase [20]. Using an excess of base and in the presence of chlorides, the oxidation reaction is fast, especially at temperatures ranging around 60 °C [21].

The spent regenerant solution contains chromates and aluminates; separation can be easily achieved by lowering the pH to around 8. Al(OH)$_3$ precipitates, thus separating it from the chromate solution. Both chemicals can be reused.

No significant degradation of the resin performance has been observed, working with hydrogen peroxide, at least at the end of the present laboratory investigation. More cycles will be run in order to evaluate resin reliability (in terms of fouling and/or degradation of the resin matrix) in the medium and the long range. Anyway, it is too early at this stage to make an economic evaluation of the process.

It is worth noting that the Italian Ministry of Environment has provided the IRSA with about 800 000 dollars in funds for building a pilot plant (2.4 m^3/d) and running experiments utilizing the above mentioned ion-exchange processes both for segregated tannery baths and acid extracts of tannery sludges. The pilot plant is under construction and data will be available in 1993.

4. Future Development

Among the different areas for future development, the category of selective recovery technology seems to be the most promising, but the development and the practical application of selective processes requires more research in terms of better understanding of chemical speciation and kinetic control.

Several extraction, separation and concentration processes are potentially applicable to metals recovery from industrial wastewaters. These technologies can be applied to soluble (wastewaters) or insoluble (solid matrices, sediments, sludges) metals. The selective adsorption of specific metals employing mineral oxides, sand, polymeric materials or inactivated microorganisms is a fast growing sector [22]. However the application in full scale plants has not been demonstrated.

The non-selective extraction of various metals from waters is generally performed via solvent extraction. Several applications are described in the literature; as examples, Cu, Ni, Fe, Cd, Zn, Pb, V, Ti are extracted by using 8-hydroxyquinoline as chelating agent [23]. More interesting should be the development of specific organic extractants for selective separation of a single metal in a mixture, as is the case with the dimethyl glyoxime for Ni.

Ion exchange is a relatively non-selective process, taking into account the limitations when using chelating resins [24]. It is most likely applicable in a facility having automated systems to isolate individual waste streams containing single metals. This was the case for a metal plating industry producing printed circuit boards [25], recovering Cu, Pb, Ni and Zn separately by ion exchange, thus decreasing the sludge generation in the wastewater treatment plant by 70 %.

Metals in solid matrices (sludges, sediments, contaminated lands and dusts) can be recovered by extractive metallurgy, mainly when one of the metals is present as a major component. It is possible to use pyrometallurgy (thermal techniques), hydrometallurgy (leaching techniques) and biohydrometallurgy (microbiological enhancement of leaching). These technologies have potentials not yet well established. More fundamental research is needed to study surface phenomena using techniques such as X-ray microprobe, ESCA, XPS and so on [26] to interpret and model ion migration into solid materials.

5. Conclusions

The physico-chemical treatment processes can be applied to inorganic contaminants, not biodegradable. These technologies can give us a great help in minimizing waste production, especially with reference to toxic heavy metals, up to now generally disposed as sludges in landfills, for this reason, due to the biorofractory nature of the metals, physico-chemical conservative technologies for reuse are not an option.

More research is needed in this sense and it is hoped that the chemical engineering sciences will greatly improve future recycling and reuse of industrial wastes in order to preserve the environment.

Acknowledgements

The author wishes to thank UNIC (Unione Nazionale Industria Conciaria) for cooperation and Mrs. G. Bagnuolo, R. Ciannarella, N. Limoni and Miss N. Rapanà for fruitful technical assistance.

References

[1] Griffin, F.A., Frost, R.R., Au, A.K., Robinson, G.D., Shimp, N.F.: Attenuation of Pollutants in Municipal Landfill Leachate by Clay Minerals. Part II: Heavy Metal Adsorption. Environmental Geology Notes *79* (1977)

[2] Hoeks, J.: Pollution of Soil and Groundwater from Land Disposal of Solid-Wastes. Inst. Land Water Management Research Wageningen, The Netherlands. Tech. Bull. *96* (1976)

[3] Patterson, J.W.: Metals Separation and Recovery. In: Metal Speciation Separation and Recovery, J.W. Patterson and R. Passino (eds.). Lewis, Chelsea, MI, USA, 1987

[4] Nemerow, N.L.: Theories and Practices of Industrial Waste Management. Addison Wesley, Reading, MA, 1963

[5] Eckenfelder, W.W., Jr.: Industrial Waste Pollution Control. McGraw-Hill, New York, NY, 1966

[6] Dillman, T.R.: Metal Ion Removal and Recovery. In: Ion Exchange for Pollution Control, C. Calmon (ed.). C.R.C. Press, Boca Raton, FL, 1979

[7] Gold, H.: Metal Finishing Wastes. In: Ion Exchange for Pollution Control, C. Calmon (ed.). C.R.C. Press, Boca Raton, FL, 1979

[8] Clifford, D., Subramanian, S., Song, T.: Recovery of Dissolved Inorganic Contaminants from Wastes. Env. Sci. Tech. *20* (11) (1986) 1072

[9] Smith, R.W.: Relations Among Equilibrium and non Equilibrium Aqueous Species of Aluminum Hydroxy Complexes. Advances in Chemistry Series no. 106, Washington, Amer. Chem. Soc., 1971

[10] Argo, G.G., Culp, G.L.: Heavy Metals Removal in Wastewaters Treatment Processes. Wat. Sewage Works *119* (1972) 62–65, 128–312

[11] Liberti, L.: Scambio Ionico, Quaderno IRSA *71* (1986) 171–191 (in Italian)

[12] Marani, D., Mezzana, M., Passino, R., Tiravanti, G.: Treatment of Industrial Effluents for Heavy Metal Removal Using the Water Soluble Starch Xanthate Process. Proc. Int. Conf. Heavy Metals in the Environment, Amsterdam, Sept. 1981, pp. 92–95

[13] Petruzzelli, D., Liberti, L., Passino, R., Tiravanti, G.: Specific Resins for Metal Ion Separation. In: Recent Developments in Ion Exchange, P.A. Williams (ed.). Elsevier, London 1990, pp. 265–275

[14] Swanson, C.L., Wing, R.E., Doane, W.M., Russel, R.R.: (1973): Mercury Removal from Wastewater with Starch Xanthate – Cationic Polymer Complex. Env. Sci. Tech. *7* (1973) 614–619

[15] Tiravanti, G., Marani, D., Passino, R., Santori, M.: Removal of Heavy Metals from Wastewaters by Cellulose Xanthate Chelating Exchangers. Env. Protect. Eng. *15* (1989) 25–33

[16] Tiravanti, G., Di Pinto, A.C., Macchi, G., Marani, D., Santori M., Wang, Y.: Heavy Metals Removal. Pilot Scale Research on the advanced MEXICO precipitation Process. In: Metal Speciation, Separation and Recovery, J.W. Patterson and R. Passino (eds.). Lewis, Chelsea, MI, 1987

[17] Tiravanti, G., Ciannarella, R., Passino, R., Santori, M.: Recovery of Elemental Mercury from Chlor-Alkali Sludge by Cementation with Aluminum. 2nd Int. Conference on Environ. Protection, S. Angelo, Ischia, Oct. 5–7, 1988, 4A, 47–54

[18] Simoncini A., Ummarino G.: Possibilità reali di recupero di sottoprodotti dell'industria conciaria. Cuoio Pelli Materiali Concianti. *64* (6) (1988) 528 (in Italian)

[19] Petruzzelli, D., Santori, M., Passino, R., Tiravanti, G.: Cr(III) Recovery and Separation from Spent Tannery Baths by Carboxylic Ion Exchange Resins. Proc. ICIE '91, Tokyo, Japan, Oct. 1–4, 1991

[20] Helfferich, F.,: Ion Exchange. McGraw Hill, New York, NY, 1962

[21] Macchi,G., Pagano, M., Pettine, M., Santori, M., Tiravanti, G.: A Bench Study for Cr(III) Recovery from Tannery Sludge. Wat. Res. *25* (1991) 1019–1026

[22] Edwards, M., Benjamin, M.M.: Adsorption Filtration Using Coated Sand: a New Approach for Treatment of Metal-Bearing Wastes. WPCF Annual Conf., Dallas, Texas, 1988

[23] Clevenger, T., Novak, J.: Recovery of Metals from Electroplating Wastes Using Liquid. Liquid Extraction. J. WRCF *55* (1983) 984–989

[24] Millar, J.R., Petruzzelli, D., Tiravanti, G.: Some Problems in the Use of Chelating Resins for Environmental Protection from Toxic Metals, In: Recent Developments in Separation Science, P.A. Williams and M.J. Hudson (eds.). Elsevier, 1990, pp. 337–346

[25] Anonymous: Automated Waste Recovery Cuts Materials and Disposal Costs. Chem. Eng. *1* (1989) 50

[26] Desimoni, E., Marcone, A., Tiravanti, G., Zambonin, P.G.: Metal Speciation in Sludges from Wastewaters Treatment by Bulk and Surface (XPS) Analysis. In: Metal Speciation Separation and Recovery, J.W. Patterson and R. Passino (eds.). Lewis, Chelsea, MI, 1990, pp. 45–68

G. Tiravanti
Istituto di Ricerca sulle Acque, CNR
Via De Blasio n. 5
I-70123 Bari
Italy

Nitrogen Elimination from Digester Supernatant with Magnesium-Ammonium-Phosphate Precipitation

H. Siegrist, D. Gajcy, S. Sulzer, P. Roeleveld, R. Oschwald, H. Frischknecht, D. Pfund, B. Mörgeli, and E. Hungerbühler

Abstract

Ammonium nitrogen was eliminated from digester supernatant (0.6–0.7 g NH_4-N/l) by magnesium-ammonium-phosphate (MAP) precipitation in laboratory and pilot-scale tests (three 0.5 m^3 reactors in series). MAP precipitation was carried out by adding phosphoric acid and magnesium oxide in a Mg : P : N ratio of 1.3 : 1 : 1. Prior to MAP precipitation, the remaining suspended solids in the digester supernatant were eliminated by flocculation. Phosphoric acid was added to the inlet of the first reactor, where the CO_2 produced was stripped. Magnesium oxide was added to the second reactor where 70 % of the ammonium was eliminated. In the third reactor, pH was adjusted to 9.0. 85–90 % of the ammonium was eliminated at a hydraulic load of 0.5 m^3/d. The MAP slurry (12–14 g TS/l) was directly dewatered with a decanter centrifuge to 50 % dry solids.

The recirculation of magnesium and phosphate was investigated, because the phosphate input during MAP precipitation is high. By combining MAP precipitation with biological phosphate elimination, phosphate released during mesophilic digestion can be used for MAP precipitation. The quality (heavy metal and TOC concentration) and utilization of MAP crystals as a crude product for fertilizers and addition to compost and dry sewage sludge were also investigated.

Introduction

The European community's new regulations on the treatment of municipal wastewater from 21 May 1991 [1] require a 70 to 80 % nitrogen removal for treatment plants above 10,000 population equivalents for sensitive zones (e.g. eutrophic coastal waters). This measure would reduce the nitrogen load of the Rhine River in the Netherlands by 25–30 %. In Switzerland it would only result in a nitrogen reduction of 10–15 %, since the nitrogen input from diffuse sources (mainly atmosphere and agriculture) is about 70–80 % [2]. Investment costs of about 1.2 billion Swiss francs [3] are needed for the installation of a permanent denitrification zone of 40 % of the total activated sludge volume in treatment plants greater than 10 000 p.e. and nitrification/denitrification for treatment

plants, where Swiss law would not demand nitrification, and would reduce the
nitrogen load in waters by about 10 %. This effort would only be beneficial if
the nitrogen input from diffuse sources could also be reduced.

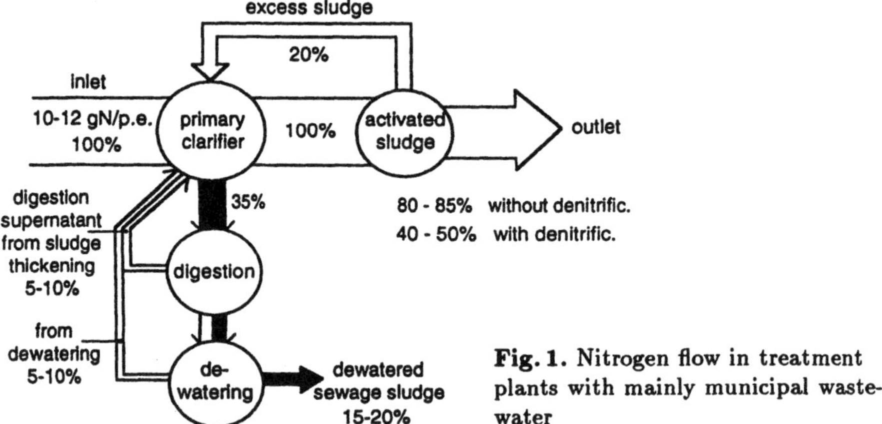

Fig. 1. Nitrogen flow in treatment plants with mainly municipal waste-water

Figure 1 illustrates the nitrogen flow through a wastewater treatment plant.
The incorporation of nitrogen into the sewage sludge reduces nitrogen load in
the effluent to about 15 %. With only a summer denitrification, the nitrogen
elimination would be increased to 25 %. A permanent denitrification zone with
40 % of the activated sludge volume would increase nitrogen elimination up
to 60 %. More than 70 % nitrogen removal by denitrification would hardly be
reached with the mostly diluted Swiss wastewater without adding a readily
degradable substrate, in other words, acidification of the raw sludge, and reduction of the voluminous primary sedimentation tanks (hydraulic detention
time mostly above 3 hours). In addition to denitrification, 70 to 80 % nitrogen
removal could also be achieved by eliminating the ammonium nitrogen from
the digester supernatant. The nitrogen load of the digester supernatant of a
municipal treatment plant is 10–20 % of the total load, depending on the degree of sludge dewatering. A separate treatment of the supernatant might be
advantageous if a substantial amount of sewage sludge from other treatment
plants is also dewatered and the sludge water partly inhibits nitrification.

Ammonium Elimination from Digester Supernatant

Ammonium can be eliminated from digester supernatant by ammonia stripping
or by magnesium-ammonium-phosphate (MAP) precipitation. Because steam
is usually not available in a treatment plant, ammonia stripping is done by air
stripping with a stripping and absorber column (recirculation of air). In the
adsorber column, the free ammonia is bound with sulfuric acid and an approx.

40 % $(NH_4)_2SO_4$ solution is produced. The pH of the supernatant is increased above 11 by adding NaOH or $Ca(OH)_2$. $Ca(OH)_2$ addition would be cheaper if the $CaCO_3$ sludge which is produced (ca. 9 kg TS/kg N) could be easily utilized in agriculture. With NaOH addition the elimination of 1 kg N requires 24 kg NaOH (30 %), 9.6 kg H_2SO_4 and yields 12 kg of $(NH_4)_2SO_4$ solution [4].

In analytical chemistry, the precipitation of magnesium, ammonium and phosphate to the insoluble MAP is a well known reaction to analyze magnesium:

$$Mg^{2+} + PO_4^{3-} + NH_4^+ \iff MgNH_4PO_4 \cdot 6\,H_2O \qquad pKs = 12.6$$

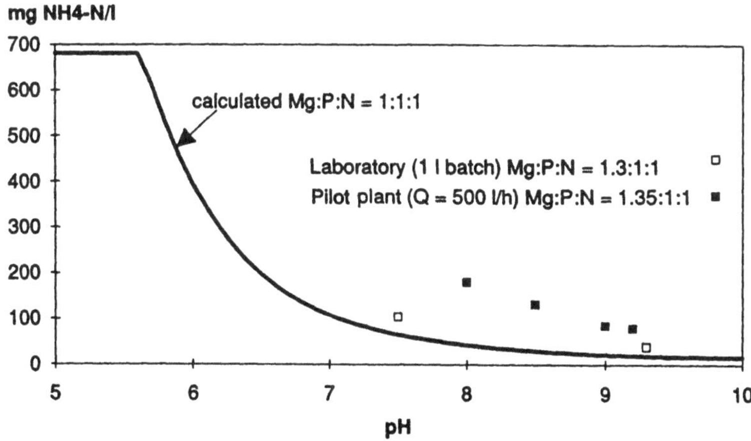

Fig. 2. The efficiency of MAP precipitation depends on pH. Optimum values are reached at pH 9. Average ammonium concentration in supernatant was 680 mg NH_4-N/l. At pH 7.5 no NaOH was added for pH control

MAP crystals contain 6 Moles of cristall water, the molecular weight is 245 g and theoretically 17.5 g MAP/g NH_4-N is formed. MAP is a basic salt and is soluble in acid. The precipitation is therefore more efficient with increasing pH (Fig. 2). The salt which is produced will smell strongly of free ammonia, however, if the pH exceeds 9.5.

The excess in magnesium is necessary to lower the equilibrium concentration of NH_4, to save NaOH, and to prevent recirculation of phosphate to the treatment plant by hazardous overdosage of phosphoric acid.

The particulate MgO dissolves slowly and needs a longer reaction time than dissolved $MgCl_2$. The addition of $MgCl_2$, however, requires a lot more NaOH (24 kg 30 % NaOH/kg NH_4-N) and increases the chloride concentration of the wastewater substantially.

MAP precipitation was carried out by adding phosphoric acid and magnesium oxide in laboratory and pilot-plant tests. The pilot plant consists of three reactors in series, each reactor having a 0.5 m^3 volume (Fig. 3).

Prior to MAP precipitation, the remaining suspended solids in the digester supernatant are eliminated by flocculation with a high cationic polyelectrolyte.

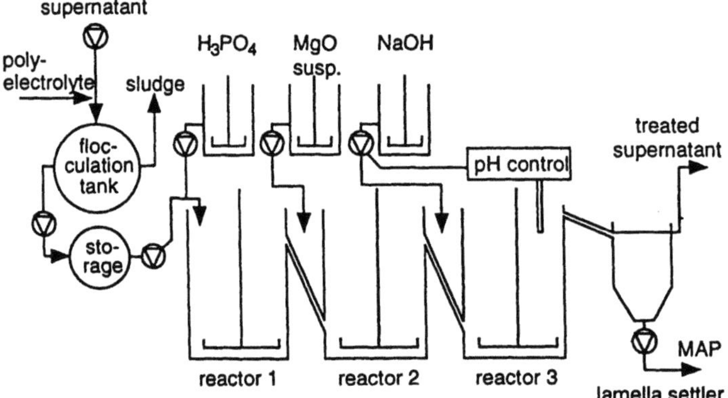

Fig. 3. Pilot plant for MAP precipitation (three 0.5 m³ reactors)

Phosphoric acid is added to the inlet of the first reactor, where the CO_2 produced is stripped. To save reactor volume, CO_2 stripping might also be carried out in a short shallow channel. Magnesium oxide (fine grained caustic: active surface 30 m²/g, average diam. 20 μm) is added dry or as a slurry to the second reactor where about 70 % of the ammonium is eliminated. In the third reactor, pH was adjusted to 9.0. 85–90 % of the ammonium was eliminated at a hydraulic load of 0.5 m³/h (hydraulic detention time 3 hours) with a Mg : P : N ratio of 1.35 : 1 : 1. Batch experiments showed that hydraulic detention time in the second and third steps can be reduced 20 % during peak load. The MAP crystals were separated either with a lamella settler and dewatered by normal gravity to 30 % total solids (TS) in porous 1 m³ plastic bags or were dewatered directly without sedimentation to 50 % TS with a decanter centrifuge.

MAP precipitation with MgO

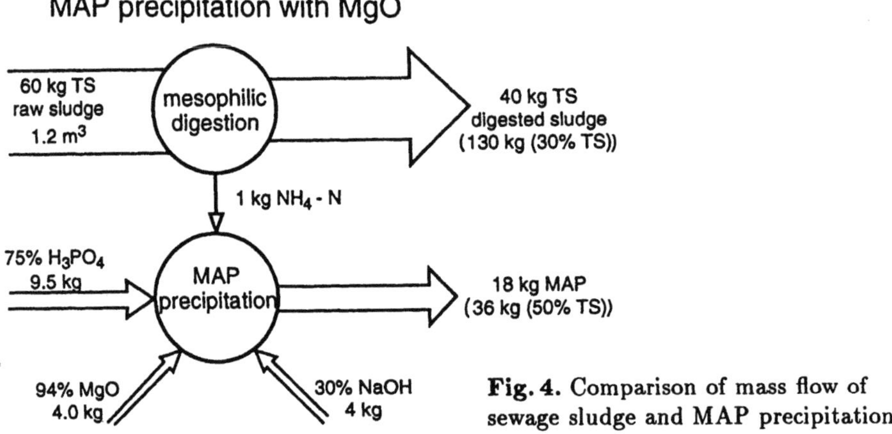

Fig. 4. Comparison of mass flow of sewage sludge and MAP precipitation

In Fig. 4, the mass flow of MAP precipitation is compared to the total solids flow in sewage sludge treatment. During digestion, 60 kg total solids with 4 %

nitrogen content will be reduced to about 40 kg of digested sludge. About 40 %
of the organic nitrogen (1 kg N) will be mineralized and appears as ammonium
in the supernatant. By adding 9.5 kg H_3PO_4 (70 %), 4 kg MgO and about 4 kg
NaOH (30 %) to 1 kg of ammonium nitrogen, 18 kg of MAP is produced. As seen
from Fig. 4, a substantial amount of dewatered MAP salt is produced compared
to the dewatered digested sludge: 0.27 kg MAP (50 % TS)/1 kg digested sludge
(30 % TS). The addition of phosphoric acid to MAP precipitation increases the
phosphate input to the sewage sludge treatment from about 0.035 kg P/kg TS
digested sludge to 0.08 kg P/kg TS. This process is thus only possible if the
phosphate in the MAP replaces actual P fertilizers.

Since the input of phosphorus to a treatment plant is more than doubled
with MAP precipitation, additional experiments with recirculation of magnesium and phosphate were carried out. With a thermal treatment at 110 °C and
previous addition of NaOH (Na : P ratio equal 1 : 1), nitrogen will escape to
more than 90 % as free ammonia and a 15–20 % NH_3-solution can be produced:

$$MgNH_4PO_4 + NaOH \longrightarrow MgNaPO_4 + NH_3$$

If NaOH is not added, ammonia will only partly escape. By adding the recycling product $MgNaPO_4$ to the digester supernatant, MAP is slowly produced
at pH 9 without adding NaOH (Fig. 5). The thermal treatment step and recirculation of magnesium and phosphate has not yet been tested on a pilot scale.
Probably only 70–80 % of the magnesium and phosphate will be recyclable because of increasing calcium content in the recycled product. The NH_3 solution
may be utilized as N fertilizer [5] or for NO_x-reduction in flue gas cleaning.

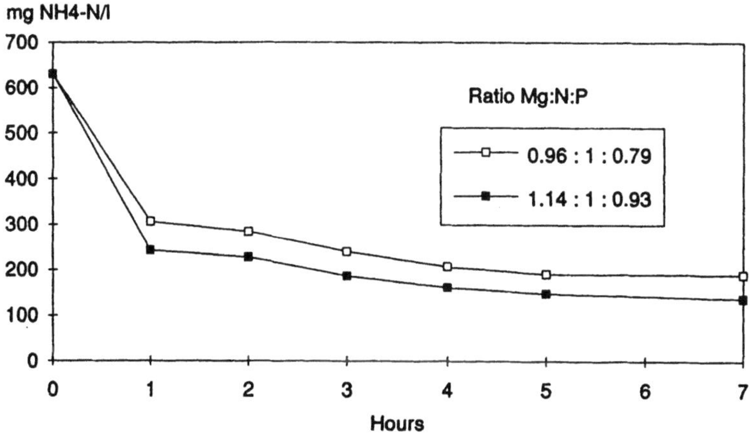

Fig. 5. NH_4 profile after addition of the recycled product ($MgNaPO_4$) to the digester
supernatant in a one liter batch reactor (pH = 9)

The MAP process can be combined with biological phosphate elimination,
since the polyphosphate of the phosphate-storing microorganisms will be hydrolysed and phosphate will partly be released to the supernatant during di-

gestion. A maximum of 10–15 % of ammonium will be eliminated by MAP-precipitation with the biologically-released phosphate. Simultaneous to phosphate release, mainly the opposite ions magnesium and potassium are released. MAP will therefore to some extent precipitate within the digester. By adding MgO or $MgCl_2$ to the supernatant, MAP precipitation can be optimized for P removal to prevent recirculation of phosphate to the activated sludge system.

Quality and Utilization of MAP

The heavy metal and organic carbon concentration in the MAP salt and the treated supernatant depends on the chemicals used for MAP precipitation and the digester supernatant (Tab. 1). By using Dorr (green) phosphoric acid, the concentration of chromium, zinc, and cadmium in MAP and digested sludge is increased (Tab. 1). With thermal (technical grade) phosphoric acid, the heavy metal inputs are very low, but the price for MAP precipitation is substantially increased (4 Fr./kg N). The organic carbon content (< 1 g TOC/kg MAP, $<$ 0.5 g DOC/kg MAP) of the product is mainly in the supernatant and depends on the efficiency of the flocculation/sedimentation process and washing of the crystals during dewatering.

Tab. 1. Heavy metal concentration of MAP and estimated increase of heavy metal concentration in the digested sewage sludge by use of Dorr (green) phosphoric acid

	H_3PO_4 (70 %) techn. green (detected) ppm	MgO (94 %) (detect.) ppm	MAP mg/kg TS (detect.) ppm	MAP mg/kg P (detect.) ppm	Increase in digest. sludge (estimat.) ppm
Cd	16	1	5	45	1
Cr	400	5	200	1 800	20
Cu	2	6	10	90	< 1
Mo	5	5	10	90	< 1
Ni	60	6	50	450	3
Pb	10	50	< 10	< 90	< 2
Zn	440	20	150	1 300	40
Hg	< 0.05	< 0.05	< 1	< 9	< 0.01

A direct addition of the dewatered MAP to the sewage sludge will hardly be possible because of a substantial increase in the phosphate concentration and the total solids concentration (Fig. 4). Agricultural use of sewage sludge is limited mainly because of phosphate input.

The availability of Mg, P and N from MAP in soil is 77, 77, and 89 %, respectively, compared to commercial fertilizers [4]. Due to the ready availability of MAP, it can be utilized as an addition to compost, garden soil or dried sewage sludge and as a raw material for fertilizer production.

For addition to compost, soil and sewage sludge MAP has to be dried for easier handling (regular addition). To minimize loss of ammonia, drying should be at a temperature below 50 °C [4]. A considerable demand for these organic-mineral fertilizers exists for recultivation and grassland cultivation. Addition of MAP and other nutrients to compost can result in a valuable product, which could replace organic-mineral fertilizers based on peat [5].

The sales prospects of MAP as final product are rather low. As a raw product for fertilizer production, MAP might partly replace imported phosphate. But for mixing with other salts, the cristall water content of MAP is too high (44 %). The release of cristall water from $MgNH_4PO_4 \cdot 6H_2O$ to $MgNH_4PO_4 \cdot 1H_2O$ will only be possible at higher drying temperatures where ammonia also escapes to some extent. The income from the nutrients in MAP hardly covers the cost for drying.

Comparison of Operation Cost

Tab. 2. Operation and investment cost of stripping and MAP precipitation for a treatment plant for 100 000 p.e.

	Air stripping sFr./kg NH₄-N	MAP precipitation sFr./kg NH₄-N
Chemicals	6.50	8.50–12.50
Operation (electricity, control, mainten.)	3.50	2.50
Investment (amortiz. 12 years, 4 % interest)	5.00	5.00
Processing of product (minus income)	1.00	–
Total	16.00	16.00–20.00

The cost of ammonium elimination from digester supernatant will be in the range of 16 to 20 SFr./kg N (Tab. 2). With Dorr phosphoric acid, the cost of MAP precipitation (16.-SFr./kg N) is equal to the stripping cost. If thermal phosphoric acid has to be used for MAP precipitation, air stripping will be about 20 % cheaper than MAP precipitation (20.-SFr./kg N). One must consider, however, that the cost of NaOH will rise in the future, which will also increase the cost of air stripping substantially. Adding $Ca(OH)_2$ is only advantageous if the $CaCO_3$ sludge which is produced can be utilized in agriculture.

Compared to nitrification/denitrification, treatment of digester supernatant will only be economical if more than 60 % nitrogen removal is demanded (Fig. 6) or if positive effects on plant operation (e.g. no inhibition of nitrification) will result. Denitrification rates are calculated with A 131 [6] and verified with the calibrated mathematical model of the IAWPRC task group on mathematical modelling [7] for summer and winter conditions. In Switzerland, the denitrification rate might be slightly reduced due to wastewater dilution and, to some extent, oxidation of the readily degradable organics in the sewer system.

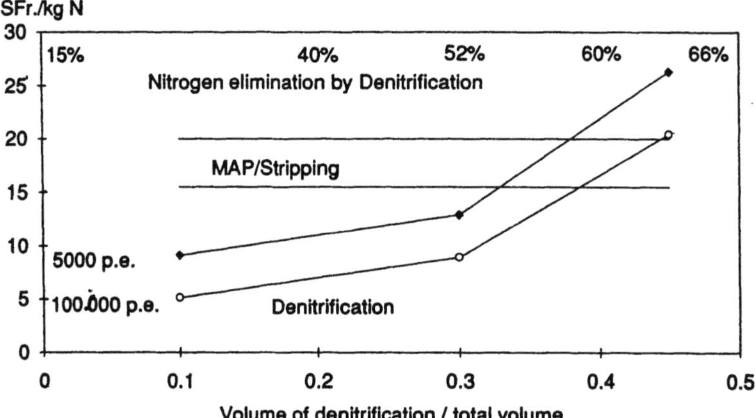

Fig. 6. Comparison of estimated cost of NH_3 stripping, MAP precipitation and nitrification/denitrification. Only the cost of denitrification (additional volume) is included. The cost for the enlargement from BOD removal to nitrific./denitrification is in the range of 15–20 SFr./kg N eliminated

Conclusions

A 60 % reduction of the nitrogen load at treatment plants above 10 000 p.e. will reduce the nitrogen load in Swiss waters by about 10 % and require investment costs of 1.2 billion Swiss francs. This measure is therefore only sensible if nitrogen input from diffuse sources (atmosphere, agriculture) is also reduced.

With a permanent denitrification volume of 40 % of the total activated sludge volume, a nitrogen elimination of 60 % should be obtained. Ammonium elimination in the digester supernatant will increase nitrogen elimination another 10 to 20 % depending on the degree of sludge dewatering. Ammonia stripping and MAP precipitation are both feasible processes for eliminating ammonium from digester supernatant. MAP precipitation will be about 20 % more expensive than stripping, if thermal phosphoric acid has to be added to keep heavy metal input low. The phosphate input to the treatment plant is about doubled with MAP precipitation. Recirculation of phosphate and magnesium after ammonia elimination by thermal treatment of MAP would reduce phosphate input and consumption of chemicals, but has not yet been tested in a pilot stage.

The cost of eliminating ammonium from the supernatant is higher than eliminating nitrogen with denitrification. The treatment of the digester supernatant is only economically feasible if more than 60 % nitrogen removal is demanded or if treatment of the supernatant will have positive effects on nitrification at treatment plants with a high percentage of ammonium in the supernatant. MAP precipitation might be a favorable process to prevent recirculation of phosphate from digester supernatant, if biological phosphate elimination is installed in the activated sludge system.

References

[1] Richtlinie des Rates vom 21. Mai 1991 über die Behandlung von kommunalem Abwasser. Amtsblatt der Europäischen Gemeinschaften, Nr. L 135/40.

[2] Zobrist J., Bührer, H., Davis, J.S.: Stickstoff in Wasser und Luft, Implikationen für den Gewässerschutz. Informationsveranstaltung 1990 "Von der Forschung zur Praxis". EAWAG News *30* (1990)

[3] Bundesamt für Umwelt, Wald und Landschaft (BUWAL), 1991 (personal communication)

[4] Gesamtbericht "Ammoniumrückgewinnung aus Schlammwasser", 1992 (in preparation)

[5] Moll, W.: Studie zur Verwertung nährstoffhaltiger Schlämme. Inst. für Bodenkunde der Justus-Liebig-Universität, Giessen, BRD, 1991

[6] Regelwerk A 131: Bemessung von einstufigen Belebungsanlagen ab 5000 Einwohnerwerten. Abwassertechnischen Vereinigung (ATV), 1991

[7] IAWPRC task group on mathematical modelling for design and operation of biological wastewater treatment: Activated sludge model Nr. 1. Scientific and technical report No. 1, IAWPRC, London, 1987

H. Siegrist, D. Gajcy, S. Sulzer, and P. Roeleveld
Swiss Federal Institute for Water Resources and Water Pollution
Control
CH-8600 Dübendorf
Switzerland

R. Oschwald, H. Frischknecht, and D. Pfund
Stadtentwässerung Zürich
CH-8064 Zürich
Switzerland

B. Mörgeli and E. Hungerbühler
Aqua System AG
CH-8400 Winterthur
Switzerland

Zeolite Filters
in Wastewater Treatment Plants

H. Witte and M. Keding

1. Introduction

German legislation has introduced minimum effluent standards for total nitrogen at 18 mg N/l, measured in a short time sample (< 2 h). A summary of the presently valid requirements is presented in Table 1 [1].

Tab. 1. Minimum requirements for wastewater discharges from public wastewater treatment plants [1]

Class PE	COD mg/l	BOD$_5$ mg/l	NH$_4$-N mg/l	N$_{tot}$ mg/l	P$_{tot}$ mg/l
> 1 000	150	40	–	–	–
1 000–5 000	110	25	–	–	
5 000–20 000	90	20	10	18	–
20 000–100 000	90	20	10	18	2
> 100 000	75	15	10	18	1

Depending on the region or into which river the treated wastewater is discharged, additional limits are set, for example, ammonia values of 2 mg/l or 4 mg/l, even at times when temperatures reach down to 5 °C. What the biological processes are able to perform is given in the following Table 2 [2] in accordance with ATV (German Association for Water Pollution Control) publications. Following general research, in Germany effluent values up to 3 mg NH$_4$-N/l can be reached safely down to wastewater temperatures of 8 °C in short time samples.

Because of these values, wastewater treatment plants must be equipped with large volume nitrification and denitrification stages and thus also produce high operating costs. Furthermore operating safety is lacking the nearer one comes to the limits of performance of such activated sludge plants. This is all the more important as, with the increase in the number of biological wastewater treatment stages, the performance limit sinks exponentially [3, 4]. Thus it is necessary to look for a technology which might assist with the establishment of a more economical plant design. A possible answer appears to be given by ion exchange filters which replace sand filtration, which itself, in any case, is necessary to keep back suspended solids to obtain low COD values.

Tab. 2. Efficiency of various wastewater treatment processes (extract according to [2])

Treatment process	BOD$_5$		COD		NH$_4$-N		P$_{tot}$		A$_s$
	A[1]	M[2]	A[1]	M[2]	A[1]	M[2]	A[1]	M[2]	A[1]
	mg/l	mg/l	mg/l	mg/l	mg/l	mg/l	mg/l	mg/l	mg/l
Activated sludge process with $B_{DS} = 0.15$ kg/(kg × d)	12	20	60	90	5	10	*	–	20
Activated sludge process with series pre-denitrification (200 % return feed)	12	20	60	90	5	10	*	–	20
Activated sludge process with $B_{DS} = 0.15$ kg/(kg × d) and rapid sand filtration	6	12	45	70	5	10	*	–	7
Activated sludge process with $B_{DS} = 0.3$ kg/(kg × d) and simultaneous precipitation	12	20	60	90	*	–	1	2	18
Activated sludge process with $B_{DS} = 0.15$ kg/(kg × d)	10	15	55	75	5	10	1	2	18
Activated sludge process with $B_{DS} = 0.3$ kg/(kg × d) simultaneous precipitation and flock filtration	6	12	45	70	*	–	0.3	0.5	5
Activated sludge process with $B_{DS} = 0.15$ kg/(kg × d) simultaneous precipitation and flock filtration	4	10	40	60	5	10	0.3	0.5	5

[1] A = Average; [2] M = Monitoring

2. Zeolite Filter as Ion Exchanger

In the Federal Republic of Germany research work on this topic is somewhat sparse. In the eighties trials were carried out with natural ion exchangers within the scope of thesis work at the RWTH [Rheinisch-Westfälische Technische Hochschule] Aachen and assessed as too expensive [5]. In the USA the properties of natural zeolites has been investigated since around 1960. The properties of natural zeolites as ammonium selective ion exchangers was discovered for the first time by Ames [6]. After this first application investigations on ammonium extraction from wastewater were undertaken. Koon and Kaufmann carried out detailed investigations in a filter column and dealt with the regeneration conditions [7]. Similar investigations were also undertaken by Liberti and colleagues [8, 9] in pilot plants with 10 and 240 m^3 wastewater per day. The effects of zeolites on the other wastewater treatments were investigated when zeolite A was used in washing agents [10–12]. Adsorption isotherms for ammonium ions on suspended clinoptilolit, which was dosed together with iron sulphate, were published by Ishii and Kaji [13]. Along with the American research intensive investigations were also carried out in Hungary [14–16]. Supplementary investigations were also carried out in Italy [17].

The criteria presented in Table 3 were determined as essential influence factors. From these, dimensioning recommendations were deduced with which, using US American zeolite, efficiencies of over 95 % with regard to the elimination of ammonium nitrogen were achieved on a semi-commercial scale with average outflow concentrations < 1 mg/l. The results are presented in detail in a manual of the US EPA (Environmental Protection Agency).

Tab. 3. Results of the EPA study – Dimensioning recommendations

Loading:	15 bed volumes/h
	Grain size 0.2–2.0 mm
	pH value 4–8
	Filter height 1–2 m
Regeneration:	15 bed volumes/h
	Different combinations from
	pH value/NaCl conc./volumes
Performance achieved:	95.7 % efficiency for NH_4-N
	with 0.75 mg NH_4-N/l in outflow

3. The Ion Exchange Process

For general understanding it is pointed out that no oxidation of ammonium nitrogen, as with the biological nitrification with subsequent denitrification, takes place, but rather a direct elimination of ammonium nitrogen in one process step. The process is shown in principle in Fig. 1.

The left half of the diagram describes the loading process while the right half shows the regeneration process. Loading and regeneration form a complete ion exchange cycle. In the loading process the ammonium ions present in the wastewater are exchanged in the filter bed with the sodium ions in the example, this means held back. The sodium ions leave the filter with the wastewater. This process is continued until the ion exchange capacity of the material is exhausted. Finally, the regeneration follows in which a regeneration solution is

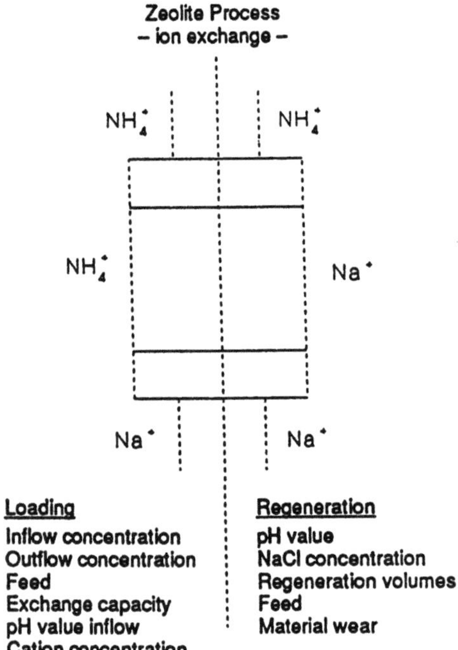

Fig. 1. Ion exchange with zeolite – process and influence quantities

pumped, under certain process conditions, through the filter. With this, a removal of the ammonium ions from the filter bed and, in turn, their replacement by sodium ions takes place. The ammonium ions are now available in the regeneration solution in concentrated form. As the regeneration solution remains in circulation, there must be a subsequent process which finally removes the ammonium ions from the regeneration solution. This can, for example, be the well- known process of AMP precipitation [5].

Loading and regeneration are determined by numerous influencing factors. For loading, an essential criterion is the service life up to achievement of an exhaustion point. This depends, according to the above mentioned research results, in turn, on the exchange capacity of the material, the hydraulic feed, the inflow concentration as well as the desired outflow concentration. The exchange capacity is, for example, dependent on the cation concentration, the grain size and the height of the filter bed [18]. Furthermore a pH value between 4 and 8 is to be maintained in the inflow.

For the regeneration the necessary regeneration volume is, in particular, an important dimension. This, in turn, depends on the pH value and the NaCl concentration in the regeneration solution. A certain hydraulic feed is not to be exceeded. The pH value in the regeneration solution, moreover, influences the wear of the filter and ion exchange material employed.

4. Objective

Following perusal of the literature it is clear that it is practical to look for an inexpensive ion exchanger which, in Germany, is offered for instance by the natural Hungarian zeolite. The assessment of the quality of such a material has been, with regard to the wastewater composition, of significance for process technology as well as the requirements of the Federal Republic of Germany. Trials have, principally, been carried out in two trials phases at two different wastewater treatment plants. With this, it was to be determined at the first wastewater treatment plant, the Bergheim wastewater treatment plant of the Erftverband (Erft Association) in Germany, to what extent zeolite is in a position to reduce the ammonium concentration of some 40 mg/l to the required discharge level. The process diagram of the Bergheim wastewater treatment plant is shown in Fig. 2. This wastewater treatment plant is operated with a sludge loading of $B_{DS} = 0.3$–0.5 kg BOD_5/kg DS and day. The wastewater for the trials plant was taken from the outflow of the filtration plant which, up until now, had served to ensure an appropriate COD reduction to the desired outflow value.

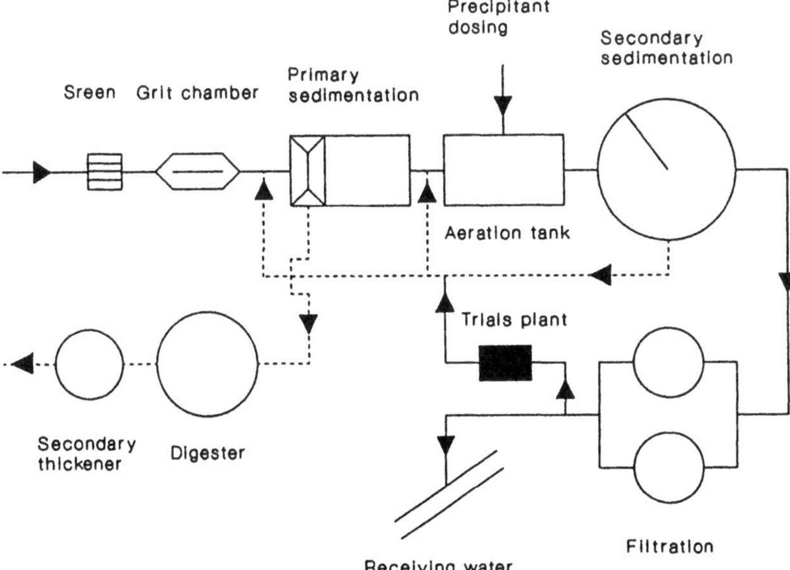

Fig. 2. Process diagram, wastewater treatment plant Bergheim

In the second case it was to be determined what results are to be achieved with a zeolite filter with NH_4-N input concentrations of ≤ 10 mg/l.

The second trials phase took place at the Lindlar wastewater treatment plant of the Aggerverband. The process diagram is shown at Fig. 3.

The wastewater treatment plant is a twin route biological filter plant whose outflow is extensively nitrified. Typical for the plant operation there is the relatively low wastewater temperature in winter which, in particular, inhibits the

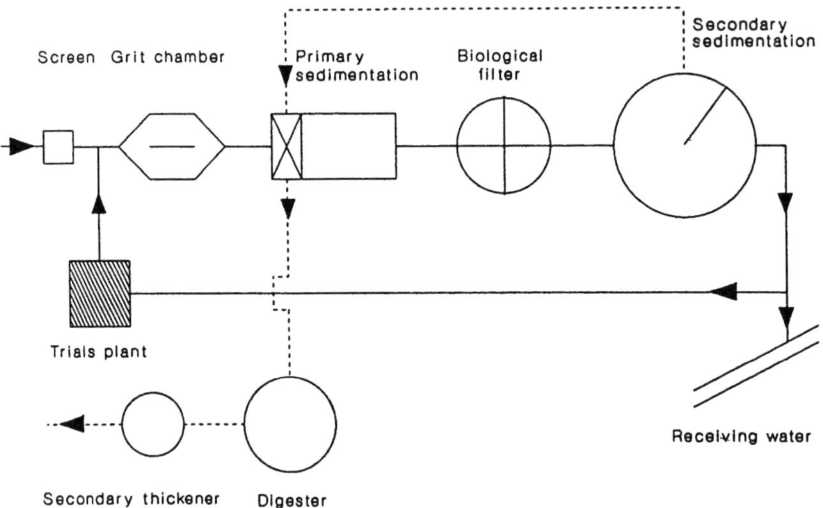

Fig. 3. Process diagram, Lindlar wastewater treatment plant

nitrification. Here the wastewater in the outflow of the secondary sedimentation stage was taken and fed into the trials plant. The aim of the process in the zeolite plant was an elimination of the residue of NH_4-N at low wastewater temperatures to 5 °C in biological reactors.

5. Description of the Trials Plant

A schematic process diagram of the trials plant is shown in **Fig. 4**.

Steel posts with a height of 2.0 m and an external diameter of 0.40 m were employed as filter. The internal structure of each post consisted of nozzle plate, filter bed and outflow facility. Hungarian zeolite was employed as filter material with a clinoptilolit component of some 50 % and a grain size of 1–2 mm. The filter bed volume was 100 l with a height of 0.9 m as well as an area of 0.11 m^2. The wastewater from the outflow of the sand filter plant was extracted, via a plastic weave hose with check valve, using a centrifugal pump and fed to the filter. Flow control was by means of a flow meter together with an adjustable plastic valve. Further piping consisted of synthetic material (PVC) and was so designed that the filter could be operated in parallel as well as in series.

For regeneration initially a 100 l synthetic material container was planned for the storage of the NaCl solution. For process technology reasons the 100 l tank was replaced, in the course of the trials adjustment, by two 1 m^3 synthetic material tanks. The transfer of the regeneration liquid took place via a further centrifugal pump. In addition the backwash water for the backwashing of the filter bed was also transferred via this pump.

The trials plant and its structure as well as the zeolite material employed are shown in the next two figures.

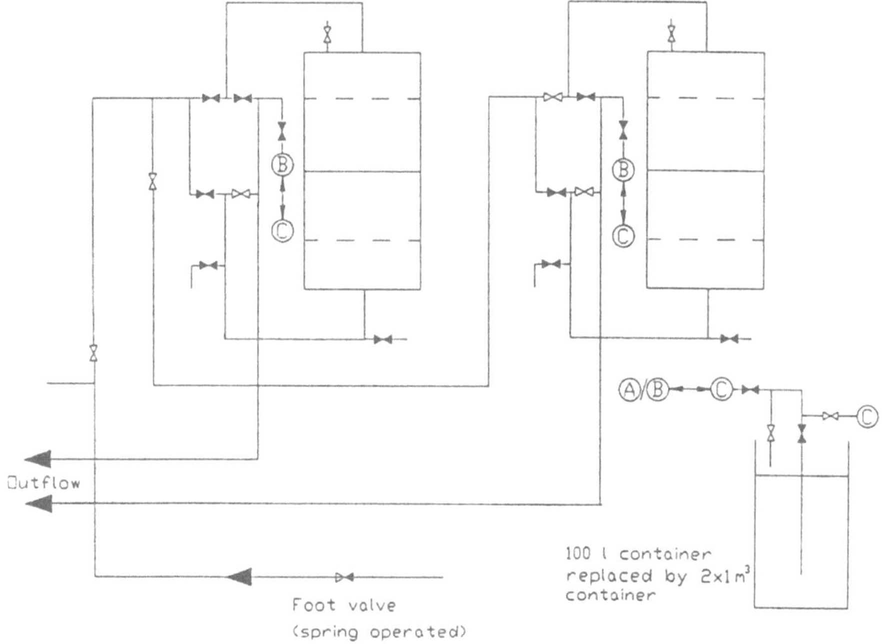

Fig. 4. Schematic trials plant process diagram – Lindlar wastewater treatment plant

Fig. 5. Trials plant

Fig. 6. Filtermaterial

6. Description of the Trials

6.1 Trials in Bergheim

The time plan of the trials is summarized in Table 4.

Tab. 4. Time plan

Time period	Trials phase
12.08.1991	Trials plant taken into operation
12.08.–27.09.1991	Running-in and modification phase
30.09.–17.10.1991	Trials Phase 1
16./17.10.1991	Determination of daily curve
21.10.–22.11.1991	Rebuild and run-in again
25.11.–29.12.1991	Trials Phase 2
27.11.1991	Determination of one daily curve
04.12.1991	Determination of one daily curve
10.12.–12.12.1991	Dismantle

The trials in Bergheim were carried out in two trials phases.

Trials Phase 1. The filters were hydraulically fed relatively low with 300 l/h corresponding to a surface feed of 3 m/h. Feeding took place in turn over 24 h. The sampling covered the complete running time, i.e., also 24 h. Additionally, from 16. to 17.01.1991, a daily curve in the form of 12 × 2 h hour composite samples was produced.

Trials Phase 2. After Phase 1 the trials plant was rebuilt in a way that two columns could be operated in series. Furthermore the filter material was exchanged with new material of the same grain size. Then the plant was running again with a higher hydraulic load of 500 l/h corresponding to 5 m/h. The trials took place from November 25th until December 9th under the conditions shown in Table 6.

6.2 Trials in Lindlar

The trials at the Lindlar wastewater treatment plant ran for six weeks from the beginning of January until the middle of February.

The trials conditions are given in Table 7.

Various combinations of parameters were set which are described in more detail in the following assessments (see also Fig. 11)

Tab. 5. Zeolite trial Bergheim – Trials Phase 1 – loadings

Trials location:	Sand filter outflow
Trials plant:	2 alternative fed single layer filters Filter bed height 1.0 m Diameter 0.35 m Filter material zeofloc 1–2 mm
Loading:	Feed 300 l/h ($= 3$ m/h) Running time 24 hours
Regeneration:	pH value 11.0 NaCl 30 g/l Feed 800 l/h ($= 8$ m/h) Regeneration volume 3 000 l ($= 30$ loading vols (BV))

Tab. 6. Zeolite trial Bergheim – Trials Phase 2 – loadings

Trials location:	Sand filter outflow
Trials plant:	2 alternatively fed single layer filters Filter bed height 1.0 m Diameter 0.35 m Filter material zeofloc 1–2 mm
Loading:	Feed 500 l/h ($= 5$ m/h) Running time 8 hours
Regeneration:	pH value 11.0 NaCl 30 g/l Feed 800 l/h ($= 8$ m/h) Regeneration volume 3 000 l ($= 30$ loading vols (BV))

Tab. 7. Zeolite trial Lindlar – loadings

Trials location:	Outflow secondary sedimentation stage
Trials plant:	2 series/alternatively fed single layer filters Filter bed height 1.0 m Diameter 0.35 m Filter material zeofloc 1–2 mm
Loading:	Feed 500–1 000 l/h Running time 24–72 hours
Regeneration:	pH value 11.0 NaCl 30 g/l Feed 800 l/h Regeneration volume 3 000 l ($= 30$ loading vols (BV))

7. Summary and Evaluation of Trials Results

7.1 Curves

7.1.1 Wastewater Treatment Plant Bergheim – Results of Trials Phase 1.
The curves for the ammonium concentration at the inflow and outflow in the first trials phase are shown in Fig. 7. Figure 8 shows the daily curve for 16./17.10.1991.

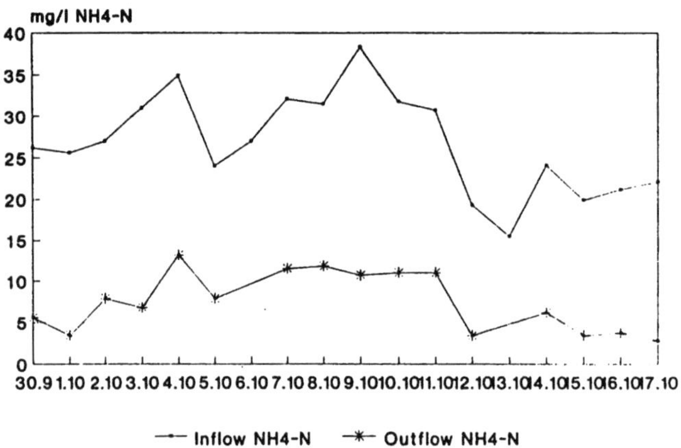

Fig. 7. Outflow, inflow (24 h composite sample) – ammonium nitrogen curves

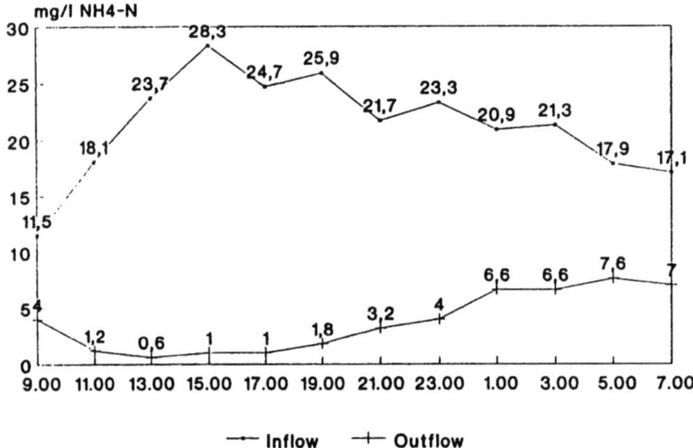

Fig. 8. Daily curves for ammonium nitrogen (24 h composite sample) for the investigations on 16./17.10.1991

The average degree of elimination for ammonium nitrogen is 72 %. A considerable reduction of the ammonium nitrogen load can also be determined. In part very low values below 5 mg/l in the 24 hour average were measured, however, in part the values lay also above 10 mg/l. Overall the figure shows a clear inflow dependency of the outflow concentration in progress over the trials period. This result is in contrast to the previous research results found in the literature. More accurate details can be seen in the daily curves from 16. and

17.10.1991 on the basis of 2 hourly composite samples. The average concentration for the period of time shown was 21.2 mg NH_4-N/l in the inflow, 3.7 in the outflow. The figure shows no inflow dependency but, on the contrary, at first a phase with constantly low values < 2 mg NH_4-N/l with simultaneously increasing or constantly high inflow concentrations. This phase corresponds with the loading process up to a certain breakthrough point which is achieved at approximately 1900 hrs, where the outflow value of 2 mg NH_4-N/l is exceeded. Finally the outflow concentrations increase continuously due to the material loading and approach the simultaneously sinking inflow concentrations. In this phase the exchange capacity of the ion exchange material is still not completely exhausted but no longer sufficient to guarantee the maintenance of a certain outflow value. According to this, it is possible over a certain time to achieve very low outflow values. The time up to achieving a maximum permissible outflow concentration is dependent on the exchange capacity of the material, the inflow concentration, the outflow concentration as well as the hydraulic feed rate (contact time).

7.1.2 Wastewater Treatment Plant Bergheim – Results of the Second Trials Phase. In order to take into account the results of the first trials phase it was attempted to increase the reaction volume through series coupling of two filters of the same size and thus achieve a higher efficiency. At the same time an attempt was to be made in this second trials phase to match the surface feed to that in practice with a normal filter plant.

The NH_4 curves of Phase 2 corresponding with Phase 1 are shown in Figs. 9 and 10.

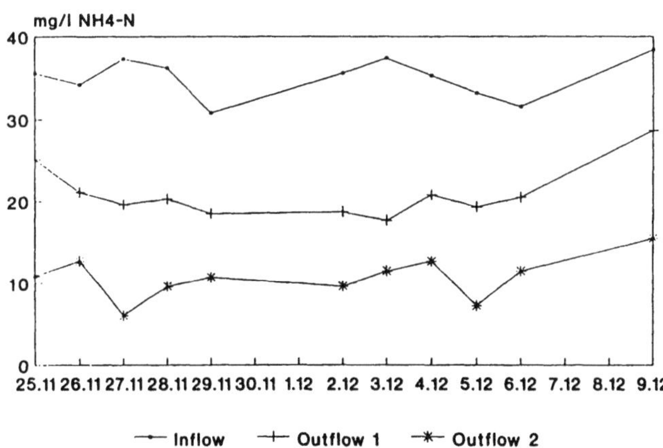

Fig. 9. Trials Phase 2: curves – (8 h composite samples) NH_4 inflow and outflow

7.1.3 Wastewater Treatment Plant Lindlar – Results of the Second Investigation Section. The trials phase in Lindlar was divided into areas with different filter bed sizes and inflow quantities. The individual phases are presented together in Fig. 11 with the inflow and outflow concentrations of NH_4-N from 2 hourly composite samples.

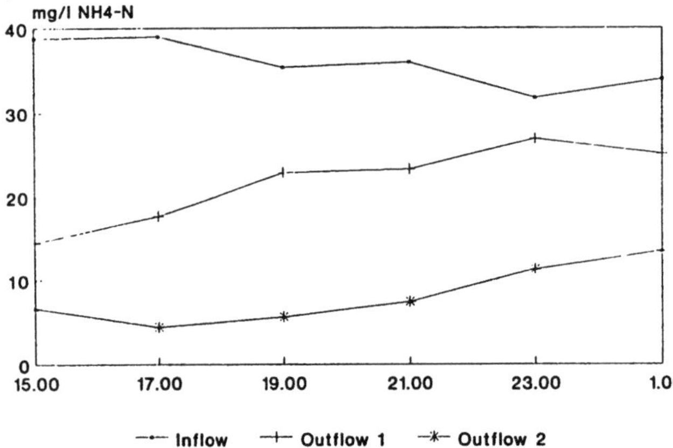

Fig. 10. Trials Phase 2 – daily curves (2 h composite samples) from 27.11.1991

The average degree of elimination for ammonium nitrogen in the second section was 55 %. The inflow concentration varied between 0.4 and 12.2 mg/l, on average 6.9 mg/l. The outflow portion varied between 0.3 and 9.2 mg/l, on average 3.1 mg/l The wastewater temperatures in the filter varied between 5 and 10 °C.

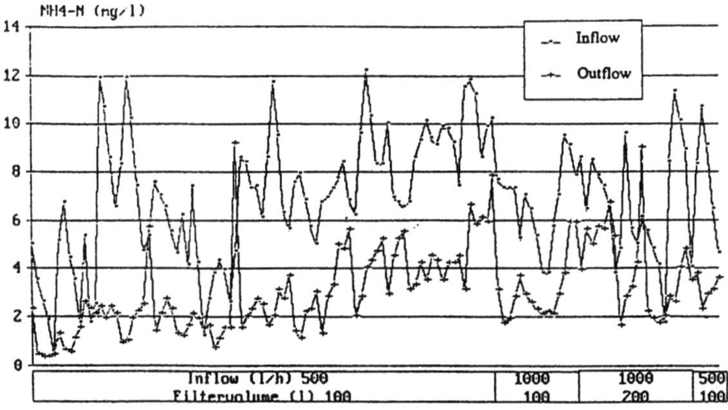

Fig. 11. Curves for NH$_4$-N trials phase in Lindlar and boundary conditions

7.2 Further Evaluation of the Trials

The trials were evaluated according to various parameters. Now, in the evaluation, only 2 hr composite samples were included in order to be able to record the influence of the zeolite loading during this time. With this, it was clear that a relationship between the ammonium volumetric loading and the outflow concentration existed. All other compared parameters (inflow concentration, inflow load, hydraulic feed) were seen as secondary in the area investigated with regard to the clarity of the results and the objective, if possible, of providing dimensioning values for a zeolite filtration.

The ammonium volumetric loading was calculated as potential dimensioning quantity for fixed bed reactors from inflow load and filter bed volume according to the following formula:

$$V_L = (c_{in} \times q_{in})/V_{FB} \quad [\mathrm{g\,NH_4\text{-}N/m^3 \cdot h}]$$

c_{in} Inflow concentration [mg/l]
q_{in} Inflow quantity [l/h]
V_{FB} Fixed bed volume [l]

The reference to the available surface of the filter material per unit volume might be seen as actual dimensioning quantity. However, as here one worked with the same material throughout the complete trial, no influence is to be expected through this. Two series connected filters were calculated as one filter with the appropriate filter height and the volume of the individual filters added.

The relationship between volumetric loading in g NH_4-N/m^3 · h and outflow concentration in mg/l is shown in Fig. 12 with the correlation characteristic values. The relationship is of great significance and shows, with increasing volumetric loading, an increasing outflow concentration. With this a first dimensioning criterion for zeolite filters is available.

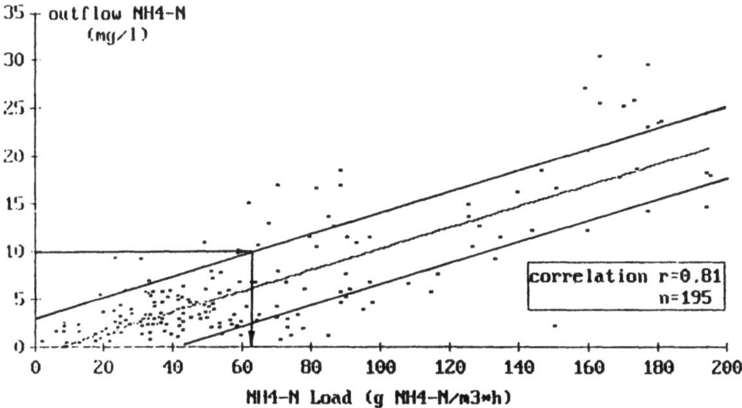

Fig. 12. Relationship NH_4-N outflow and volumetric loading

The standard prediction error is of interest with regard to the maintenance of monitoring values. The confidence interval is also shown in the diagram. From this one can deduce that ammonium nitrogen outflow values of, for instance, 10 mg/l can be maintained with a volumetric loading of some 65 g ammonium nitrogen/m^3 · h in 2 hourly composite samples. Lower monitoring values are equally possible.

The relationship between temperature and outflow values is shown in Fig. 13.

Figure 13 substantiates the fact that no recognisable relationship exists between the change in temperature and the outflow value. This statement is of

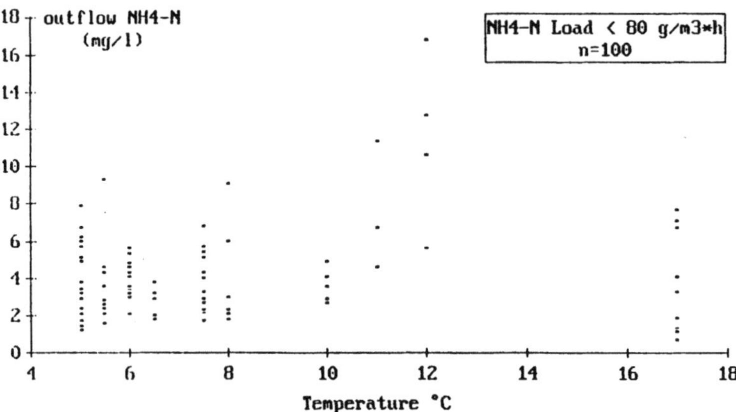

Fig. 13. Relationship between temperature and outflow values

importance insofar that it is to be deduced from this that the process leads to
constant success even at low temperatures. The possibilities are considerably
increased through this if the wastewater treatment plant is to be operated in
cold regions. The dimensioning diagram for the volumetric loading remains the
same under these circumstances.

A further important criterion for the planning of a filter plant is the running time. From the trials there resulted an exchange capacity up to complete
material loading of some 1.4–2.3 g NH_4-N/kg zeolite. This result is identical
with the results of Hungarian trials. The influence of material loading on the
relationship shown in Fig. 13 has to be investigated furthermore.

8. Operating Resources

Along with the technical process assessment the question of operating resources
is also of fundamental significance for the planning engineer. It is not possible
to come to a final statement on the basis of the trials carried out here, as
only the first part of the process - the loading of the ion exchanger - was part
of the investigation. Therefore here only a basic summary shall be given on
the individual cost factors. Insofar as usable statements are available from the
execution of the trial, these are included at the appropriate position.

8.1 Investment and Capital Costs

The investment costs are made up, according to current estimates, of the following individual items:

- investment costs for filter plant,
- investment costs for the regeneration cycle,
- investment costs for the regeneration preparation.

No statements can yet be made for these items. According to previous estimates the costs will depend particularly on:

- the ion exchange, if required to be coupled with the well-known technical process of flocculation filtration,
- the regeneration (filter regeneration and regeneration agent preparation) process to be technically so optimised that a continuous, automatic process is achieved, this in particular to minimise construction volume and operating costs.

8.2 Operating Costs

The operating costs are made up of:
- the capital services,
- the operating costs for the filter plant,
- the operating costs for the regeneration,
- the operating costs for the regeneration agent preparation,
- the costs for the disposal of residues.

For the filtration costs it should be noted that, during the trials phase, no noteworthy wear was determined in the zeolite filter material. From this one can expect long times of service life for the filter bed filling under the given conditions.

In the regeneration cycle the adjustment of the pH value to a value of about 11 in the regeneration fluid is, in particular, a continuous cost factor. The use of sodium hydroxide solution (33 percent) in the trial on average gave 0.53 l $NaOH/m^3$ regenerate (minimum 0.3, maximum 1.1 l $NaOH/m^3$). A relationship between ammonium elimination and NaOH usage was not determined.

No statements can yet be made with regard to the preparation of the regeneration agent, e.g. using AMP precipitation. However, it can basically be recorded that the regeneration fluid in the circulation remains and only the ammonium is eliminated. A continuous consumption of regeneration fluid is therefore not to be expected.

Magnesium ammonium phosphate, in more or less heavily dewatered form, appears as end-product of the ammonium elimination process. Its final destination is also to be included as credit or debit in the cost considerations.

9. Ecological Considerations

As a basic argument against the employment of the zeolite process it can be said that although a removal of ammonium ions from the wastewater takes place, at the same time, however, the content of sodium ions in the wastewater is increased.

In the wastewater in the outflow of the ion exchanger ammonium is present as monovalent cation. During the ion exchange process ammonium is exchanged

for the equally monovalent sodium cation. The ammonium cation is then located in the regeneration agent; the sodium cation in the wastewater. The latter
is in the respective receiving water via the wastewater treatment plant runoff.
The ammonium cation is eliminated from the regenerate using, for instance,
the AMP precipitation process (Precipitation as magnesium ammonium phosphate at pH 9). The loss of cations in the regeneration agent is balanced by
the addition of sodium hydroxide solution.

The question of the effects of the increase in sodium concentration in the
wastewater treatment plant runoff can only be assessed taking into account
the quantitative relationships. In the trials Phase 1 the average ammonium
nitrogen inflow was, for instance, 24.9 mg/l, the outflow on average 6.5 mg/l.
This results in an increase in the sodium concentration in the wastewater,
taking into account the different masses, of some 30 mg/l Na^+/l.

As in practice various future application cases are feasible, a quantification is given by example in Table 8 for various cases. Three different application cases were differentiated: non-, partial and extensive nitrified wastewater
treatment plants. Correspondingly different are the process aims with the employment of ion exchangers. According to the pretreatment of the wastewater
different ammonium nitrogen dimensioning values for the ion exchanger result.
For the receiving waters it is, however, not the peak load but the average ratio
which is of interest. Taking into account a variation factor of approx. 2, the
elimination performances and the resulting increase in Na^+ in the outflow are
calculated from the average inflow concentration taking into account an outflow
value of 2 mg ammonium nitrogen/l.

Tab. 8. NH_4-N elimination and increase in Na^+ concentration

	Plant type pretreatment		
	Carbon reduction (e.g. $B_{DS} = 0.30$) mg/l	Nitrifikation (e.g. $B_{DS} = 0.15$) mg/l	Extensive nitrifikation (e.g. $B_{DS} = 0.07$) mg/l
Process aim ion-exchange	Complete NH_4-N-elimination	Partial NH_4-N-elimination	Residual NH_4-N-elimination
NH_4-N-dimensioning	40	10	5
Average inflow concentration	20	5	2.5
Elimination with NH_4-N in outflow $= 2$ mg/l	18	3	0.5
Na^+-increase in outflow	30	5	0.8

For the assessment of the calculated increase it is pointed out that sodium
is the alkali metal ion most frequently contained in freshwater. It is essential

in larger quantities for humans. The recommended daily intake lies between 1.1 and 3.3 g. However, excessive amounts of sodium may have a possible connection with coronary artery disease as a result of a sodium damaged kidney.

According to Koppe [20] the average concentrations in Ruhr water are some 50 mg sodium/l. According to the origin of the raw water the sodium concentrations in the drinking water can, however, be very different and up to 250 mg/l. An increase in sodium concentration in the receiving waters is therefore not necessarily to be feared due to the employment of the zeolite ion exchange process. However, even when this occurs with receiving waters with low sodium contents it is to be seen as relatively without problem. In any case the increases with the employment for only partial ammonium nitrogen elimination are negligible, as Table 8 shows.

10. Outlook

The trials described here present us with an introduction to the subject. The trials results confirm that it is fundamentally possible to achieve, with the process described, with appropriate dimensioning, very extensive removal performances for ammonium nitrogen even at low wastewater temperatures. Compared with biological processes the zeolite process represents a possibility, if the occasion arises, of saving considerable building costs. On the other hand, the process demands an increase in apparatus resources which, at least in Germany, has not up until now, been usual. For the future it will therefore depend on developing the process, particularly on the technical apparatus side, so far that the necessary volumetric as well as the operating resources are minimised. Furthermore, one must consider, for example, how the exchange capacity of the material can be increased, for instance through pretreatment, in order to lengthen the service life of the process. The corresponding trials will be continued.

Bibliography

[1] Anonym: Leistungstabelle über Verfahren der weitergehenden Abwasserreinigung nach biologischer Behandlung beim Belebungsverfahren. Arbeitsbericht des ATV-Fachausschusses 2.8. Korrespondenz Abwasser *9* (1988) 937

[2] Anonym: Technischer Leitfaden zur Nitrifikation und Denitrifikation in kommunalen Kläranlagen. Merkblatt Nr. 6 des Landesamtes für Wasser und Abfall Nordrhein-Westfalen, Düsseldorf 1990

[3] Böhnke, B.: Bemessung der Stickstoffelimination in der Abwasserreinigung. Korrespondenz Abwasser *9* (1989) 1046 ff.

[4] Witte, H.: Verfahrensvarianten zur Erfüllung der Anforderungen an die Abwasserwirtschaft, Vortrag 3. Siegener Fachkolloquium Siedlungswasserwirtschaft, 4./5.11.1991. Abwassertechnik *1* (1992)

[5] Schulze-Rettmer, R.: Kritische Gedanken zur Denitrifikation und Alternativen. Korrespondenz Abwasser *3* (1987) 218–222

[6] Ames, L.L.: The Cation Sieve Properties of Clinoptilolith. American Mineralogist *45* (1960) 689–700

[7] Koon, J.H., Kaufman, W.J.: Optimisation of Ammonia Removal by Ion Exchange Using Clinoptilolit. US Report for the EPA 1971

[8] Liberti, L. et al.: Water Supply *1* (1983) 169–176

[9] Liberti, L. et al.: The $10^3 \, h^{-1}$ RIM-NUT Demonstration at West Bari for Removing and Recovering N and P from Wastewater. Wat. Res. *20* (1986) 735–739

[10] Carrondo, M.S.T., Perry, R., Lester, J.N.: Type A Zeolite in the Activated Sludge Process – Part I: Treatment Parameters. J. Wat. Pollut. Control Federation *52* (1980) 2796–2806

[11] Holman, W.F., Hopping, W.D.: Treatability of Type A Zeolite in Wastewater – Part II: Treatment Parameters. J. Wat. Pollut. Control Federation *52* (1980) 2887–2905

[12] King, J.E., Hopping, W.D., Holman, W.F.: Treatability of Type A Zeolite in Wastewater – Part I: Treatment Parameters. J. Wat. Pollut. Control Federation *52* (1980) 2875–2886

[13] Ishii, M., Kaji, K.: Gypsum and Lime *186* (1983) 8–16

[14] Olah, J. et al.: Simultaneous Separation of Suspended Solids, Ammonium and Phosphate Ions from Wastewater by Modified Clinoptilolit. In: Celithes as Catalysts, Adsorbant and Detergent Builders, J. Weidkamp and H.G. Karge (eds.). Elsevier Science Publishers, Amsterdam 1989

[15] Czaran, E. et al.: Ion Exchange with Clinoptilolit. Akta Chemika Hungarika *125* (1) (1988) 201–210

[16] Czaran, E. et al.: Separation of Ammonia from Wastewater Using Clinoptilolit as Ion Exchanger. Nuclear and Chemical Waste Management *8* (1988) 107–113

[17] N.N.: Abstraktion von Ammonium durch "zeonyme" Filter. HydroTrade, Achern 1991

[18] N.N.: Optimisation of Ammonia Removal by Ion Exchange Using Clinoptilolit. US Environm. Protection Agency, Washington 1971

[19] Jahresbericht. Erftverband, Bergheim 1990

[20] Koppe, P., Skozek, A.: Kommunales Abwasser. Vulkan Verlag, Essen 1996

H. Witte and M. Keding
Gemeinnütziges Institut Wasser und Boden e.V.
Mülldorfer Str. 25
D-5205 St. Augustin
Germany

Rapid Sewage Clarification with Magnetite Particles

N.A. Booker, A.J. Priestley, and C.B. Ritchie

Introduction

As discharge standards for wastewaters have become increasingly stringent, the need for a diverse range of treatment technologies has become more urgent. Physico-chemical processes based on coagulation chemistry have been applied to wastewater treatment, and have been found especially effective for suspended solids removal. This paper describes a novel way to engineer a process based on coagulation/flocculation chemistry. The process is based on the use of very fine magnetite particles, which not only remove the necessity for an extended period of flocculation but also provide very rapid solid-liquid separation by utilisation of their magnetic properties.

The process was originally used for the treatment of drinking water (Anderson and Priestley, 1983; Anderson et al., 1983; Maloney et al., 1989), but in recent years has also been applied to the treatment of sewage and industrial wastes (Booker et al., 1991). This latter application required significant changes to the process design because of the different nature of the material being removed. In water treatment the process removes colloidal clay particles and humic acids, which cause the water to be turbid and coloured respectively. These materials carry a negative charge which has to be neutralised before effective flocculation can occur. Pretreatment of the magnetite particles with a dilute caustic soda solution removes silica based impurities from their surface and increases the point of zero charge (iso-electric point) of the particles from a pH level of about 4 to at least 7. At pH levels less than the iso-electric point, the magnetite particles will have an overall positive charge and are then able to neutralise and adsorb the negatively charged colloidal impurities. For the treatment of highly turbid and/or coloured waters, the capacity of the magnetite particles is enhanced by the addition of a cationic polyelectrolyte. No inorganic coagulant is required.

When applied to wastewater treatment this simple mechanistic picture of the mode of action of the magnetite particles must be modified to account for the more complex nature of the feed. Wastewaters can contain a wide spectrum of materials, ranging from millimetre sized particles, emulsified oil and grease to colloidal and soluble matter of both organic and inorganic origin. A variety of mechanisms must be employed to remove this diverse range of materials from water. Coagulation and flocculation mechanisms still play a predominant role, but other mechanisms such as hydrophobic interactions and simple physical enmeshment also play an active part. For most wastewaters the role of the magnetite particles as a primary coagulant is reduced, compared to water treatment, simply because of capacity limitations. An inorganic coagulant, such as alum, can be added in conjunction with the magnetite particles to increase charge neutralisation and thus increase the capacity of the magnetite particles.

The original discovery of the coagulating properties of alkali treated magnetite was reported by Kolarik (1983). Fine particles of both magnetic and non-magnetic nature had been previously used in coagulation/flocculation processes (Demeter & Galgaczi, 1967, De Latour & Kolm, 1976) but only as a floc weighting agent and usually on a once through basis. In contrast, the alkali treated magnetite particles were not only able to act as a coagulant in their own right, but were also able to be readily regenerated and recovered for reuse.

As indicated earlier, the original process development work was targeted at the treatment of drinking water and, for this application, the technique is now a fully commercial process with seven plants either installed or under construction around the world. Development work has continued on the water treatment process with the aim of improving its cost effectiveness (Anderson et al., 1991). However, the major recent development has been the application of the technique to the treatment of sewage and industrial wastewaters.

After laboratory jar tests had established the feasibility of successfully treating raw or primary settled sewage, attention turned to the design of a pilot plant (Booker et al., 1991). These studies indicated that the main advantage of the process was speed, with coagulation/flocculation times being reduced to around one minute and clarifier upflow rates of around 10 m/hr being achievable. Two fully continuous pilot plants have been built and operated since these early studies. The first pilot plant, of capacity 60 litres per minute, was built and operated at the Lower Plenty laboratories of CSIRO in Melbourne, while the second, of capacity 140 litres per minute, was constructed for the Sydney Water Board and was operated at their Malabar ocean outfall sewage treatment works.

The Lower Plenty pilot plant treated a high rate primary sedimented sewage of purely domestic origin, while the Malabar pilot plant was fed screened and degritted raw sewage of mixed industrial and domestic source. Both pilot plants have been operated for a period of two years and the results were sufficiently encouraging to convince the Sydney Water Board to proceed with construction of a 5 ML/day prototype plant. This plant is due for commissioning in September 1992, and should provide the basis for confidently designing single stage units of capacity up to 100 ML/day.

This paper describes the results of pilot plant operations while treating sewage at both the Malabar and Lower Plenty sites and concentrates on the process design changes necessary to adapt the process to this application.

Process Description

A flow diagram for the process as applied to the treatment of sewage or industrial wastewaters, is shown in Fig. 1. Screened degritted raw sewage is mixed with cleaned recycled magnetite particles at a concentration of about 10 g/L. An acidic recycle stream from the sludge separator is immediately mixed with the magnetite suspension and fresh inorganic coagulant added to achieve the optimum conditions for coagulation/flocculation. A very rapid process of colloid destabilisation and adsorption onto the magnetite surface occurs. If the magnetite surface is sufficiently clean, no free flocs are formed and the coagulated material is bound to the magnetite surface. Kinetic studies have demonstrated that this adsorption process is complete within 1 to 2 minutes and is sensitive to the agitation conditions. Proper design of the mixing conditions is essential for effective adsorption of destabilised material. Laboratory experiments (Kolarik, 1983) have clearly demonstrated the importance of the fine particle size in achieving optimum interaction with colloidal material in the water.

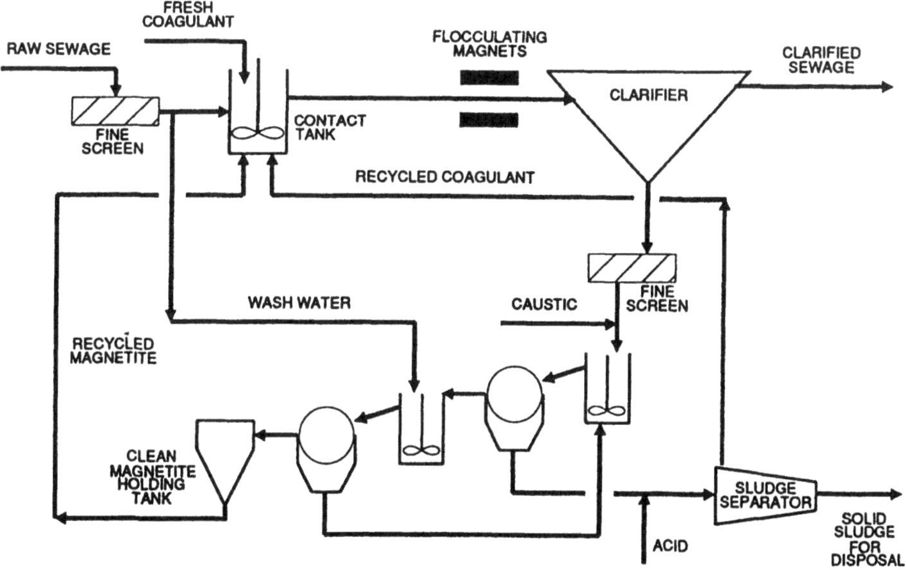

Fig. 1. SIROFLOC sewage treatment process – process flow scheme

When the adsorption step is complete, the fine magnetite particles can be separated from the treated water by a magnetic flocculation technique. Although dense (S.G. 5.1), the fine magnetite particles by themselves settle very slowly, but when magnetised by passage through a moderate magnetic field they floc together almost instantaneously under their own magnetic attraction to form large flocs which settle rapidly at rates up to 100 m/hr. However, because of the wide distribution of floc sizes formed, clarifier upflow rates used are about an order of magnitude lower than this figure in order to achieve low carryover of magnetite particles (< 10 mg/L). Clarifier upflow rates of 24 m/hr were tested in pilot plant trials on a high turbidity river water and yielded magnetite carryovers of between 10 to 20 mg/L. This magnetite can be effectively removed on a high rate sand filter. Thus the combination of clarification and filtration may allow the attainment of higher upflow rates in clarification. The magnetic properties of magnetite particles allow the use of very fine particle sizes of high surface area, while at the same time achieving rapid solid/liquid separation by the magnetic flocculation technique.

After settlement, the magnetite particles are regenerated and reused. The regeneration process involves raising the pH of the thickened magnetite slurry to around 10.5 with caustic soda. At this pH the magnetite surface becomes negatively charged and all of the attached impurities can be rapidly stripped from the surface. This charge reversal mechanism is a considerable oversimplification of the surface cleaning process which occurs when a heavily-contaminated wastewater is being treated. However, it has been firmly established that, for all wastewaters tested, stripping of material from the magnetite surface is rapid and virtually complete within twenty seconds (Booker et al., 1991). Lime cannot be used to raise the pH in the surface stripping operation as the calcium ions induce precipitation onto the magnetite surface.

After regeneration, the magnetite particles are separated from the stripped sludge by means of permanent magnetic wet drum separators and washed in a two stage countercurrent flow operation. The washed magnetite particles are recycled, while a dilute sludge of approximately 0.6 per cent solids is produced from the first magnetic drum separator. Flowrate of the dilute sludge is about 3 per cent of the raw sewage flow. This dilute sludge has a pH around 9.5 and contains the impurities removed from the sewage and all of the inorganic coagulant added to the sewage with the magnetite. To recover the coagulant and produce a concentrated sludge, acid is added to the dilute sludge to reduce the pH to between 2 and 3. Addition of the acid results in precipitation of most of the organic matter into a thick sludge, which can be easily separated by flotation or centrifugal techniques. A clear acidic solution, which contains most of the inorganic coagulant, remains and this solution is recycled to the start of the process where value is obtained from both the acid and recycled coagulant. Around 10 per cent of the organic material originally present in the dilute sludge is also recycled with this solution.

Pilot Plant Operating Experience

Lower Plenty, Melbourne

This pilot plant commenced operation in June 1989, treating domestic sewage which had been passed through a macerator pump and a high rate primary sedimentation tank. The process flowscheme was as shown in Fig. 1 but did not include fine screening of either the raw sewage or settled magnetite slurry from the bottom of the clarifier. Initial operation was aimed at determining a number of key operating factors:

(i) would the magnetite particles retain their effectiveness after many cycles of regeneration and reuse?

(ii) what were the effects of key operating variables such as magnetite dose, regeneration pH, washwater volume and sewage/magnetite contact times.

The main process parameters measured to evaluate process performance were turbidity, suspended solids (SS), chemical oxygen demand (COD) and biochemical oxygen demand (BOD). For process control purposes, however, turbidity was used as the principal quality parameter.

Early operation confirmed that, provided the regeneration pH was kept above 9.5, the magnetite particles could be successfully regenerated and reused. This result was established not only by the continued good performance of the pilot plant but also by monitoring the concentration of organic carbon remaining on the magnetite surface after regeneration. This level reached a stable concentration early in the pilot plant operation. Once steady operation had been achieved, the key operating parameters were varied to evaluate their effect on the process performance. A magnetite dose of around 10 g/l was found to be optimal; little improvement was achieved by going above this concentration, while capacity limitations began to appear below about 5 g/l. Washwater volumes were varied in the range 2–6 % of the main sewage flow. Satisfactory process performance was maintained down to a washwater volume of 3 per cent. Below this level the process performance slowly began to deteriorate as impurities built up on the magnetite surface.

Contact of the magnetite with the sewage was initially achieved in a series of three stirred tanks, with total contact time being in the range 6–8 minutes. However, samples from the first tank showed that adsorption of material from the sewage was already virtually complete and the remaining tanks were achieving very little. Subsequently, the stirred tanks were replaced by a simple 25 mm I.D. pipeline contactor in which the contact time varied from 0.5 to 2 minutes. Process performance improved marginally as contact time increased in this range. For the same contact time, higher pipe velocities led to better clarification performance, possibly as a result of better mixing within the contactor. With continued operation of the pilot plant, the process developed some sensitivity to shearing conditions during contact and magnetic flocculation. This effect became more marked once coagulant recovery and recycle from the sludge were initiated. To provide optimum mixing conditions while maintaining a short residence time a new magnetite/sewage contactor was designed, based on vertical upward flow of the mixture through an agitated baffled cylindrical vessel. The best performance from the pilot plant was achieved while this contactor was in operation, emphasizing the importance of achieving optimum mixing conditions.

Some representative results from the pilot plant operation at Lower Plenty utilising the pipeline contactor are given in Table 1. The results are given as the average $(\bar{x})\pm$ the standard deviation (σ) and represent the averaged performance taken from a six week period (26.7.1990 to 5.10.1990) where the pilot plant was operated for seven hours per day, five days per week under reasonably steady operating conditions.

Tab. 1. Averaged results from extended period of pilot plant operation (Lower Plenty)

	Raw $(\bar{x} \pm \sigma)$	Treated $(\bar{x} \pm \sigma)$
Turbidity [NTU]	95 ± 13	20 ± 5
COD [mg/L]	406 ± 49	121 ± 23

Turbidity and chemical oxygen demand were the parameters measured on a routine basis because of ease of analysis. Some measurements of suspended solids and BOD were carried out early in the pilot plant operation and the results correlated with the turbidity and COD measurements (Priestley et al., 1989). Averaged results for suspended solids and BOD removal, based on these correlations, are given in Table 2. It should be pointed out that the pilot plant operation was controlled to achieve a treated effluent turbidity of 20 NTU, with a better performance being achievable simply by lowering the operating pH or increasing the coagulant dose. However, this improved process performance could only be achieved with a significant increase in operating cost. A summary of the key operating variables during the period of the test is given in Table 3.

Tab. 2. Averaged results for suspended solids and BOD removal (obtained by correlation)

	Raw	Treated
Suspended solids [mg/L]	190	30
BOD [mg/L]	180	40

Tab. 3. Averaged plant operating conditions $(\bar{x} \pm \sigma)$

Plant throughput	42 ± 3 litres/min
Contact time in pipe	58 ± 4 secs.
Coagulant dose*	25 ± 4 mg/l as Al
Magnetite dose	18 ± 4 g/L
Operating pH	6.0 ± 0.3
Operating temperature	$17.0 \pm 1.0\,°C$

* No coagulant recycle

The results given in Tables 1 and 2 show that the process achieved the following percentage removals: turbidity $\sim 79\,\%$; suspended solids $\sim 84\,\%$; COD

$\sim$ 70%; BOD $\sim$ 78%. Other measurements demonstrated that phosphorus and organic nitrogen were effectively removed from the sewage while ammonia passed through the process unaffected.

Operation of the pipeline contactor clearly demonstrated the speed of the coagulation process, with one minute of contact being adequate for most situations. Coagulants contributed to most of the operating cost and further research work has been targeted at maximising the recovery and reuse of coagulant while producing a sludge of high solids density.

Malabar, Sydney

A second pilot plant of capacity 200 m^3/day (140 l/min) was designed and constructed for installation at the Malabar ocean outfall of the Sydney Water Board. This plant was commissioned in January 1990, and operated for the next eighteen months. The feed to the plant was screened degritted raw sewage and operation was designed to provide information on the following points:

– plant operating characteristics on the Malabar sewage,
– treated effluent quality,
– operating costs,
– long term operability,
– ability to recover and reuse coagulant,
– sludge quality and production rates.

In contrast to the Lower Plenty experience, physical operating problems were immediately experienced on the Malabar pilot plant. The raw sewage contained significant levels of fine paper fibre and while these fibres were effectively removed from the sewage they were only partially separated from the magnetite during regeneration. Consequently, the loading of fibre on the magnetite particles slowly built up and so changed their settling characteristics that the process became inoperable. It quickly became evident that either some form of pretreatment was required to remove most of the fibre from the raw sewage before contact with the magnetite or an in-process step was required to separate fibre from magnetite particles. In fact, both approaches were implemented in the pilot plant and, as shown in Fig. 1, both the raw sewage feed and the settled magnetite slurry were subjected to fine screening. Either approach was capable of solving the fibre handling problem but operating experience indicated that the best process performance for minimum cost was achieved with an optimal choice of both screens.

Results from Malabar confirmed the Lower Plenty conclusion that magnetite could be recovered and reused in the long term. Operating data were collected for two major process configurations:

1) screening of the settled magnetite slurry only,
2) both prescreening and in-process screening.

Tab. 4. Pilot plant operating results from 50 day trial at Malabar, (Configuration 1)

	Raw sewage		Treated sewage		% Removal	
	$\bar{x}$	σ	$\bar{x}$	σ	$\bar{x}$	σ
Suspended solids [mg/L]	161	39	33	9	80	6
Oil& grease [mg/L]	38	14	6	4	84	16
Turbidity [NTU]	152	57	12	5	92	2
BOD [mg/L]	200	27	103	13	48	2
COD [mg/L]	439	99	217	38	50	8
Chromium [mg/L]	0.12	0.32	0.01	0	90	
Copper [mg/L]	0.18	0.06	0.02	0.01	90	
Nickel [mg/L]	0.07	0.09	0.02	0.02	70	
Zinc [mg/L]	0.31	0.10	0.13	0.09	60	

An averaged set of operating results from a 50 day trial for process Configuration 1 is given in Table 4.

As analytical support was available a comprehensive range of process variables was monitored on a twice daily basis. The numbers quoted in Table 4 provide an interesting comparison with the performance of the process at Lower Plenty. Suspended solids removal at around 80 % was similar to the Lower Plenty performance although final turbidities were lower at around 12 NTU. Oil and grease removal was also good (84 % removal) with final levels averaging 6 mg/L. However, the removal of both BOD and COD was significantly lower than at Lower Plenty. Removals averaged 50 % with only small deviations and it appeared that the Malabar sewage had a much higher fraction of purely soluble BOD/COD.

As shown in Table 4, measurements of heavy metal concentrations were also carried out. Although the levels of heavy metals in the raw sewage were quite low, the process still proved capable of removing up to 90 % of the incoming metal. Based on simple plate count measurements, the process was also capable of removing greater than 99 % of bacteria present in the raw sewage.

When prescreening of the raw sewage was also included in the process flowscheme, the quality of the treated effluent improved even further. Results from this latter period of operation are given in Table 5. While the removal of BOD/COD remained essentially unchanged, both suspended solids and oil and grease removal improved, with final concentrations averaging 25 and 3.3 mg/L respectively.

This improved process performance was a result of better fibre removal, a significant fraction of the suspended solids in the treated effluent comprised fibre particles with magnetite particles attached. Laboratory experiments have demonstrated that these particles can be easily removed in a simple sand filter producing an effluent with a significantly reduced suspended solids level.

Tab. 5. Process performance after introduction of sewage prescreening (Configuration 2)

	Treated sewage		% Removal
	$\bar{x}$	σ	$\bar{x}$
Suspended solids [mg/L]	25	5	84
Oil & grease [mg/L]	3.3	1.1	91
BOD [mg/L]	100		

Operating conditions on the Malabar pilot were similar to those at Lower Plenty, e.g. operating pH $\sim$ 6.0; magnetite dose $\sim$ 10 g/L; regeneration pH $\sim$ 10.5. The major difference was the significantly lower coagulant dose required, with the Malabar requirements being half that of Lower Plenty.

A summary of the chemical dose requirements for the pilot plant operation at Malabar is given in Table 6.

Tab. 6. Chemical dosage requirements for Malabar pilot plant operation

Coagulant*	10 to 18 mg/L as Al
Acid	50 mg/L of H_2SO_4
Caustic soda	22 mg/L of NaOH

* Level quoted is with no coagulant recycle

Further Process Development

Pilot plant operation at both Lower Plenty and Malabar has clearly demonstrated the ability of the magnetite particle process to clarify raw sewage on a continuous basis. Major development work on the process is now targeted at optimising the recovery and recycle of inorganic coagulant from the regeneration effluent and maximising the solids density of the final sludge produced. As outlined in Fig. 1, these objectives are presently achieved by acidifying the alkaline regeneration effluent to achieve a pH between 2 and 3, whereupon the organic material in the regeneration effluent precipitates. The precipitated organics can be readily separated leaving a clear acidic stream which can be recycled to the magnetite contact stage. This recycle stream contains not only most of the acid added to the regeneration effluent but also a significant fraction of the inorganic coagulant. All of the coagulant is stripped from the magnetite surface during regeneration. Depending on the coagulant (either iron or aluminium salts), an optimum pH exists which maximises coagulant recovery while minimising the solubilisation and recycle of organic material and other inorganic contaminants such as heavy metals (Zn, Cu, Cr, etc).

Both pilot plants have been operated in this coagulant recovery mode and both have demonstrated reductions in demand for fresh coagulant of around 50 %. Up to 95 % of the coagulant and 10 % of organic material is solubilised and recycled, the presence of the organic material reducing the effectiveness of the recovered coagulant. Acid costs are not excessive, as most of the acid used to acidify the regeneration effluent can be used to adjust the pH of the incoming raw sewage. Long term operating tests have so far not detected any detrimental effect of the coagulant recycle on treated sewage quality.

The other major advantage of recovering and recycling the coagulant is its effect on sludge production. Because the precipitated sludge contains significantly less coagulant and also because of the low pH, it has proven relatively easy to produce a high density sludge at the solids separation stage. Both dissolved air flotation and centrifugation have proven to be suitable sludge separation techniques in pilot plant trials. Sludges of up to 10 % w/w dry solids have been produced by dissolved air flotation, while centrifugation has displayed the ability to produce a 20 % w/w dry solids sludge directly from the 0.6 % w/w dry solids regeneration effluent. No polymer addition is required at any stage.

Research work is presently investigating at cost effective techniques for removing soluble organics and phosphate from the acidic recycle stream leaving the coagulant largely free from contaminants. Laboratory experiments have shown that such a process greatly increases the effectiveness of the recycled aluminium and the potential exists to reduce the demand for fresh coagulant to 10 % of that required without coagulant recycle.

Optimisation of the magnetite contact and clarification stages is also being studied. As mentioned earlier, improved results have been achieved by paying close attention to agitation conditions during magnetite contact, an agitated upflow contactor proving excellent in this role. A prototype plant based on the design shown in Fig. 1 and of capacity 5 ML/day, is presently being constructed at the Malabar site. This plant should be commissioned in September 1992, and will prove out major aspects of the process design on a proper engineering scale. It is hoped that further developments on coagulant recovery and recycle will be incorporated into the design of this plant.

Final Discussion and Conclusions

While the technical feasibility of the process has been clearly demonstrated on the pilot plant scale, its commercial viability depends upon a number of other factors. Costs, both capital and operating, are obviously a major determinant and a comparison of the magnetite based process with conventional chemical coagulation clearly needs to be made. A comparison of design parameters for chemical flocculation of sewage, discussed in an earlier paper (Booker et al., 1991), points to the speed of the magnetite based process, with flocculation times of around 1 minute and clarifier overflow rates of 10 m/hr. This relative speed should translate directly into capital cost savings, even though some cost is associated with recovery and reuse of the magnetite particles. In situations of limited land availability the advantages of the magnetite based process should be increased.

Operating costs depend very much on the level of coagulant dose required e.g. the coagulant dose required at Malabar was half that required at Lower Plenty. The use of caustic soda for magnetite regeneration also adds to the operating cost, although it has major advantages in minimising odour production, significantly reducing pathogen levels in the sludge and greatly facilitating the recovery and reuse of coagulant. Indeed, this latter ability, combined with the ability to produce a final sludge of 20 % w/w dry solids should significantly reduce total operating costs compared with conventional chemical coagulation. In many sewage treatment processes costs associated with treatment and disposal of the sludge can represent up to 50 % of the total operating cost.

Apart from costs, the suitability of a chemical coagulation/flocculation process for sewage treatment also needs to be addressed. The conventional approach to sewage treatment has relied almost entirely on microbiological processes such as activated sludge or trickling filters. These processes are designed to reduce effluent BOD levels to less than 20 mg/L but have long residence times, usually of around eight hours. As has been illustrated in this paper, a chemical coagulation process based on the use of fine magnetite particles will remove suspended solids, oil and grease, phosphorus, heavy metals and bacteria to levels equal to or better than those achieved by conventional secondary treatment. BOD removals of 50 to 75 % are less than conventional secondary treatment but the remaining BOD is virtually all in the soluble form and, as pointed out by Ødegaard (1989), should have a very high rate of biological oxidation. This point raises the real prospect of an optimal combination of physico-chemical and microbiological techniques for treating sewage, especially where removal of both phosphorus and nitrogen is required.

A major operating feature of the magnetite based process is the speed of start up and shut down. Good quality effluent can be achieved within ten minutes of start up, making the process suitable as a peak load plant or as a compact plant for treating sewer overflows. For such applications the magnetite regeneration equipment can be significantly reduced in size further reducing capital costs.

Another factor of potential significant advantage to the magnetite based process is the scope for odour reduction. Both pilot plants were constructed and operated inside a standard shipping container. One feature of the pilot plant operation was the complete lack of odour within this enclosed space, even when there was a lot of odour emanating from the large conventional treatment plant nearby. The lack of odour from the process may be a result of adsorption of the odour causing compounds onto the magnetite surface, although this is hard to quantify given the very sensitive nature of the nose as an odour detection device.

All of the potential advantages of the magnetite based process will be fully tested by the construction and operation of the 5 ML/day prototype plant at Malabar. Research work is now targeted at the critical areas of coagulant recovery and recycle and sludge treatment and disposal, as these areas provide the greatest potential for cost reduction. It is planned that future advances in these two fields will be incorporated into the prototype plant operation.

Acknowledgements

The authors would like to record their thanks and appreciation to Dr Brian Bolto, Chief Research Scientist with the Division for his continued support and encouragement during the course of this work. Thanks are also due to Gunseli Akyel and Chris Hallam for their valued technical assistance. Financial and technical assistance from the Sydney Water Board to support the pilot plant operation is gratefully acknowledged.

References

Anderson, N.J., Priestley, A.J.: Colour and Turbidity Removal with Reusable Magnetite Particles – V Process Development. Wat. Res. *17* (10) (1983) 1227–1233

Anderson, N.J., Bolto, B.A., Blesing, N.V., Kolarik, L.O., Priestley, A.J., Raper, W.G.C.: Colour and Turbidity Removal with Reusable Magnetite Particles – VI Pilot Plant Operation. Wat. Res. *17* (10) (1983) 1235–1243

Anderson, N.J., Chin. C.T., Kolarik, L.O.: Colour and Turbidity Removal by Magnetite and Recycled Inorganic Coagulant, Vol. 2. Proc. 14th Fed. Conv., Aust. Water and Wastewater Ass., Perth 1991, pp. 373–380

Booker, N.A., Keir, D., Priestley, A.J., Ritchie, C.B., Sudarmana, D.L., Woods, M.A.: Sewage Clarification with Magnetite Particles. Wat. Sci. Tech. *23* (1991) 1703–1712

De Latour, C., Kolm, H.H.: High Gradient Magnetic Separation – a Water Treatment Alternative. J. Am. Wat. Wks. Ass. *68* (1976) 325–327

Demeter, L., Galgaczi, D.: US Patent No 3 350 302, 1967

Kolarik, L.O.: Colour and Turbidity Removal with Reusable Magnetite Particles – IV, Alkali Activated Magnetite. Wat. Res. *17* (1983) 141–147

Maloney, R.J., Horne, G.P., Stockley, M.: The SIROFLOC process:. Development, Construction and Operating Experience on a 20 000 m^3/day Potable Water Plant. Inst. Chem. Engrs, Symposium Series, No. 113 (1989) 93–104

Ødegaard, H.: Appropriate Technology for Wastewater Treatment in Coastal Tourist Areas. Wat. Sci. Tech., Vol. 21, No. 1 (1989) 1–17

Priestley, A.J., Sudarmana, D.L., Woods, M.A.: Sewage Treatment Combining Physico-Chemical Clarification and Anaerobic Digestion. Proc. 13th Fed. Conv. Aust. Water and Waste Water Ass. Canberra 1989, pp. 396–401

N.A. Booker, A.J. Priestley, and C.B. Ritchie
CSIRO Division of Chemicals & Polymers
Private Bag 10
Clayton Victoria 3168
Australia

Magnetic Floc Separation
in Chemical Phosphate Removal

U. Rott

1. Introduction

In chemical water and wastewater treatment, separation of flocs is an essential process unit. Different separation systems are in use, including sedimentation, filtration, centrifugation and flotation. Magnetic separation of flocs is a rather new technology and has not been used much up to now because chemical wastewater flocs do not have magnetic properties.

Magnetic floc separation is a method for separating undissolved particles from a fluid and can be applied in phosphate removal, separation of heavy metals and other flocculated pollutants or in chemical water softening. Because all of the flocs removed by this method must possess magnetic properties, it is necessary to attach to them a magnetic carrier material. A much used magnetic carrier in water treatment is granular magnetite (Fe_3O_4). The magnetite is incorporated in the floc under addition of polymer to achieve strong binding forces between floc and magnetic carrier. Afterwards the magnetic flocs are separated from the fluid by means of a magnet with a high gradient magnet field. A metal matrix filter is situated within the magnet field. Magnetic forces fix the flocs on the matrix while the wastewater passes through the high gradient area.

When the matrix is filled up to 70–80 % of the separation chamber volume, the magnet is switched off and the matrix is backwashed with water and compressed air. The loading time of the matrix is 20–40 minutes while the complete backwashing procedure may take less than one minute.

The magnetite is recovered from the backwash fluid by means of strong mechanical shear forces. Thus the flocs are separated into their original chemical compounds and into the magnetic carrier. The magnetite is separated by means of a magnetic drum separator and is directly reused in the main process.

In the research sewage treatment plant of Stuttgart University, tests of magnetic separation of phosphate flocs were carried out, applying magnetic wastewater treatment in the field of phosphate elimination from communal wastewater. Phosphate elimination from the mechanically and biologically treated wastewater was carried out through chemical post-precipitation. An additional sedimentation unit was not required.

The tests used iron- and alum salts from different manufacturers, lime and polyelectrolyte as coagulant aid. The influent concentration was 2.5–4 mg P/l.

The effluent concentration was less than 0.7 mg P/l and less than 0.2 mg ortho P/l.

2. Physical Basis of Magnetic Separation

Some basic information on the physical basis of magnetic separation of flocs from water was given by Van Velsen [1] and Van Velsen et al. [2].

Magnetic separation is a method for separating particles from a fluid on the basis of the magnetic properties of the particles. It has to be emphasized that only particulate material can be separated off magnetically. This can be achieved by using preliminary treatment, for example, adsorption, coagulation, flocculation and precipitation. Furthermore, the particles to be separated must possess magnetic properties. Most wastewater components, however, are only slightly magnetic, if at all. These components and particles have to be attached to a magnetic carrier material. A much used magnetic carrier material is magnetite (Fe_3O_4).

An important design parameter for separation performance in a magnetic field is the magnetic velocity, V_m, defined by

$$V_m = \frac{V \cdot M \cdot \operatorname{grad} B}{3\pi \cdot \eta \cdot d}$$

V Volume of the particle [m^3]
M Magnetisation per unit volume of the particle [A/m]
B Magnetic induction [T]
η Viscosity of the liquid [Pa·s]
d Diameter of the particle [m]

Obviously, the magnetic velocity must be sufficiently high to separate off the particles when the water passes the magnet. When the velocity is too high, however, some of the particles will not reach the matrix and are not trapped. As a result, the separation efficiency decreases.

In wastewater treatment a magnetic carrier material is used and it is essential that the binding forces between the particles be removed and the carrier material exceed the hydrodynamic forces which are exerted on the particle by the fluid. If not, the components to be removed will be freed from the magnetic carrier and pass the magnet.

The attachment depends on the components to be eliminated and is governed by colloidal and chemical forces. The literature indicates that there are various auxiliary substances that enhance the attachment, e.g. montmorillonite, aluminium salts, ion exchangers and polymers.

3. Separation Magnets

Most magnets used for separation processes in water technology are of the flat cylindrical type. The separation chamber is situated inside the cylindrical coil. Coil and chamber are enclosed by the magnetic circuit in the shape of a cylindrical box with flat bottom and top (see Fig. 1).

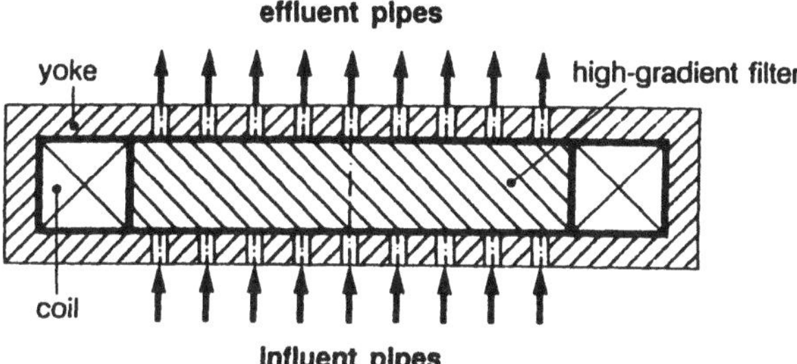

Fig. 1. Flat box magnetic separation system

The bottom and top plates of the magnetic circuit both have to be perforated in order to leave a passage for the water to be treated. In designing such a system it is necessary to avoid some of the typical disadvantages of a flat box magnet system as there are:

– clogging of the plates
– high local velocities in the scparation chamber
– even flushing of the matrix
– large dimensions of the coil compared to the total magnetic flux.

A different magnet type has been developed especially for wastewater treatment to meet the following demands:

– high separation efficiency
– high velocities in the separation chamber to reduce the magnet dimensions
– low energy consumption
– open structures to prevent clogging
– large storage capacity for separated particles.

This magnet called Aquamag (see Fig. 2) has two main construction principles:

1. The use of a magnetic flux generator, the dimensions of which are not directly governed by the volume of the separation chamber. The flux generator consists of a core and a coil. The coil is always magnetised up to saturation, which leads to minimum coil dimensions and, consequently, a low power consumption. The separation chamber is situated outside the coil, giving freedom in the choice of dimensions and in the choice of the field strength.
2. Integration of the flushing medium storage volume inside the magnetic circuit.

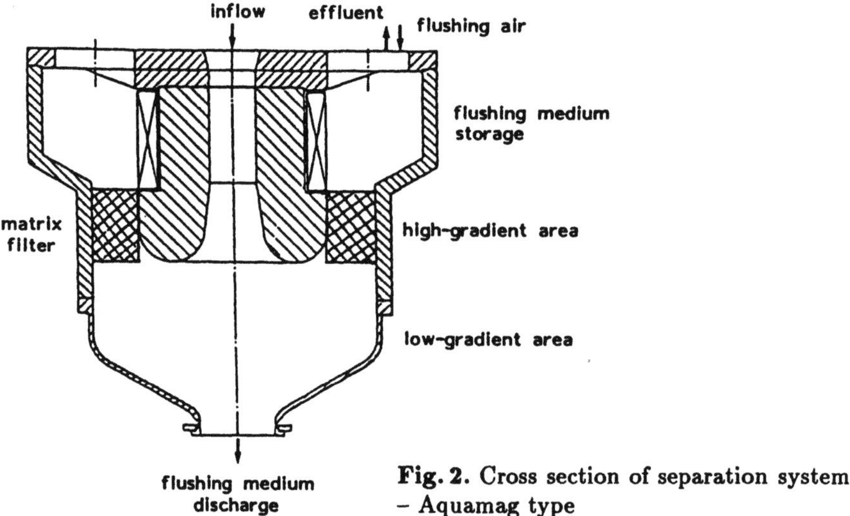

Fig. 2. Cross section of separation system
– Aquamag type

Since the separation chamber is also situated inside the magnetic circuit, there are no obstacles between the flushing medium and the matrix inside the separation chamber; this has a clear positive effect on the cleaning efficiency. The metal enclosed coil is situated around the upper part of the core. The magnetic flux passes via the cover to the housing. The main part of the flux crosses over to the lower part of the core through the separation chamber, which is equipped with a matrix.

Pretreated wastewater is pumped into the magnet via a central hole in the cover and the core and is forced to pass the separation chamber upwards. The water flows into the flushing storage, from where it is discharged through pipe connections in the cover. In the separation chamber with the high-gradient matrix, the magnetic particles are eliminated from the water.

When the matrix is filled up to 70–80 % of the separation chamber volume, the coil is switched off and the entry and discharge valves are closed. By applying compressed air the flushing water is pressed through the matrix with a high velocity and discharged through the rapid-action valve at the bottom of the magnet.

Loading the matrix will take 20 to 40 minutes; the complete flushing procedure takes less than one minute.

4. Flocculation and Magnetic Separation System

The separation of flocs has been tested on a complete system of magnetic phosphate removal for domestic sewage. In this system the phosphates are precipitated with lime, different trivalent iron salts and aluminium salts.

The predominant consideration in using lime as the precipitation agent is the production of calcium phosphate/carbonate, a product that can be used for agricultural purposes or as a raw material in industry. The precipitated phosphate is attached to magnetite, the magnetic carrier material. After magnetic separation the magnetite is recovered selectively and re-used in the process. The complete magnetic phosphate removal plant is shown in Fig. 3.

FLOCCULATION AND MAGNETIC SEPARATION SYSTEM

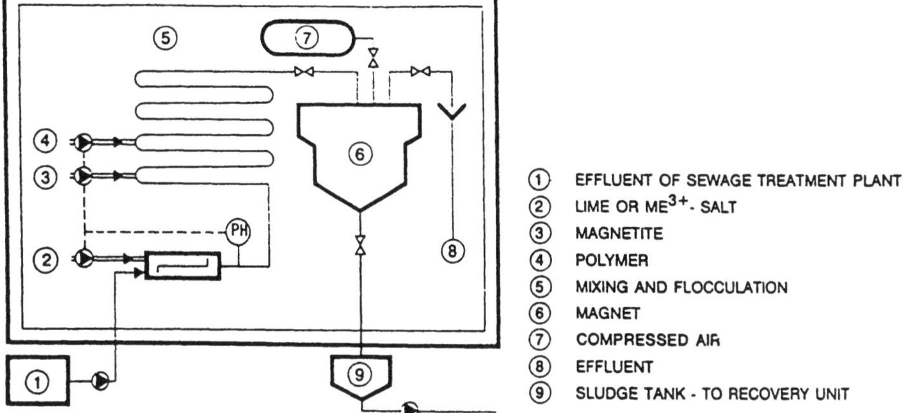

Fig. 3. Flocculation and magnetic separation system

The effluent of the sewage treatment plant is pumped into a pretreatment unit, consisting of a mixing tube with in-line dosage of lime or iron or aluminium salts, magnetite and polymer, and a flocculation tube. The lime dosage up to a pH of 10.5–11 is controlled by continuous pH measurements. Polymer addition is necessary to attach the phosphate-containing particles to magnetite.

The flocculation tube has a diameter of 110 mm and a length of 39 m. With a hydraulic capacity of 15 m^3/h, a flow velocity of 1580 m/h is achieved and the flocculation time is about 2 min.

Since flocculation takes place in only one flocculation tube in this system, flocculation cannot be carried out in two different steps, namely, perikinetic flocculation and orthokinetic flocculation with different velocity gradients.

From the mixing/flocculation tube the water flows directly into the magnet. Intermediate pumping between the pretreatment unit and the magnet is avoided to prevent a disruption of the conglomerate particles.

After saturation of the filter matrix, the magnet is backwashed. The magnet flushing medium is collected in a small tank under the magnet. From this collection tank the flushing medium is pumped to the magnetite recovery unit.

5. Magnetite Recovery

The function of the magnetite recovery unit is to recover magnetite for reuse in the magnetic separation process, and to produce calcium phosphate/carbonate residues with as low a concentration of magnetite as possible.

The magnetite is uncoupled from the phosphate-containing particles by means of shear forces, e.g. generated with an ultrasonic transducer element. Thereafter, magnetite is selectively separated from the medium by means of a magentic drum separator (see Fig. 4).

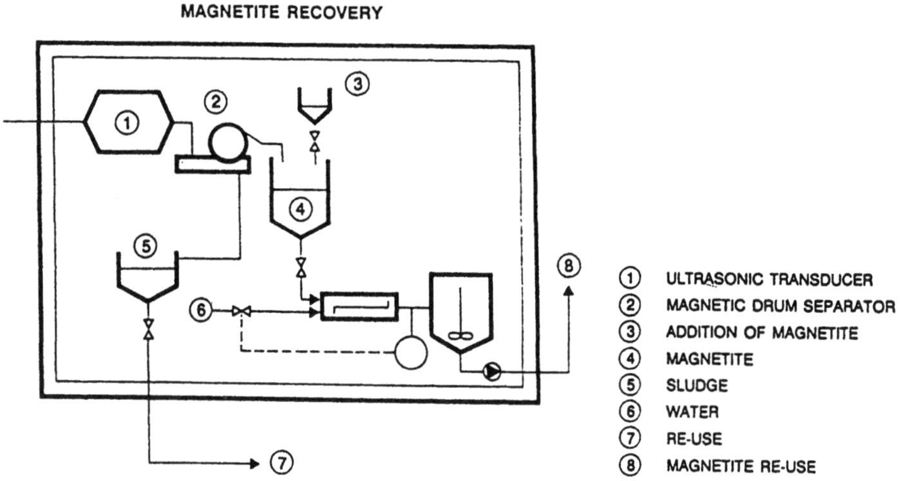

Fig. 4. Magnetite recovery system

The remaining product, a suspension of calcium phosphate-carbonate in water or a sludge of metal-phosphates, is handled depending on its final re-use.

6. Phosphate Removal Tests

Phosphate removal tests were carried out in a mobile pilot plant for magnetic separation with a hydraulic capacity of 15–20 m^3/h [3]. The pilot plant was installed at the research sewage treatment plant of Stuttgart University at Stuttgart-Büsnau.

Some quality parameters of the effluent of the sewage treatment plant after mechanical and biological treatment are given in Table 1.

Tab. 1. Quality parameters of treated sewage

COD	35 mg/l	SS	5–10 mg/l
BOD	7 mg/l	pH	7
NH_4^+-N	1 mg/l	P_{total}	3.5 mg/l
NO_3^--N	20 mg/l		

Pretests showed that the optimum capacity of the pilot plant is 20 m^3/h with lime dosage and 15 m^3/h with Me^{3+}-salt dosage. The optimum length of a filtration cycle was found to be 12.5 min. Longer cycles cause breakthrough of magnetite, which would mean a loss of magnetite. Up to a cycle time of 12.5 min, the loss of magnetite was less than 1 %. The magnet of the pilot plant was a high gradient type with a magnetic induction of $0.2\,T$. Flushing of the metal matrix took 2 minutes with compressed air of 6 bar.

Different coagulants were used during the test:

- Ferric chloride $FeCl_3$
- Ferric chloride sulphate $FeClSO_4$
- AVR (mixed product of $Al^{3+}Fe^{3+}$-sulphate)
- Lime $Ca(OH)_2$.

The flocculant in all tests was Praestol 2540, a copolymer of acrylamide and acrylate. The magnetic carrier was magnetite Fe_3O_4 with a constant dosage of 1 g/l.

The best results were obtained with FeCl3 and AVR and a flocculant dosage of 1 mg/l (see Table 2).

Tab. 2. Results of phosphate removal tests

Flocculant	Effluent		COD Reduction
	mg P/l	mg oP/l	%
$Ca(OH)_2$	0.46	0.09	12–54
$FeCl_3$	0.6	0.2	20–30
$FeClSO_4$	0.89	0.6	30
AVR	0.3	0.05	50

The good results with AVR were found on the basis of a molecular ratio of 5, which is not yet sufficient and the AVR-phosphate sludge required a longer sedimentation time. The results of $FeClSO_4$ flocculation were not as satisfactory, with effluents of 0.6–1.8 mg P/l and 0.3–0.9 mg oP/l. Flocculation with lime showed good phosphate removal results, but insufficient removal of suspended solids, due to incomplete building of lime flocs.

The results of suspended solids analyses are given in Table 3:

Tab. 3. Suspended solids in phosphate removal tests

Flocculant	Influent	Effluent
	mg ss/l	mg ss/l
$Ca(OH)_2$	5.9	80
$FeCl_3$	6.5	8.5
$FeClSO_4$	5.7	8.7
AVR	7.5	7.5

In general, the concentration of suspended solids in the influent is relatively low due to effective sedimentation in the sewage treatment plant. The concentration of suspended solids in the effluent of the magnetic separation system, especially in $Ca(OH)_2$ tests, is higher than in the influent due to the absence of an orthokinetic flocculation phase as mentioned above.

The magnetic carrier Fe_3O_4 had a grain size of 0–60 μm. The use of a grain size of 20–60 μm is recommended to avoid the inevitable loss of the smaller grains. The magnetite recovery unit was not part of the mobile test plant.

7. Conclusion

Water quality management of surface waters requires low phosphate concentrations in sewage water effluents. In Germany sewage treatment plants with a population equivalent above 100 000 must maintain a phosphate limit value of less than 1 mg P/l, and plants with a population equivalent above 20 000 must maintain a limit value of less than 2 mg P/l. Even in future the phosphate concentration of raw sewage water will exceed 3 mg P/l and traditional mechanical-biological sewage treatment will reduce this value not more than 20–30 %. Therefore additional (tertiary) treatment of sewage is necessary.

Chemical phosphate removal usually consists of coagulation and flocculation of phosphate and of liquid-solid separation by means of sedimentation or filtration. This separation step can be replaced by magnetic separation.

Because phosphate flocs do not have magnetic properties, however, a magnetic carrier, granulated magnetite (Fe_3O_4), must be added and is incorporated in the phosphate flocs. Tests were carried out with a mobile pilot plant for magnetic separation using the effluent of a mechanical-biological sewage treatment plant. The pilot plant was designed for a capacity of 10–20 m^3/h. The phosphate concentration of the treated sewage was 2.5–4.0 mg P/l.

Various concentrations of coagulants such as ferric chloride, ferric chloride sulphate, AVR and lime, with a constant concentration of magnetite, were used in the tests, as well as a flocculant based on acrylamide.

The best results were obtained using $FeCl_3$ and AVR with effluent values below 0.7 mg/l total phosphate and 0.2 mg/l orthophosphate. It is possible to attain the required values of less than 1 mg P/l for sewage discharge.

Magnetic separation of phosphate flocs can be recommended as a replacement for sedimentation or filtration units in the course of physical-chemical phosphate removal in sewage water. It is an economical and space-saving alternative, and is well suited to supplement existing sewage treatment plants.

8. References

[1] Van Velsen, A.F.-M.: High Gradient Magnetic Filtration for Wastewater Treatment. 5th International Filtration Conference, Nice, June 1990
[2] Van Velsen, A.F.-M., Boersma, R., de Reuver, J.L.: Magnetic Separation in Wastewater Treatment. ENS-Conference, Stavanger, August 1991

[3] Rott, U., Lamberth, B.: Untersuchungsbericht zur Phosphatabscheidung mit einer Magnetseparationsanlage. SMIT-NYMEGEN, Nijmegen, NL, April 1990

Ulrich Rott
Institut für Siedlungswasserbau, Wassergüte-
und Abfallwirtschaft der Universität Stuttgart
Bandtäle 1
D-7000 Stuttgart 80
Germany